D1203078

Rüdiger Memming

Semiconductor Electrochemistry

Related Titles from WILEY-VCH

C. H. Hamann / A. Hamnett / W. Vielstich
Electrochemistry
1998. XVII, 420 pages with 242 figures and 29 tables.
Hardcover. ISBN 3-527-29095-8
Softcover. ISBN 3-527-29096-6

A. E. Kaifer / M. Gomez-Kaifer
Supramolecular Electrochemistry
1999. XIV, 255 pages with 98 figures and 2 tables.
Hardcover. ISBN 3-527-29597-6

G. Hadziioannou / P. F. van Hutten (Eds.)
Semiconducting Polymers
1999. XXVII, 632 pages with 350 figures and 19 tables.
Hardcover. ISBN 3-527-29507-0

Rüdiger Memming

Semiconductor Electrochemistry

Weinheim · New York · Chichester
Brisbane · Singapore · Toronto

Professor Dr. Rüdiger Memming
Wildbret 3
D-25462 Rellingen

This book was carefully produced. Nevertheless, author and publisher do not warrant the information contained therein to be free of errors. Readers are advised to keep in mind that statements, data, illustrations, procedural details or other items may inadvertently be inaccurate.

1st Reprint 2002

Library of Congress Card No. applied for

A catalogue record for this book is available from the British Library

Die Deutsche Bibliothek – CIP Cataloguing-in-Publication-Data
A catalogue record for this publication is available from
Die Deutsche Bibliothek

ISBN 3-527-30147-X

Printed on acid-free paper

Composition: Mitterweger & Partner Kommunikationsgesellschaft mbH, D-68723 Plankstadt
Printing: betz-druck GmbH, D-64291 Darmstadt
Bookbinding: Wilh. Osswald + Co., D-67433 Neustadt
Printed in the Federal Republic of Germany

Preface

Several books on classical electrochemistry had already appeared about 30 to 40 years before the present book was written, for example, *Electrochemical Kinetics* by K. Vetter, in 1958, and *Modern Electrochemistry* by O. Bockris and A. Reddy in 1970. In the latter book a wide-ranging description of the fundamentals and applications of electrochemistry is given, whereas in the former the theoretical and experimental aspects of the kinetics of reactions at metal electrodes are discussed. Many electrochemical methods were described by P. Delahay in his book *New Instrumental Methods in Electrochemistry*, published in 1954. From the mid-1950s to the early 1970s there was then a dramatic development of electrochemical methodology. This was promoted by new, sophisticated electronic instruments of great flexibility. About 20 years ago, in 1980, Bard and Faulkner published the textbook *Electrochemical Methods*, which is an up-to-date description of the fundamentals and applications of electrochemical methods.[1]

The modern work on semiconductor electrodes dates back to mid-1950s when the first well-defined germanium and silicon single crystals became available. Since then many semiconducting electrode materials and reactions have been investigated and most processes are now well understood. Since charge transfer processes at semiconductor electrodes occur only at discrete energy levels, i.e. via the conduction or valence band or via surface states, and since they can be enhanced by light excitation, detailed information on the energy parameters of electrochemical reactions has been obtained. Corresponding investigations had greatly contributed to the understanding of electrochemical processes at solid electrodes in general. In this area, a major role was played by the modern theories on electron transfer reactions, developed by Marcus, Gerischer, Levich and Dogonadze. During the last quarter of the 20th century, models and experimental results have been described and summarized in various review articles and books. In this context, *Electrochemistry at Semiconductor and Oxidized Metal Electrodes*, by S. R. Morrison (1980), and *Semiconductor Photoelectrochemistry* by Y. V. Pleskov and Y. Gurevich (1986), should be mentioned.

Semiconductor electrochemistry has various important applications, such as solar energy conversion by photoelectrochemical cells, photo-detoxification of organic waste, etching processes in semiconductor technology and device fabrication, photoplating and photography. In particular, the first-mentioned application provided a great impetus to research in the field of semiconductor electrochemistry.

Therefore, there is a need for a textbook for teaching the fundamentals and applications of semiconductor electrochemistry in a systematic fashion. This field has interdisciplinary aspects insofar as semiconductor physics and also, in part, photochemistry are involved, as well as electrochemistry. Thus, one can expect that students and scientists with backgrounds in semiconductor physics on the one hand and in metal electrochemistry on the other will become interested in this area. A physicist will have no problem with the concept of band model and little difficulty with the energy concept of electron

[1] The second edition is in preparation.

transfer, but the ion interactions in the solution and at the interface may present obstacles. On the other hand, an electrochemist entering the field of semiconductor electrochemistry may have some problems with energy bands and the Fermi level concept and in thinking of electrode reactions in terms of energy levels. In the present book, these difficulties are taken into account by including appropriate chapters dealing with some of the fundamentals of semiconductor physics and classical electrochemistry. Accordingly, it is the intention of this textbook to combine solid state physics and surface physics (or chemistry) with the electrochemistry and photoelectrochemistry of semiconductors. It is not the aim of the book to cover all results in this field. The references are limited, and are selected primarily from an instructional point of view.

I have been helped by several people in preparing this work. The basic parts of some chapters were prepared during several extended visits to the National Renewable Energy Laboratory (NREL) in Golden, Colorado, USA. I am mainly indebted to Dr A. J. Nozik and Dr B. B. Smith (NREL), for many stimulating discussions, essential advice, and support. I would like to thank Prof. B. Kastening, Institute of Physical Chemistry, University of Hamburg and Dr D. Meissner of the Forschungszentrum Jülich, for discussions and for reading some chapters. I would also like to acknowledge the help of Dr R. Goslich of the Institute for Solar Energy Research (ISFH), Hannover, in solving computer and software problems. Finally, I wish to thank my wife for her support and for affording me so much time for this work.

October 2000 Rüdiger Memming

Contents

1 Principles of Semiconductor Physics

The understanding of electrochemical processes at semiconductor electrodes naturally depends on a knowledge of semiconductor physics. This chapter presents a brief introduction to this field; only those subjects relevant to semiconductor electrochemistry are included here. For detailed information, the reader is referred to the standard textbooks on semiconductor physics by C. Kittel [1], R. A. Smith [2], T. S. Moss [3] and Pankove [4].

1.1 Crystal Structure

A crystalline solid can be described by three vectors $\boldsymbol{a}$, $\boldsymbol{b}$ and $\boldsymbol{c}$, so that the crystal structure remains invariant under translation through any vector that is the sum of integral multiples of these vectors. Accordingly, the direct lattice sites can be defined by the set

$$\boldsymbol{R} = m\boldsymbol{a} + n\boldsymbol{b} + p\boldsymbol{c} \tag{1.1}$$

where m, n and p are integers [1]

Various unit cells of crystal structures are shown in Fig. 1.1. Most of the important semiconductors have diamond or zincblende lattice structures which belong to the tetrahedral phases, i.e. each atom is surrounded by four equidistant nearest neighbors. The diamond and zincblende lattices can be considered as two interpenetrating face-centered cubic lattices. In the case of a diamond lattice structure, such as silicon, all the atoms are silicon. In a zincblende lattice structure, such as gallium arsenide (a so-called III-V compound), one sublattice is gallium and the other arsenic. Most other III-V compounds also crystallize in the zincblende structure [5]. Various II-VI compounds, such as CdS, crystallize in the wurtzite structure, and others in the rock salt structure (not shown). The wurtzite lattice can be considered as two interpenetrating hexagonal close-packed lattices. In the case of CdS, for example, the sublattices are composed of cadmium and sulfur. The wurtzite structure has a tetrahedral arrangement of four equidistant nearest neighbors, similar to a zincblende structure. The lattice constants and structures of the most important semiconductors are given in Appendix C.

It is common also to define a set of reciprocal lattice vectors $\boldsymbol{a}^*$, $\boldsymbol{b}^*$, $\boldsymbol{c}^*$, such as

$$\boldsymbol{a}^* = 2\pi\frac{\boldsymbol{b}\cdot\boldsymbol{c}}{\boldsymbol{a}\cdot\boldsymbol{b}\cdot\boldsymbol{c}}; \quad \boldsymbol{b}^* = 2\pi\frac{\boldsymbol{c}\cdot\boldsymbol{a}}{\boldsymbol{a}\cdot\boldsymbol{b}\cdot\boldsymbol{c}}; \quad \boldsymbol{c}^* = 2\pi\frac{\boldsymbol{a}\cdot\boldsymbol{b}}{\boldsymbol{a}\cdot\boldsymbol{b}\cdot\boldsymbol{c}} \tag{1.2}$$

so that $\boldsymbol{a}\cdot\boldsymbol{a}^* = 2\pi$; $\boldsymbol{a}\cdot\boldsymbol{b}^* = 0$ and so on. The general reciprocal lattice vector is given by

$$\boldsymbol{G} = h\boldsymbol{a}^* + k\boldsymbol{b}^* + l\boldsymbol{c}^* \tag{1.3}$$

where h, k, l are integers.

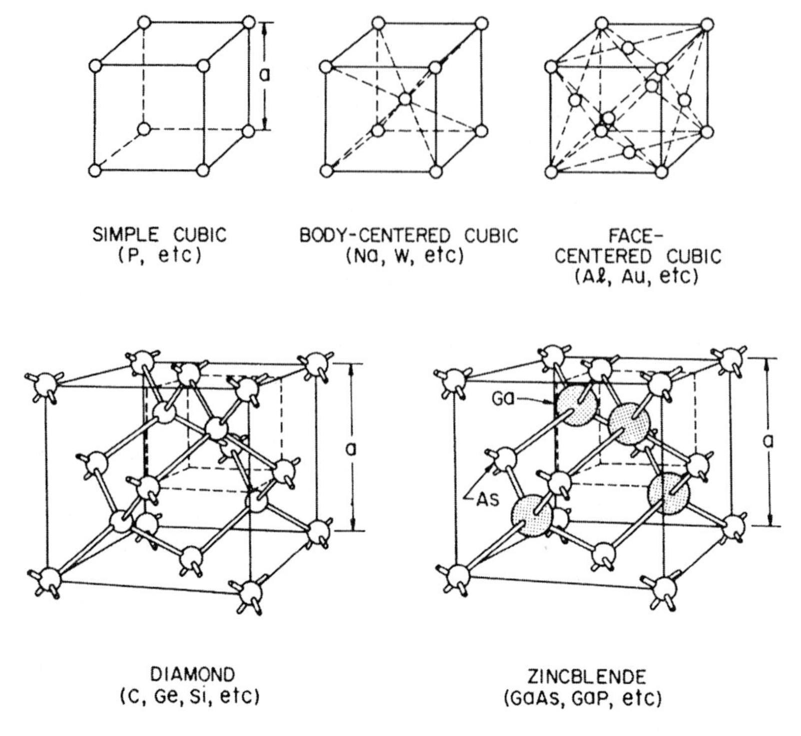

Fig. 1.1 Important unit cells. (Taken from ref. [7])

According to the definitions given by Eqs. (1.1) to (1.3), the product $\boldsymbol{G} \cdot \boldsymbol{R} = 2\pi \times$ integer. Therefore each vector of the reciprocal lattice is normal to a set of planes in the direct lattice, and the volume $V_c{}^*$ of a unit cell of the reciprocal lattice is related to the volume of the direct lattice V_c by

$$V_c = (2\pi)^3 / V_c \tag{1.4}$$

where $V_c = \boldsymbol{a} \cdot \boldsymbol{b} \cdot \boldsymbol{c}$

It is convenient to characterize the various planes in a crystal by using the Miller indices h, k, l. They are determined by first finding the intercepts of the plane with the basis axis in terms of the lattice constants, and then taking the reciprocals of these numbers and reducing them to the smallest three integers having the same ratio. The three integers are written in parentheses (hkl) as Miller indices for a single plane or a set of parallel planes. One example is given in Fig. 1.2 where the Miller indices of some planes in a cubic crystal are shown. Planes that intercepted, for example, the x-axis on the negative side would be characterized by $(\bar{h}kl)$. For directions perpendicular to the corresponding planes, one uses the Miller indices in brackets, i.e. $[hkl]$.

Some physical properties of semiconductor electrodes depend on the orientation of the crystal, and surface properties vary from one crystal plane to the other. It is therefore very important in studies of surface and interface effects that the proper surface is selected. A semiconductor crystal can be cut by sawing or by cleavage. Cleavage in

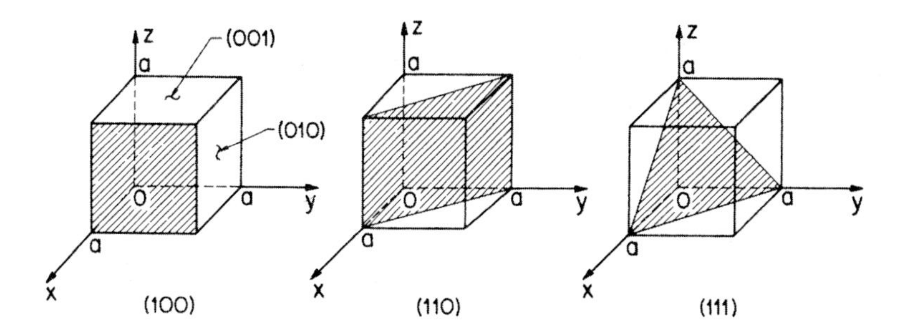

Fig. 1.2 Miller indices of some important planes in a cubic crystal

particular is a common technique for preparing clean surfaces in an ultrahigh vacuum. Unfortunately, however, only a few surface planes can be exposed by cleavage. The easiest planes in silicon and germanium are (111) and their equivalents. In contrast, gallium arsenide cleaves on (110) planes. Accordingly, the most interesting planes, which consist of a Ga surface (111) or an As surface ($\bar{1}\bar{1}\bar{1}$), cannot be produced by cleavage.

1.2 Energy Levels in Solids

Before the energy bands of semiconductors can be described, the following basic quantities must be introduced.

A free electron in space can be described by classical relations as well as by quantum mechanical methods. Combining both methods, the wavelength λ of the electron wave is related to the momentum p by

$$\lambda = h/p = h/mv \tag{1.5}$$

in which h is the Planck constant, m the electron mass and v the electron velocity. The electron wave can also be described by the wave vector defined by the relation

$$k = 2\pi/\lambda \tag{1.6}$$

Combining Eqs. (1.5) and (1.6) one obtains

$$k = \frac{2\pi}{h} p \tag{1.7}$$

The kinetic energy of a free electron is then given by

$$E = \tfrac{1}{2} m v^2 = \frac{h^2}{8\pi^2 m} k^2 \tag{1.8}$$

The parabolic relation between the energy and the wave vector k is illustrated in Fig. 1.3.

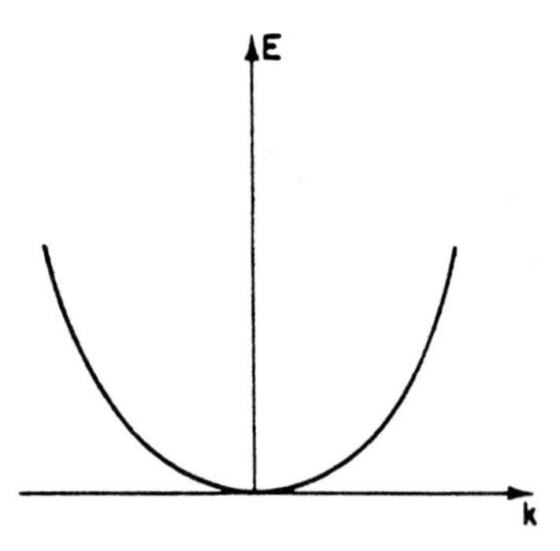

Fig. 1.3 Parabolic dependence of the energy of a free electron vs. wave vector

In a metal, the electrons are not completely free. A quantum mechanical treatment of the problem leads to the consequence that not all energy values are allowed. The corresponding wave vectors are now given by

$$k = \pi n/L \tag{1.9}$$

in which L is the length of a metal cube and n is any non-zero integer. Inserting Eq. (1.9) into (1.8), one obtains

$$E = \frac{h^2}{8mL^2}n^2 \tag{1.10}$$

The relation between energy and wave vector is still parabolic but the energy of the electron can only attain certain values. Since, however, the range of the allowed k values is proportional to the reciprocal value of L, the range of the energy values is very small for a reasonable size of metal, so that the E–k dependence is still a quasi-continuum.

The band structure of crystalline solids is usually obtained by solving the Schrödinger equation of an approximate one-electron problem. In the case of non-metallic materials, such as semiconductors and insulators, there are essentially no free electrons. This problem is taken care of by the Bloch theorem. This important theorem assumes a potential energy profile $V(\boldsymbol{r})$ being periodic with the periodicity of the lattice. In this case the Schrödinger equation is given by

$$E = \frac{h^2}{8mL^2}n^2 \tag{1.11}$$

The solution to this equation is of the form

$$\psi_{\mathbf{k}}(r) = e^{\mathbf{jkr}}U_{\mathrm{n}}(\boldsymbol{kr}) \tag{1.12}$$

where $U_{\mathrm{n}}(\boldsymbol{k},\boldsymbol{r})$ is periodic in $\boldsymbol{r}$ with the periodicity of the direct lattice, and n is the band index. Restricting the problem to the one-dimensional case then the lattice constant is a, b or c (see Eq. 1.1). If N is an integral number of unit lattice cells, then $k = \pi/a$ is the maximum value of k for $n = N$. This maximum occurs at the edge of the socalled Brillouin zone. A Brillouin zone is the volume of k space containing all the values of k up to π/a. Larger values of k lead only to a repetition of the first Brillouin zone.

Accordingly, it is only useful to determine the band structure within the first Brillouin zone. The solution of the Schrödinger equation (see Eqs. 1.11 and 1.12) leads to two energy bands separated by an energy gap, as shown in Fig. 1.4. The energy profile of the conduction band (upper curve) still appears parabolic (at least near the minimum) but it may deviate considerably from a parabolic *E–k* relation. In order to continue to use the relation derived for free electrons (Eq. 1.8), the electron mass is adjusted to provide a good fit. We then have, instead of Eq. (1.8),

$$E = \frac{h^2}{8\pi^2 m^*} k^2 \tag{1.13}$$

in which m* is the effective mass. Differentiating this equation, the effective mass is given by

$$m^* = \frac{h^2}{4\pi^2} \frac{1}{\dfrac{d^2E}{dk^2}} \tag{1.14}$$

This means that the effective mass is determined by the second derivative of the *E–k* curve, i.e. by its curvature. From this follows that the width of an energy band is larger for a small value of m^* and smaller for a large m^* value. The width can be determined by optical investigation and the effective mass by cyclotron resonance measurements.

According to Eq. (1.14), negative curvature of the valence band would mean a negative electron mass, which is physically not acceptable. It has therefore been concluded that occupied orbitals in the valence band correspond to holes. A hole acts in an applied electric or magnetic field as though it were a particle with a positive charge. This concept has been experimentally proved by Hall measurements (see Section 1.6). However, it only makes sense if nearly all energy states are filled by electrons. It should be further mentioned that the effective mass of holes may be different from that of electrons. A selection of values is listed in Appendix D.

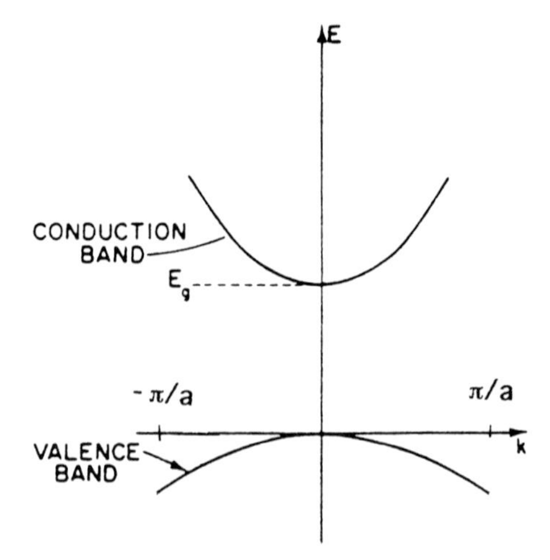

Fig. 1.4 Electron energy vs. wave vector in a semiconductor

The band structure of solids has been studied theoretically by various research groups. In most cases it is rather complex as shown for Si and GaAs in Fig. 1.5. The band structure, $E(k)$, is a function of the three-dimensional wave vector within the Brillouin zone. The latter depends on the crystal structure and corresponds to the unit cell of the reciprocal lattice. One example is the Brillouin zone of a diamond type of crystal structure (C, Si, Ge), as shown in Fig. 1.6. The diamond lattice can also be considered as two penetrating face-centered cubic (f.c.c.) lattices. In the case of silicon, all cell atoms are Si. The main crystal directions, $\Gamma \rightarrow$ L ([111]), $\Gamma \rightarrow$ X ([100]) and $\Gamma \rightarrow$ K ([110]), where Γ is the center, are indicated in the Brillouin zone by the dashed lines in Fig. 1.6. Crystals of zincblende structure, such as GaAs, can be described in the same way. Here one sublattice consists of Ga atoms and the other of As atoms. The band structure, $E(k)$, is usually plotted along particular directions within the Brillouin zone, for instance from the center Γ along the [111] and the [100] directions as given in Fig. 1.5.

In all semiconductors, there is a forbidden energy region or gap in which energy states cannot exist. Energy bands are only permitted above and below this energy gap. The upper bands are called the conduction bands, the lower ones the valence bands. The band-gaps of a variety of semiconductors are listed in Appendix D.

According to Fig. 1.5, the conduction as well as the valence band consists of several bands. Some valence bands are degenerated around $k = 0$ (the Γ point). Since the curvature differs from one band to another, each band is associated with a different effective mass (see also Appendix D). Rather flat energy profiles correspond to heavy holes

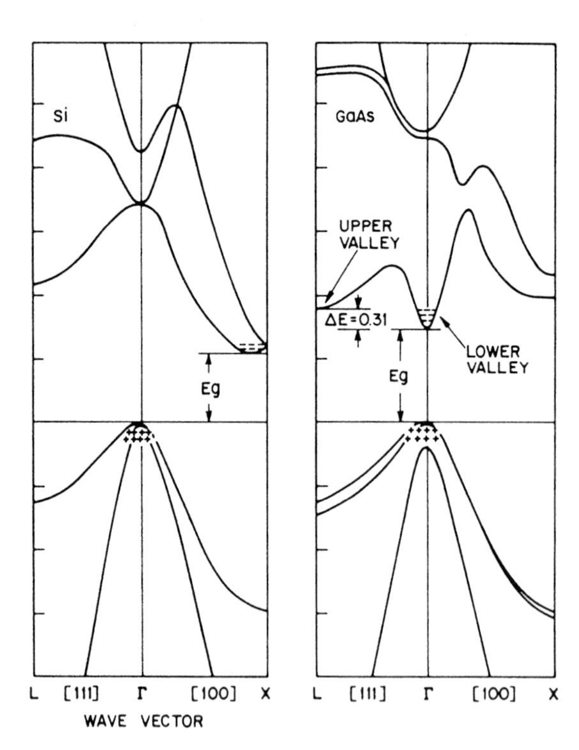

Fig. 1.5 Energy band structure of Si and GaAs. Compare with Fig. 1.4. (After ref. [11])

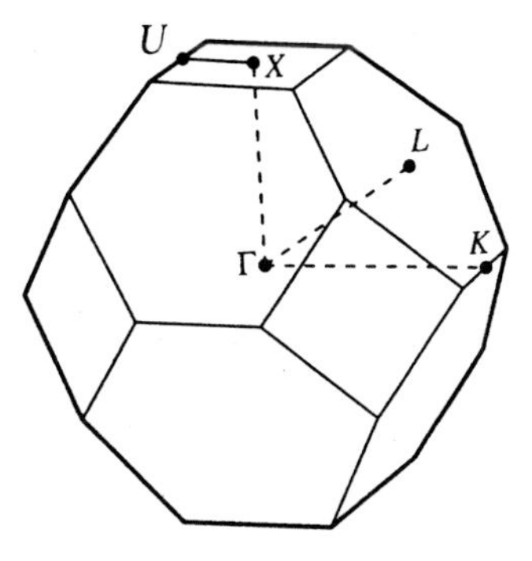

Fig. 1.6 Brillouin zone for face-centered cubic lattices, with high symmetry points labelled. (After ref. [6])

ass), and steep profiles to light holes (small effective mass). In the case of GaAs, the maxima of all valence bands and the minimum of the lowest conduction band occur at $k = 0$, i.e. in the center of the Brillouin zone (Γ point) (Fig. 1.5). The corresponding bandgap (1.4 eV for GaAs) is indicated. In the band structure of many semiconductors, however, the lowest minimum of the conduction band occurs at a different wave vector ($k \neq 0$) from the maximum of the valence band ($k = 0$). For instance, in the case of silicon the lowest minimum of the conduction band occurs at the edge of the Brillouin zone (X point) (Fig. 1.5). If both the conduction band minimum and the valence band maximum occur at $k = 0$, the energy difference, E_g, is a so-called direct bandgap. If the lowest conduction band minimum is found at $k \neq 0$, E_g is termed an indirect bandgap. The consequences of these differences in band structure will be discussed in Chapter 1.3. During the course of this book, only the lowest edge of the conduction band (E_c) and the upper edge of the valence band (E_v) are considered (as illustrated in Fig. 1.9 in Section 1.3).

1.3 Optical Properties

The simplest method for probing the band structure of semiconductors is to measure the absorption spectrum. The absorption coefficient, α, is defined as

$$\alpha = \frac{1}{d} \ln \frac{I_0}{I} \tag{1.15}$$

in which d is the thickness of the sample, and I and I_0 the transmitted and the incident light intensities, respectively. Since the refractive index of semiconductors is frequently quite high, accurate measurements require the determination of the transmission coefficient, T, as well as the reflection coefficient, R. For normal incidence they are given by

$$T = \frac{(1 - R^2)\exp(-4\pi d/\lambda)}{1 - R^2 \exp(-8\pi d/\lambda)} \tag{1.16}$$

$$R = \frac{(1 - n)^2 + k^2}{(1 + n)^2 + k^2} \tag{1.17}$$

in which λ is the wavelength, n the refractive index and k the absorption constant. The latter is related to the absorption coefficient α by

$$\alpha = 4\pi k/\lambda \tag{1.18}$$

By analyzing the T and λ or the R and λ data at normal incidence, or by measuring R and T at different angles of incidence, both n and α can be obtained.

The fundamental absorption refers to a band-to-band excitation which can be recognized by a steep rise in absorption when the photon energy of the incident light goes through this range. Since, however, optical transitions must follow certain selection rules, the determination of the energy gap from absorption measurements is not a straightforward procedure.

Since the momentum of photons, h/λ, is small compared with the crystal momentum, h/a (a is the lattice constant), the momentum of electrons should be conserved during the absorption of photons. The absorption coefficient $\alpha\,(h\nu)$ for a given photon energy is proportional to the probability, P, for transition from the initial to the final state and to the density of electrons in the initial state as well as to the density of empty final states. On this basis, a relation between absorption coefficient α and photon energy E_{ph} can be derived [2, 4]. For a direct band–band transition, for which the momentum remains constant (see Fig. 1.7), it has been obtained for a parabolic energy structure (near the absorption edge):

$$\alpha \sim (E_{ph} - E_g)^{1/2} \tag{1.19}$$

in which E_g is the bandgap. Accordingly, a plot of $(\alpha\ E_{ph})^2$ vs. E_{ph} should yield a straight line and E_g can be determined from the intercept. However, this procedure does not always yield a straight line. Therefore some scientists define E_g at that photon energy where $\alpha = 10^4$ cm^{-1}. High α values of up to 10^6 cm^{-1} have been found for direct transitions. Electrons excited into higher energy levels of the conduction band (transition 1a in Fig. 1.9) are thermalized to the lower edge of the conduction band within about 10^{-12} to 10^{-13} s.

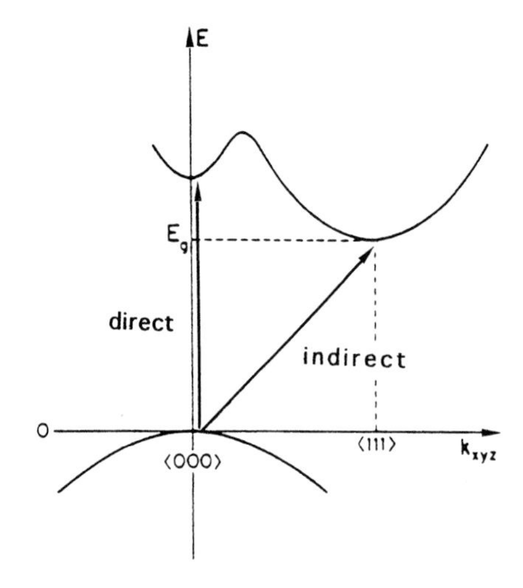

Fig. 1.7 Optical transitions in semiconductors with an indirect bandgap

As already mentioned in the previous section, the lowest minimum in the conduction band energy frequently occurs not at $k = 0$ but at other wave numbers as shown for silicon in Fig. 1.5. The law of conservation of momentum excludes here the possibility of the absorption of a photon of an energy close to the bandgap. A photon absorption becomes possible, however, if a phonon supplies the missing momentum to the electron as illustrated in Fig. 1.7. Since such an indirect transition requires a "three-body" collision (photon, electron, phonon) which occurs less frequently than a "two-body" collision, the absorption coefficient will be considerably smaller for semiconductors with an indirect gap. This becomes obvious when the absorption spectra of semiconductors are measured; a selection is given in Fig. 1.8. For instance, GaAs and $CuInSe_2$ provide examples of a direct bandgap, i.e. the absorption coefficient rises steeply near the bandgap and reaches very high values. Si and GaP provide typical examples of an indirect transition. In the case of Si one can recognize in Fig. 1.8 that α remains at a very low level for a large range of photon energies (GaP not shown). For indirect transitions, the relation between α and E_{ph} is given in refs. [2] and [7] as

$$\alpha \sim (E_{ph} - E_g)^2 \tag{1.20}$$

The interpretation of the interband transition is based on a single-particle model, although in the final state two particles, an electron and a hole, exist. In some semiconductors, however, a quasi one-particle state, an exciton, is formed upon excitation [4, 8]. Such an exciton represents a bound state, formed by an electron and a hole, as a result of their Coulomb attraction, i.e. it is a neutral quasi-particle, which can move through the crystal. Its energy state is close to the conduction band (transition 3 in Fig. 1.9), and it can be split into an independent electron and a hole by thermal excitation.

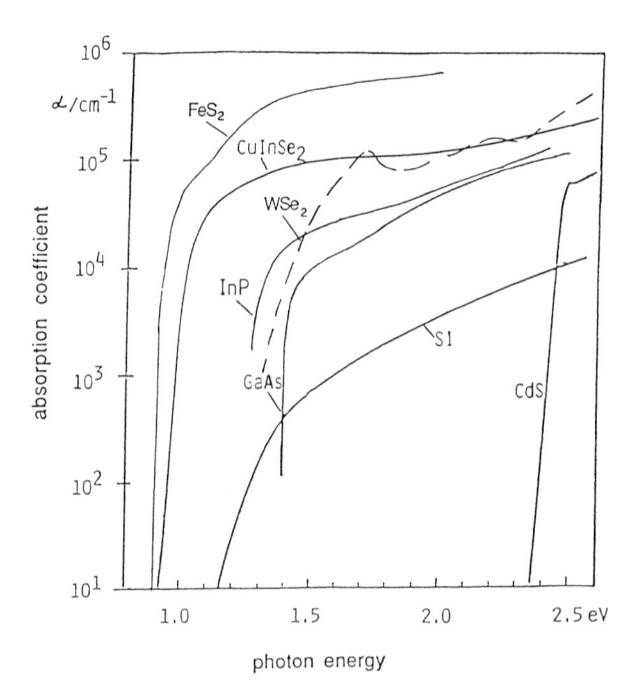

Fig. 1.8 Absorption spectra of various semiconductors

Therefore, a sharp absorption peak just below the bandgap energy can usually only be observed at low temperatures, whereas at room temperature only the typical band–band transition is visible in the absorption spectrum. The situation is different in organic crystals [9] and also for small semiconductor particles (see Chapter 9).

Various other electronic transitions are possible upon light excitation. Besides the band–band transitions, an excitation of an electron from a donor state or an impurity level into the conduction band is feasible (transition 2 in Fig. 1.9). However, since the impurity concentration is very small, the absorption cross-section and therefore the corresponding absorption coefficient will be smaller by many orders of magnitude than that for a band–band transition. At lower photon energies, i.e. at $E_{ph} \ll E_g$, an absorption increase with decreasing E_{ph} has frequently been observed for heavily doped semiconductors. This absorption has been related to an intraband transition (transition 4 in Fig. 1.9), and is approximately described by the Drude theory [4]. This free carrier absorption increases with the carrier density. It is negligible for carrier densities below about 10^{18} cm^{-3}.

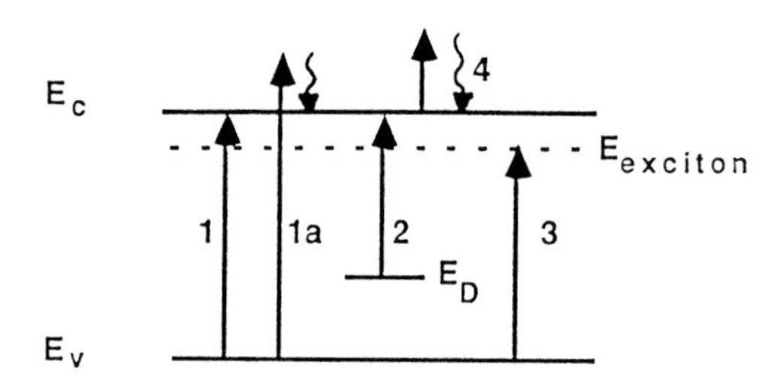

Fig. 1.9 Optical transitions in a semiconductor

1.4 Density of States and Carrier Concentrations

Semiconductor single crystals grown from extremely pure material exhibit a low conductivity because of low carrier density. The latter can be increased by orders of magnitude by doping the material. The principal effect of doping is illustrated in Fig. 1.10, taking germanium as an example. Fig. 1.10a shows intrinsic Si which contains a negligibly small amount of impurities. Each Ge atom shares its four valence electrons with the four neighboring atoms forming covalent bonds. By doping the material with phosphorus, n-type germanium is formed (Fig. 1.10a). The phosphorus atom with five valence electrons has replaced a Ge atom and an electron is donated to the lattice. This additional electron occupies one level in the conduction band. Similarly, p-type germanium is made by doping with a trivalent atom such as boron. This atom with three valence electrons substitutes for a Ge atom, an additional electron is transferred to boron leaving a positive hole in the Ge lattice (Fig. 1.10b). In principle, compound semiconductors are doped in the same way. In this case, however, doping can also occur by unstoichiometry, as illustrated for n-type CdS in Fig. 1.11. The bonding is partly ionic and additional free electrons occur if a sulfur atom is missing; if a sulfur vacancy, V_S, is formed the material becomes n-type.

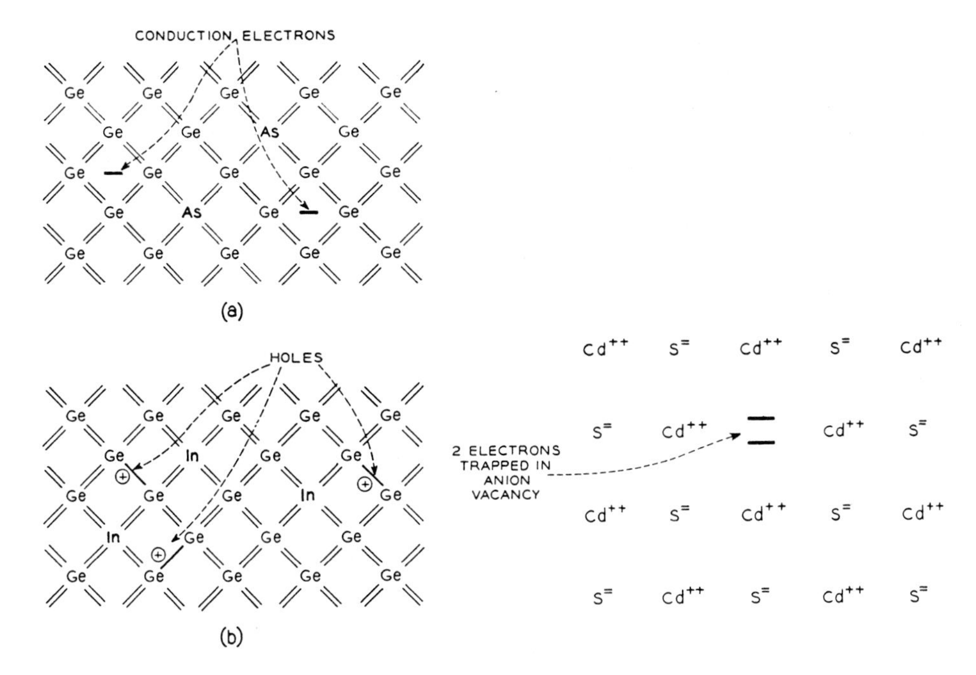

Fig. 1.10 Doping of a semiconductor crystal (Ge): (a) n-type doping; (b) p-type doping

Fig. 1.11 Imperfections in a compound semiconductor (CdS)

The additional electrons and holes occupy energy states in the conduction and valence bands, respectively. Before discussing the rules of occupation of energy levels, the energy distribution of the available energy states must first be derived, as follows.

In momentum space, the density of allowed points is uniform. Assuming that the surfaces of constant energy are spherical, then the volume of k space between spheres of energy E and $E + \Delta \mathrm{E}$ is $4\pi k^2\,\mathrm{d}k$ (see ref. [4]). Since a single level occupies a volume of $8\pi^3/V$ (V = crystal volume) in momentum space and there are two states per level, the density of states is given by

$$N(E)\,\mathrm{d}E = \frac{8\pi k^2}{8\pi^3}\mathrm{d}k = \frac{k^2}{\pi^2}\mathrm{d}k \tag{1.21}$$

It has been assumed here that the volume is unity (e.g. 1 cm^3). Inserting Eq. (1.13), one obtains

$$N(E)\,\mathrm{d}E = \frac{1}{2\pi^2 h^3}(2m^*)^{3/2} E^{1/2}\,\mathrm{d}E \tag{1.22}$$

E is measured with respect to the band edge. This equation is valid for the conduction and valence bands. The energy states can be occupied by electrons in the conduction band and by holes in the valence band. According to Eq. (1.22) the density of states per energy interval increases with the square root of the energy from the bottom of the

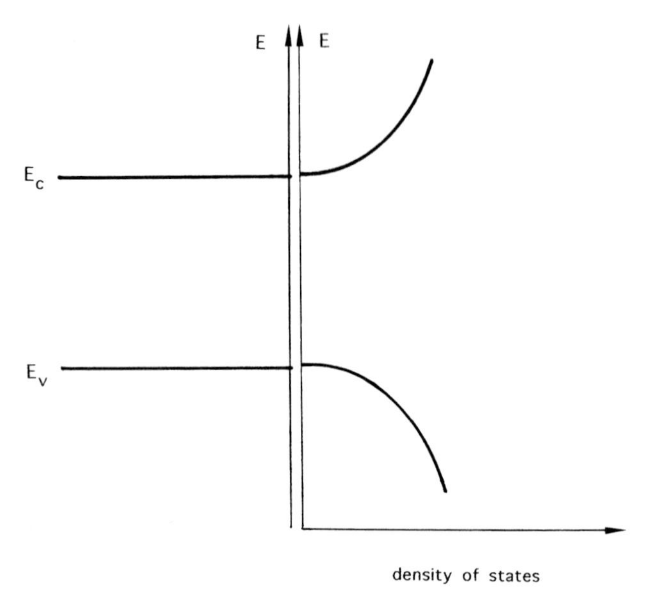

Fig. 1.12 Density of energy states near the band edges of a semiconductor vs. energy

corresponding band edge as illustrated in Fig. 1.12 (electron and hole energies have opposite signs). Since the reduced masses may be different for electrons and holes, the slopes of the curves are also different. These curves are based on the parabolic shape of the E–k relation as assumed near the minimum. The density of states looks very different when it is measured over a much larger energy range as shown in Fig. 1.13.

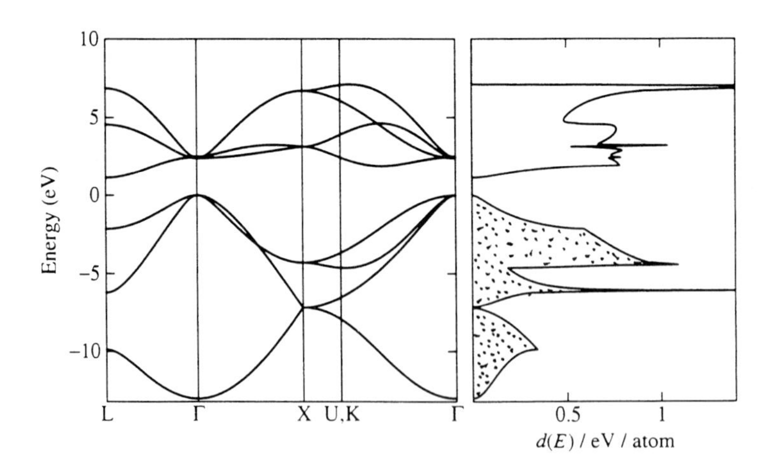

Fig. 1.13 Band structure and energy distribution of the density of states of silicon, as calculated for a large energy range. Dotted distribution is occupied by electrons. (From ref. [6])

The total density of energy states up to a certain energy level is obtained by integration of Eq. (1.22). The result is

$$N(E) = \frac{1}{3\pi^2 h^3}(2m^*)^{3/2} E^{3/2} \tag{1.23}$$

1.4.1 Intrinsic Semiconductors

The number of electrons occupying levels in the conduction band is given by

$$n = \int_{E_c}^{\infty} N(E) f(E)\,\mathrm{d}E \tag{1.24}$$

in which f(E) is the Fermi–Dirac distribution as given by

$$f(E) = \frac{1}{1 + \exp\left(\frac{E - E_F}{kT}\right)} \tag{1.25}$$

where E_F is the Fermi level. The integral in Eq. (1.24) cannot be solved analytically. Nevertheless, the integral must have a limited value because the density of states increases with increasing energy whereas $f(E)$ decreases. Eq. (1.24) can only be solved by assuming that $(E - E_F)/kT \gg 1$. In this case, the following has been obtained [2]

$$n = N_c \exp\left(-\frac{E_c - E_F}{kT}\right) \tag{1.26}$$

in which N_c is the density of energy states within few kT above the conduction band edge and is given by

$$N_c = \frac{2(2\pi m_e^* kT)^{3/2}}{h^3} \tag{1.27}$$

According to Eq. (1.27) one obtains $N_c \approx 5 \times 0^{19}$ cm^{-3} for the density of states within 1 kT above the lower edge of the conduction band, assuming an effective mass of $m^* = 1 \times m_0$ (m_0 = electron mass in free space). Since semiconductors with doping of less than 1×10^{19} cm^{-3} are used in most investigations and applications, the majority of the energy levels remain empty.

Similarly, we can obtain the hole density near the top of the valence band. We have, then

$$p = \int_{-\infty}^{E_V} N(E)\big(1 - f(E)\big)\,\mathrm{d}E \tag{1.28}$$

Using the same approximations as above, we obtain

$$p = N_v \exp\left(\frac{E_v - E_F}{kT}\right) \tag{1.29}$$

where the density of states, N_v, around the top of the valence band is given by

$$N_V = \frac{2(2\pi m_h^* kT)^{3/2}}{h^3} \tag{1.30}$$

in which m_h^* is the effective hole mass.

In order to preserve charge neutrality in an intrinsic semiconductor, the electron and hole densities must be equal. The position of the Fermi level can then be calculated from Eqs. (1.26) and (1.29). We then have

$$\begin{aligned} E_F &= \frac{E_c + E_v}{2} + \frac{kT}{2} \ln\left(\frac{N_v}{N_c}\right) \\ &= \frac{E_c + E_v}{2} + \frac{kT}{2} \ln\left(\frac{m_h^*}{m_e^*}\right)^{3/2} \end{aligned} \tag{1.31}$$

Accordingly, the Fermi level E_F is close to the middle of the energy gap, or for $m_e^* = m_h^*$ it is exactly at the middle of the gap. The intrinsic carrier density can be obtained by multiplying Eqs. (1.26) and (1.29), i.e.

$$np = N_c N_v \exp\left(-\frac{E_g}{kT}\right) = n_i^2 \tag{1.32}$$

The product of n and p is constant and the corresponding concentration is $n = p = n_i$, i.e. is the intrinsic electron density. Eq. (1.32) is called the "mass law" of electrons and holes, in comparison with chemical equilibria in solutions. The intrinsic concentration can be calculated from Eq. (1.32) if the densities of states are known. Assuming that $m_e^*/m_0 = 1$, then $n_i \approx 10^{11}$ cm^{-3} for a bandgap of $E_g = 1$ eV, i.e. n_i is a very small quantity. In the case of intrinsic material, the electron hole pairs are created entirely by thermal excitation. Since this excitation becomes very small for large bandgaps, n_i decreases with increasing bandgaps as proved by Eq. (1.32). Eq. (1.32) which is also valid for doped semiconductors, is of great importance because when one carrier density (e.g. n) is known then the other (here p) can be calculated. Examples are given in Appendix E.

1.4.2 Doped Semiconductors

When a semiconductor is doped with donor or acceptor atoms (see Fig. 1.10), then corresponding energy levels are introduced within the forbidden zone, as shown on the left side of Fig. 1.14. The donor level is usually close to the conduction band and the acceptor level close to the valence band. A donor level is defined as being neutral if filled by an electron, and positive if empty. An acceptor level is neutral if empty, and negative if filled by an electron. Depending on the distance of the donor and acceptor levels with respect to the corresponding bands, electrons are thermally excited into the conduction band and holes into the valence band.

In the presence of impurities, the Fermi level must adjust itself to preserve charge neutrality. The latter is given for an n-type semiconductor by

$$n = N_D^+ + p \tag{1.33}$$

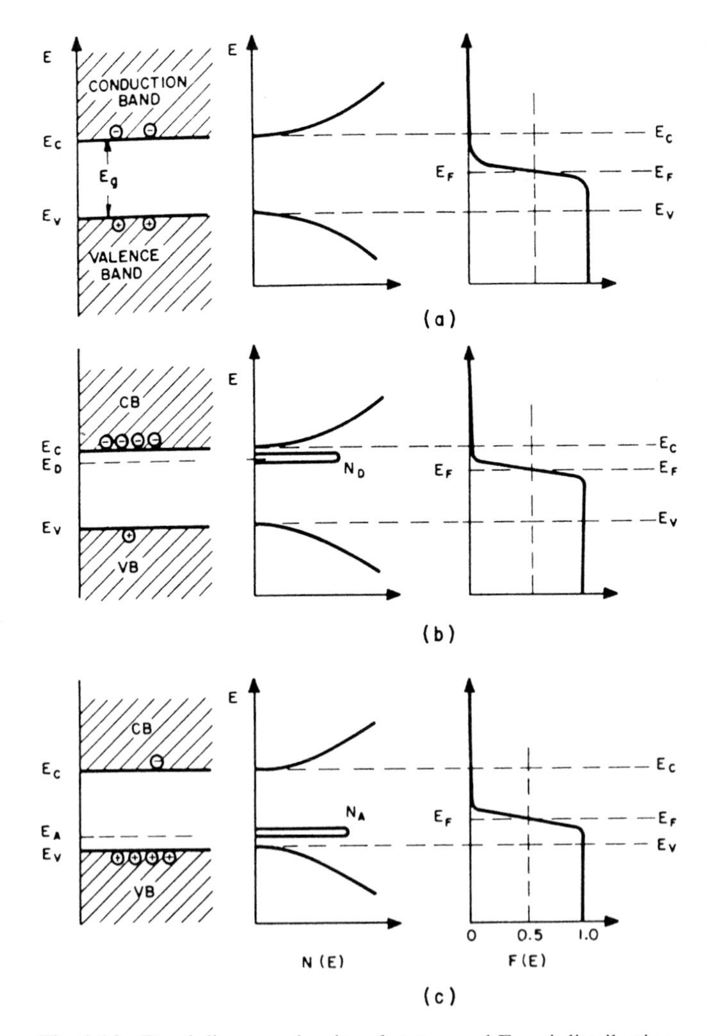

Fig. 1.14 Band diagram, density of states, and Fermi distribution

in which N_D^+ is the density of ionized donors. The latter is related to the occupied donor density N_D by the Fermi function, i.e.

$$N_D^+ = (1 - f)N_D = N_D\left[1 - \frac{1}{1 + \exp\left(\dfrac{E_D - E_F}{kT}\right)}\right] \tag{1.34}$$

Introducing Eqs. (1.28), (1.30) and (1.34) into (1.33), the Fermi level, E_F, can be calculated. According to Eq. (1.34), it is clear that all donors are completely ionized if the Fermi level occurs below the donor level, as shown on the right side of Fig. 1.14. On the other hand, if the donor concentration is increased then the electron density also rises.

In this case E_F may be located between E_c and E_D, but then not all of the more highly concentrated donors are ionized. Similar relations can be derived for acceptor states in a p-type semiconductor.

At extremely high impurity concentrations, the Fermi level may pass the band edge. In this case the semiconductor becomes degenerated, and most of the relations derived above are no longer applicable. The semiconductor then shows a metal-like behavior.

1.5 Carrier Transport Phenomena

When an electric field of strength $\mathscr{E}$ is applied across a crystal, electrons and holes are forced to move in the material. The corresponding current density is given by

$$j = \sigma \mathscr{E} \tag{1.35}$$

in which σ is the conductivity, the reciprocal value of the resistivity P. For semiconductors with both electrons and holes as carriers, the conductivity is determined by

$$\sigma = e(\mu_n n + \mu_p p) \tag{1.36}$$

in which e is the elementary charge, and μ_n and μ_p are the mobilities of electrons and holes, respectively. For doped semiconductors, the first or second term within the brackets dominates. According to Eq. (1.36) the conductivity can be varied by many orders of magnitude by increasing the doping.

The mobility is a material constant. Values for some typical semiconductors are given in Appendix E. Electron and hole mobilities are typically in the range between 1 and 1000 $cm^2\ V^{-1}s^{-1}$ These values are many orders of magnitude higher than the mobility of molecules and ions in solution ($\sim 10^{-4}$–10^{-3} $cm^2\ V^{-1}\ s^{-1}$). The presence of acoustic phonons and ionized impurities leads to carrier scattering which can significantly affect the mobility. The mobility μ_i (μ_n or μ_p), determined by interaction with acoustic phonons, as shown in refs. [2] and [7], is given by

$$\mu_i \sim (m^*)^{-5/2} T^{-3/2} \tag{1.37}$$

Accordingly, the mobility decreases with temperature. The mobility influenced by scattering of electrons (or holes) at ionized impurities can be described [2, 7] by

$$\mu_i \sim (m^*)^{-1/2} N_i^{-1} T^{3/2} \tag{1.38}$$

in which N_i is the density of impurity centers (N_D or N_A). In contrast with phonon scattering, mobility increases with temperature for impurity scattering. This makes it possible to distinguish experimentally between these two scattering processes. For the complete equations and their derivation, the reader is referred to refs. [2] and [7].

The carrier diffusion coefficient, D_n for electrons and D_p for holes, is another important parameter associated with mobility. It is given by

$$D_n = \frac{kT}{e}\mu_n\,; \quad D_p = \frac{kT}{e}\mu_p \tag{1.39}$$

It should be emphasized that a carrier transport can only be described by Ohm's law (Eq. 1.35) if sufficient empty energy levels exist in the corresponding energy band and a minimum carrier density is present in the material. On the other hand, in the case of an intrinsic high bandgap semiconductor, the carrier density may be negligible so that only those carriers carry the current which are injected into the crystal via one contact. In this case we have a space charge limited current which is proportional to $\mathscr{E}^2$ (Child's law).

The most common method for measuring the conductivity is the four-point probe technique [10]. Here a small current I is passed through the outer two probes and the voltage V is measured between the inner two probes (s is the distance between two probes). When such a measurement is performed with a semiconductor disk of diameter 2r and a thickness w, the resistivity is given by

$$\rho = \frac{\pi}{(\ln 2)} \frac{V}{I} w \tag{1.40}$$

provided that $2r \gg s$. The advantage of this method is that the conductivity can be measured without there being ohmic contacts between the semiconductor and the outer probes.

In order to measure the carrier concentration directly, a method is applied which uses the Hall effect. The simplest set-up is shown in Fig. 1.15. Here a voltage is applied to a semiconducting sample in the x-direction and a magnetic field is applied along the z-direction. The resulting Lorentz effect causes a force on the charge carriers, in the y-axis; this leads to an accumulation of electrons at the top side of an n-type sample and of holes at the bottom of a p-type sample. This effect causes a voltage V_H in the y-direction which is given by

$$V_H = R_H I_x \mathscr{B}_z w \tag{1.41}$$

in which $\mathscr{B}_z$ is the magnetic field, w the thickness of the sample and R_H the Hall coefficient.

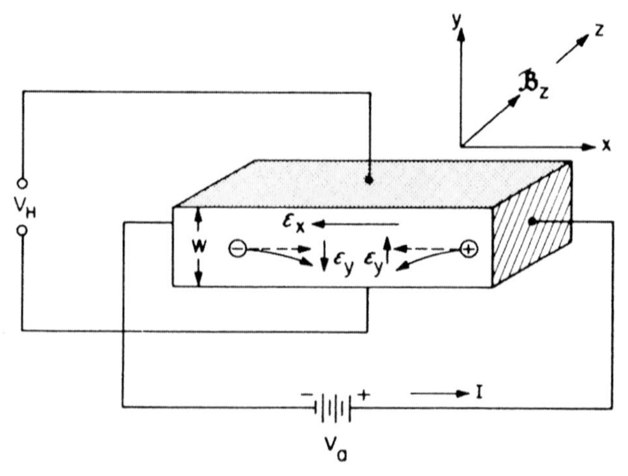

Fig. 1.15 Arrangement for measuring carrier concentrations by the Hall effect. (From ref. [7])

The latter is defined by

$$R_H = -r\frac{1}{en} \text{ (n-type)}; \quad R_H = r\frac{1}{ep} \text{ (p-type)} \tag{1.42}$$

where r is a constant depending on the scattering mechanism [2]. The corresponding mobility values can be obtained by using Eq. (1.36).

1.6 Excitation and Recombination of Charge Carriers

If the equilibrium of a semiconductor is disturbed by excitation of an electron from the valence to the conduction band, the system tends to return to its equilibrium state. Various recombination processes are illustrated in Fig. 1.16. For example, the electron may directly recombine with a hole. The excess energy may be transmitted by emission of a photon (radiative process) or the recombination may occur in a radiationless fashion. The energy may also be transferred to another free electron or hole (Auger process). Radiative processes associated with direct electron–hole recombination occur mainly in semiconductors with a direct bandgap, because the momentum is conserved (see also Section 1.2). In this case, the corresponding emission occurs at a high quantum yield. The recombination rate is given by

$$R_{np} = C_0 np \tag{1.43}$$

in which C_0 is a constant. During excitation the carrier density is increased by Δn and Δp, where $\Delta n = \Delta p$. Taking an n-type material as an example ($n_0 \gg p_0$), and using light intensities which are such that $\Delta n \ll n_0$ and $\Delta p \gg p_0$, then we have

$$R_{np} = C_0 n_0 \tag{1.44}$$

The lifetime is defined as $\tau = \Delta n / R_{np}$, so that

$$\tau = \frac{1}{C_0 n_0} \tag{1.45}$$

Accordingly the recombination rate as well as the lifetime of a band–band recombination process depends strongly on the carrier density.

In the case of semiconductors with an indirect gap, recombination occurs primarily via deep traps (Fig. 1.16). Here, an electron is first captured by the trap; in a second

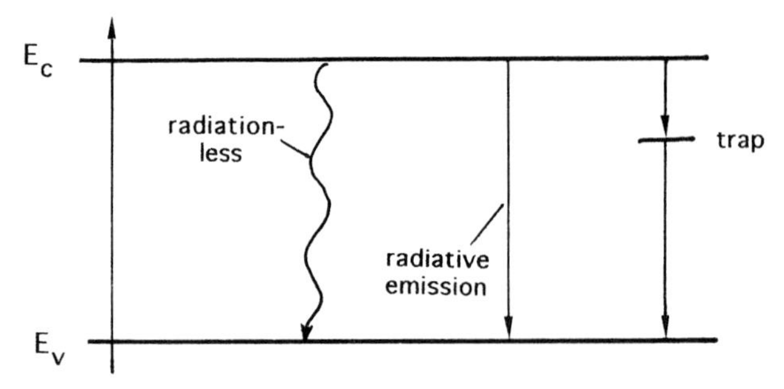

Fig. 1.16 Excitation and recombination of electrons

step the trapped electron recombines with the hole. It has been found that the recombination probability is much higher for a two-step process than for a single recombination process. This two-step process can be analyzed as follows.

The trapping rate for electrons from the conduction band into traps is proportional to the electron density in the conduction band and to the number of empty traps. We have then

$$R_C = C_n(1 - f_t)N_t n \tag{1.46}$$

in which N_t is the trap density, f_t denotes the fraction of traps occupied by electrons and C_n is given by

$$C_n = \gamma_n v_{th} \tag{1.47}$$

where γ_n is the electron capture cross-section and v_{th} the carrier thermal velocity equal to $(3kT/m^*)^{1/2}$. The rate of excitation of electrons from the trap into the conduction band is given by

$$R_e = C'_n f_t N_t \tag{1.48}$$

C'_n can be related to C_n by analyzing the equilibrium state which is determined by $R_c = R_e$. Applying this condition we have

$$C'_n = C_n \frac{n_0(1 - f_t^0)}{f_t^0} = C_n n_1 \tag{1.49}$$

in which n_0 and f_t^0 are the electron density and the fraction of occupied traps at equilibrium, respectively, with f_t^0 being given by

$$f_t^0 = \frac{1}{1 + \exp\left(\dfrac{E_t - E_F}{kT}\right)} \tag{1.50}$$

The carrier density for a Fermi level located just at the trap level ($E_F = E_t$) is given by

$$n_1 = N_c \exp\left(\frac{E_c - E_t}{kT}\right) \tag{1.51}$$

Combining Eqs. (1.28) and (1.51) one can show that n_1 is equal to the second term in Eq. (1.49), i.e.

$$n_1 = n_0\left(1 - f_t^0\right)\left(f_t^0\right)^{-1} \tag{1.52}$$

Using Eqs. (1.46) to (1.49), we can derive the overall flow of electrons into the traps as given by

$$R_n = \gamma_n v_{th} N_t\left[(1 - f_t)n - f_t n_1\right] \tag{1.53}$$

By analogy, a similar expression can be derived for the net capture rate of holes, R_p. We then have

$$R_p = \gamma_p v_{th} N_t\left[f_t p - (1 - f_t)p_1\right] \tag{1.54}$$

In the case of stationary illumination, the electron and hole flow must be equal ($R = R_n = R_p$). Applying this condition to Eqs. (1.53) and (1.54), f_t can be determined. Inserting the resulting equation into Eq. (1.50), one obtains

$$R = \frac{\gamma_n \gamma_p v_{th} \left(np - n_i^2 \right) N_t}{\gamma_n (n + n_1) + \gamma_p (p + p_1)} \tag{1.55}$$

This is the so-called Shockley–Read equation describing recombination via traps. It also plays an important role in the description of recombination processes via surface states, as discussed in Chapter 2. In the above equation one may also replace n_1 and p_1 by the relations

$$n_1 = n_i \exp\left(\frac{E_t - E_i}{kT} \right) \tag{1.56a}$$

$$p_1 = n_i \exp\left(-\frac{E_t - E_i}{kT} \right) \tag{1.56b}$$

which can be derived using Eqs. (1.28) and (1.30).

There are various techniques for measuring the lifetime of excited carriers, which cannot be described here. Details are given by Sze [7].

1.7 Fermi Levels under Non-Equilibrium Conditions

At equilibrium, the Fermi level, i.e. the electrochemical potential is constant throughout the semiconductor sample (Fig. 1.17a). In addition, the density of electrons and holes can be calculated simultaneously from Eqs. (1.28) and (1.30) if the position of the Fermi level within the bandgap is known. If the thermal equilibrium is disturbed, for instance by light excitation, then the electron and hole densities are increased to above their equilibrium value and we have $np > n_i^2$. Accordingly, the electron and hole density are not determined by the same Fermi level. It is useful to define quasi-Fermi levels, $E_{F,n}$ and $E_{F,p}$, one for electrons and another for holes, as given by

$$E_{F,n} = E_c - \ln\left(\frac{N_c}{n} \right) \tag{1.57}$$

$$E_{F,p} = E_v + \ln\left(\frac{N_v}{p} \right) \tag{1.58}$$

so that formally the original relations between carrier densities and Fermi level remain the same.

Let us consider a light excitation of electrons and holes ($\Delta n = \Delta p$) within a doped n-type semiconductor so that $\Delta n \ll n_0$ and $\Delta p \gg p_0$. Then the Fermi level of electrons, $E_{F,n}$, remains unchanged with respect to the equilibrium case, whereas that of holes, $E_{F,p}$, is shifted considerably downwards, as illustrated in Fig. 1.17b. In many cases, however, the excitation of electron–hole pairs occurs locally near the sample surface because the penetration of light is small. Then the splitting of the quasi-Fermi levels is

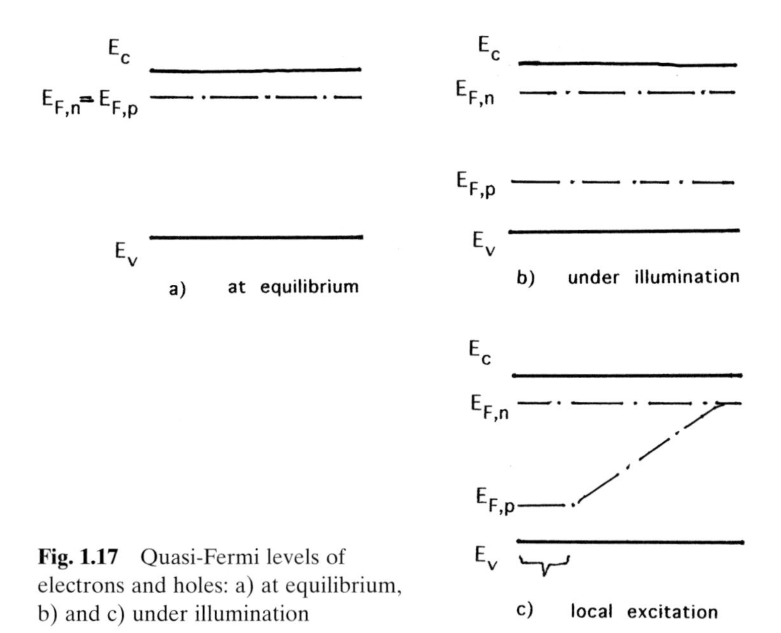

Fig. 1.17 Quasi-Fermi levels of electrons and holes: a) at equilibrium, b) and c) under illumination

large near the surface. Since the carriers diffuse out of the excitation range and recombine, the quasi-Fermi level of holes varies with distance from the excitation area (Fig. 1.17c).

The quasi-Fermi levels play an important role in processes at the semiconductor–liquid interface, because the relative position of the quasi-Fermi level with respect to that in solution yields the thermodynamic force which drives an electrochemical reaction (see Section 7.4).

2 Semiconductor Surfaces and Solid–Solid Junctions

2.1 Metal and Semiconductor Surfaces in a Vacuum

Clean metal and semiconductor surfaces can be produced, for instance, by cleavage of a crystal in vacuum. Assuming that there is no charge on the surface then the essential energies for a metal and for n- and p-type semiconductors in a vacuum are given in Fig. 2.1. These energy diagrams look fairly similar for metals and semiconductors. Differences occur primarily with respect to the Fermi level. The energy position of the Fermi level is determined by the work function $e\phi_M$. The vacuum E^{∞}_{vac} is taken as a reference level for the electrochemical potential of electrons where the electron energy is zero. In the case of a metal, $e\phi$ is identical to the ionization energy I and to the electron affinity E_A. Semiconductors differ from metals insofar as the position of the Fermi level depends on the doping (compare n- and p-type doping in Fig. 2.1) and there are no energy states at the Fermi level. In addition the ionization energy now corresponds to the energy required to excite an electron from the valence band to the vacuum, and the electron affinity corresponds to the energy gained if an electron is transferred from the vacuum into the conduction band. thus, the work function and therefore the Fermi level in semiconductors is not directly accessible. Only the ionization energy is directly measurable, by photoelectric methods. From these data the position of the Fermi level can be calculated if the distance between E_F and E_c or E_v is known.

With respect to E^{∞}_{vac} the position of the Fermi level is composed of two parts, namely the chemical potential μ_e and an electrostatic term $e\chi$, i.e. we then have (compare with Fig. 2.1)

$$E_F = \mu_e - e\chi = -e\phi \tag{2.1}$$

Provided that there is no additional surface charge, μ_e is a pure bulk term which is independent of any electrostatic potential. The term $e\chi$ is the contribution of surface dipoles [1, 2] (Fig. 2.1). Such a dipole can be caused by an unsymmetrical distribution of charges at the surface because there is a certain probability for the electrons to be located outside the surface. In the case of compound semiconductors, dipoles based on the surface structure caused by a particular ionic charge distribution occur. These effects depend on the crystal plane and on the reconstruction of the surface atoms [3, 4]. These dipole effects also influence the electron affinity and ionization energy. In the case of metals, the work function is a directly measurable quantity, and for semiconductors it is calculable from ionization measurements. However, the relative contributions of μ and $e\chi$ are not accessible experimentally and data given in the literature are based on theoretical calculations (see e.g. ref. [1]).

The situation becomes even more complex if surface states, i.e. additional energy levels within the bandgap, are present as illustrated in Fig. 2.2. In general, two types of

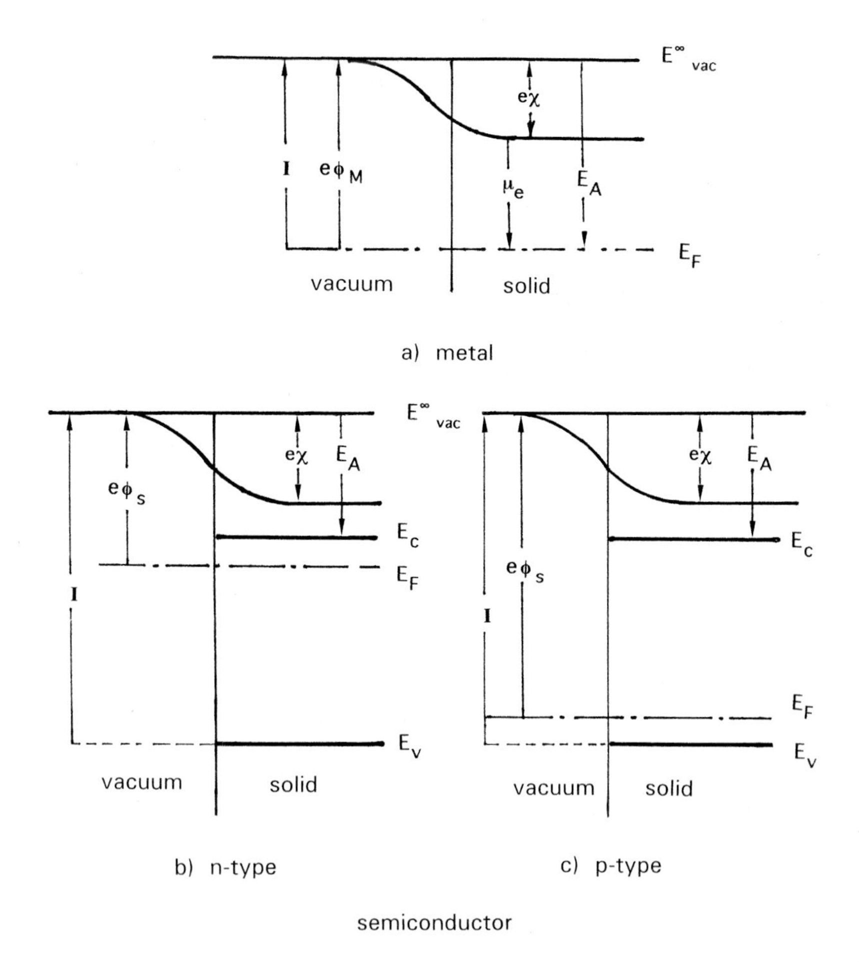

Fig. 2.1 Schematic presentation of different surface potentials at the solid–vacuum interface (details in the text). a) Metal; b) n-type; c) p-type semiconductor. (After ref. [1])

surface state are distinguished, i.e. intrinsic and extrinsic. Intrinsic states are associated with the original semiconductor surface, whereas extrinsic surface states result from the interaction with an external ambient, such as a strongly adsorbed species (see also Section 2.2). Even in ultrahigh vacuums, surface states are formed on clean surfaces as a result of the termination of the periodic crystal at the surface. Intrinsic surface states are categorized into two types: ionic states (Tamm states) and covalent surface states (Shockley states) [5]. The first type results from a heteronuclear splitting of bonding interaction. Since these energy states occur close to the band edges, they will not cause any band bending. Tamm states are mainly found with semiconductors, the bonding of which hasve a large ionic contribution. Examples include oxide semiconductors and wide bandgap chalcogenides such as CdS and ZnS.

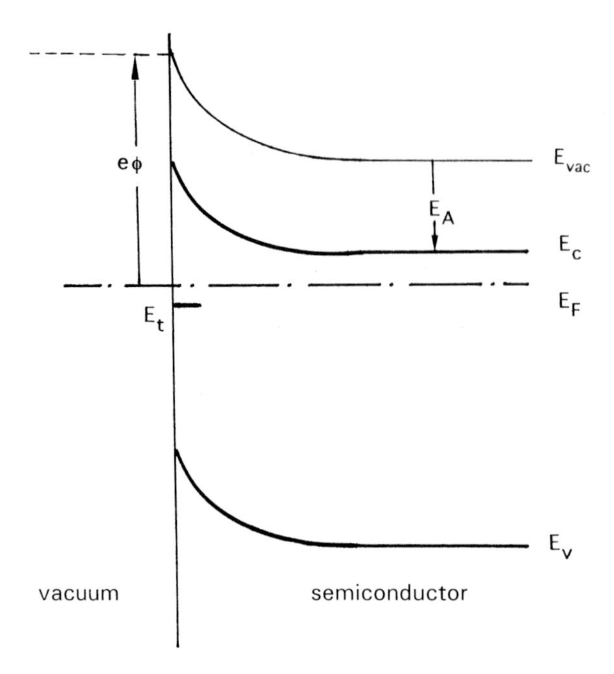

Fig. 2.2 Energy diagram of the semiconductor–vacuum interface in the presence of surface states

Shockley states correspond to unsaturated radical states which occur mainly at the surface of covalently bonded semiconductors. In the literature they are also described as dangling bonds. Typical examples are found with silicon and germanium surfaces. According to the analysis of the surface structure, the surface atoms rearrange, resulting in new bonds between surface atoms. In this case only a few dangling bonds are left, i.e. surface state densities in the order of 10^{12} cm^{-2} have been reported (see e.g. ref. [6]). These surface states are mainly located near the middle of the bandgap. In order to achieve electronic equilibrium between the surface and the bulk, a positive space charge is formed below the surface of an n-type semiconductor because some electrons are transferred to the surface states. A positive space charge means that fewer electrons are in this range. Accordingly, the energy distance between the Fermi level and the conduction band is increased at the surface, leading to a corresponding band bending as shown in Fig. 2.2. A corresponding energy scheme can be drawn for a p-type semiconductor. Intrinsic surface states have been found for many semiconductors, and in most cases the surface state energies have been determined experimentally (see e.g. refs. [7–11]).

2.2 Metal–Semiconductor Contacts (Schottky Junctions)

This section is of special interest because at first sight there are certain similarities between semiconductor–metal junctions and semiconductor–liquid interfaces. This will be discussed in more detail in Chapter 7.

2.2.1 Barrier Heights

When a contact is made between a semiconductor and a metal, the Fermi levels in the two materials must coincide at thermal equilibrium. We will consider here two limiting cases as shown in Fig. 2.3. In the first, we have a somewhat ideal situation (Fig. 2.3a). The semiconductor is assumed to be free from surface states so that the energy bands are flat as far as the surface before contact. In addition, any dipole layer on the metal and on the semiconductor surface is neglected. In the example shown in Fig. 2.3a, the Fermi level of the semiconductor occurs at a higher energy than that of the metal. Accordingly, some electrons are transferred from the semiconductor to the metal after close contact is made between them. This leads to a positive space-charge layer after

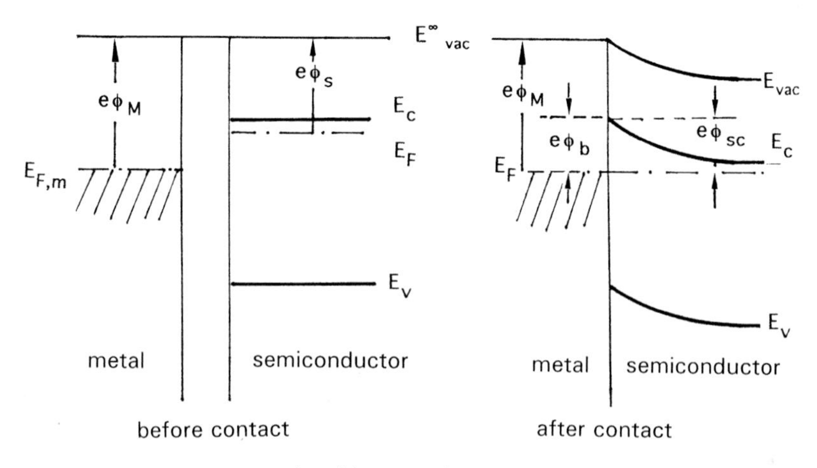

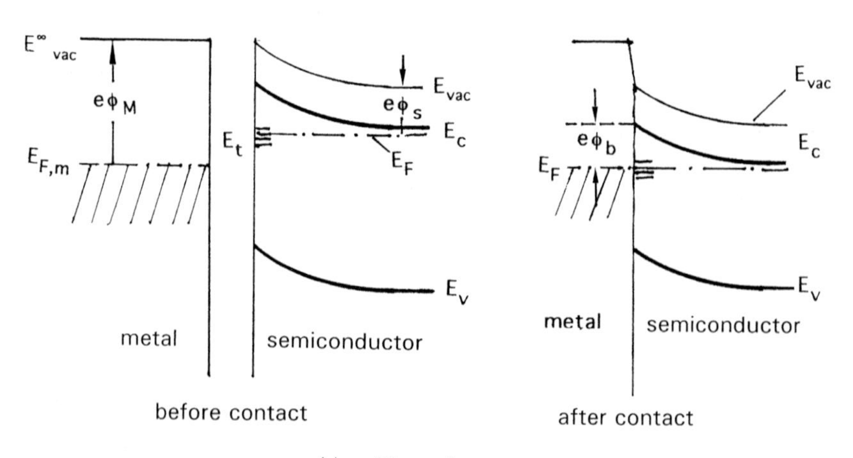

Fig. 2.3 Energy diagram of a metal–n-type semiconductor interface before and after contact. a) Without surface states; b) with surface states

thermal equilibrium has been achieved. Since the carrier density is small in a semiconductor, the positive space charge is distributed over a certain range below the surface. The thickness of this region depends on the doping, as described in more detail in Chapter 5. As a consequence of the positive space charge, an upward band bending occurs at the surface (Fig. 2.3a). The potential difference is called the contact potential, V_k. It is given by the difference between the work functions of the contacting materials, i.e.

$$V_k = \phi_m - \phi_s \tag{2.2}$$

This cannot be measured because there are other metal–metal and metal–semiconductor contacts in the measuring circuit, including those of the voltmeter, and the sum of all contact potentials is zero. As can easily be deduced from Fig. 2.3, the barrier height $e\phi_b$ at the metal–semiconductor contact is given by

$$e\phi_b(n) = e\phi_m - E_A \tag{2.3}$$

and the potential across the space charge layer is

$$e\phi_{sc} = e\phi_b - (E_c - E_F) \tag{2.4}$$

where E_A is the electron affinity of the semiconductor. It is also clear from Fig. 2.3 that the final position of the Fermi level at the surface and the potential across the space charge layer, ϕ_{sc}, depend on the relative position of the two Fermi levels before the materials come into contact. This is illustrated in Fig. 2.4. In the case of a metal with a very small work function, electrons are transferred from the metal to the semiconductor, leading to a negative space charge, i.e. the energy bands are bent downwards (Fig. 2.4a). In an n-type semiconductor this is called an accumulation region. On the other hand, if $\phi_m > \phi_s$, a positive space charge (upward band bending) is formed. In this case we have a depletion layer (Fig. 2.4b). In the case of a very large ϕ_m, one obtains a large band bending. If the hole density dominates at the surface, the space charge region is called an inversion layer (Fig. 2.4c). The border between depletion and inversion will be specified in more detail in Chapter 5 in relation to semiconductor–liquid interfaces.

It should be mentioned that for p-type materials, usually, a negative space charge is formed because the work function of the semiconductor is below that of the metal. Again assuming an ideal contact, the energy barrier is given by

$$e\phi_b(p) = E_g - (e\phi_m - E_A) \tag{2.5}$$

For a given semiconductor the sum of the two barrier heights, $e\phi_b(n)$ and $e\phi_b(p)$, is expected to be equal to the bandgap.

In the other extreme case, the formation of contact between a metal and a semiconductor with a high density of surface states is shown in Fig. 2.3b. The Fermi level occurs at the energy of the surface states before making the contact, indicating that the surface levels are about half-filled. If the concentration of the surface states is sufficiently large to take further charges without much change of the occupation level (E_F remains constant at the surface), the space charge below the semiconductor surface remains unchanged upon contact between semiconductor and metal. Accordingly, if some electrons are transferred from the surface to the metal, the barrier height remains unaffected. It is entirely determined by the properties of the semiconductor surface and the

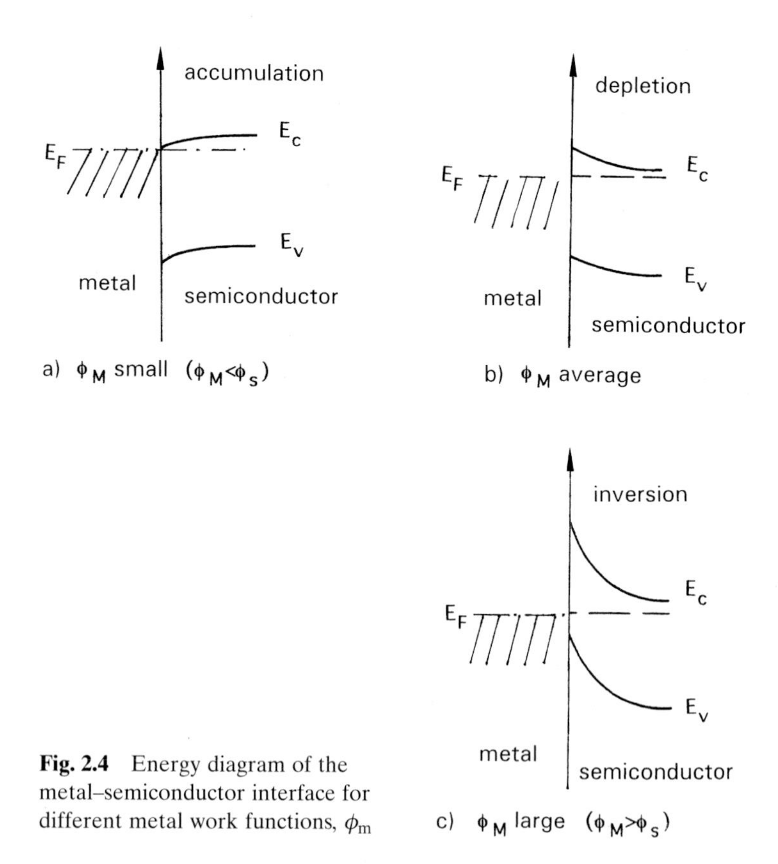

Fig. 2.4 Energy diagram of the metal–semiconductor interface for different metal work functions, ϕ_m

contact potential occurs across a very small range at the interface. Some authors describe this situation as "Fermi level pinning" [12–15].

A large number of semiconductor metal junctions have been studied. There are various experimental techniques for measuring barrier heights, such as photoelectric and capacity measurements and current–voltage investigations. The first-mentioned technique seems to be the most accurate. These methods are not described here; some of them are discussed in Chapters 4 and 5 (see also refs. [7, 12, 16]).

Many experimental values for barrier heights at semiconductor–metal junctions have been obtained. Many researchers have also measured the barrier height as a function of the work function of the metal, and have mostly obtained a straight line, as expected from Eq. (2.3). However, in many cases the slope, $d\phi_b/d\phi_m$, of a corresponding plot was much smaller than unity. In 1969 it was shown by Kurtin et al that the sensitivity of a barrier height to different metals increases with the ionicity of the semiconductor [17]. In order to obtain a better characterization of the experimental data, they defined an index of interface behavior, S, which they introduced into Eq. (2.3) as

$$e\phi_B = S(\phi_m - E_A) + C \tag{2.6}$$

in which C is a constant. S is obtained experimentally as the linear slope of an $e\phi_b$ versus $e\phi_m$ or X_m plot (X_m = the electronegativity of the metal). From $e\phi_b$ vs. X_m plots of

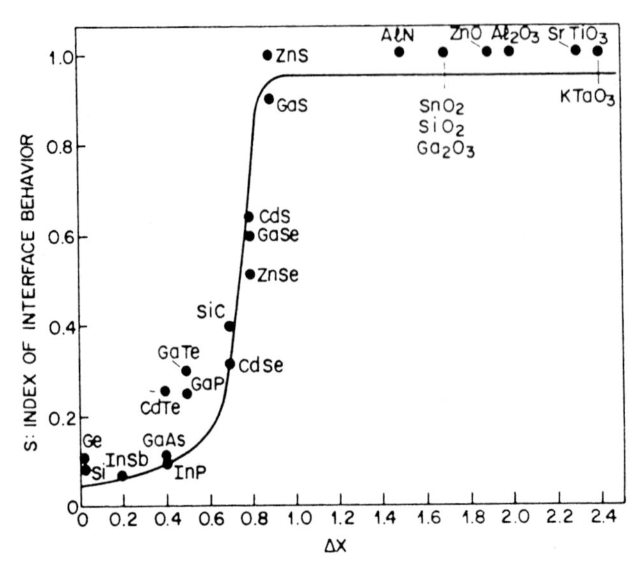

Fig. 2.5 Index of interface behavior, S, as a function of the electronegativity differences of semiconductors (After ref. [17])

various metals on a given vacuum-cleaved semiconductor surface, these authors developed a well-accepted curve of interface behavior. A corresponding plot of S vs. semiconductor ionicity, the electronegativity difference ΔX, is given in Fig. 2.5. This curve displays a marked increase in S between more covalent and ionic semiconductors. In the case of ionic semiconductors, S reaches values close to unity. The increase of S with ΔX was widely interpreted as evidence for intrinsic surface states [12, 18]. Investigations of microscopic details by surface science techniques have shown, however, that the interface is much more complex [1, 7, 12, 16, 19, 20]. For instance, the surface dipoles cannot be expected to remain unchanged when a contact between metal and semiconductor is made in a vacuum. Changes may result from the reconstruction of the surfaces, from changes in electron polarization and in surface relaxation. These effects make it more or less impossible to predict the barrier height at a metal–semiconductor interface. Such a concept is quite common in semiconductor electrochemistry because the formation of a double layer at a solid–liquid interface controls the position of the band edges (see Chapter 5).

The contact between metals and III-V semiconductors has been studied in detail [21]. These semiconductors are of special interest here because many electrochemical experiments have been performed with these materials. In the case of metal–semiconductor junctions, the barrier heights were found to be nearly independent of the type of metal, which was interpreted as strong Fermi level pinning. Various models have been proposed for interpreting the pinning of the barrier height at the surface of several III-V compounds. Besides the "metal induced gap state" (MIGS), the "unified defect model" (UDM) has been used to interpret the pinning of the Fermi level at the interface (see e.g. refs. [22, 23]). The main argument in favor of the unified defect model was the experimental result that the pinning of the Fermi level is relatively independent of the type of foreign atom deposited on the surface. Results consistent with this were not only obtained with GaAs but also with InP. It has been suggested that the pinning is caused by defects located in the semiconductor near the interface produced during the deposition of foreign atoms [21]. In the case of GaAs, two defects are even considered, namely an acceptor level (As vacancy) at 0.7 eV and a donor level at 0.9 eV below the conduction band [21], which would explain a constant barrier height of 0.8 eV for n-GaAs.

The question arose, however, as to whether other treatments could lead to different values. Concerning this problem, very interesting data were reported by Aspnes et al. [24]. These authors deposited thin metal layers onto n-GaAs and p-InP and exposed the junctions alternatively to air or hydrogen. In both cases the distance between Fermi level and conduction band at the interface decreased upon hydrogen exposure. They found a small change of 0.1–0.2 eV for GaAs [24] and a very large change of 0.5 eV for InP [24]. This effect was reversible. It was interpreted as a variation in the dipole component of the metal work function upon changing the ambient gas. This result can only be understood in terms of the UDM model if the density of defects is not too large. In a more recent investigation, in which metals such as Cu, Au and Pt were deposited electrochemically on n-GaAs, barrier heights of 1.1–1.2 eV were found [26]. This result also indicates that the Fermi level pinning is not primarily caused by intrinsic surface states but depends on the surface dipole layer. It has been discussed whether an electrochemical deposition of noble metals on a hydride layer on the GaAs surface is responsible for these effects.

2.2.2 Majority Carrier Transfer Processes

The mechanism most often used to describe the electron transfer across a semiconductor–metal Schottky junction, is the thermionic emission model. The theory, derived by Bethe [27], is based on the assumptions that: (1) the barrier height is larger than kT: (2) thermal equilibrium exists at the plane which determines emission, and (3) the net current does not affect this equilibrium [16]. Accordingly, the current flow depends only on the barrier height. Considering an n-type semiconductor, the current density $j_{\mathrm{s}\rightarrow\mathrm{m}}$ from the semiconductor to the metal is given by the concentration of electrons (majority carriers in n-type semiconductors) with energies sufficient to overcome the energy barrier. The basic model of an electron transfer from the conduction band of an n-type semiconductor to a metal is illustrated in Fig. 2.6. At first the electron density, dn, in a small energy interval dE at energies above $E_{\mathrm{c}} + e\phi_{\mathrm{b}}$ in the bulk is calculated. Postulating that all this energy is kinetic energy, the electron velocity v_x towards the surface will be derived and finally the corresponding current is obtained. Following the derivation given by Sze [16] we have

$$j_{\mathrm{s}\rightarrow\mathrm{m}} \rightarrow \int_{E_F+e\phi_{\mathrm{b}}}^{\infty} e v_x \,\mathrm{d}n \tag{2.7}$$

where $E_{\mathrm{F}} + e\phi_{\mathrm{b}}$ is the lowest energy required for an electron transfer into the metal if no external voltage is applied, and $\mathrm{v_x}$ is the electron velocity in the x-direction (x-direction is perpendicular to the surface). The electron density dn in an incremental energy interval is given by

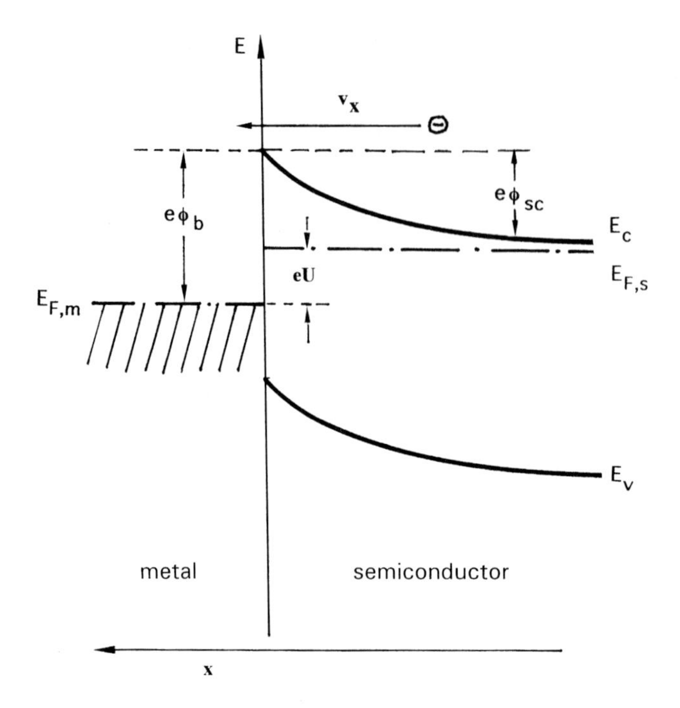

Fig. 2.6 Electron transfer at the semiconductor–metal interface according to the thermionic emission model

$$\mathrm{d}n = N(E)f(E)\,\mathrm{d}E$$
$$= \frac{4\pi(2m^*)^{3/2}}{h^3}(E - E_c)^{1/2}\exp\left[-\frac{(E - E_c) + (E_c - E_F)}{kT}\right]\mathrm{d}E \tag{2.8}$$

in which $N(E)$ is the density of states (see Eq. 1.22) and $f(E)$ the distribution function (Eq. 1.25), respectively. For $f(E)$ the approximation of $E - E_F \gg kT$ is used.

As already mentioned above, it is assumed that the energy of the electrons in the conduction band is entirely kinetic energy. We have then

$$E - E_c = \tfrac{1}{2}m^*v^2 \tag{2.9a}$$
$$\mathrm{d}E = m^*v\,\mathrm{d}v \tag{2.9b}$$
$$\left(E - E_c\right)^{1/2} = v\left(m^*/2\right)^{1/2} \tag{2.9c}$$

in which E_c is the conduction band edge and m^* the effective mass. Inserting Eq. (2.9) into Eq. (2.8) one obtains

$$\mathrm{d}n = 2\left(\frac{m^*}{h}\right)^3 \exp\left(\frac{E_c - E_F}{kT}\right)\exp\left(-\frac{m^*v^2}{2kT}\right)\left(4\pi v^2\,\mathrm{d}E\right) \tag{2.10}$$

Since the electrons can move everywhere, Eq. (2.10) describes the density of electrons having velocities between v and dv distributed over all directions. Considering the speeds in the three main directions we have

$$v^2 = v_x^2 + v_y^2 + v_z^2 \tag{2.11}$$

Using the transformation $4\pi v^2\, dv = dv_x\, dv_y\, dv_z$ one obtains from Eqs. (2.7), (2.10) and (2.11)

$$j_{s\to m} = 2e\left(\frac{m^*}{h}\right)^3 \exp\left(-\frac{E_c - E_F}{kT}\right) \times$$

$$\int_{v_x^0}^{\infty} \exp\left(-\frac{m^* v_x^2}{2kT}\right) dv_x \int_{-\infty}^{+\infty} \exp\left(-\frac{m^* v_y^2}{2kT}\right) dv_y \int_{-\infty}^{+\infty} \exp\left(-\frac{m^* v_z^2}{2kT}\right) dv_z$$

$$= \left(\frac{4\pi e m^* k^2}{h^3}\right) T^2 \exp\left(-\frac{E_c - E_F}{kT}\right) \exp\left(-\frac{m^*(v_x^0)^2}{2kT}\right) \tag{2.12}$$

in which v_x^0 is the minimum electron speed at equilibrium. If a negative voltage U is applied to the semiconductor with respect to the metal, then the barrier for the electrons in the conduction band becomes smaller. Using Eq. (2.9) we have then

$$\tfrac{1}{2} m^* (v_x^0)^2 = e(\phi_{sc} - U) \tag{2.13}$$

Substituting this equation into Eq. (2.12) one obtains

$$j_{s\to m} = A T^2 \exp\left(-\frac{e\phi_b}{kT}\right) \exp\left(\frac{eU}{kT}\right) \tag{2.14}$$

in which ϕ_b is the barrier height and A is given by

$$A = \frac{4\pi e m^* k T}{h^3} \tag{2.15}$$

This is the effective Richardson constant for thermionic emission. In the case of isotropic effective mass one can rewrite Eq. (2.15) as

$$A = \frac{4\pi e m_0 k T}{h^3}\left(\frac{m^*}{m_0}\right) = 120\ \mathrm{A\,cm^{-2}}\left(\frac{m^*}{m_0}\right) \tag{2.16}$$

in which m_0 is the free electron mass. The first term contains only physical constants and can be calculated as given in Eq. (2.16). For further details concerning unisotropic effective mass see ref. [16].

At equilibrium ($U = 0$), the total current must be zero. Accordingly, there is a reverse current, $j_{m\to s}$, from the metal to the semiconductor. Since the externally applied voltage occurs only across the space charge layer of the semiconductor, the barrier height also remains constant for reverse bias so that $j_{m\to s}$ is independent of the voltage U. Therefore, $j_{m\to s}$ must be equal to the $j_{s\to m}$-value at equilibrium, i.e.

$$j_{m\to s} = j_0 = j_{s\to m} = -A T^2 \exp\left(-\frac{e\phi_b}{kT}\right) \quad \text{at } U = 0 \tag{2.17}$$

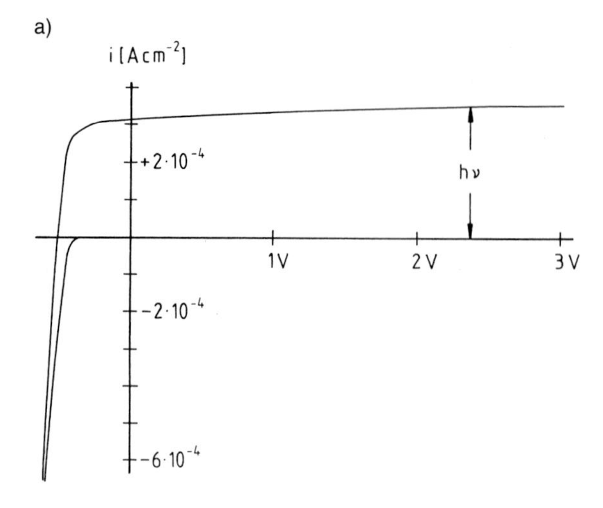

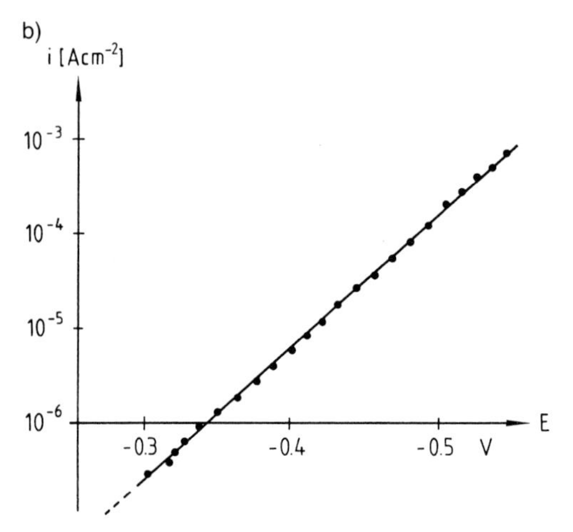

Fig. 2.7 a) Current–voltage curve for an n-type GaAs–Au Schottky junction. Au was deposited electrolytically.
b) Semilogarithmic plot of the forward dark current.
(After ref. [26]).

The total current, $j = j_{s \to m} + j_{m \to s}$, is then given by

$$j = j_0 \left[\exp\left(\frac{eU}{kT} \right) - 1 \right] \tag{2.18}$$

with

$$j_0 = AT^2 \exp\left(-\frac{e\phi_b}{kT} \right) \tag{2.19}$$

It is important to realise that this derivation is based on the assumption that all electrons reaching the surface with a speed v_x are transferred. This is of interest especially in comparison with processes at the semiconductor–liquid interface. As a consequence of this assumption, the forward currents attain large values at relatively low voltages. Taking for instance a relative effective mass of one, then $AT^2 \approx 10^7$ A cm^{-2}. Assuming further a barrier height of $\phi_b = 0.8$ eV as found for GaAs ($E_g = 1.4$ eV), because of Fermi level pinning by surface states, then the pre-exponential factor in Eq. (2.18) amounts to $j_0 \approx 10^{-7}$ A cm^{-2} (calculated from Eq. 2.19). A typical current–voltage curve is given in Fig. 2.7.

Quite a large number of systems have been studied and most of the current–voltage curves follow the thermionic emission model [12, 16]. Frequently, there is some difference in the slope; for example, instead of a theoretical slope of 60 mV per decade in current, slopes of 70–75 mV were found. This deviation may either be due to an image-force-induced lowering of the barrier or to tunneling through the space charge layer, as has been quantitatively studied for Au/Si barriers [28]. These two effects have been treated in detail by Sze [16].

2.2.3 Minority Carrier Transfer Processes

A system in which only majority carriers (electrons in n-type) carry the current, is frequently called a "majority carrier device". On the other hand, if the barrier height at a semiconductor–metal junction reaches values close to the bandgap then, in principle, an electron transfer via the valence band is also possible, as illustrated in Fig. 2.8a. In this case holes are injected under forward bias which diffuse towards the bulk of the semiconductor where they recombine with electrons ("minority carrier device"). It is further assumed that the quasi-Fermi levels are constant across the space charge region; i.e. the recombination within the space charge layer is negligible. In addition Boltzmann equilibrium exists so that we have according to Eqs. (1.57) and (1.58)

$$n = N_c \exp\left(\frac{E_c - E_{F,n}}{kT}\right) \tag{2.20a}$$

$$p = N_v \exp\left(\frac{E_v - E_{F,p}}{kT}\right) \tag{2.20b}$$

The pn product becomes

$$np = N_c N_v \exp\left(-\frac{E_c - E_v}{kT}\right) \exp\left(\frac{E_{F,n} - E_{F,p}}{kT}\right) \tag{2.21}$$

and with Eq. (1.32)

$$np = n_i^2 \exp\left(\frac{E_{F,n} - E_{F,p}}{kT}\right) \tag{2.22}$$

As already mentioned, we have a forward bias if the n-type semiconductor is made positive with respect to the metal. Under these conditions holes are injected into the semiconductor and we have $pn > n_i^2$. In this case the quasi-Fermi level of holes, $E_{F,p}$, occurs below that of electrons, $E_{F,n}$, in Fig. 2.8, i.e. it is closer to the valence band which is equivalent to the fact that the minority carrier density is increased. The externally applied voltage is then determined by

$$eU = E_{F,n} - E_{F,p} \tag{2.23}$$

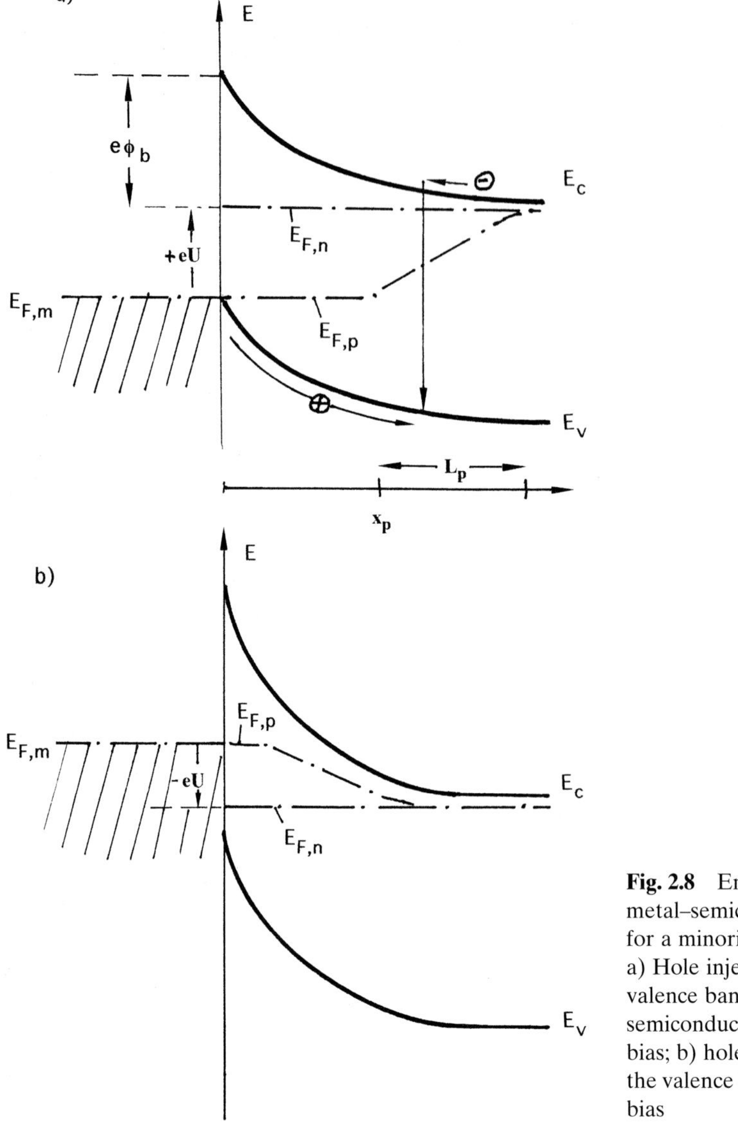

Fig. 2.8 Energy diagram of a metal–semiconductor junction for a minority carrier device. a) Hole injection into the valence band of an n-type semiconductor under forward bias; b) hole extraction from the valence band under reverse bias

If there is a strong coupling between the metal and the semiconductor, $E_{F,p}$ is close to E_F in the metal at all potentials. For a reverse bias (negative U), there is an extraction of holes and $E_{F,p}$ occurs above $E_{F,n}$ as shown in Fig. 2.8b. The resulting current–voltage curve can be derived as follows.

In the case of minority carrier injection, the interfacial current is not only determined by the electric field but also by diffusion of the carriers. The hole current j_p is then

$$j_p = e\mu_p p\mathscr{E} + eD_p\nabla p \tag{2.24}$$

in which μ_p is the mobility of holes and $\mathscr{E}$ the electric field at the interface. The second term in Eq. (2.24) determines the hole diffusion where D_p is the diffusion constant and ∇p the gradient of the hole density (Fick's law). Using the Einstein relation (Eq. (1.39) one obtains from Eq. (2.22) and from the fact that $\mathscr{E} = -\nabla E_c$ (here taking the conduction band as a measure for the potential)

$$\begin{aligned} j_p &= e\mu_p \left(p\mathscr{E} + \frac{kT}{e} \nabla p \right) \\ &= \mu_p \left[-\nabla(E_c) + e\mu_p \frac{kT}{e} \left\{ \frac{p}{kT} (\nabla(E_c) - \nabla(E_{F,p})) \right\} \right] \\ &= \mu_p \nabla(E_{F,p}) \end{aligned} \tag{2.25}$$

According to this equation the hole current is determined by the gradient of the hole quasi- Fermi level as indicated in Fig. 2.8a. This gradient is caused by a gradient of the hole concentration which decreases with distance from the surface because of recombination. Since it has been assumed that this gradient occurs in a range which is much larger than the space charge thickness, i.e. in a field-free region, the hole current is governed by diffusion. Accordingly, the first term in Eq. (2.24) can be neglected. The exact concentration profile can be derived from the continuity equation (see e.g. ref. [16]). Here we assume that there is a linear concentration profile over a distance which corresponds to the diffusion length L defined by

$$L = (D\tau)^{1/2} \tag{2.26}$$

in which τ is the lifetime of the carrier (see also Section 1.6). This is an essential quantity in semiconductor systems which is also of importance for charge transfer processes at semiconductor–liquid interfaces.

The diffusion current is then given by

$$j_p = eD_p \nabla p = eD_p \,\mathrm{grad}(p - p_0) \qquad |_{x=x_p \approx 0} \qquad = eD_p \frac{(p - p_0)}{L_p} \tag{2.27}$$

in which D_p and L_p are the diffusion constant and length of holes, respectively, whereas p_0 is the hole density at $x = x_p \approx 0$ at equilibrium. According to Eq. (1.29) we have

$$p = N_v \exp\left(\frac{E_v - E_{F,p}}{kT} \right) \tag{2.28}$$

$$p_0 = N_v \exp\left(\frac{E_v - E_F}{kT} \right) \tag{2.29}$$

Dividing Eq. (2.28) by (2.29) one obtains

$$\frac{p}{p_0} = N_v \exp\left(\frac{E_F - E_{F,p}}{kT} \right) \tag{2.30}$$

Inserting Eqs. (2.30) and (2.23) into (2.27) leads to

$$j_p = j_0 \left[\exp\left(\frac{eU}{kT} \right) - 1 \right] \tag{2.31}$$

with

$$j_0 = \frac{eD_p p_0}{L_p} \tag{2.32a}$$

or when substituting p_0 by Eq. (1.32):

$$j_0 = \frac{eD_p n_i^2}{n_0 L_p} \tag{2.32b}$$

Eq. (2.31) is identical to Eq. (2.18) derived for a majority carrier device (thermionic emission model). Accordingly, the same type of current–voltage curve is expected as that given in Fig. 2.7. The characteristics of the models occur only in the pre-exponential factors, which indeed are different in both cases (compare Eqs. 2.17 and 2.30). As mentioned before the j_0 of the majority carrier device is only determined by the barrier height and some physical constants (Eq. 2.19), whereas the j_0 of the minority carriers depends on material-specific quantities such as carrier density, diffusion constant and diffusion length.

It is interesting to ask at which barrier height of a given system the minority injection process should dominate over the thermionic emission of majority carriers. Taking n-type GaAs (E_g = 1.4 eV) as an example, one can calculate the barrier height at which the two j_0 are equal. Using typical values for GaAs, such as D_p = 6.3 cm^2 s^{-1}, n_i^2 = 3.1 × 10^{13} cm^{-6}, n_0 = 3 × 10^{16} cm^{-3}, L_p = 10^{-4} cm, one obtains j_0 = 1 × 10^{-17} A cm^{-2} for the hole injection current. The same j_0 value is obtained by applying the thermionic emission model (Eq. 2.19) with m^*/m_0 = 0.05 (light holes) and a barrier height of ϕ_b = 1.3 eV! According to this result, the hole injection process would only dominate for a system with a barrier height which almost reaches the bandgap of 1.4 eV. This result shows again the efficiency of a majority carrier transfer. In practice, one would never obtain a barrier height sufficient for a hole injection process because of Fermi level pinning as discussed in Section 2.2.2. Nevertheless, the minority carrier process has been treated here in detail because it plays an important role in charge transfer processes at the semiconductor–liquid junction.

2.3 p–n Junctions

In accordance with the definitions given above, a p–n junction is a pure minority carrier device. In this case the forward current is entirely determined by the injection of holes into the n-doped region and of electrons into the p-doped region, as illustrated by the energy diagram for an abrupt p–n junction in Fig. 2.9. This diagram is very similar to that derived for the metal–semiconductor device, where minority carrier injection also determines the forward current, as discussed in Section 2.2.3. The only difference here is that we have two space charges on both sides of the interface: a positive in the n-doped region and a negative in the p-doped region. In the simplest case, it is also assumed here that the recombination of the injected minority carriers with the majority carriers occurs in the bulk of the semiconductor, i.e. outside the space charge

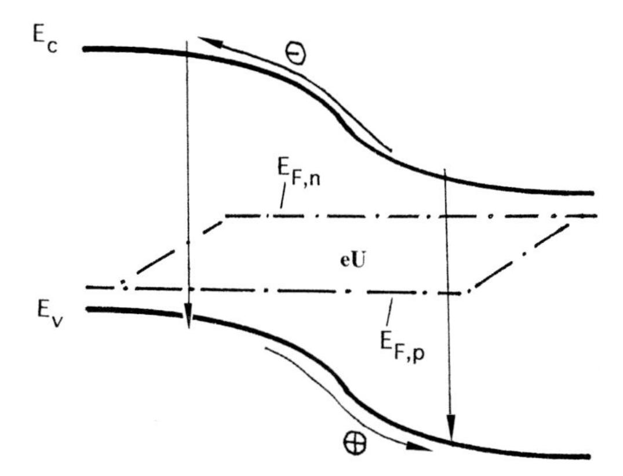

Fig. 2.9 Energy diagram of a p–n junction under forward bias

regions. Accordingly, it is assumed that the quasi-Fermi levels remain constant within the space charge regions (Fig. 2.9).

In principle the same derivation can be applied as that used in Section 2.2.3. The only difference is that the minority carrier current is not only determined by the hole injection and recombination in the n-type region but also by the injection of electrons into the p-type area. Thus, we now have instead of Eq. (2.31)

$$j_{\mathrm{p}} = \frac{eD_{\mathrm{p}}n_{\mathrm{i}}^2}{n_0 L_{\mathrm{p}}}\left[\exp\left(\frac{eU}{kT}\right) - 1\right] \tag{2.33}$$

and

$$j_{\mathrm{n}} = \frac{eD_{\mathrm{n}}n_{\mathrm{i}}^2}{p_0 L_{\mathrm{n}}}\left[\exp\left(\frac{eU}{kT}\right) - 1\right] \tag{2.34}$$

and the total current is given by

$$j = j_{\mathrm{n}} + j_{\mathrm{p}} = j_0\left[\exp\left(\frac{eU}{kT}\right) - 1\right] \tag{2.35}$$

in which j_0 is given by

$$j_0 = en_{\mathrm{i}}^2\left(\frac{D_{\mathrm{n}}}{p_0 L_{\mathrm{n}}} + \frac{D_{\mathrm{p}}}{n_0 L_{\mathrm{p}}}\right) \tag{2.36}$$

Eq. (2.36) is the famous Shockley equation which is the ideal diode law [16]. According to Eq. (2.35) we have obtained again the same basic current–voltage dependence as already derived for majority and minority carrier devices with semiconductor–metal junctions (see Eqs. 2.18 and 2.31). As already mentioned, the physical difference occurs only in the pre-exponential factor j_0. The general shape of a complete j–U curve in a linear and semilog plot has already been given in Fig. 2.7. Concerning the slope of the j–U dependence in the semilog plot, one would expect, at room temperature, a current rise by one decade if the voltage changes by about 59 mV (= 2.3 kT/e). Frequently, however, a considerably higher slope, i.e. $\gg$ 59 mV, has been found with real metal–semiconductor and p–n junctions. This is empirically expressed by a diode equation such as

$$j = j_n + j_p = j_0 \left[\exp\left(\frac{eU}{nkT} \right) - 1 \right] \tag{2.37}$$

in which the n in the exponent is the so-called quality or ideality factor. In the case of a minority carrier device, such as a p–n junction, this can be explained by an additional recombination process within the space charge region, i.e. the recombination does not occur only in the bulk of the semiconductor. This means, in terms of quasi-Fermi levels, that the latter do not remain constant across the space charge region. In this case the recombination within the space charge region can be expressed by an additional recombination current as given in ref. [30]

$$j_{rec} = \int_0^{d_{sc}} eR\,\mathrm{d}x \tag{2.38}$$

in which d_{sc} is the thickness of the space charge region and R is the recombination rate given by Eq. (1.55) (see Chapter 1). It is helpful to use Eq. (1.55) because it describes the recombination of electron–hole pairs via energy states within the gap and their occupation is determined by the banding and therefore by the applied voltage U. Inserting Eq. (2.22) into (1.55), and assuming that the capture cross-sections for electrons and holes are identical ($\sigma_n = \sigma_p = \sigma$), and further assuming that the recombination centers are located in the middle of the gap ($E_t = E_i$), then one obtains

$$j_{rec} \approx \frac{e}{2} e d_{sc} v_{th} N_t n_i \exp\left(\frac{eU}{2kT} \right) \tag{2.39}$$

in which N_t is the density of recombination centers, v_{th} the thermal velocity and d_{sc} the thickness of the space charge region. The definition of d_{sc} will be derived in relation to semiconductor–liquid junctions in Chapter 5. For details of the derivation of Eq. (2.39) see ref. [16]. Accordingly, the slope of the semilog plot of the current due to recombination within the space charge layer versus voltage exhibits an ideality factor of 2. The total forward current is then given by

$$j_{tot} = j_0 \left[\exp\left(\frac{eU}{kT} \right) - 1 \right] + \frac{e}{2} e d_{sc} v_{th} N_t n_i \exp\left(\frac{eU}{2kT} \right) \tag{2.40}$$

The factor n equals 2 when the recombination current dominates and n equals 1 when the diffusion current dominates. When both currents are comparable, n has a value between 1 and 2.

Sometimes an ideality factor of greater than 1 is also reported for a majority carrier device. In this case, however, there is no physical basis for an ideality factor of $n > 1$ and any deviation from $n = 1$ must have technological reasons.

As mentioned above, electrons are injected into the p-doped area and holes into the n-doped area of the semiconductor if a p–n junction is under forward bias (Fig. 2.9). In the case of some semiconductors, for example, GaAs or GaP, the subsequent recombination of the minority carriers occurs with light emission. This effect is used in so-called "light-emitting diodes" (LED). Emission effects have also been observed with semiconductor–liquid junctions and corresponding recombination effects in semiconductors are treated in Section 7.7.

2.4 Ohmic Contacts

Whenever charge transfer processes across solid–solid or solid–liquid junctions are studied by measuring currents, it is necessary that the current-carrying cables form an ohmic contact to the semiconducting specimen, i.e. any energy barrier for electrons and holes has to be avoided at such a contact. According to the derivations for rectifying contacts given in Section 2.2, an energy diagram for an ohmic barrierless contact can be easily developed for n- and p-type semiconductors as shown in Fig. 2.10.

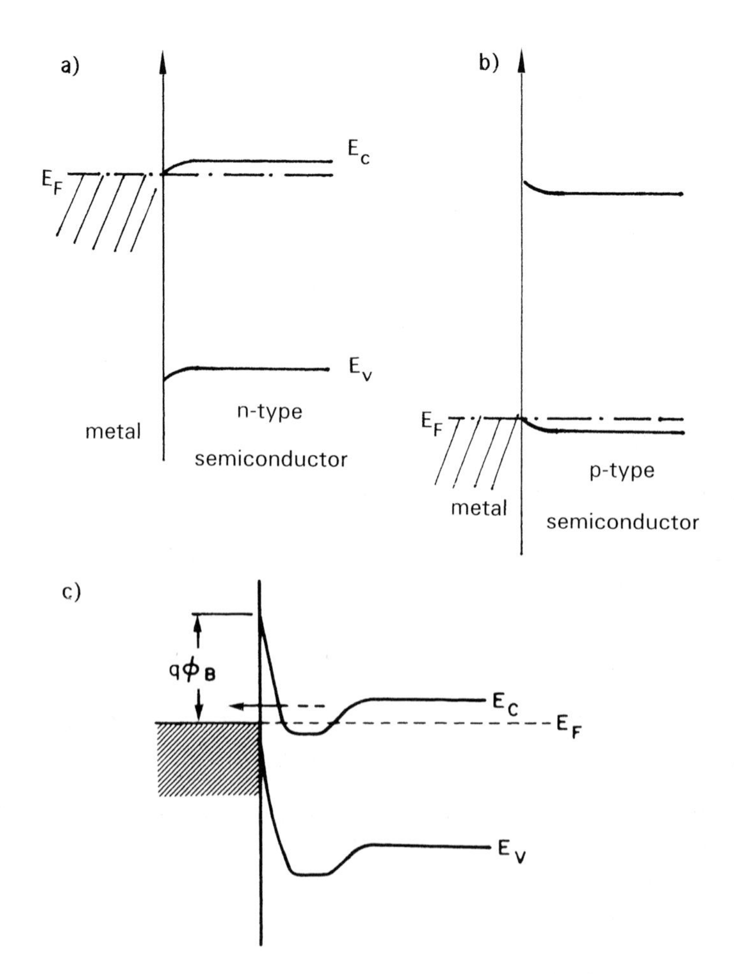

Fig. 2.10 Ohmic contacts between metal and semiconductor. a) and b) Using metals forming low barriers with n- and p-type semiconductors; c) using a metal forming a high barrier (see text)

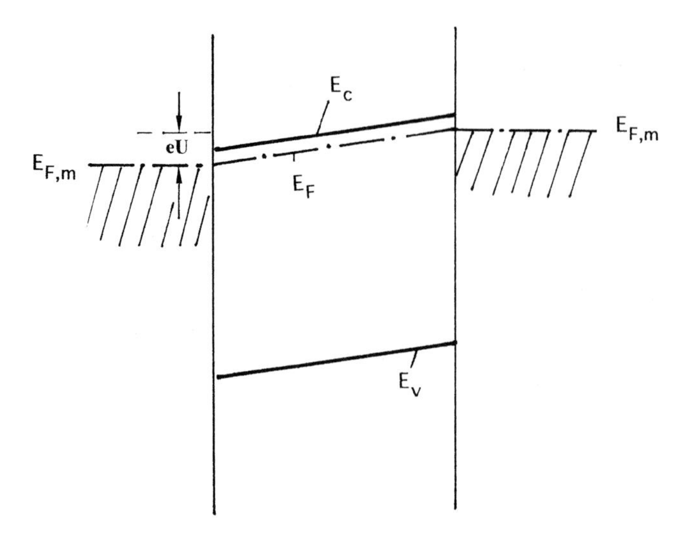

Fig. 2.11 Energy diagram of a metal–semiconductor–metal system under bias (ohmic contacts)

Theoretically, there are two ways of making a good ohmic contact: (1) deposition of a metal of a sufficiently low work function on an n-type crystal (or of high work function on a p-type material); (2) application of a metal which can act as an electron donor when used as a doping material in the semiconductor. In the first case, an accumulation layer is formed within the semiconductor at the interface (Figs. 2.10a and b). This method is not applicable for wide-gap semiconductors because generally a metal does not exist with a low enough work function to yield a small barrier height. Therefore, the second method has to be applied. Here, the sample is heated up so that the metal diffuses into it forming a highly doped area (n^+ in n-type and p^+ in p-type) just below the semiconductor surface. The Fermi level passes then the semiconductor–metal interface (degenerated surface) very close to the corresponding band edge, or the electrons tunnel through the very thin n^+ region (Fig. 2.10c). In practise, however, problems frequently arise in the technology of ohmic contacts and many of these are solved empirically. The quality of ohmic contacts can easily be tested by forming two contacts and applying a voltage across them, and measuring a current–voltage curve which should not only exhibit a linear dependence but also a slope which corresponds to the resistivity of the material.

In terms of the Fermi level no step should occur at two ohmic metal–semiconductor contacts if a voltage is applied between the contacts. This voltage occurs across the whole specimen between the contacts leading to a linear rise of Fermi level and energy bands (constant electric field) as illustrated in Fig. 2.11. In the case of a rectifying contact, however, the externally applied voltage occurs only across the space charge region (Figs. 2.6 and 2.8), as already discussed in detail in Section 2.2. Consequently the Fermi level of the majority carriers here remains constant within the semiconductor.

2.5 Photovoltages and Photocurrents

Typical current–voltage characteristics have been derived for metal–semiconductor and p–n junctions in Sections 2.2 and 2.3. The forward current, rising exponentially with voltage, was determined either by majority carrier transfer or by minority carrier injection. The reverse current, on the other hand, is entirely limited by minority extraction which leads to extremely small currents as shown in Fig. 2.7. The small minority carrier concentration can be increased by light excitation within the semiconductor, as illustrated for a majority carrier device (metal–semiconductor junction) in Fig. 2.12. Assuming that the light enters the system through the metal layer (i.e. the metal layer is made sufficiently thin), and assuming that nearly all the light is absorbed within the

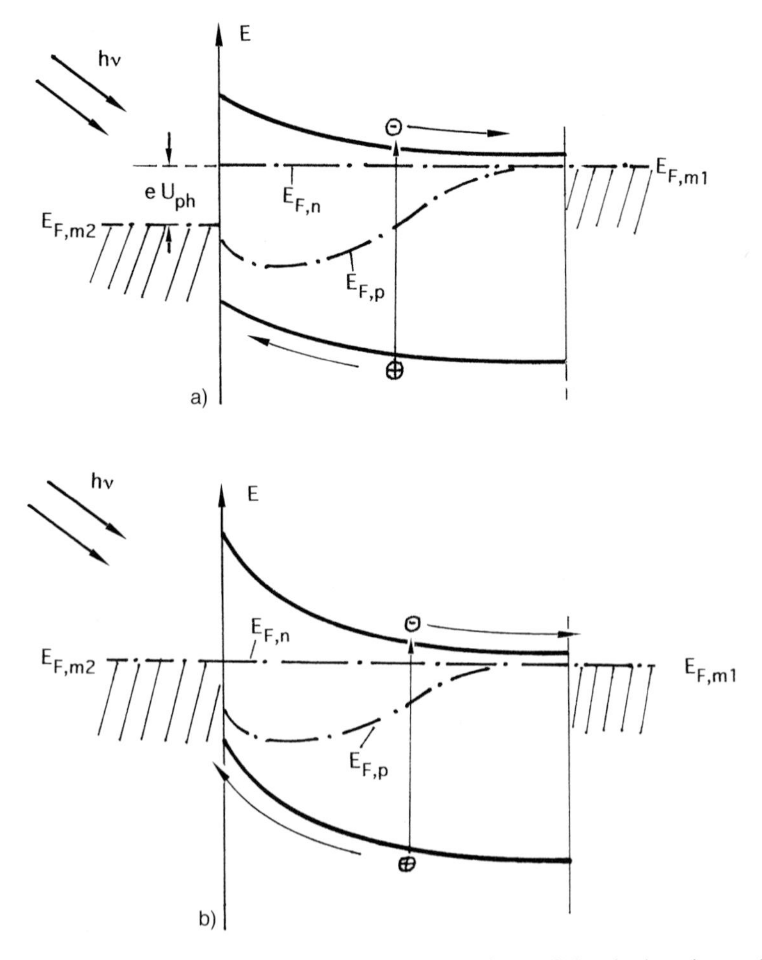

Fig. 2.12 Energy diagram of a metal–semiconductor Schottky junction under illumination. a) Open circuit condition, b) short circuit condition

space charge layer (high absorption coefficient), then all electron–hole pairs created by this process are separated by the electric field within the space charge layer. According to the sign of the field, negative on the metal side and positive within the space charge region, the electrons are pushed towards the bulk of the semiconductor whereas the holes created by light excitation move toward the interface (Fig. 2.12). These holes can easily be captured by an electron in the metal because a large density of occupied energy states exists on the metal side just at the edge of the valence band of the semiconductor. This hole transfer leads to an additional current, the sign of which is the same as that of the reverse dark current. A corresponding current-potential curve as would be expected for an illuminated system, is shown in Fig. 2.13. For comparison, the dark current is also given.

In principle the same process occurs in minority carrier devices. In all cases the photocurrent is proportional to the light intensity and is independent of the applied potential. Accordingly, the photocurrent occurs in the diode equation (Eqs. (2.18) or (2.31) or (2.37)) as an additive term, so that we have

$$j = j_0 \left[\exp\left(\frac{eU}{kT}\right) - 1 \right] - j_{ph} \tag{2.41}$$

in which j_{ph} is the photocurrent, and j_0 depends on the type of system and is given by the corresponding Eqs. (2.19), (2.32) and (2.36).

We can distinguish between two cases: first the short circuit condition and secondly the open circuit condition. The first is defined as a condition where the system is short circuited, i.e. $U = 0$. According to Eq. (2.41) the total current is then only determined

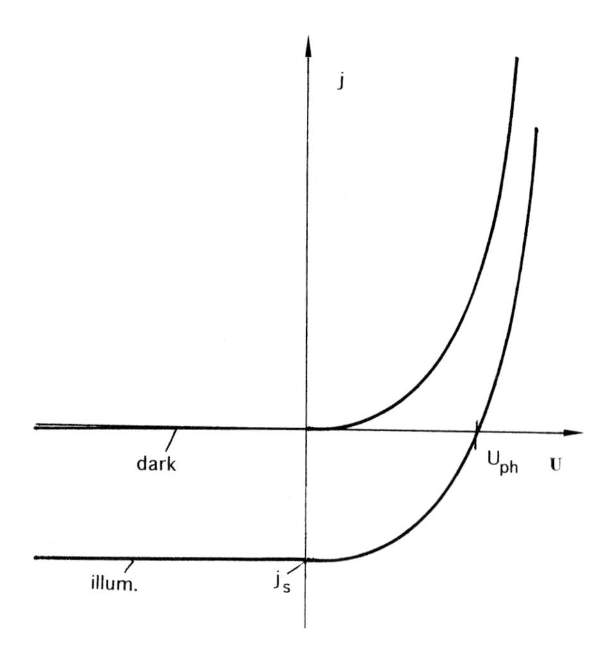

Fig. 2.13 Current–voltage curve for a metal–semiconductor junction in the dark and under illumination

by the photocurrent ($j = j_{ph}$). This case is also illustrated by the energy diagram in Fig. 2.12b. $U = 0$ means that the quasi-Fermi level of the majority carriers is equal across the whole system, from the metal which forms the rectifying contact, through the semiconductor up to the metal forming the ohmic contact. Although $U = 0$ there is no equilibrium between electrons and holes because of light excitation ($np \gg n_i^2$). Accordingly, the quasi-Fermi level of holes (minority carriers in the n-type semiconductor) occurs closer to the valence band and at the interface below the Fermi level $E_{F,m2}$ of the metal. The latter statement means that a hole transfer is thermodynamically possible.

The second case, the open circuit condition, means that the total current is zero ($j = 0$). Inserting this value into Eq. (2.41), one obtains the so-called photovoltage U_{ph} as given by

$$U_{ph} = \frac{kT}{e} \ln\left(\frac{j_{ph}}{j_0} + 1\right) \tag{2.42}$$

as also indicated in the current–voltage curve in Fig. 2.13. Since this photovoltage can be measured between the two metal contacts their Fermi levels cannot be equal here; they just differ by eU_{ph} as shown in the corresponding energy diagram (Fig. 2.12a). At first sight it may seem surprising that the quasi-Fermi level of the electrons ($E_{F,n}$) in the semiconductor occurs here above that in the metal ($E_{F,m2}$), whereas the Fermi level of the holes ($E_{F,p}$) is below $E_{F,m2}$. This is caused by the fact that we have actually two partial currents of opposite signs at $U = U_{ph}$, a positive dark current and a negative photocurrent.

The discussion of photoeffects in terms of quasi-Fermi levels may seem to be rather pointless because there is no way of determing the quasi-Fermi levels at solid–solid junctions experimentally. It has been introduced here, however, because it is possible to obtain experimental information on quasi-Fermi levels in the case of semiconductor–liquid junctions, and it will be shown in Chapter 7 that the same principles can be applied for semiconductor–liquid and solid–solid junctions.

Returning to Eq. (2.42), it should be mentioned that the photovoltage U_{ph} depends only on the ratio of j_{ph}/j_0. Since the j_0 value is very different for majority and minority carrier devices, there can be huge differences in U_{ph}. Particularly in the case of the majority carrier devices where the forward current is derived by the thermionic emission model, large j_0 values can be obtained (see Eq. 2.19) leading to very small photovoltages. This would be a great disadvantage for an application for a photovoltaic device (see Chapter 11). In addition it should be mentioned that the quantum yield of the photocurrent depends strongly on the penetration depth of the incident light and also on the diffusion length of minority carriers. A quantum yield of unity can be expected only in cases where the diffusion length is larger than the penetration depth. A quantitative relation between photocurrent or quantum yield on the one hand and diffusion length, diffusion constant, absorption coefficient and thickness of the space charge layer on the other, has been derived by Gärtner [31]. This problem will be treated quantitatively for semiconductor–liquid junctions in Section 7.3.3.

2.6 Surface Recombination

Electrons and holes may not only recombine via energy states in the bulk of the semiconductor but also via surface states. Surface recombination can occur at all surfaces of a semiconconductor device but in the case of pure solid and semiconductor–liquid junctions, we are mainly interested in the recombination via surface states at the interface. Assuming the surface states to be associated with single charged centers having a single, discrete level within the energy gap of the semiconductor, then the Shockley-Read model [32] can be applied to surface recombination, as derived at first by Stevenson and Keyes [33]. Using Eq. (1.55) which was derived for bulk recombination, the rate of surface recombination is given by

$$R = \frac{\gamma_n \gamma_p v_{th} (n_s p_s - n_i^2) N_t}{\gamma_n (n_s + n_1) + \gamma_p (p_s + p_1)} \tag{2.43}$$

This equation differs from Eq. (1.55) insofar as we have introduced the carrier densities at the surface, namely n_s and p_s. We further assume that during light excitation, quasi-equilibrium exists for the distribution of electrons and holes in the space charge layer, i.e. the surface densities of electrons and holes at the surface are related to the bulk densities, n_b and p_b, via the Boltzmann factor. We have then

$$n_s = n_b \exp\left(-\frac{e\phi_{sc}}{kT}\right) \tag{2.44a}$$

$$p_s = p_b \exp\left(\frac{e\phi_{sc}}{kT}\right) \tag{2.44b}$$

Accordingly, the product of the densities is given by

$$n_s p_s = n_b p_b > n_i^2 \tag{2.45}$$

This assumption does not include the condition that equilibrium is achieved between electrons and holes ($n_s p_s > n_i^2$).

The electron and hole densities are changed from their thermal equilibrium δn and δp when electron–hole pairs are created by light excitation. In the bulk, these deviations must be equal to preserve electrical neutrality. Therefore we may write

$$n_s = n_0 + \delta n \tag{2.46a}$$

$$p_s = p_0 + \delta n \tag{2.46b}$$

Inserting Eqs. (2.44) and (2.46) into (2.43) and assuming $\delta n \ll n_0$ and $\delta n \ll p_0$, one obtains the so-called surface recombination velocity (dimension, cm s^{-1})

$$s = \frac{R}{\delta n} = \frac{\gamma_n \gamma_p v_{th} (n_0 + p_0) N_t}{\gamma_n \left[n_p \exp\left(-\frac{e\phi_{sc}}{kT}\right) + n_1 \right] + \gamma_p \left[p_0 \exp\left(\frac{e\phi_{sc}}{kT}\right) + p_1 \right]} \tag{2.47}$$

The definition of $s = R/\delta n$ is useful because s is a quantity which is independent of δn and consequently of the light intensity. According to Eq. (2.47), s must pass a maxi-

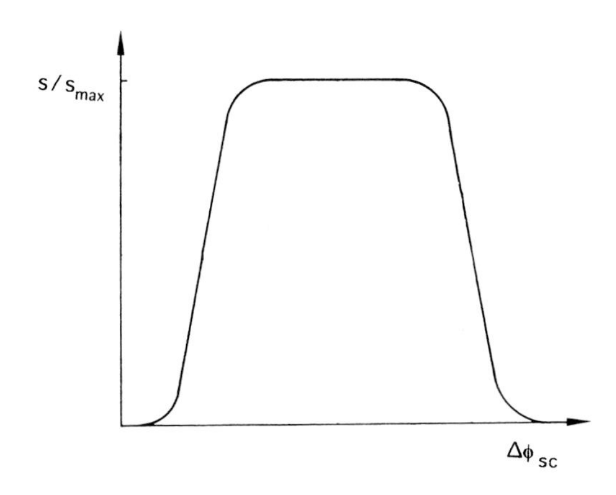

Fig. 2.14 Surface recombination, s, vs. potential across the space charge layer $\Delta\phi_{sc}$ (qualitative description)

mum when the potential is varied across the space charge layer, ϕ_{sc}, as illustrated qualitatively in Fig. 2.14. One branch of the curve is determined by the majority carriers, the other by the minority carriers. The position of the maximum on the ϕ_{sc} scale and the width of the curve depend on the doping of the semiconductor and on the capture cross-sections, γ_n and γ_p. Physically speaking, Eq. (2.47) illustrates that for electron-hole recombination at the surface, both holes and electrons must be able to flow into the surface [34]. If ϕ_{sc} is such as to strongly repel either electrons or holes from the surface, s will be low. The validity of Eq. (2.47) has been proved at least qualitatively in many experiments (see e.g. ref. [5]).

In the case of doped semiconductors, the assumption $< n_0$ or $\delta n < p_0$ is almost impossible to fulfill for minority carriers under experimental conditions. An increase of the light intensity then leads to a decrease of the surface recombination velocity along the branch determined by the minority carrier density (for details see ref. [5]).

3 Electrochemical Systems

Many of the topics discussed here have also been presented in other classical textbooks (see for instance refs. [1–5]).

3.1 Electrolytes

3.1.1 Ion Transport in Solutions

In contrast to solid electronic conductors such as metals and semiconductors, the current in liquid electrolytes is carried by ions. These ions are formed by dissociation of salts in a polar solvent, for example KCl or Na_2SO_4 in water. The dissociation leads to the formation of positively as well as negatively charged ions and both types participate in the conduction process. In contrast to the situation with semiconductors, where there are also two types of charge carrier (electrons and holes) but one type usually dominates because of doping, both types of carrier are always present in equal concentrations in the electrolyte. The conductivity is, in principle, given by Eq. (1.36), i.e. by the same equation used for electronic conductivity in solids. Employing the usual electrochemical terminology, we then have for one type of dissociated molecule (compare with Eq. 1.36)

$$\sigma = F(z_1\mu_+ + z_2\mu_-)c/1000 \tag{3.1}$$

in which z_1 and z_2 are the charges of the ions; μ_+ and μ_- are the mobilities of the positively and negatively charged ions, having the dimension $cm^2\ V^{-1}\ s^{-1}$; c is the concentration of the ions, given here in moles per cm^3; and F is the Faraday constant, which is given by the product of elementary charge and of Avogadro number (eN_{avo}), so that we have $F = 96500$ A s $mole^{-1}$. With these definitions the dimension of σ is again $(\Omega\ cm)^{-1}$ as already defined in Section 1.5.

When different salts are dissolved in the electrolyte then the conductivity can be expressed in more general terms

$$\sigma = F\sum_i |z_i|\mu_i c_i \tag{3.2}$$

(In most books on electrochemistry κ and u_i are used as symbols for conductivity and mobility, respectively. In order to have the same symbols throughout this book, we are using here those which were introduced in the solid state physics chapter).

The mobility of an ion is related to the velocity v and the electric field $\mathscr{E}$ in which an ion is moving. The corresponding electric force, $|z_i|e\mathscr{E}$, accelerates the ion until the frictional drag exactly counterbalances the electric force. The frictional drag can be approximated from Stokes' law as $6\pi\eta rv$, in which η is the viscosity of the solution,

r the radius of the ion and v the velocity. Using these equations and the definition of mobility, namely $\mu_i = v/\mathscr{E}$, then one obtains

$$\mu_i = \frac{v}{\mathscr{E}} = \frac{|z_i|e}{6\eta r} \tag{3.3}$$

in which r is actually the hydrodynamic radius which accounts for the "solvation" (see Section 3.1.2) of the ion. Eq. (3.3) makes it possible to calculate theoretical values of mobility. According to Eq. (3.1), conductivity in solutions should be proportional to the concentration of the ions. Since deviations from linearity are frequent, it is useful to introduce a so-called equivalent conductivity, Λ, defined as

$$\Lambda = \frac{\sigma}{c_{eq}} \tag{3.4}$$

where c_{eq} is the concentration of positive or negative charges in moles per cm^3. Thus Λ expresses the conductivity per unit concentration of charge. Since $|z|c = c_{eq}$ for either ionic species, one finds from Eqs. (3.2) and (3.4) that

$$\Lambda = F(\mu_+ + \mu_-) \tag{3.5}$$

This equation also implies the definition of individual equivalent conductivities, so that we have according to Eq. (3.5)

$$\Lambda = \Lambda_+ + \Lambda_- \tag{3.6}$$

in which

$$\Lambda_+ = F\mu_+; \qquad \Lambda_- = F\mu_- \qquad \text{or} \qquad \Lambda_i = \mu_i \tag{3.7}$$

In addition it is common in electrochemistry to use transference numbers defined as

$$t_i = \frac{\Lambda_i}{\Lambda} \tag{3.8}$$

or alternatively, when using Eqs. (3.6) and (3.7), as

$$t_i = \frac{\mu_i}{\mu_+ + \mu_-} \tag{3.9}$$

Transference numbers have been introduced because they can be determined experimentally. On the other hand, the individual mobilities cannot be determined independently as in the case of electrons and holes in a semiconductor; conductivity measurements yield only the sum of the cation and anion mobilities (see Eq. 3.1). Accordingly, the mobilities can be evaluated from measurements of the conductivity and the corresponding transference numbers. There are various methods for measuring transference numbers which are not described here (for details, see for example ref. [2]). Since both Λ_i and t_i depend on the concentration of the ions, data are usually evaluated from measurements in very dilute solutions because of interactions between ions. According to many investigations, the transference numbers of most ions are not far from $t_i = 0.5$; i.e. the mobilities of cations and anions of dissolved salts are about the same. Large t_i values are only obtained with acids and alkaline solutions. This is the result of the special transport mechanism of hydrogen and hydroxyl ions in water in which hydrogen bonds play an important role.

According to the experimental data, the mobility of most ions is around 10^{-4} cm^2 V^{-1} s^{-1}. These values are about 5–6 orders of magnitude smaller than the mobility of electrons and holes in a semiconductor. In order to achieve sufficient conductance in an electrochemical cell, electrolyte conductivities of $\sigma > 10^{-2}$ $(\Omega\ \text{cm})^{-1}$, and therefore ion concentrations of $c > 10^{-1}$ mole per liter, are required. In investigations of electrode processes it is important that the solution is made sufficiently conductive by the addition of ions which are not involved in the electrode reaction. Such a solution is usually called a "supporting electrolyte".

The conductivity of an electrolyte describes the transport of ions within an electrical field in the solution. Ions and molecules can also move in the solutions via diffusion. This becomes important in the electrode reactions of molecules or ions added to the supporting electrolyte. Provided that the concentration of these ions is much smaller than that of the supporting electrolyte, then the electric field does not affect the movement of the ions, i.e. these ions, as well as uncharged molecules, reach the electrodes only by diffusion. If the rate of the electrode process becomes large then the concentration of the reacting species decreases and that of the generated species increases near the electrode, which leads to a concentration profile. The corresponding diffusion process can be described by Fick's law, as will be discussed in Section 7.1.2. The diffusion process is essentially characterized by the diffusion constant D. Similarly to electrons and holes, the diffusion constant is related to the mobility by the Einstein relation (see Eq. 1.39). We have then

$$D_{\mathrm{i}} = \frac{kT}{e}\mu_{\mathrm{i}} \tag{3.10}$$

Since the mobility of ions and molecules in solution is very low the diffusion constants are also low. With $\mu_{\mathrm{i}} \approx 5 \times 10^{-4}$ cm^2 V^{-1} s^{-1} one obtains $D \approx 2 \times 10^{-6}$ cm^2 s^{-1} for $z_{\mathrm{i}} = 1$.

3.1.2 Interaction between Ions and Solvent

The ions in the solution interact with other ions and molecules but mostly with the solvent molecules. This overall interaction is termed the "solvation". It is interesting to recognize that the heat involved when a salt such as KCl is dissolved in water, is rather small (of the order of 10 kJ mole^{-1}) although the lattice energy or the dissociation energy is very large, typically in the range of 200–1000 kJ mole^{-1} (8–20 eV). The reason for this small value is that the large dissociation energy is compensated by the solvation or hydration energy which is of the same order of magnitude. Various attempts have been made to calculate the hydration energy theoretically, assuming an electrostatic ion–dipole attraction. Basically there are two essential contributions to calculation of the solvation enthalpies. One is a "near order" model in which the interaction between the ion and the solvent dipoles (e.g. water dipoles) are calculated as given by

$$\Delta H_{\mathrm{id}} = \frac{N_{\mathrm{avo}} n |z_{\mathrm{i}}| e_0 P_{Lm}}{4\pi\varepsilon_0 (r_{\mathrm{i}} + r_{\mathrm{sol}})} \tag{3.11}$$

in which N_{avo} is the Avogadro number; n is the number of solution molecules which are in direct contact with the ion; z_i is the charge of the ions; P_{Lm} is the dipole moment; r_i and r_{sol} the radii of the ion and the solution molecules, respectively, and ε_0 is the permittivity of free space.

This equation actually describes the interaction between the ion and the first solvation shell. Ions frequently form a very stable "aquo complex" with H_2O molecules, for example an $Fe(H_2O)_6^{3+}$ complex. Besides this "inner sphere" interaction there is also an "outer sphere" one, leading to a corresponding arrangement of the H_2O dipoles around the ions. The outer sphere interaction is given by the energy required if an ion with the inner solvation shell (radius $r_i + r_{sol}$) is transferred from a vacuum into the solution, as derived by Born using the continuum model:

$$\Delta G_{Born} = -\frac{N_{avo} z_i^2 e_0^2}{4\pi \varepsilon_0^2 (r_i + 2r_{sol})} \left(1 - \frac{1}{\varepsilon_{sol}}\right) \tag{3.12}$$

in which ε is the dielectric constant of the solution. The corresponding enthalpy is obtained by using

$$\Delta H_{Born} = \Delta G_{Born} + T\Delta S \tag{3.13}$$

Inserting (3.13) into (3.12) one obtains

$$\Delta H_{Born} = -\frac{N_{avo} z_i^2 e_0^2}{4\pi \varepsilon_0^2 (r_i + 2r_{sol})} \left(1 - \frac{1}{\varepsilon_{sol}} - \frac{T}{\varepsilon_{sol}^2} \frac{\partial \varepsilon_{sol}}{\partial T}\right) \tag{3.14}$$

Evaluations of both contributions yield values of the same order of magnitude, i.e. several hundreds of kJ $mole^{-1}$. For an exact evaluation some further contributions must be considered, namely the ion quadrupole interaction and an additional ion dipole interaction induced by the electric field of the ion. Further details of these derivations are given in ref. [2].

The solvent structure around the ions depends heavily on the charge of the ion. Accordingly, it changes during or after an electron transfer. Such a "rearrangement" or "reorganization" of the solvent molecules and the corresponding energy change play an important role in the theory of electron transfer, as described in detail in Section 6.1.2.

Since the dielectric constants of polar solvents have rather high values ($\varepsilon = 80$ for H_2O) the solvation can be a very energetic process for a low ion radius and a high charge. Typical values are in the range from 2 to 20 eV. Although more accurate models for calculating the hydration energy have been developed, the order of magnitude of the energy has been about the same as that obtained from the Born equation. The best results are obtained from Eq. (3.14) for cases where mainly the outer sphere is involved. This occurs for ions which are already surrounded by an inner sphere complexing compound before they are dissolved. One example is the $Fe(CN)_6^{4-}$ complex, for which the net charge number $z = 4$ is effective when the Born equation is applied.

3.2 Potentials and Thermodynamics of Electrochemical Cells

3.2.1 Chemical and Electrochemical Potentials

The thermodynamics of solutions and solid–liquid interfaces can be well described in terms of the chemical and electrochemical potentials of the system. The basic definition of the chemical potential [6] is

$$\mu_i^\alpha = \left(\frac{\partial G}{\partial n_i}\right)_{T,p,n_{j \neq i}} \tag{3.15}$$

where G is the Gibbs free energy and n_i is the number of moles of the ith species in phase α if the temperature T, the pressure p and the concentration of the species are kept constant. For an ideal solution, the chemical potential of an ion is related to its concentration c_i (see e.g. ref. [6]) by

$$\mu_i = \mu_i^0 + RT \ \ln c_i/c_i^0 \tag{3.16}$$

in which μ_i^0 is the standard chemical potential for $c_i = c_i^0$. For real solutions, this equation is not sufficiently accurate because of ion–ion interactions in the solution. Therefore a correction term, the so-called activity coefficient f, has been introduced. We have then

$$\mu_i = \mu_i^0 + RT \ \ln\left(fc_i/fc_i^0\right) = \mu_i^0 + RT \ \ln\left(a_i/a_i^0\right) \tag{3.17}$$

in which a_i is the activity. In many applications, i.e. for dilute solutions, it can be assumed that $a_i \approx c_i$.

The chemical potential or free energy of a species, the latter being a component in a solid or in a solution, depends on the chemical environment. In the case of a charged species, such as an ion or an electron, we have to consider in addition the electrical energy required for bringing a charge to the site of the species. Accordingly, an electrochemical potential μ_i is defined instead of the chemical potential. Both are related by

$$\overline{\mu_i} = \mu_I + z_i F\phi \tag{3.18}$$

in which $z_i F\phi$ is the potential energy of the charged species, with F as the Faraday constant (96 500 Coulomb mole^{-1}); and z_i is the charge number of the species, and ϕ is the potential in volts. Instead of F one can also use the elementary charge $e = 1.6 \times 10^{-19}$ A s. The reference point of ϕ is that at which $\overline{\mu}_i = \mu_i$; we return to the problem of the reference point later. It should be emphasized once more that the potential, ϕ, depends only on the large-scale environment of a given phase, and the potential energy $z_i F\phi$ on the charge number z_i while it is independent of the chemical nature of the species; for example, the potential energies of Cu^{2+} and Zn^{2+} in a given (i.e. aqueous) phase are identical.

It is important to realize that at equilibrium in a system the electrochemical potential, μ_i, is constant over all contacting phases as far as the ith substance is exchangeable between these phases. Accordingly we have

$$\overline{\mu_i} = \text{const.} \tag{3.19a}$$

Hence, the following relation results for such an exchange under equilibrium conditions:

$$\Delta\overline{\mu} = \Delta G = 0 \tag{3.19b}$$

Considering for instance a reaction in a single phase we have

$$\nu_1 A_1 + \nu_2 A_2 \Leftrightarrow \nu_3 A_3 + \nu_4 A_4 \tag{3.20}$$

where the A symbols represent the species in a reaction, and the v symbols are integers. The rate of the reaction must be equal in both directions at equilibrium, and it can be shown in accordance with Eq. (3.19) that the electrochemical potentials of the reactants and the products must then be related by

$$\textstyle\sum_{\mathrm{i}} \nu_{\mathrm{i}}\overline{\mu_{\mathrm{i}}} = 0 = \Delta G \tag{3.21}$$

The v_{i} values of the product have to be taken as positive and those of the educt negative. According to Eqs. (3.15) to (3.21) the properties of the electrochemical potential can be classified as follows.

a) Reactions in a Single Phase

Within a single conducting phase, the potential ϕ has the same value everywhere within this phase and has no effect on the chemical equilibrium. This is frequently illustrated using the dissociation of acetic acid in aqueous solutions

$$\mathrm{CH_3COOH} = \mathrm{H^+} + \mathrm{CH3COO^-} \tag{3.22}$$

According to Eq. (3.21) we have

$$\overline{\mu}(\mathrm{CH_3COOH}) = \overline{\mu}(\mathrm{H^+}) + \overline{\mu}(\mathrm{CH_3COO^-}) \tag{3.23a}$$

Taking into account that $\overline{\mu}_{\mathrm{i}} = \mu_{\mathrm{i}}$ for an uncharged species we obtain, by using Eq. (3.18),

$$\overline{\mu}(\mathrm{CH_3COOH}) = \mu(\mathrm{CH_3COOH}) = \mu(\mathrm{H^+}) + F\phi + \mu(\mathrm{CH_3COO^-}) - F\phi \tag{3.23b}$$

Since the ϕ terms cancel, only the chemical potentials remain, i.e.

$$\mu(\mathrm{CH_3COOH}) = \mu(\mathrm{H^+}) + \mu(\mathrm{CH_3COO^-}) \tag{3.23c}$$

Since the equilibrium constant K is defined by

$$\ln K = \ln a_{\mathrm{i}}^{\nu_{\mathrm{i}}} \tag{3.24}$$

by inserting this equation into Eq. (3.17) one obtains

$$\ln K = \left(-\textstyle\sum \nu_{\mathrm{i}}\mu_{\mathrm{i}}^{0}\right)/RT = -\Delta G^{0}/RT \tag{3.25}$$

Again, the v_{i} terms of the product have to be taken as positive and those of the educt as negative.

b) Reactions Involving Two Phases

Considering two phases which are in direct contact, equilibrium is again achieved if the electrochemical potential is identical in the two phases. Taking the system of a metal electrode in contact with a metal ion electrolyte as an example, we have the equilibrium

$$M \Leftrightarrow M^{2+} + 2e^- \tag{3.26}$$

Applying Eq. (3.21) to this reaction, one obtains

$$\overline{\mu}(M) = \overline{\mu}(M^{2+}) + 2\overline{\mu}_e \tag{3.27}$$

where $\overline{\mu}_e$ bar is the electrochemical potential of electrons in the electrode. We can calculate $\overline{\mu}$ bar from Eq. (3.27) by also using Eq. (3.18)

$$\overline{\mu_e} = \tfrac{1}{2}\left[\mu(M) - \mu(M^{2+})\right] - F\phi^{sol} \tag{3.28}$$

According to Eq. (3.17), the chemical potential of the metal ions, $\mu(M^{2+})$, depends on the M^{2+} concentration in the solution whereas that of the metal, $\mu(M)$, is constant (= $\mu^0(M)$), so that Eq. (3.28) becomes

$$\overline{\mu_e} = \tfrac{1}{2}\left[\mu^0(M) - \mu^0(M^{2+})\right] - RT/2\left\{\ln\left[a(M^{2+})\right]\right\} - F\phi^{sol} \tag{3.29}$$

Accordingly, the electrochemical potential of the electrons, $\overline{\mu}_e$, can be varied by changing the ion concentration; in this case of $v(M^{2+}) = 1$ and $n = 2$ by 30 meV if the activity a (or the concentration c for very dilute solutions) is changed by one order of magnitude.

On the right side of Eq. (3.28) we have replaced the electrochemical potential by the chemical potential. One also could express the electrochemical potential $\overline{\mu}_e$ of the electrons in terms of the corresponding chemical potential μ_e which leads to (see Eq. 3.18)

$$\overline{\mu_e} = \mu_e + nF\phi^M \tag{3.30}$$

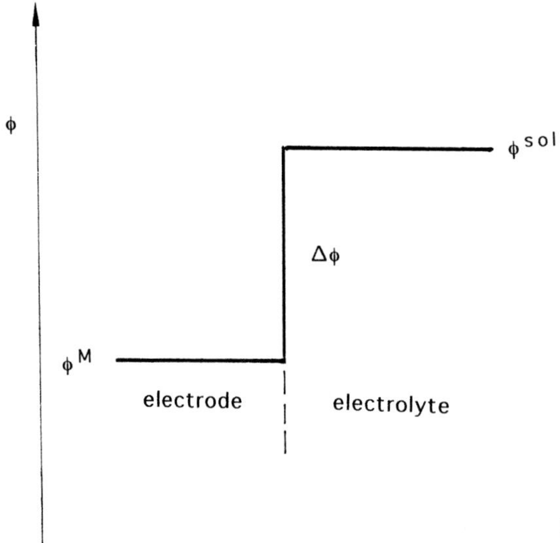

Fig. 3.1 Potential profile across a metal–liquid interface

Accordingly, we have a potential difference $\Delta\phi = \phi^{M} - \phi^{sol}$ across the interface as illustrated in Fig. 3.1. According to Eqs. (3.29) and (3.30), the potential difference $\Delta\phi$ between the metal electrode and the electrolyte is also expected to vary by 30 mV when the concentration of the divalent metal ions is changed by one decade. This has been verified experimentally over a large concentration range. At lower metal ion concentrations, however, such a measurement can be influenced by other factors. Then a potential drop across the electrode–electrolyte interface is mainly determined by the interaction between electrode and electrolyte, as will be discussed in detail in Chapter 5, or by impurities such as a redox system or oxygen.

Another essential aspect to be considered here, is the fact that the potential $\Delta\phi$ across the metal–electrolyte interface (Galvani potential) is not measurable. The reason is that an additional probe, i.e. a second electrode, is required for measuring any potential difference. If one used identical electrodes for potential measurements, then the potential difference between both electrodes would be zero because identical Galvani potentials would occur at both of them.

3.2.2 Cell Voltages

If two different electrodes are used in an electrolyte, for instance Cu and Zn in H_2O, then a two-compartment cell is required in order to separate the ions and to prevent a direct chemical reaction from occurring, as schematically shown in Fig. 3.2. The reactions involved here are given by

$$\mathrm{Cu} \Leftrightarrow \mathrm{Cu}^{2+} + 2\mathrm{e}^{-} \tag{3.31a}$$

$$\mathrm{Zn} \Leftrightarrow \mathrm{Zn}^{2+} + 2\mathrm{e}^{-} \tag{3.31b}$$

The Cu^{2+} ions are separated from the Zn^{2+} ions by a membrane which is permeable for the corresponding counter ions, such as SO_4^{2-} ions, but not for the metal ions. Accordingly, an electrical connection across the membrane is achieved by the transport of the SO_4^{2-} ions. An equilibrium throughout the whole cell does not exist because exchange of the metal ions between the two partial systems has been made impossible. On the other hand, equilibrium still exists in the two half-cells, i.e. between the Cu electrode and the Cu^{2+} in the left compartment and between the Zn electrode and the Zn^{2+} in the right one. However, the electrochemical potentials of the electrons in the two electrodes are different. The electrochemical potentials of the electrons for the reactions (3.31a) and (3.31b) can be derived by applying Eq. (3.29). Their difference is then given by

$$\begin{aligned}\Delta\mu_e &= \mu_e(\mathrm{Cu}) - \mu_e(\mathrm{Zn})\\ &= \tfrac{1}{2}\left[\mu^0(\mathrm{Cu}) - \mu^0(\mathrm{Cu}^{2+})\right] - (RT/2)\ln\left[a(\mathrm{Cu}^{2+})\right] - F\phi_1^{sol}\\ &\quad - \tfrac{1}{2}\left[\mu^0(\mathrm{Zn}) - \mu^0(\mathrm{Zn}^{2+})\right] + (RT/2)\ln\left[a(\mathrm{Zn}^{2+})\right] + F\phi_2^{sol}\end{aligned} \tag{3.32}$$

In the case of $a(Cu^{2+}) = a(Zn^{2+})$ we have

$$\Delta\mu = \Delta\mu^0 = \tfrac{1}{2}\left\{\left[\mu^0(\mathrm{Cu}) - \mu^0(\mathrm{Cu}^{2+})\right] - \left[\mu^0(\mathrm{Zn}) - \mu^0(\mathrm{Zn}^{2+})\right]\right\} \tag{3.33}$$

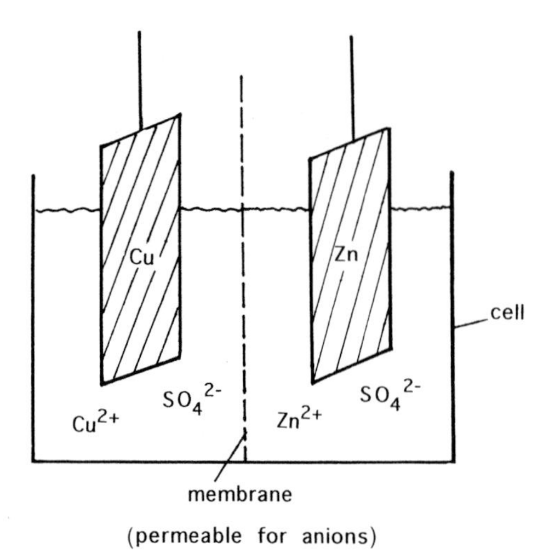

Fig. 3.2 A copper–zinc cell with a permeable membrane

Rewriting Eq. (3.32) by using (3.33) one obtains

$$\Delta\overline{\mu} = \Delta\mu_0 + \frac{RT}{2} \ln\left[\frac{a(\mathrm{Zn}^{2+})}{a(\mathrm{Cu}^{2+})}\right] \tag{3.34}$$

Since $\Delta\mu = \Delta G$ and $\Delta\mu^0 = \Delta G^0$, Eq. (3.34) can also be written in terms of free energy. The difference in electrochemical potential can be measured as a voltage U between the Cu and Zn electrode. It is related to the electrochemical potential by

$$U = \Delta\overline{\mu_e}F = -\Delta G/F \qquad \text{and} \qquad U^0 = \Delta\mu^0/F = -\Delta G^0/F \tag{3.35}$$

The corresponding potentials and cell voltage are illustrated in Fig. 3.3. Certainly some kind of equilibrium would be reached if the membrane was removed. However, no useful information would be obtained because the more noble Cu^{2+} would oxidize Zn to Zn^{2+}, leading to a deposition of Cu on the Zn electrode. Since no equilibrium exists in the cell containing the membrane, a current flows across the system if the two electrodes are short-circuited. On the other hand, equilibrium can be achieved and the current made zero if a counter voltage which balances the original cell voltage, is applied to the cell. Such equilibrium cell voltages can be used for determining corresponding ΔG values.

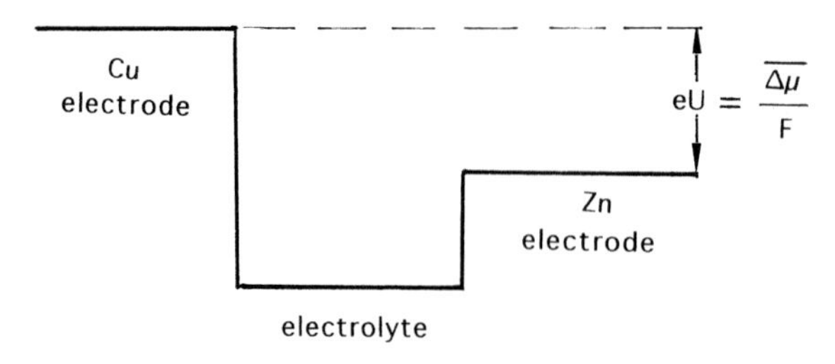

Fig. 3.3 Potential profile for a Cu–electrolyte–Zn system

3.2.3 Reference Potentials

Each electrochemical couple exhibits a characteristic electrochemical potential and a characteristic $\Delta\phi$ value between electrode and electrolyte. The cell voltage *U*, defined above, depends on the couples combined in a cell and on their concentrations. In order to quantify the properties of electrochemical couples reference electrodes are used. The reference electrode which is primarily used is the standard hydrogen electrode (SHE), or normal hydrogen electrode (NHE). In this case it is an inert Pt electrode around which hydrogen is flushed (see Fig. 3.4). The reaction involved is given by

$$H_2 \Leftrightarrow 2H^+ + 2e^- \tag{3.36}$$

Following the procedure derived in Section 3.3.1 and using Eqs. (3.17), (3.18), and (3.27) to (3.29), the electrochemical potential of the electrons is then given by

$$\overline{\mu_e} = \tfrac{1}{2}\mu^0(H_2) - \mu^0(H^+) - RT\ \ln\left(\frac{a(H^+)}{p^{1/2}(H_2)}\right) - F\phi^{sol} \tag{3.37}$$

In the case of gases, the activity is given in terms of pressure, *p*. The electrochemical potential of electrons is an energy. One can express it in terms of a (not directly measurable) potential as defined for a H_2/H^+(Pt) half-cell by

$$U_{ref}(H_2/H^+) = -\overline{\mu_e}F \tag{3.38}$$

A standard condition can be defined for $p^{1/2}(H_2) = a(H^+) = 1$. We have then

$$U^0_{ref} = -F^{-1}\left[\tfrac{1}{2}\mu^0(H_2) - \mu^0(H^+)\right] \tag{3.39}$$

The latter condition is internationally accepted as a reference point (standard hydrogen electrode). If another electrode or another half-cell is now combined with this reference system then the corresponding cell voltage is given by (compare with Eqs. 3.34 and 3.35):

$$U_E = U_{syst} - U^0_{ref} = \Delta\mu/F = \Delta G/zF \tag{3.40}$$

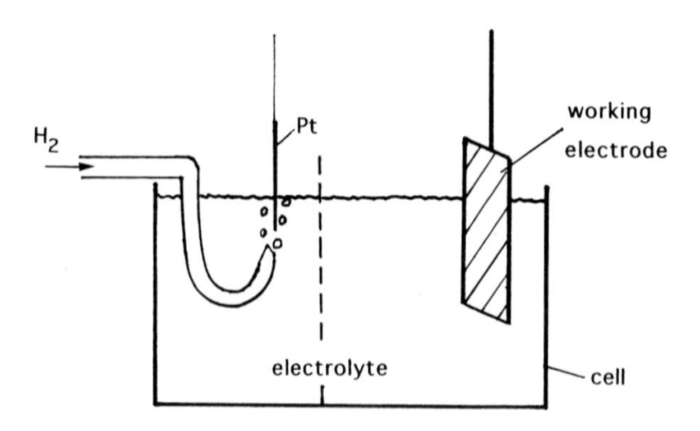

Fig. 3.4 $Pt(H_2/H^+)$ reference electrode (normal hydrogen electrode, NHE)

Such a cell voltage measured with respect to a reference electrode is called the "electrode potential" and is symbolized in the following text by the term U_E. Accordingly, the reference potential U^0_{ref} (H_2/H^+) is actually set to zero. A scale and data for various systems are given in Appendix G. It should be mentioned here, that the symbol E rather than U_E is generally used in the classical electrochemical literature. We prefer to use U_E for electrode potential because the symbol E is used for electron energies (see also ref. [7]).

According to Eqs. (3.37) to (3.39) the potential of a hydrogen reference electrode $U_{ref}(H_2/H^+)$ depends on the activity (or concentration) of protons, $a(H^+)$, i.e. it depends on the pH of the solution. $U_{ref}(H_2/H^+)$ varies by 60 mV when the pH is changed by one unit. Therefore all standard potentials refer to pH = 0 ($a(H^+) = 1$). Since the hydrogen electrode is not a convenient reference electrode for experimental work, other reference electrodes, such as the saturated calomel electrode (SCE) or a silver/silverchloride electrode, are mainly used in electrochemical experiments. The first consists of a Hg/Hg_2Cl_2 electrode in KCl (saturated in H_2O) which has a potential of +0.242 V vs. NHE. This reference electrode exhibits a very constant potential and it is stable against cell currents. In addition, its potential is independent of the pH of the solutions. Further details about reference electrodes are given in Chapter 5. A corresponding scale with SCE as the reference electrode is also given in Appendix G.

3.2.4 Standard Potential and Fermi Level of Redox Systems

Sections 3.2.1 and 3.2.2 dealt with electrochemical reactions at metal electrodes where metal ions were transferred across the metal–electrolyte interface. Redox couples are characterized by molecules or ions in a solution which can be reduced and oxidized by a pure electron transfer. The corresponding reaction is given by

$$\text{Red} \Leftrightarrow \text{Ox} + e^- \tag{3.41}$$

in which Red is the reduced and Ox the oxidized species (for example Red: Fe^{2+}, Ox: Fe^{3+}). It is important to note here that in an electrochemical reaction both the Red and Ox species remain in the solution and only the electron is transferred between the redox system and the electrode. Usually corresponding processes are investigated at inert metal (e.g. Pt) or semiconductor electrodes, or at least in a potential range where the electrode remains stable. In principle the thermodynamics of the half-cell electrode/redox system can be handled in the same way as described in Section 3.2.1. With respect to semiconductor electrochemistry, it is useful, however, to use a somewhat different approach as given below.

First we consider only the redox couple itself, dissolved in an electrolyte without any electrode. We again use reaction (3.41) but assume that the electron is still located in the solution. The electrochemical potential of this electron is then given by

$$\overline{\mu}_{e,\text{redox}} = \overline{\mu}_{\text{red}} - \overline{\mu}_{\text{ox}} \tag{3.42}$$

Inserting Eqs. (3.17) and (3.18) into (3.42) one obtains

$$\overline{\mu}_{e,\text{redox}} = \mu^0_{\text{red}} - \mu^0_{\text{ox}} - RT \ln \frac{c_{\text{ox}}}{c_{\text{red}}} + zF\phi^{\text{sol}} - (z+1)F\phi^{\text{sol}} \tag{3.43}$$

The question arises concerning what is a reasonable reference level. This problem becomes clear by means of another "thought experiment". Let us assume that an electron e_∞^- from the vacuum level far outside the electrolyte is captured by a redox ion $M^{(z+1)}$(liq) (oxidized species) in the electrolyte, which leads to the formation of M^{z+}(liq) (reduced species). We have then the reaction

$$M^{(z+1)}(\text{liq}) + e_\infty^- \rightarrow M^{z+}(\text{liq}) \tag{3.44}$$

Assuming that the electrochemical potential of the electron in the vacuum is zero, one obtains essentially the same equation as Eq. (3.43), namely

$$\overline{\mu}_{e,\text{redox}} = \mu^0_{e,\text{redox}} - RT \ln\left[\frac{C_{M^{(z+1)+}}}{C_{M^{z+}}}\right] \tag{3.45}$$

in which

$$\overline{\mu}^0_{e,\text{redox}} = \mu^0_{M^z} + \mu^0_{M^{(z+1)+}} + zF\phi^{\text{sol}} - (z+1)F\phi^{\text{sol}} \tag{3.46}$$

As already mentioned, the electrochemical potential of the electrons in the redox system is measured against the vacuum level. This is similar to the semiconductor-vacuum and semiconductor–metal contacts discussed in Sections 2.1 and 2.2. A corresponding energy scheme for a semiconductor–liquid–vacuum junction is shown in Fig. 3.5. Accordingly, the redox system has here the same reference point as is used in solids

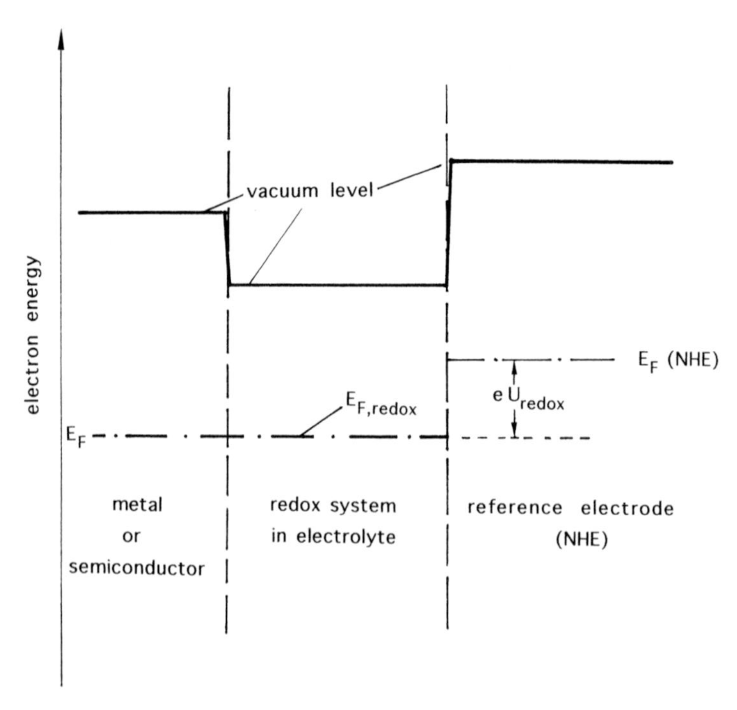

Fig. 3.5 Energy diagram for a metal (or semiconductor)–redox system–reference electrode system

(see Chapters 1 and 2). Since the electrochemical potential of electrons in a metal or a semiconductor is called the Fermi level, E_F, we also term the electrochemical potential of a redox system $\overline{\mu}_{e,redox}$ as the Fermi level ($E_{F,redox}$) as first introduced by Gerischer (see ref. [8]). All chemical and electrochemical potentials are usually given in units of joule mole^{-1}, whereas Fermi energies are usually given in units of electron volts (eV) and refer to single electrons, so that

$$E_{F,redox} = (e/F)\overline{\mu}_{e,redox} \tag{3.47}$$

Applying this to Eq. (3.45) and using Red and Ox again, instead of M^{z+} and $M^{(z+1)}$, respectively, we have

$$E_{F,redox} = E^0_{F,redox} - kT \ln\left(\frac{c_{ox}}{c_{red}}\right) \tag{3.48}$$

A contact is now made between the electrolyte containing a redox system and a metal or a semiconductor electrode, and equilibrium between the two phases is achieved; i.e.

$$E_F = E_{F,redox} \quad \text{at equilibrium} \tag{3.49}$$

These electron energies are schematically illustrated in Fig. 3.6. Such a contact leads to a certain potential difference or at least a potential change across this interface, as will

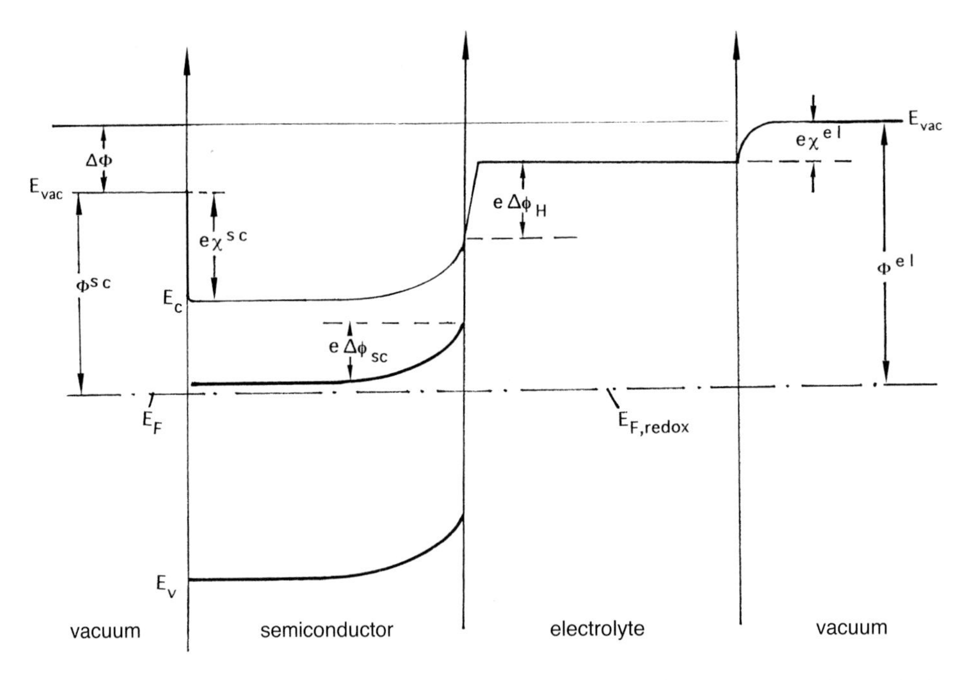

Fig. 3.6 Energy diagram for the semiconductor–vacuum, semiconductor–liquid and liquid-vacuum interfaces; Φ^{sc}, Φ^{el}, work functions; $e\chi^{sc}$, $e\chi^{el}$ surface dipole contributions (neglected at the semiconductor–liquid junction); $e\Delta\phi_{sc}$ $e\Delta\phi_H$ potentials across the space charge layer and Helmholtz layer, respectively. (Compare with Figs. 2.1 to 2.4)

be discussed in more detail in Chapter 5. In practise, however, one would not use the vacuum level as a reference but would take again a classical reference electrode such as a hydrogen electrode or any other reference electrode, for instance the saturated calomel electrode (SCE). In this case one can treat the reference electrode in the same way as was been derived for a redox system. Thus, one can also derive a Fermi energy for the normal hydrogen electrode (E_F[NHE]), as indicated in Fig. 3.5. Since there is no equilibrium between the working electrode or the redox system on the one hand and the reference electrode on the other hand, a corresponding potential difference, U_{redox}, between the inert working electrode and the reference electrode can be experimentally measured (Fig. 3.5). Accordingly, the redox potential U_{redox} as measured against a reference electrode, is then given by

$$U_{redox} = -\frac{E^0_{redox}}{e} = U^0_{redox} + kT \ln\left(\frac{c_{ox}}{c_{red}}\right) \tag{3.50}$$

which is the famous Nernst equation, with U^0_{redox} as the standard redox potential at which $c_{ox} = c_{red}$. It should be emphasized here that the actual redox potential as defined above can only be measured if an inert metal electrode is in contact with the redox system and if the latter two are in electronic equilibrium.

Although the vacuum level is of no practical reference, the question arises of whether the standard potential or energy of the hydrogen reference electrode (NHE) can be quantitatively related to the vacuum level. The first quantitative approach, which was a straightforward one, was published by Lohmann [9]. He examined free energy changes associated with the reduction of silver:

$$Ag(g) = Ag(s) \qquad -7.64\,eV \tag{3.51a}$$

$$Ag^+(g) + e^- = Ag(g) \qquad -2.60\,eV \tag{3.51b}$$

$$Ag^+(aq) = Ag^+(g) \qquad +4.96\,eV \tag{3.51c}$$

$$Ag(s) + H^+(aq) = Ag^+(aq) + \tfrac{1}{2}H_2 \qquad +0.80\,eV \tag{3.51d}$$

$$H^+ + e^- = \tfrac{1}{2}H_2 \qquad +4.48\,eV \tag{3.51e}$$

in which g indicates the gas phase, s the solid, and aq the solution (here an aqueous phase). The determination of such a reference electrode has been the subject of various further calculations. In semiconductor electrochemistry, the value of $E_{ref} = -4.5$ eV for the NHE, as derived by Lohmann [9] is used mostly. More recently Trasatti recalculated E_{ref} and obtained –4.31 eV [10]. The calculations are based on a reaction cycle (see ref. [11]) which contains energy terms such as sublimation energy, ionization energy, and the single hydration energy. The latter quantity is rather difficult to determine because it contains a term describing the potential difference, $\chi(H_2O)$, between the solution surface and its interior. Trasatti estimated this potential, whereas Lohmann neglected it, which explains the difference between the two E_{ref} values. Both authors used experimental values obtained by Randles [12]. A few years later Gomer and co-workers took up this problem again [13]. They determined absolute potentials experimentally using the vibrating condenser method, and finally obtained $E_{ref} = -4.73$ eV for NHE. This is considerably larger than Trasatti's value. According to Gomer the deviation from the other values is due to an incorrect value for the work function of Hg in air, the latter having being used by Randles. A further value of $E_{ref} = -4.85$ eV was published by Kötz et al. [14].

Because of these discrepancies we also continue to use Lohmann's value, so that the electron energy of a redox couple is given by

$$E_{abs} = -4.5\,\text{eV} - eU_{redox} \tag{3.52}$$

The Fermi level concept is very useful in the quantitative description of reactions at semiconductor electrodes, as described in Section 7.4. Other energy states of a redox system besides the Fermi level can also be defined. This problem is discussed in detail in Chapter 5.

4 Experimental Techniques

4.1 Electrode Preparation

Most electrochemical measurements have been performed using semiconductor single crystals. The latter were oriented using x-ray analysis in order to select the correct surface for the electrochemical investigation. In most cases, the crystal face does not play a dominant role because other faces are developed during the experiment (see Chapter 8). The semiconductor specimens are then provided with an ohmic contact usually at the rear of the electrode. The technology for producing the ohmic contact depends on the semiconducting material and on its doping. The procedure can be found in the appropriate literature (see also Section 2.4). To test the quality of the ohmic contact it is useful to have two contacts at the rear, between which the resistance of the semiconductor at different voltages can be measured. There are different techniques for mounting the crystal in the cell, depending mainly on the size of the crystal. In experiments where defined diffusion conditions toward the electrode in the liquid are required, it is useful to take circular slices which can be prepared by ultrasonic cutting. The specimens are then mounted in an electrode holder as illustrated in Fig. 4.1. The advantage of this technique is that the sample can easily be taken out of the solution and put back again. In addition, such an electrode can be rotated as described in Section 4.2.3. Before each measurement, an electrode is usually etched.

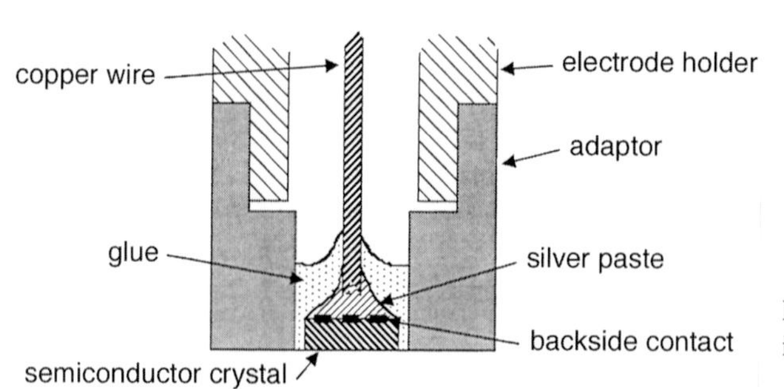

Fig. 4.1 A holder for a rotating semiconductor electrode

4.2 Current–Voltage Measurements

4.2.1 Voltametry

Current–potential measurements at semiconductor electrodes are usually performed in a cell with three electrodes under potentiostatic conditions. The cell is illustrated schematically in Fig. 4.2. It consists of the working electrode WE (a semiconductor electrode as described in Fig. 4.1), a large counter electrode CE (usually a Pt electrode) and a reference electrode RE. The latter is usually separated from the actual cell by a salt bridge (see Fig. 4.2) in order to avoid interference with the electrolyte in the cell.

There are several choices for a reference electrode such as a normal hydrogen electrode (NHE), a saturated calomel electrode (SCE) or a silver/silver chloride electrode (Ag/AgCl). The first consists of a Pt electrode around which H is flushed. This reference electrode is not very convenient to use. The saturated calomel electrode is the most commonly used reference electrode in aqueous solutions. It consists of a Hg/Hg_2Cl_2 system as indicated in the left part of Fig. 4.2. This is a highly suitable reference electrode because its potential is very constant because of the low solubility of Hg_2Cl_2. Its standard potential is U°(SCE) = 0.27 V vs. NHE. Another advantage is its independence of pH. Similar conditions are met for the couple Ag/AgCl with a standard potential of U° = 0.23 V vs. NHE. In the case of water-free organic electrolytes, frequently a simple redox system of known potential is used, which is also separated from the main solution by a salt bridge [1–5].

To avoid artifacts in impedance measurements when modulating the input voltage of the potentiostat at high frequencies (>10 kHz), the reference electrode should be short-circuited via a capacitance of 10 nF and a Pt wire dipped into the solution in the main part of the cell (Fig. 4.2). The surface of the counter electrode should be sufficiently large so that its interface with the electrolyte does not influence the current-potential curve. Usually a platinized Pt sheet is used as a counter electrode. The electrolyte is made conductive by adding an inert salt of a concentration in the range of 10^{-3}–10^{-1} M.

The external voltage is supplied by a corresponding voltage generator as schematically shown in Fig. 4.3. In this way the electrode potential can be scanned over several volts at a selected scan rate. The resulting current–potential curve can be displayed by a recorder or by using a computer.

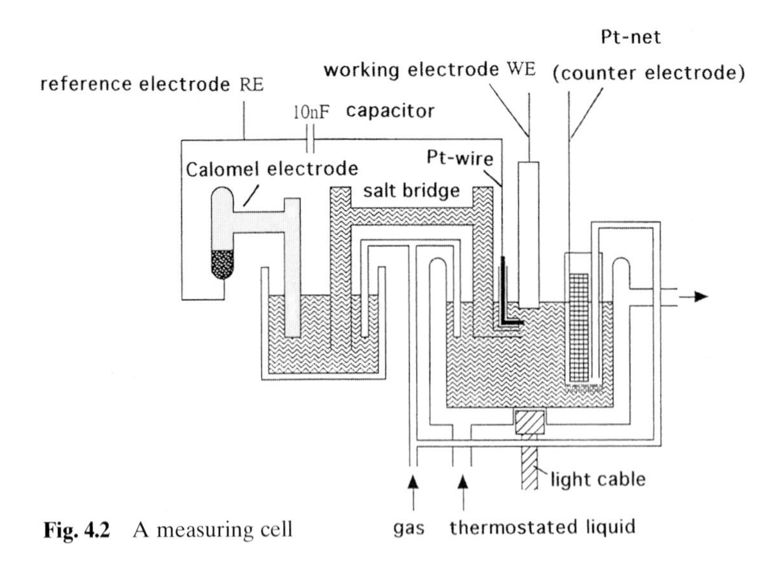

Fig. 4.2 A measuring cell

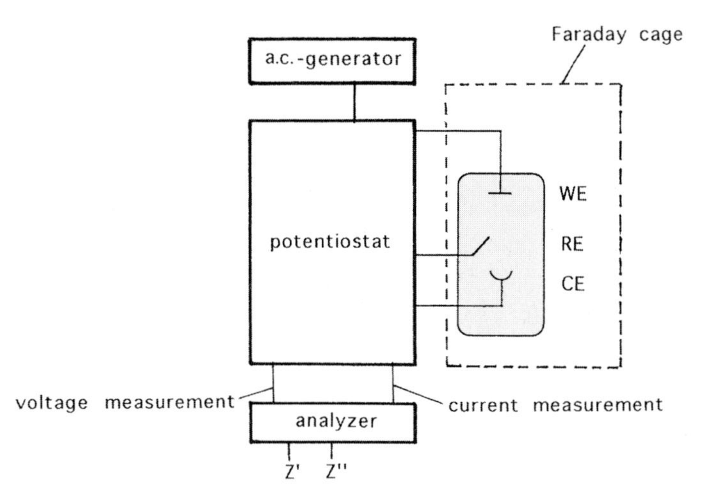

Fig. 4.3 A potentiostatic measuring system: WE, working electrode; RE, reference electrode; CE, counter electrode

4.2.2 Photocurrent Measurements

The photocurrent in the semiconductor is generated by illuminating the electrode, for instance through the bottom of the cell as indicated in Fig. 4.2. It is often very convenient to introduce the light via a light cable. Usually a xenon lamp is used as a light source which yields sufficiently high intensities when monochromatic light is used for excitation. Several research groups use chopped light for excitation leading to a corresponding a.c. photocurrent signal which can be easily amplified by using a lock-in technique. However, this is a rather dangerous procedure in many cases because valuable information can be lost. For instance, especially in the range of the photocurrent onset, photocurrent transients are frequently found when the light is turned on (see Section 7.3.3). These are only visible if the time profile of the photocurrent signal is displayed during the whole cycle, i.e. during the time interval when the light is turned on and also when the light is turned off.

4.2.3 Rotating Ring Disc Electrodes

In some cases it is of interest to determine products formed at semiconductor electrodes. If redox reactions are involved this can be done by using a rotating ring disc electrode assembly (RRDE), which has proved to be a powerful tool for investigating electrochemical reactions at metal electrodes. The technique and corresponding results as obtained with metal electrodes have been reviewed by Bruckenstein and Miller [6] and by Pleskov et al. [7].

Such an assembly consists here of a semiconductor disc and a Pt ring as illustrated in Fig. 4.4. When such an electrode assembly is rotated the solution flows upward towards the semiconductor disc similarly as with a pure rotating electrode, which leads

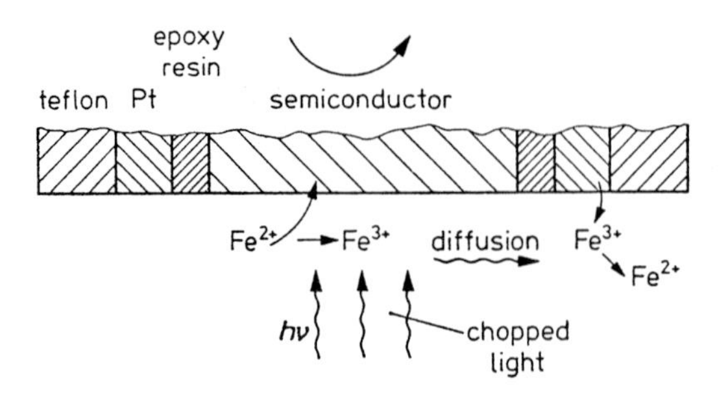

Fig. 4.4 Rotating ring disc electrode assembly

to an increase of the disc current if it is diffusion-limited (compare with Fig. 4.1). During the rotation of the electrode system, the liquid moves radially along the surface of the whole system as indicated in Fig. 4.4. Accordingly, a redox system which is, for instance, oxidized at the semiconductor disc electrode, can be reduced at the Pt ring electrode provided that the latter electrode is set to a potential at which the redox system is reduced back to its original state. Such an investigation is of special interest if holes, produced by light excitation in an n-type semiconductor, are consumed for the anodic dissolution of the semiconductor as well as for the oxidation of a redox system. The total anodic photocurrent j_{ph} is then given by

$$j_{ph} = j_{diss} + j_{ox} \tag{4.1}$$

where j_{ph} is the dissolution current and j_{diss} the current due to the oxidation of the redox system. At the Pt ring, of course, only the redox couple is oxidized but not the product produced by the anodic dissolution. Absolute values of j_{ox} at the disc depend on the distance between the disc and ring electrode. Since the hydrodynamics of the rotating system is well known, the ring current can be calculated for a given oxidation current at the disc [6] or it can be determined experimentally by using an inert Pt disc instead of a semiconductor disc electrode [8]. Such a metal disc/metal ring system of the same dimensions can also be used for calibration of a RRDE assembly [8]. The best insight into the processes can be obtained by studying light-induced reactions (see Section 8.5).

4.2.4 Scanning Electrochemical Microscopy (SECM)

A few years ago Bard and his group developed the technique called scanning electrochemical microscopy (SECM) which makes possible a spatial analysis of charge transfer processes [9]. In this method an additional 'tip' electrode of a diameter of about 2 μm is used as well as the three other electrodes (semiconductor, counter and reference electrode). Assuming that a redox system is reduced at the semiconductor, then the reduced species can be re-oxidized at the tip electrode, the latter being polarized positively with respect to the redox potential. The corresponding tip current j_T is proportional to the local concentration of the product formed at the semiconductor surface and therefore also to the corresponding local semiconductor current, provided

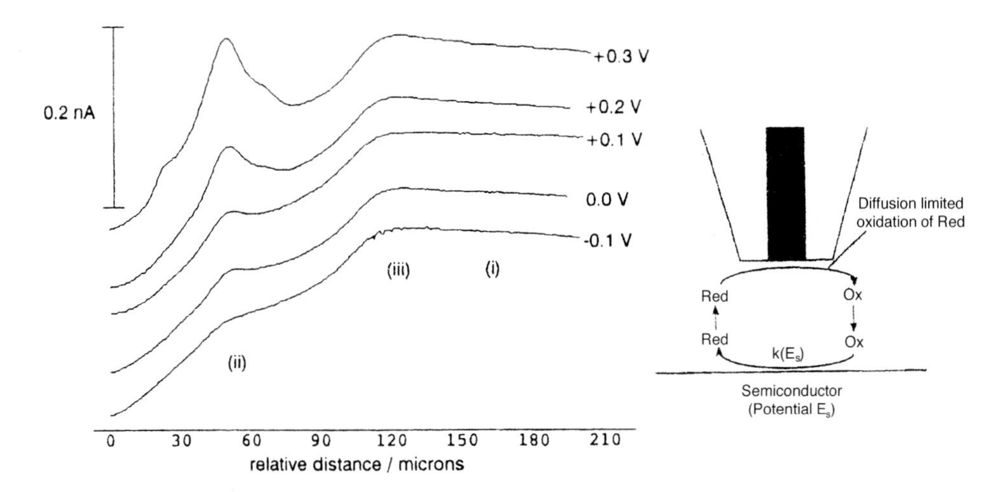

Fig. 4.5 Scanning electrochemical microscopy (SECM) line scans across a stepped portion of a WSe_2 electrode: tip current at different potentials of a WSe_2 electrode in aqueous solutions. Insert: tip arrangement; tip diameter, 2 μm. (After ref. [9])

that the tip is sufficiently close to the semiconductor electrode (insert in Fig. 4.5). This method yields a lateral photocurrent map when scanning the tip across the surface area of interest, as illustrated for a stepped portion of a p-type WSe_2 electrode in Fig. 4.5. Recently the sensitivity of the method was considerably increased by using much smaller tip electrodes of a diameter used for scanning tunneling microscopy (STM) measurements as described in Section 4.7.2.

4.3 Measurements of Surface Recombination and Minority Carrier Injection

The surface recombination is defined by Eq. (2.47) in Chapter 2. It can be measured by the "thin slice" method as described below. This method involves mainly a p–n junction at the rear of the electrode in addition to the ohmic contact mentioned in Section 4.1. Surface recombination is measured after excitation of electron–hole pairs near the electrode surface. They can diffuse towards the electrode surface but also towards the p–n junction at the rear. The electron–hole pairs are separated at the p–n junction which leads to a corresponding short-circuit current across this junction. As already discussed in Section 2.5, this current is proportional to the light intensity and therefore to the density of electron–hole pairs created by light excitation. If there is some additional surface recombination in a certain potential range, then fewer electron–hole pairs reach the rear p–n junction. A quantitative relation between the short-circuit current at the rear p–n junction j_s and light intensity and surface recombination can be derived as follows.

Taking a p-type electrode as an example, the electrode is illuminated through the electrolyte as illustrated in Fig. 4.6. The diffusion of the excess minority carrier density Δn is given by the continuity equation [10]

$$D_{\mathrm{n}} \frac{\delta^2 \Delta n}{\delta x^2} - \frac{\delta \Delta n}{\tau} + g(x) = 0 \tag{4.2a}$$

in which D_n is the diffusion constant of electrons in the p-type electrode, τ the lifetime of the electrons and $g(x)$ the generation rate of minority carriers, the latter being given by

$$g(x) = \alpha I \exp(-\alpha x) = g_0 \exp(-\alpha x) \tag{4.2b}$$

The generation rate depends on the light intensity I, and on the absorption coefficient α. The short-circuit current is defined as

$$j_{\mathrm{s}} = -D_{\mathrm{n}} \operatorname{grad} \Delta n \mid_{x=d} \tag{4.3}$$

The surface recombination velocity can be introduced by a further boundary condition. At the semiconductor–liquid interface the recombination rate is given by

$$R_{\mathrm{s}} = s\Delta n = D_{\mathrm{n}} \operatorname{grad} \Delta n \mid_{x=0} \tag{4.4}$$

Using Eqs. (4.2b) to (4.4), restricting their application to a small penetration depth of light ($\alpha \leq L_n$) and assuming that $d > L_n$, one obtains by solving Eq. (4.2a):

$$j_{\mathrm{s}} = \frac{I}{s} \frac{2eD_{\mathrm{n}}}{L_{\mathrm{n}} \exp\left(\frac{d}{L_{\mathrm{n}}}\right)} \tag{4.5}$$

in which L_n is the diffusion length of the electrons as defined by Eq. (2.26). An equivalent equation is valid for an n-type electrode. According to Eq. (2.26), the short-circuit current j_{s} is inversely proportional to the surface recombination velocity s. The experimental conditions are given in Fig. 4.6b, and i_{s} is measured when the p–n junction is short-circuited with the ohmic contact as indicated in Fig. 4.6b.

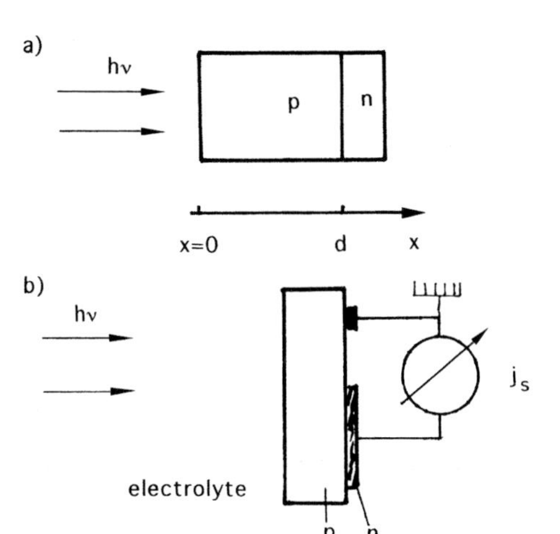

Fig. 4.6 Technique for measuring injection of minority carriers into a semiconductor electrode. a) Geometry of a p–n junction; b) cell arrangement

The same method can be applied if minority carriers are produced by injection from a species in the electrolyte, as first proposed by Brattain and Garrett [11]. Taking again a p-type electrode as an example, the short circuit current j_s is simply related to the corresponding injection current j_{inj} by

$$j_s = \exp(-d/L_n)\, j_{inj} \tag{4.6}$$

The ratio of j_{inj} and the total current j_{tot} then yields quantitative information about whether or not an electrochemical reaction is partly or fully a minority carrier process. This method has mainly been applied to reactions at germanium and silicon electrodes because here the diffusion length is in the order of some hundred microns. In other cases, such as GaAs, it is very difficult to apply this method because the diffusion length is rather small ($\approx$1 μm).

It should be mentioned that a somewhat different method was introduced by Pleskov [12]. Here, the semiconductor is in contact with a second liquid at the rear instead of a p–n junction being used. The semiconductor is under reverse bias with respect to this liquid. Any minority carrier injection at the front then leads to an increase of current at the rear contact. This method is of interest for semiconductors with which a p–n junction cannot be made.

4.4 Impedance Measurements

4.4.1 Basic Rules and Techniques

When a sinusoidal voltage is applied across the interface

$$U(t) = U_m \sin(\omega t) = U_m \exp(i\omega t) \tag{4.7}$$

then the corresponding current usually shows a phase shift Φ with respect to $U(t)$ and is given by

$$J(t) = J_m \sin(\omega t + \varphi) = J_k \exp(i\omega t) \tag{4.8}$$

In the second part of Eqs. (4.7) and (4.8) $U(t)$ and $J(t)$ are written as complex numbers in which $J_k = J_m \exp(i\Phi)$. The impedance is then defined as

$$Z(\omega) = \frac{U_m}{J_m} \exp(i\varphi) \tag{4.9}$$

and is also a complex number. It is composed of a real (Z') and an imaginary part (iZ'') and we have then

$$Z(\omega) = Z'(\omega) + iZ''(\omega) \tag{4.10}$$

This is illustrated in Fig. 4.7. In the simplest case the solid–liquid interface can be described by a charge transfer resistance R_{ct} and a capacity in parallel (Helmholtz capacity for metal electrodes, C_H, and a space charge capacity for semiconductor elec-

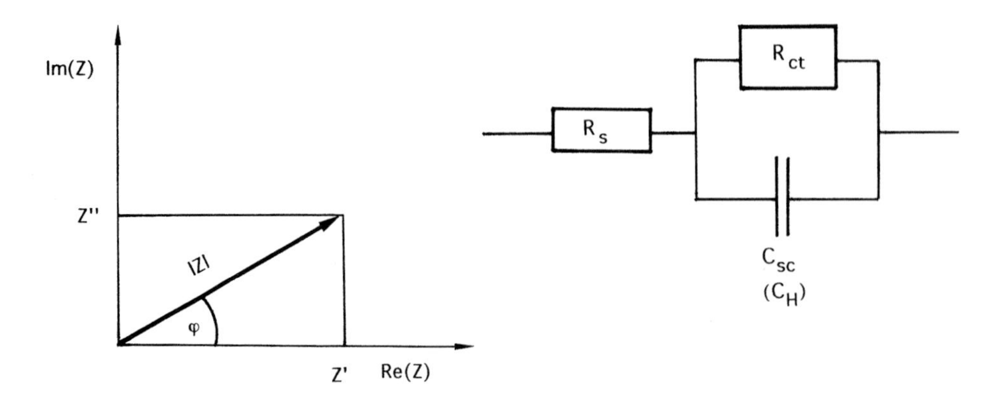

Fig. 4.7 Imaginary and real part of impedance

Fig. 4.8 Equivalent circuit

trodes, C_{sc}) including a series resistance R_s as shown in Fig. 4.8. The current flowing through the two branches is given by

$$J = J_C + J_F \tag{4.11}$$

in which J_C and J_F are the currents flowing through the capacitor and the resistance, respectively. In order to have the impedance in ohms, the currents are given in amperes, i.e. they are not current densities. Since the space charge capacity C_{sc} and the charge transfer resistance R_{ct} strongly depend on the potential, the system is nonlinear. This makes it necessary to apply only a small a.c.voltage, ΔU, which is superimposed on the d.c. voltage. Accordingly, we have

$$U(t) = U + \Delta U \exp i\omega t \tag{4.12}$$

and the corresponding current is then given by

$$J(t) = J + \Delta J \exp i\omega t \tag{4.13}$$

as illustrated in Fig. 4.9.

Since the current across a capacitor is given by $J_c = C\,dU(t)/dt$ one obtains by applying Eq. (4.12)

$$J_C(t) = i\omega C_{sc} \Delta U \exp i\omega t \tag{4.14}$$

Since most current–potential curves show usually an exponential type of dependence the small modulation of the Faradaic current, ΔJ_F, flowing only through R_{ct}, can be obtained by

$$\Delta J_F = \frac{dJ_F}{dU} \Delta U \tag{4.15}$$

Using Eq. (4.11), the total current is then given by

$$J(t) = i\omega C_{sc} \Delta U \exp i\omega t + J_F + \frac{\delta J_F}{\delta U} \Delta U \exp i\omega t \tag{4.16}$$

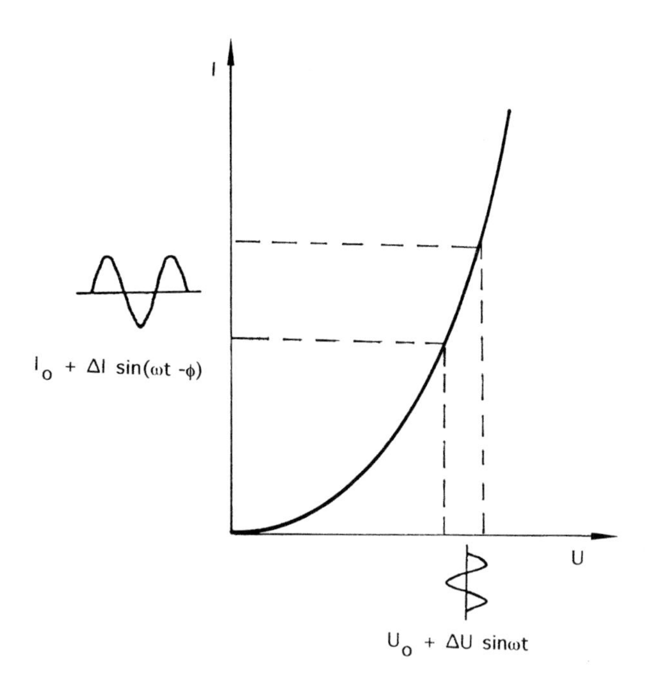

Fig. 4.9 Modulation of potential and current for a semiconductor electrode

Since the d.c. current, J, is equal to the J_F, one obtains from Eq. (4.16) after having inserted Eq. (4.13)

$$\Delta J = i\omega C_{sc}\Delta U + \frac{\delta J_F}{\delta U}\Delta U \tag{4.17}$$

The impedance is then given by

$$\Delta J = i\omega C_{sc}\Delta U + \frac{\delta J_F}{\delta U}\Delta U \tag{4.18}$$

where the series resistance is included. The derivative dJ_F/dU is the charge transfer resistance R_{ct} which is usually potential-dependent.

The principal set-up for the equipment for measuring the impedance is shown in Fig. 4.10. It is similar to that shown in Fig. 4.3. In addition there is an a.c. voltage generator by which a small voltage of less than kT/e (<0.025 V) is superimposed on the d.c. input voltage. An analyzer is used for measuring the imaginary and real components of Z. The frequency of the generator, f, can be varied. The angular frequency is given by $\omega = 2\pi f$.

4.4.2 Evaluation of Impedance Spectra

C_{sc} and R_{ct} can only be evaluated from the complex Z data if the equivalent circuit is known. One simple circuit which is valid for a one-electron redox process, has been shown in Fig. 4.8. Any equivalent circuit can be tested by measuring Z' and Z'' over a large frequency range. It is common to plot the imaginary Z' values versus the cor-

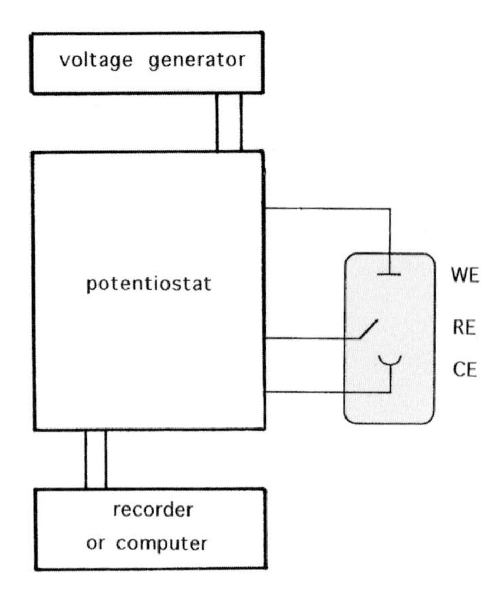

Fig. 4.10 An impedance measuring system (d.c. voltage input neglected)

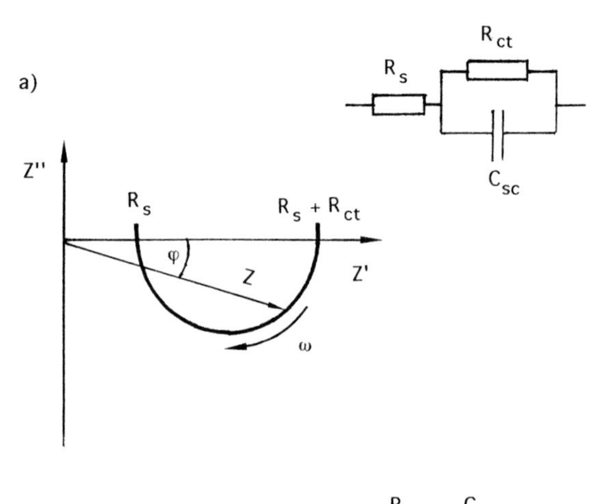

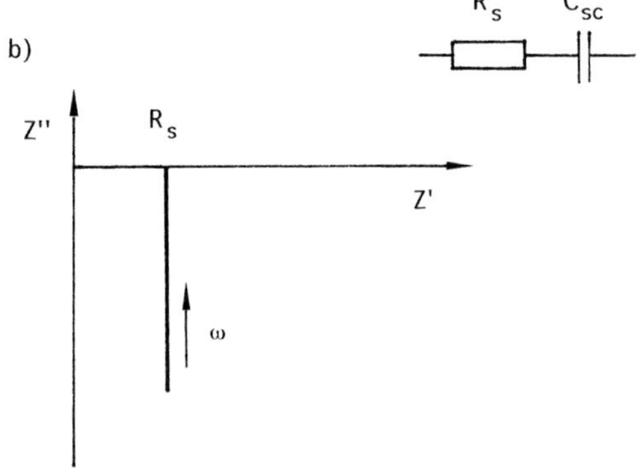

Fig. 4.11 Impedance spectra for two simple equivalent circuits

responding real Z'' component over a certain range of frequencies as shown in Fig. 4.11a. The semicircle has a radius of $\frac{1}{2}R_{ct}$, the low-frequency intercept on the real axis is $R_s + R_{ct}$ and the high frequency intercept R_s. C_{sc} can be calculated by using Eq. (4.18), whereas R_s and R_{ct} can be taken directly from Fig. 4.11. The quality of the equivalent circuit can be tested by calculating the phase shift ϕ and the absolute value of the impedance, $|Z|$, from the evaluated C_{sc}, R_s, R_{ct} values for the whole frequency range, and comparing these with the experimental ϕ and $|Z|$ values (Bode plot) as illustrated in Fig. 4.12. Such a test is useful because more complex semicircles sometimes occur in the complex plane (see below). Only this or other tests make it possible to determine reliable data, especially for obtaining C_{sc} values which are independent of ω. In the case of semiconductor electrodes, impedance measurements are frequently performed in a potential range where the interfacial current is extremely low and constant due to extraction of minority carriers. Then R_{ct} becomes extremely large and the semicircle degenerates into a vertical line at the high frequency end which is parallel to the Z' axis, as shown in Fig. 4.11b.

As already mentioned, the equivalent circuit given in Fig. 4.8 is the simplest case. If the impedance is also investigated in a potential range where large currents are observed, then one has to take into account that the concentration of the ox or red species may be lower at the surface than in the body. In such a case, the concentration of the involved species will also be modulated when a small a.c. voltage is applied. Thus, we have instead of Eq. (4.15)

$$\Delta J_F = \frac{dJ_F}{dU}\Delta U + \frac{dJ_F}{dc_{ox}}\Delta c_{ox} + \frac{dJ_F}{dc_{red}}\Delta c_{red} \tag{4.19}$$

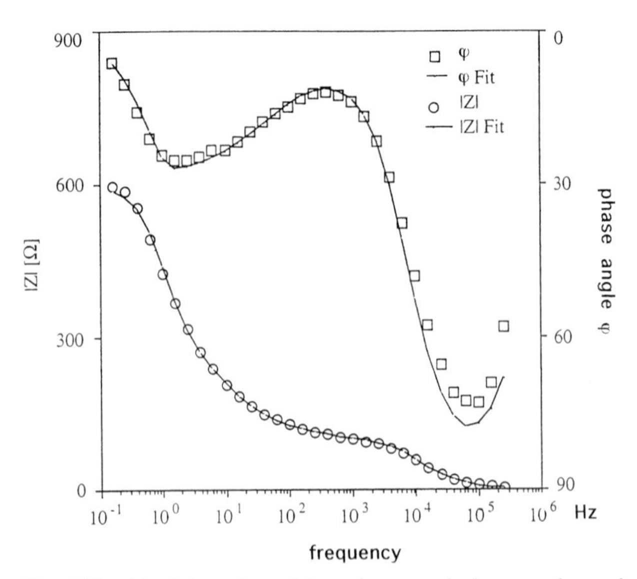

Fig. 4.12 Absolute value of impedance and phase angle vs. frequency (Bode plot). Example: p-GaAs in 7 mM Cu^+ aqueous solution at 0.4 V vs. saturated calomel electrode (SCE); the solid curve is calculated from a equivalent circuit, the squares and circles are experimental data. (After ref. [35]).

The concentrations follow the diffusion equation. The derivation of this effect leads to an additional impedance (Warburg impedance) as given by

$$Z_W = \sigma_0 \frac{\tanh\left(\delta_N \sqrt{\frac{i\omega}{D}}\right)}{\sqrt{i\omega}} \tag{4.20}$$

with

$$\delta_N = 1.6 D^{1/3} \nu^{1/6} \omega_r^{-1/2}$$

where σ_0 is a function of the rate constants involved in a redox process; D is the diffusion constant for the redox system; v is the viscosity, and ω_r the angular velocity of the electrode.

For details and an exact derivation of Z_w the reader is referred to ref. [13]. The derivation also shows that Z_w is in series with R_{ct} as shown in Fig. 4.13a. Typically, the Warburg impedance leads to a linear increase of Z' with rising Z'' and the slope is 45° as also shown in Fig. 4.13a. In this case, Z_w has been calculated assuming an infinite thickness of the diffusion layer. Any convection of the liquid limits the thickness of the diffusion layer. The latter is limited to a well defined value when a rotating disc electrode is used (see Section 4.2.3). In this case, the impedance spectrum is bent off at low frequencies as shown in Fig. 4.13b. The Z_w branch is only linear at its high frequency end where it shows a slope of 45°.

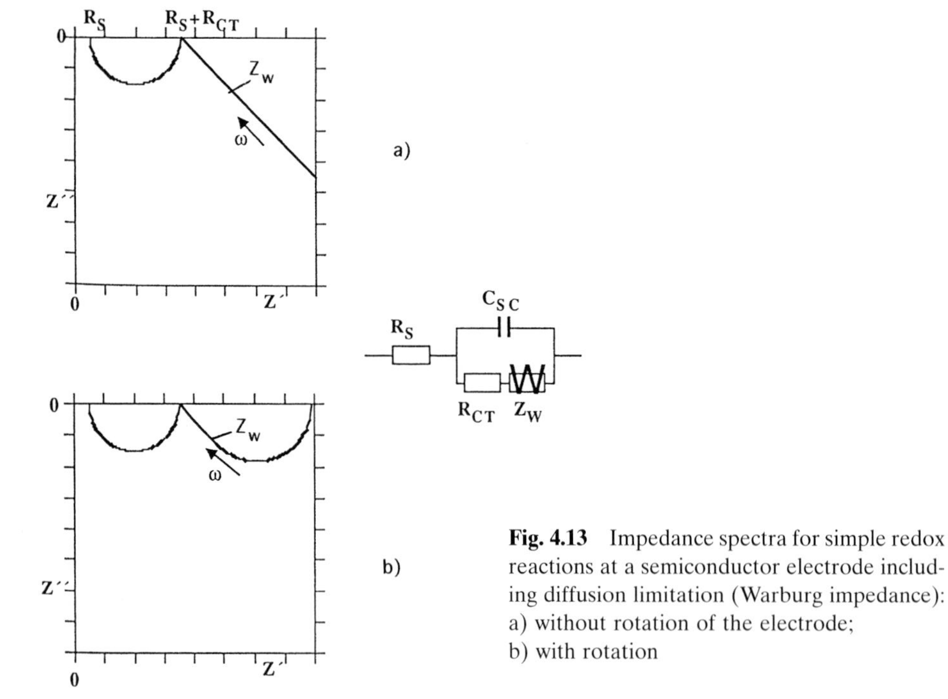

Fig. 4.13 Impedance spectra for simple redox reactions at a semiconductor electrode including diffusion limitation (Warburg impedance): a) without rotation of the electrode; b) with rotation

Sometimes more than one semicircle occurs in the impedance spectrum as well as the Warburg impedance. The origin of the second semicircle is usually due to a two-step reaction process, i.e. an intermediate state is involved. This can occur, for instance, if an adsorbed molecule participates in the reaction, or if energy states within the energy gap at the semiconductor surface are involved, or if just more than one electron occurs in the reaction. In these cases, R_{ct} becomes a complex quantity and we have to replace R_{ct} by a complex Faraday impedance Z_F, as illustrated in Fig. 4.14. Such a Faraday impedance depends on the reaction mechanism. One can derive Z_F from a kinetic model proposed for a reaction process. First we derive ΔJ, which depends finally on rate constants and on various derivatives, such as $\Delta c_{intermediates}$ and Δn_s or Δp_s where the latter are the modulations of electron and hole densities at the surface. The Faraday impedance is given by $Z_F = \Delta U/\Delta J$, which turns out to be a complex number (see e.g. refs. [13, 14]). One can also express Z_F in terms of an equivalent circuit; i.e. by resistors and capacitors (Fig. 4.14) which frequently makes the evaluation easier. In this case, the elements such as R_a, R_b and C_p depend on a set of various kinetic parameters such as rate constants and concentrations. Fairly complex equations are obtained, but this is arbitrary. The impedance measurements make it possible to determine certain kinetic parameters. In the case of anodic dissolution, very complex impedance spectra with negative and positive Z'' values have often been found which cannot be discussed here (see e.g. refs. [13, 15].

It should be emphasized that impedance measurements are mainly used for measuring space charge capacities. They are usually performed in a frequency range of 10 kHz up to nearly 1 MHz depending on the Faraday current.

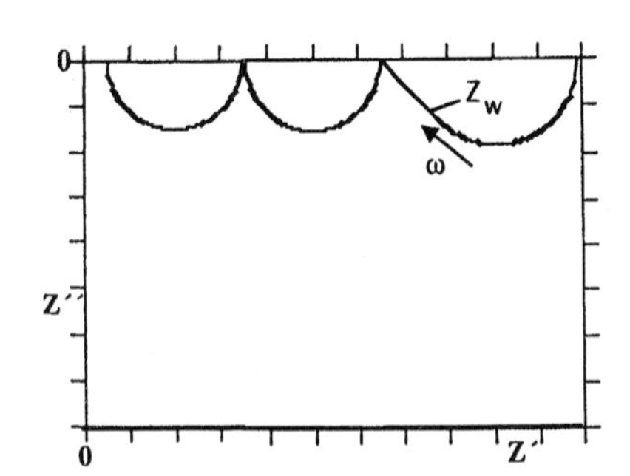

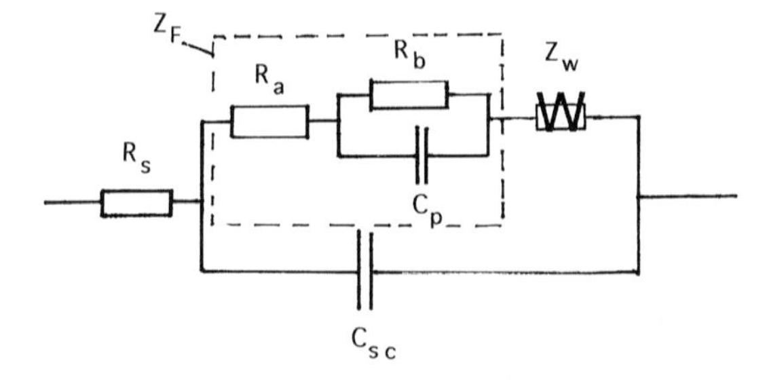

Fig. 4.14 Impedance spectrum for a semiconductor–liquid system for a redox reaction including an intermediate state

4.5 Intensity-Modulated Photocurrent Spectroscopy (IMPS)

In this technique, as first developed by Li and Peters [16], the photocurrent instead of the potential is modulated. Hence, it is only applicable for minority carrier processes. The modulation of current is achieved by modulating the exciting light intensity. The current modulation is illustrated by a current–potential curve of an n-type semiconductor electrode (Fig. 4.15). The quantum efficiency is defined as the ratio of the current and intensity modulation ($\phi = \Delta J/\Delta I$). Since the intensity is not always known it is easier to use a relative quantum yield as defined by

$$\phi_r(\omega) = \frac{\Delta J}{\Delta J_{ph}} \tag{4.21}$$

(Here also currents and not current densities are considered.)

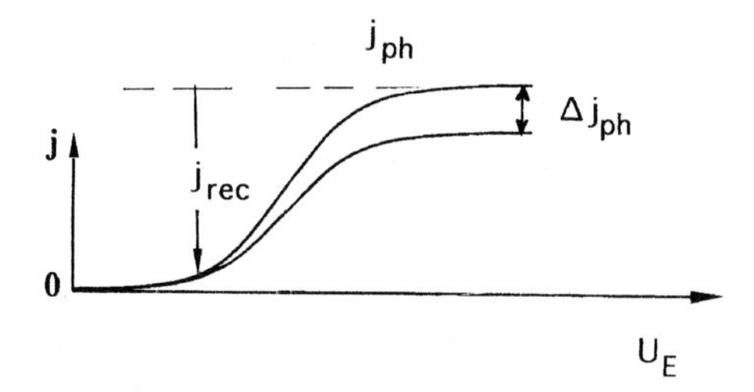

Fig. 4.15 Intensity modulated photocurrent–potential curve for an n-type semiconductor electrode (intensity-modulated photocurrent spectroscopy, IMPS)

According to this definition $\phi_r = 1$ in the range of saturated photocurrent. The current decrease observed at lower potentials, is caused by recombination of electron–hole pairs. The current is then given by

$$J_F(t) = J_{ph}(t) - J_{rec}(t) \tag{4.22}$$

and the derivative with respect to time by

$$dJ_F(t)/dt = dJ_{ph}(t)/dt - dJ_{rec}/dt \tag{4.23}$$

Similarly to the derivation in Section 4.4, the modulations of these currents are described by

$$J_F(t) = J_F + \Delta J \exp i\omega t \tag{4.24a}$$

$$J_{ph}(t) = J_{ph} + \Delta J_{ph} \exp i\omega t \tag{4.24b}$$

$$J_{rec}(t) = J_{rec} + \Delta J_{rec} \exp i\omega t \tag{4.24c}$$

Inserting Eq. (4.24) into (4.23) we have

$$\Delta J = \Delta J_{ph} + \Delta J_{rec} \tag{4.25}$$

The relative quantum yield ϕ_r plays a role similar to that of the impedance Z, derived in Section 4.4. It is frequency-dependent and ϕ_r spectra can be obtained. Examples are given in Section 7.4.

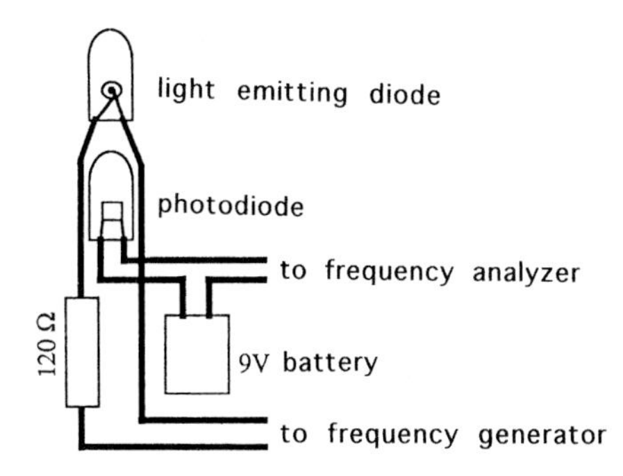

Fig. 4.16 Electrical circuit for light-emitting diode and photodiode (IMPS measurement)

The experimental equipment is quite simple and is similar to that used for impedance measurements. It is convenient to use a laser for light excitation because the light intensity is easily modulated and focused on a rotating disc electrode. When investigating a semiconductor of low bandgap ($E_g < 1.7$ eV) a light-emitting gallium arsenide/phosphide diode (LED) can be used where its intensity can be simply modulated by passing an alternating current through the diode. Since a phase shift occurs between intensity and applied voltage, a photodiode is used as a reference for the incident light, as schematically shown in Fig. 4.16. The diode is operated by the a.c. generator and the current signal is detected by an analyzer. The latter two pieces of equipment can also be used for impedance measurements. In other cases where shorter wavelengths are required, a gas laser must be used. In this case the light is modulated by an acousto-optic modulator.

4.6 Flash Photolysis Investigations

In studies of the properties of semiconductor particles or corresponding colloidal solutions, the formation of intermediates needs to be measured. This is usually done by employing a laser flash apparatus, as has been commonly used in photochemistry for many decades. The principle is as follows.

A pulse laser is used to excite a primary system such as a colloidal solution. The absorption of a possible transient formed by the laser pulse is then measured by using a continuously emitting xenon lamp. The principal set-up as used for measurements in the range of time intervals of nanoseconds, is shown in Fig. 4.17. Here, an excimer laser ($\lambda = 308$ nm) pumps a dye laser at a certain pulse frequency and the corresponding laser emission is used for excitation of the colloid. The advantage of the dye laser is that different excitation wavelengths can be selected by using the appropriate dye. The xenon light beam passes the cell perpendicularly to the laser beam, and is focused on the entrance slit of a monochromator, and is finally detected by a multiplier detection system. Such a system enables the measurement of a complete absorption spectrum of a transient and also its decay time.

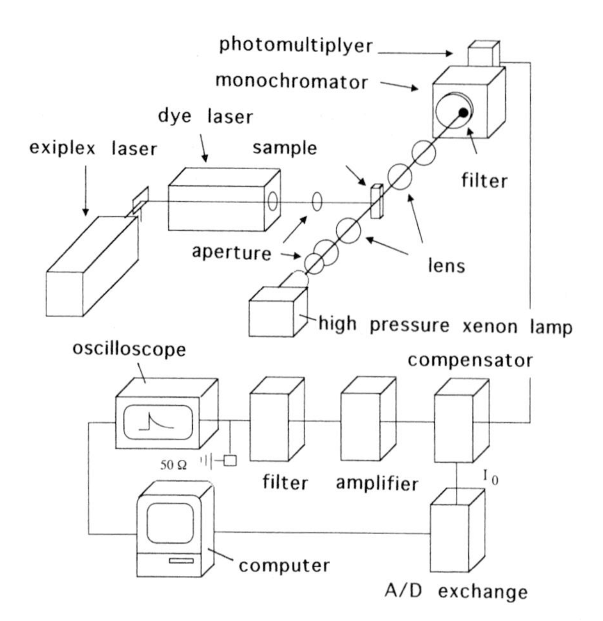

Fig. 4.17 Laser photoflash apparatus

It is also possible to determine transients in the pico- and femtosecond range. Fairly sophisticated equipment has been developed, which cannot be described here.

4.7 Surface Science Techniques

It has been shown in Chapter 3 that electrodes (metals and semiconductors) interact with the electrolyte which strongly influences the optoelectronic properties of the junction. Frequently, it is very difficult to identify the microscopic and molecular nature of the states at the interface. Better scientific understanding demands a spectroscopic identification of the surface and the interface states. Several spectroscopic methods are available which allow the analysis of the chemical, structural and, also, electronic properties of the surface.

4.7.1 Spectroscopic Methods

Infrared spectroscopy (IR) is a fairly simple in situ method. Since the absorption coefficients of molecular vibrations are rather low, it is impossible to detect the IR absorption of a molecule adsorbed or bonded to the semiconductor surface, merely by an ordinary vertical transmission measurement. This problem was solved by using attenuated total reflection (ATR) spectroscopy, as introduced by Harrick [17], and first applied to semiconductor–liquid junctions by Beckmann [18, 19]. In this technique, the incident IR light beam is introduced via a prism into a semiconductor, at such an angle that total internal reflection occurs at the semiconductor–liquid interface, as illustrated in Fig. 4.18. Total reflection is possible if the refractive index of the semiconductor is higher than that of the liquid, a condition which is usually fulfilled for semiconductors. In order to achieve a sufficiently high sensitivity, relatively large crystals are required for a large number of reflections. Since these conditions are only fulfilled with Ge and Si, appropriate IR measurements have mostly been performed with Si which is transparent over a large wavelength range.

There is one disadvantage of this method insofar as the light beam which is reflected at the Si–liquid interface penetrates into the liquid over a certain depth (up to one-tenth of the wavelength near the critical angle), so that the light could also be absorbed by molecules in the solution. The sensitivity could be further increased by modulating the electrode potential [20, 21]. The latter technique is of special interest for studying changes in surface composition.

Other methods for the spectroscopic analysis of metal and semiconductor surfaces are, however, only sufficiently sensitive in an ultrahigh vacuum, which makes in situ application impossible. Thus, ex situ techniques have primarily been applied. The question arises, however, whether an emersion of the semiconductor electrode from the liquid changes the conditions at the interface. It is generally assumed that there is a strong chemical interaction which is not changed during emersion, and also the electrochemical double layer may be preserved so that an analysis of surface species is still possible. On the other hand, certain surface transformations may occur during evaporation of the liquid. In addition, impurities, such as hydrocarbons, may be adsorbed during the transfer into the vacuum system. In order to avoid the latter problem, experimental equipment has been developed by which an electrochemical cell is coupled with an ultrahigh vacuum (UHV) system (see e.g. ref. [22]). Various methods are used for the surface analysis.

UV photoelectron spectroscopy (UPS), x-ray photoelectron spectroscopy (XPS) and low energy electron diffraction (LEED) are most commonly applied in this context. In the first method (UPS) electrons are excited by UV light (sources He I = 21.22 eV; He II = 40.82 eV) and information on the electronic structure of the valence band region is obtained. The second method (XPS) provides information about the

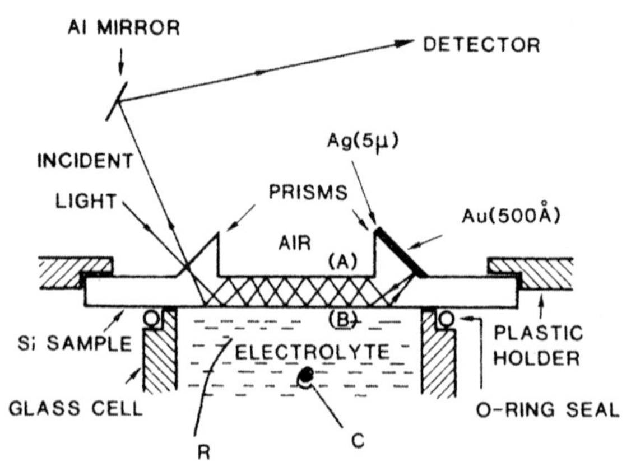

Fig. 4.18 Si sample design for attenuated total reflection (ATR) measurements. (After ref. [20])

elemental composition and the valence states of the elements. Here, x-ray excitation is used (possible radiation sources: Mg K_α = 1253.6 eV or Al K_α = 1486.6 eV). In both methods, the emitted electrons are analyzed as current densities in dependence of their kinetic energy. Since the XPS signals depend not only on elemental composition but are also sensitive to the chemical environment of specific atoms, valuable information on molecular structure can be obtained (see Chapter 8). LEED is used for the analysis of the geometric structure of the surface. Details of these and other methods applicable in combined electrochemical/UHV systems are very well discussed in a review article by Jägermann [23].

There is a small number of reports on ex situ surface analysis of semiconductor electrode surfaces. Usually products formed during an electrochemical reaction have been determined. On the other hand, there are only very few systematic studies and all of these were performed with transition metal chalcogenides. It would be beyond the scope of this book to describe these investigations and the reader is referred to the literature [23, 24].

One notable experimental difficulty is the formation of really clean surfaces; these can only be obtained and conserved under UHV conditions at a residual gas pressure in the order of 10^{-10} mbar. A very convenient technique for producing a clean surface is the cleavage of a suitable crystal under UHV. Since the cleavage plane of III-V and II-VI compounds is the (110) plane, most analytic investigations of clean surfaces in UHV have been done at that plane, whereas electrochemical measurements have generally been done with the (100) and (111) planes. In the case of layered transition chalcogenides, clean van der Waal faces can be prepared simply by stripping off one surface layer. This cleavage can easily be performed in solutions by using scotch tape to pull off a layer.

4.7.2 In Situ Surface Microscopy (STM and AFM)

Scanning tunneling microscopy (STM) is one method of investigation. Here, a metal tip is brought very close to the surface of the sample to be investigated so that a tunneling current can flow between tip and sample when a suitable voltage is applied, as illustrated in Fig. 4.19. When the tip is moved along the surface, the tunneling current can be kept constant by moving the tip up or down depending on the flatness or contour of the sample surface (constant current mode). These small up-and-down movements are achieved by means of a piezocrystal at the back end of the tip (not shown in Fig. 4.19). The corresponding voltage variations at the piezocrystal are taken as a measure of the topography of the surface [25, 26].

During the 1990s, in situ STM measurements have been successfully performed for solid surfaces in contact with an electrolyte and atomic resolution has been obtained. In the case of metal electrodes, topics such as adsorption of monomolecular layers, the structure of underpotential-deposited layers and other areas have been studied [27]. The in situ application of STM to semiconductor electrodes is a little more difficult because there exist the following limitations concerning the tunneling current. Even for low tunneling currents, there is still a liquid layer between the sample and the tip. The semiconductor–liquid junction actually determines the barrier height. As discussed in Chapter 5 in detail, the position of the energy bands at a semiconduc-

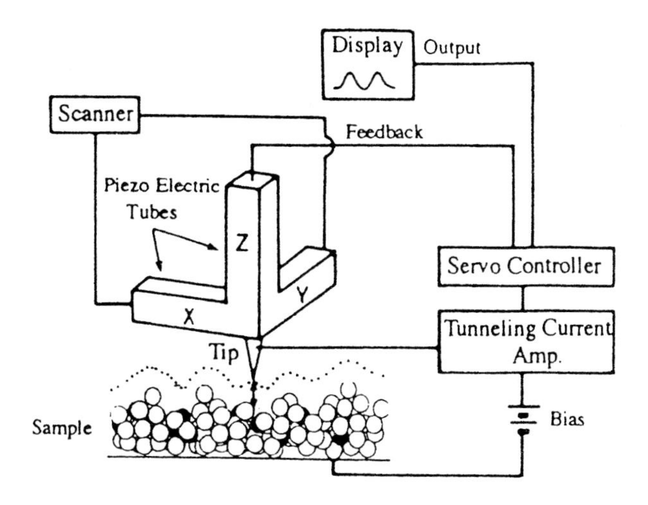

Fig. 4.19 Operation of scanning tunneling microscopy (STM). (After ref. [31])

tor-liquid interface is determined by the interaction of the semiconductor and the contacting liquid and not by the properties of the metal tip. Accordingly, a certain energy band bending occurs at the semiconductor– liquid interface at equilibrium, as shown for an n-type semiconductor in Fig. 4.20a. In this case, almost no electrons can be transferred from the semiconductor to the metal tip. It is quite clear from Fig. 4.20 that an electron transfer (tunneling current) from the semiconductor to the tip is only possible under cathodic polarization of the semiconductor electrode, as illustrated in Figs. 4.20b and c. Accordingly, STM pictures of semiconductor electrodes were only obtained at suitable electrode potentials. The STM technique has mainly been applied to corrosion studies and to the etching of semiconductor electrodes (Si, GaAs) [28–30].

It should be emphasized that the STM technique is only applicable for a conductive surface or electrode. Accordingly, problems may arise if an insulating oxide layer is

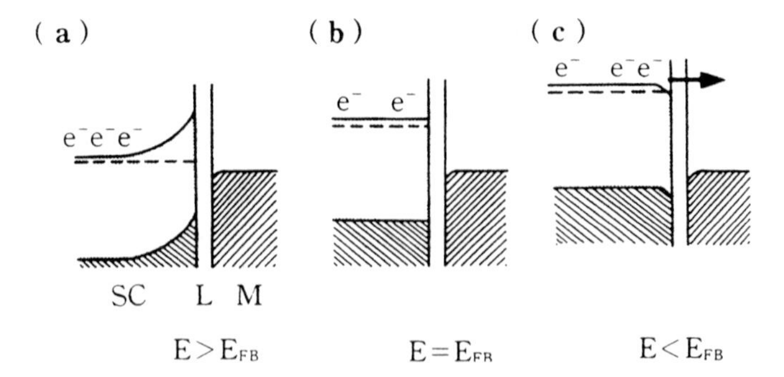

Fig. 4.20 Electron transfer at a semiconductor–metal tip tunneling barrier in STM measurements. (After ref. [28])

formed on a semiconductor surface. This problem is avoided by applying atomic force microscopy (AFM). In this case, the mass attraction and repulsion between a tip and a sample is measured, which is independent of the conductivity of insulating layers at the surface of a given sample. In order to have sufficient sensitivity the tip vibrates at a given resonance frequency and the change of this frequency is taken as a measure of distance between tip and sample [27].

5 Solid–Liquid Interface

5.1 Structure of the Interface and Adsorption

If a metal or a semiconductor comes into contact with a liquid then ions or molecules in the solution may be adsorbed at the electrode or chemical bonds may even be formed with a molecule or preferentially with the liquid molecules, depending on the type of electrode materials. One must distinguish here between an adsorption as a result of electrostatic attraction and a specific adsorption which is accompanied by bond formation. For instance many semiconductors exhibit a strong affinity for water, leading to a bond formation with hydroxyl groups or hydrogen. Examples are germanium or oxide semiconductors such as TiO_2 as illustrated in Fig. 5.1a and b. In a few cases the composition of electrode surfaces has been studied in more detail using special infrared techniques which will not be discussed here. More quantitative information have been obtained with TiO_2 powders, i.e. with TiO_2 particles. Here suitable sus-

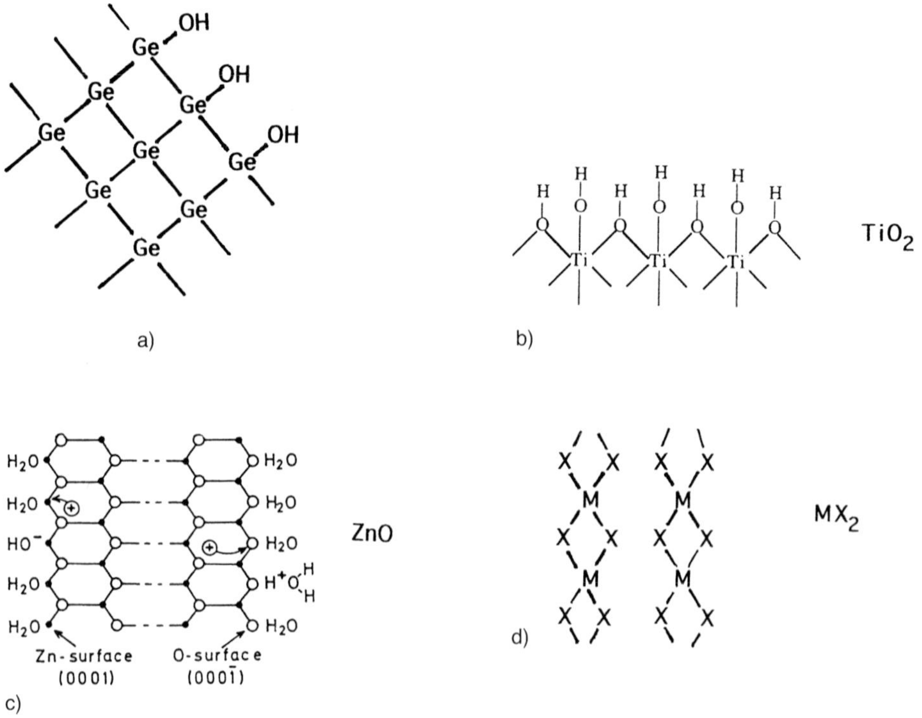

Fig. 5.1 Surface structure of various semiconductors: a) Ge; b) TiO_2; c) ZnO; d) layered dichalcogenides (MX_2)

pensions or colloidal solutions have been titrated and the following equilibria have been found [1]:

$$\mathrm{Ti_sO^- + H^+ \Leftrightarrow Ti_sO{-}H} \qquad (\mathrm{pK_1 = 8}) \tag{5.1}$$

$$\mathrm{Ti_sO{-}H + H^+ \Leftrightarrow Ti_sOH_2^+} \qquad (\mathrm{pK_2 = 4.5}) \tag{5.2}$$

in which the subscript s refers to a surface atom. According to these results two pK values have even been found.

The structure of the surface layer also depends on the orientation of the semiconductor single crystal as has been shown for ZnO. Because of the wurzite lattice of the crystal, two different surfaces can be prepared, namely one which consists only of Zn ions ((0001)surface) and a second where only oxygen is exposed to the electrolyte ($(000\bar{1})$ surface) as shown in Fig. 5.1c [2]. Other interesting examples are layered dichalcogenide crystals with the composition MX_2 (M = transition metal, X = S, Se, Te). These are characterized by sheets of covalently bound X–M–X sandwiches, the latter being kept together by van der Waals forces, i.e. the interaction between these layers is rather weak (Fig. 5.1d). Because of the layered structure, strong anisotropic properties have been observed. For example, the crystals can easily be cleaved along the van der Waal gap, leading to defined and fairly defect-free surfaces. Such a surface produced by cleavage exhibits only a very small interaction with the solvent. Only at steps perpendicular to the layer might a free valency of the metal cause bonding to a hydroxyl group of water.

Instead of a bonding with OH^-, direct bonding with a few other ions has been reported, for instance with F^- at Ge and Si surfaces, at high F^- concentrations [3].

In addition ions may also be simply adsorbed at the surface due to electrostatic forces. This occurs preferentially on hydroxylated surfaces, for example

$$\mathrm{Ti_s{-}OH_2^+ + A^- \Leftrightarrow Ti_s{-}OH_2^+A^-} \tag{5.3}$$

$$\mathrm{Ti_s{-}O^- + K^+ \Leftrightarrow Ti_sO^-K^+} \tag{5.4}$$

A corresponding model of the structure of a double layer is presented in Fig. 5.2. According to this model, the solution side of this double layer itself consists of several layers. The first layer is formed by solvent molecules and specifically adsorbed ions as discussed above. This is called the inner Helmholtz layer or plane (IHP). The center of electrical charge occurs at a distance x_1. The latter is in the order of angstroms. Solvated ions can only approach the electrode surface to a distance x_2. The center of their charge is called the outer Helmholtz plane (OHP). Since the interaction between the electrode and the solvated ions involves only long-range forces, a concentration profile of the solvated ions exists over a relatively large distance, depending on their concentration. Because of their thermal motion they are distributed in a three-dimensional region, called the diffuse layer (or Gouy layer, see Section 5.2.2) which extends from the OHP into the bulk of the solution.

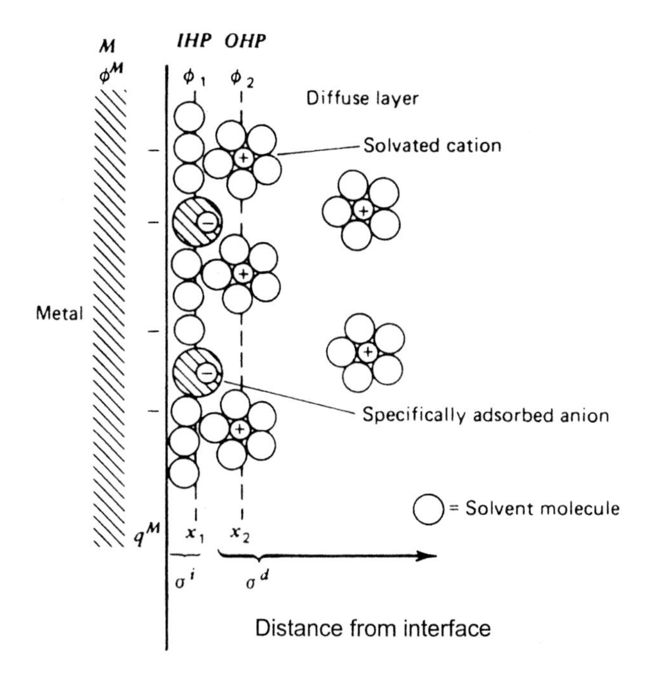

Fig. 5.2 Model for the double layer region at the metal–electrolyte interface IHP, inner Helmholtz plane; OHP, outer Helmholtz plane. (After ref. [4])

5.2 Charge and Potential Distribution at the Interface

The formation of a double layer has considerable consequences for the charge and potential distribution across the interface. In the case of a metal electrode the counter charges are located just below the surface. Since, however, the carrier density in a semiconductor is usually much smaller than in a metal electrode, the counter charges can be distributed over a considerable distance below the interface, i.e. a space charge layer is formed, similar to that in pure solid devices (compare with Chapter 2). The potential and charge distribution across the Helmholtz layer, Gouy layer and space charge region will be treated separately in the next three sections. This topic has been treated in various review articles [5–9] and books [10, 11].

5.2.1 The Helmholtz Double Layer

Considering at first a metal electrode and assuming that cations are specifically adsorbed at the electrode surface, then the equivalent negative counter charge occurs just below the metal surface, as indicated in Fig. 5.3. This charge separation causes a corresponding potential $\Delta\phi_H$ across the interface as also shown in Fig. 5.3. It is partly determined by the specifically adsorbed ions and in addition by the solvated ions

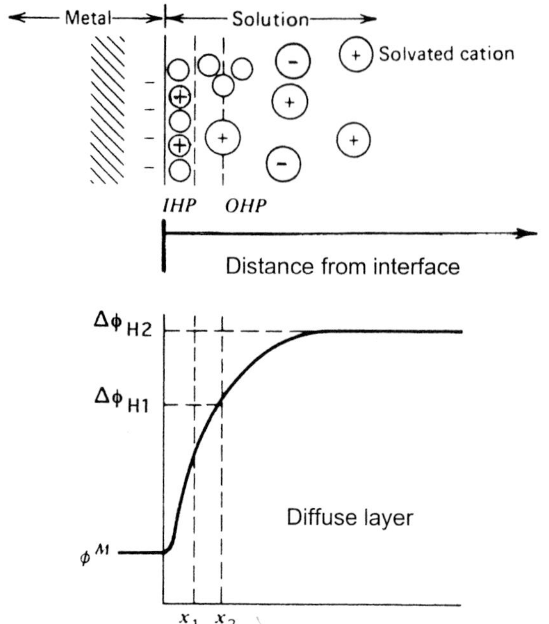

Fig. 5.3 Potential distribution at the metal–electrolyte interface

within the outer Helmholtz layer. Assuming for instance that the formation of the inner Helmholtz layer is due to the reaction of the electrode with the solvent, then an outer Helmholtz layer may be formed due to electrostatic forces on ions formed upon addition of a conducting salt to the electrolyte. At high concentrations all of these charges are concentrated within the outer Helmholtz layer, at low concentrations the charges are distributed over a much thicker diffuse layer as described in Section 5.1 (see e.g. ref. [4]). The case of the diffuse layer will be analyzed separately in Section 5.2.2.

As already discussed in Section 3.2 the potential across a single solid–liquid interface cannot be measured. One can only measure the potential of an electrode vs. a reference electrode. It has already been shown in Section 3.2 that a certain potential is produced at a metal or semiconductor electrode upon the addition of a redox system, because the redox system equilibriates with the electrons in the electrode, i.e. the Fermi level on both sides of the interface must be equal under equilibrium. It should be emphasized here that the potential caused upon addition of a redox couple to the solution occurs in addition to that already formed by the specific adsorption of, for instance, hydroxyl ions. A variation in the relative concentrations of the oxidized and reduced species of the redox system leads to a corresponding change of the potential across the outer Helmholtz layer, as required by Nernst's law (see Eq. 3.47), which can be detected by measuring the electrode potential vs. a reference electrode. However, there still exists a potential across the inner Helmholtz layer which remains unknown.

As already mentioned, the charges are concentrated within the inner and outer Helmholtz layer at high ion concentrations. In this case, the double layer acts as parallel-plate capacitor. The charge density Q_i stored in such a capacitor, is related to the voltage drop U by

$$Q_i = \frac{\varepsilon\varepsilon_0}{x_1}U \tag{5.5}$$

where ε is the dielectric constant of the medium, ε_0 is the permittivity of free space and x_1 the spacing. The differential capacity is then given by

$$C_H = \frac{dQ_i}{dU} = \frac{\varepsilon\varepsilon_0}{x_1} \tag{5.6}$$

Using $\varepsilon = 20$ and $x_1 = 5 \times 0^{-7}$ cm, one obtains $C_H \approx 3 \times 10^{-5}$ F cm^{-2}. This capacity value agrees with experimental data within an order of magnitude. This Helmholtz capacity is independent of the electrode potential, i.e. in the case of a metal electrode any external variation of the electrode potential leads only to a corresponding change of the charges on both sides of the interface.

5.2.2 The Gouy Layer in the Electrolyte

As already discussed in Section 5.1, at lower ion concentrations the charges near the interface are distributed over a diffuse layer. In this case the charge balance at the interface is given by

$$Q_{sol} = Q_H + Q_d \tag{5.7}$$

in which Q_{sol} represents the charge in the solid, Q_H that in the inner Helmholtz layer and Q_d that in the diffuse layer. In the case of the diffuse double layer the corresponding capacity is given by

$$C_d = \frac{dQ_d}{d\phi} = \left(\frac{2z^2e^2\varepsilon\varepsilon_0 c_0}{kT}\right)\cosh\left(\frac{ze\Delta\phi_0}{2kT}\right) \tag{5.8}$$

in which z is the number of ion charges, c_0 the ion concentration per cm^3 in the bulk of the solution, and $\Delta\phi_0$ the potential difference between the electrode surface and the bulk of the solution.

This is a differential capacity because it depends on the potential $\Delta\phi_0$. The total capacity for a metal–liquid interface can be considered as two capacitors, namely C_H and C_d, in series. We have then

$$\frac{1}{C_{H,d}} = \frac{1}{C_H} + \frac{1}{C_d} \tag{5.9}$$

This has been verified experimentally; examples are given in refs. [4, 12]. Evaluations of corresponding data have shown that $C_d \ll C_H$ for ion concentrations of $c < 10^{-3}$ M. Accordingly, any potential drop across the diffuse double layer can be neglected for higher ion concentrations. Since in experiments with semiconductor electrodes the occurance of a Gouy diffuse double layer complicates the analysis of the data, only solutions containing a conducting salt of high concentration ($\geq 10^{-2}$ M) are used. This is also the reason why the derivation of Eq. (5.8) (see [4, 12]) is ignored here. Its origin will be discussed again in conjunction with the space charge layer in the semiconductor.

5.2.3 The Space Charge Layer in the Semiconductor

In contrast to metals, the carrier density in a semiconductor is much smaller. Accordingly, in a semiconductor electrode the charge in the solid (counter charge to Q_{sol}) is distributed over a certain range below the surface. The resulting potential and charge distribution is shown in Fig. 5.4. As already mentioned in the previous section, we consider here an electrolyte with a high ion concentration so that a diffuse layer on the solution side can be neglected.

The potential and charge distribution within the space charge region is quantitatively described by the Poisson equation as given by

$$\frac{d^2 \Delta\phi_{sc}}{dx^2} = -\frac{1}{\varepsilon\varepsilon_0}\rho(x) \tag{5.10}$$

in which the charge density is given by

$$\rho(x) = e\,[N_d - N_a - n(x) + p(x)] \tag{5.11}$$

in which x is the distance from the surface and N_d and N_a are the fixed ionized donor and acceptor densities, respectively, which are given by the doping of the semiconducting material. The electron and hole densities, $n(x)$ and $p(x)$, vary with the distance x. According to Eqs. (1.27) and (1.30) they are given by

$$n(x) = N_c \exp\left(-\frac{E_c(x) - E_F}{kT}\right) \tag{5.12}$$

$$p(x) = N_v \exp\left(\frac{E_v(x) - E_F}{kT}\right) \tag{5.13}$$

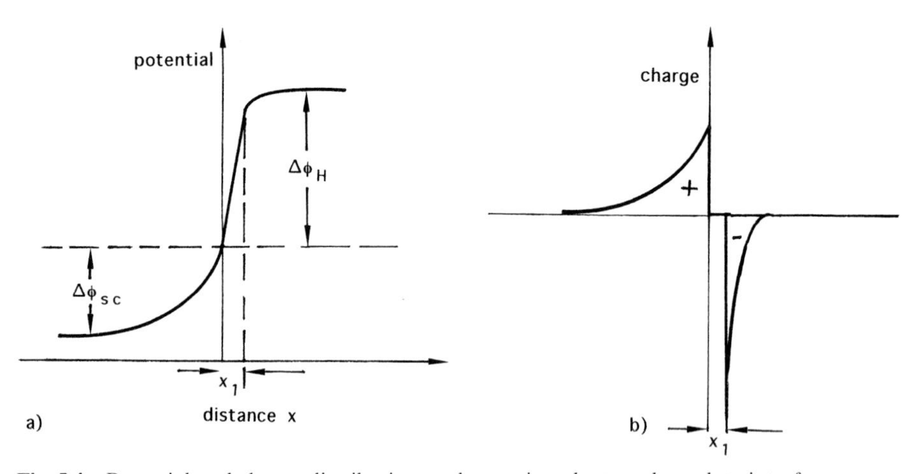

Fig. 5.4 Potential and charge distribution at the semiconductor–electrolyte interface

Since the Fermi level is expected to be constant within the space charge region the position of the energy bands $E_c(x)$ and $E_v(x)$ vary with distance. Denoting the carrier density in the bulk of the semiconductor by n_0 and p_0 one obtains

$$n(x) = n_0 \exp\left(-\frac{E_c(x) - E_c^b}{kT}\right) = n_0 \exp\left(-\frac{e\Delta\phi_{sc}(x)}{kT}\right) \tag{5.14}$$

$$p(x) = p_0 \exp\left(\frac{E_v(x) - E_v^b}{kT}\right) = p_0 \exp\left(\frac{e\Delta\phi_{sc}(x)}{kT}\right) \tag{5.15}$$

in which E_c and E_v are the energy bands in the bulk. The latter equations imply a Boltzmann distribution within the space charge layer. In the bulk of the semiconductor, far from the surface, charge neutrality must exist. Therefore,

$$N_d - N_a = n_0 - p_0 \tag{5.16}$$

Inserting Eqs. (5.11), (5.14), (5.15) and (5.16) into (5.10) we have

$$\frac{d^2(\Delta\phi_{sc})}{dx^2} = \frac{e}{\varepsilon\varepsilon_0}\left[n_0 - p_0 - n_0 \exp\left(-\frac{e\Delta\phi_{sc}}{kT}\right) + p_0 \exp\left(\frac{e\Delta\phi_{sc}}{kT}\right)\right] \tag{5.17}$$

Using the transformation

$$\frac{d^2(\Delta\phi_{sc})}{dx^2} = \frac{\delta(\Delta\phi_{sc})}{\delta x} \cdot \frac{d\left(\frac{\delta(\Delta\phi_{sc})}{\delta x}\right)}{d(\Delta\phi_{sc})} \tag{5.18}$$

Eq. (5.17) can be rewritten as

$$\int_0^{\delta(\Delta\phi_{sc})} \frac{\delta(\Delta\phi_{sc})}{\delta x} \cdot d\left(\frac{\delta(\Delta\phi_{sc})}{\delta x}\right) \tag{5.19}$$

$$= -\frac{e}{\varepsilon\varepsilon_0}\int_0^{\Delta\phi_{sc}}\left[n_0 - p_0 - n_0 \exp\left(-\frac{e\Delta\phi_{sc}}{kT}\right) + p_0 \exp\left(\frac{e\Delta\phi_{sc}}{kT}\right)\right] d(\Delta\phi_{sc})$$

After integration, the electric field $\mathscr{E}_s$ at the surface can be derived as given by

$$\mathscr{E}_s = \left.\frac{d(\Delta\phi_{sc})}{dx}\right|_{x=0} = \frac{kT/e}{L_D} G(\Delta\phi_{sc}) \tag{5.20}$$

and

$$G(\Delta\phi_{sc}) = \frac{kT/e}{L_D}\left\{\frac{p_0}{n_i}\left[\exp\left(\frac{e\Delta\phi_{sc}}{kT}\right) - 1\right] + \frac{n_0}{ni}\left[\exp\left(-\frac{e\Delta\phi_{sc}}{kT} - 1\right)\right]\right\} \tag{5.20a}$$

in which n_i is the intrinsic carrier density which is determined by the equilibrium equation Eq. (1.32), i.e. by $n_0 p_0 = n_i^2$. L_D is the so-called Debye length as given by

$$L_D = \left(\frac{\varepsilon\varepsilon_0 kT}{2n_i e^2}\right)^{1/2} \tag{5.21}$$

Similarly, by Gauss' law the space charge required to produce the field $\mathscr{E}_s$ is

$$Q_{sc} = \varepsilon\varepsilon_0 \mathscr{E}_s \tag{5.22}$$

The differential capacity of the space charge layer is defined by

$$C_{sc} = \frac{dQ_{sc}}{d(\Delta\phi_{sc})} \tag{5.23}$$

Inserting Eqs. (5.20) and (5.21) into (5.23) one obtains after differentiation

$$C_{sc} = \frac{\varepsilon\varepsilon_0}{L_D} \cosh\left(\frac{e\Delta\phi_{sc}}{kT}\right) \tag{5.24}$$

This equation shows clearly that the space charge capacity C_{sc} depends strongly on the potential $\Delta\phi_{sc}$ across the space charge layer, although in a rather complex way. Before analyzing this relation in more detail it is useful to introduce two further equations. In accordance with Eqs. (5.14) and (5.15) the electron and hole densities at the surface are given by

$$n_s = n_0 \exp\left(-\frac{e\Delta\phi_{sc}}{kT}\right) \tag{5.25a}$$

$$p_s = p_0 \exp\left(\frac{e\Delta\phi_{sc}}{kT}\right) \tag{5.25b}$$

If electrons and holes at the surface are at equilibrium we have:

$$n_s p_s = np = n_i^2 \tag{5.26}$$

Taking an n-type semiconductor as an example, the resulting C_{sc} –$\Delta\phi_{sc}$ curve is illustrated in Fig. 5.5. According to Eq. (5.24), the potential range can be divided in three sections as follows.

(1) If $\Delta\phi_{sc}$ is negative then the electron density at the surface becomes much larger than that in the bulk ($n_s \gg n_0$). This range is the accumulation region, in which C_{sc} rises exponentially with increasing negative $\Delta\phi_{sc}$ values (left side of Fig. 5.5).

(2) The center range, where the capacity curve is rather flat, is characterized by a majority carrier density being depleted with respect to the bulk concentration, i.e. $n_s < n_0$ (depletion region) whereas $n_s \gg p_s$. In this range the capacity is mainly determined by the linear term in Eq. (5.24) which originates from the ionized donors.

(3) As soon as the hole density increases above the electron density at the surface ($p_s > p_0$), C_{sc} increases exponentially again. Since C_{sc} is here entirely determined by the minority carriers, this range is called the inversion region. Using a p-type electrode instead, the resulting capacity curve would be a mirror image to that obtained for the n-type electrode, i.e. the accumulation range would occur at positive $\Delta\phi_{sc}$ values.

As will be shown below, the depletion region plays a dominant role in the analysis of the potential distribution of the semiconductor–electrolyte interface. As already mentioned, in this range the linear term in Eq. (5.24) dominates whereas the exponential terms can be neglected. In this case one obtains

$$\frac{1}{C_{sc}^2} = \left(\frac{2L_{D,eff}}{\varepsilon\varepsilon_0}\right)^2 \left(\frac{e\Delta\phi_{sc}}{kT} - 1\right) \tag{5.27}$$

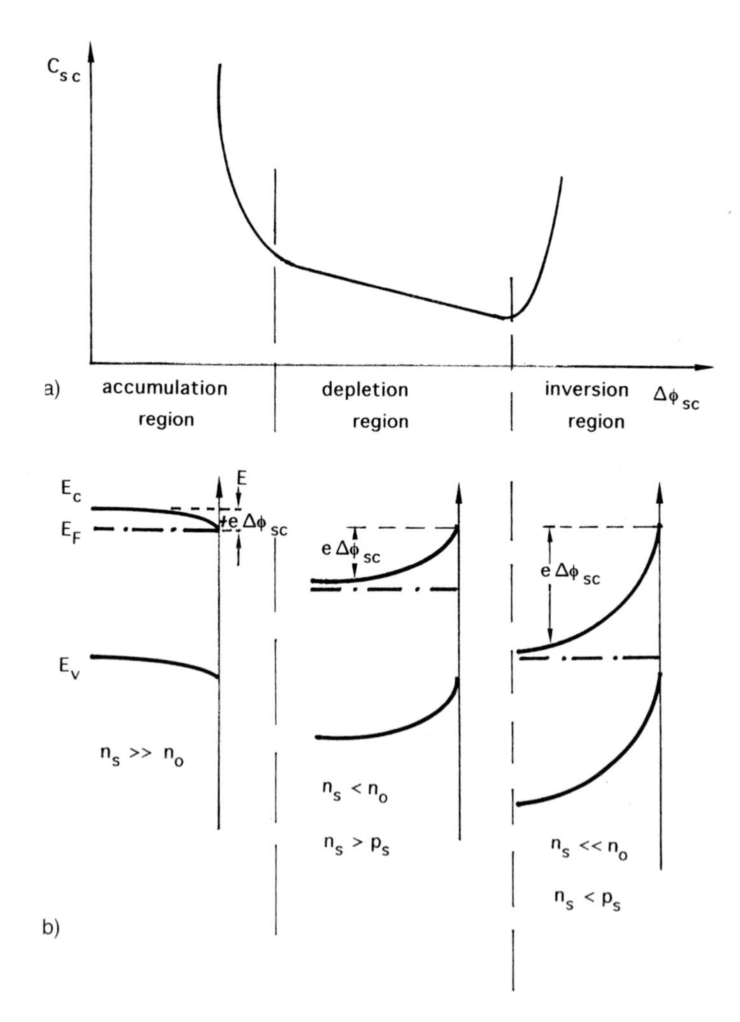

Fig. 5.5 Accumulation, depletion and inversion layer at the semiconductor–electrolyte interface: a) space charge capacity C_{sc} vs. potential across the space charge layer $\Delta\phi_{sc}$; b) energy model

in which $L_{D,eff}$ is the effective Debye length as given by

$$L_{D,eff} = \left(\frac{\varepsilon\varepsilon_0 kT}{2n_0 e^2} \right)^{1/2} \tag{5.28}$$

The latter equation is in principle identical to Eq. (5.21), only n_i is replaced by n_0. A plot of $1/C_{sc}^2$ vs. $\Delta\phi_{sc}$ for an n-type electrode is presented in Fig. 5.6. At $\Delta\phi_{sc} = kT/e$ which is nearly zero ($kT/e = 0.025$ V), the energy bands are flat and $1/C_c^2 = 0$. The slope is determined by the Debye length and therefore by the doping. In the literature Eq. (5.27) is usually known as the Mott–Schottky relation.

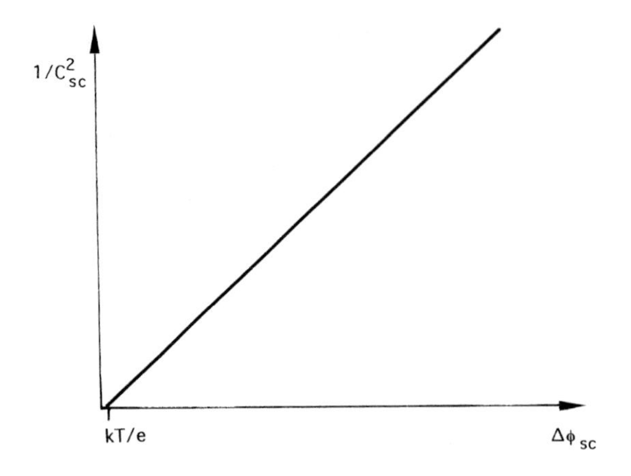

Fig. 5.6 Mott–Schottky plot of the space charge capacity vs. potential across the space charge layer (C_{sc}^{-2} vs. $\Delta\phi_{sc}$) for an n type semiconductor electrode (theoretical curve).

Eq. (5.24) becomes also very simple in the case of an intrinsic semiconductor in which the electron and hole densities are equal ($n_0 = p_0 = n_i$). One obtains then:

$$C_{sc} = \frac{\varepsilon\varepsilon_0}{L_D} \cosh\left(\frac{e\Delta\phi_{sc}}{kT}\right) \tag{5.29}$$

A corresponding plot of log C_{sc} vs. $\Delta\phi_{sc}$ is shown in Fig. 5.7. The minimum occurs at $\Delta\phi_{sc} = 0$. At first sight it may be surprising that C_{sc} is not zero at the minimum because there is then no space charge. The reason for a finite C_{sc} value at the minimum is that C_{sc} is a differential capacity and $dQ_{sc}/d(\Delta\phi_{sc})$ is not zero at $\Delta\phi_{sc} = 0$.

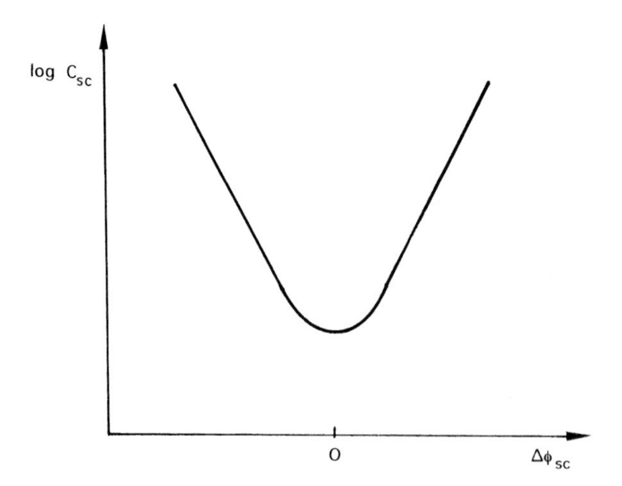

Fig. 5.7 Space charge capacity C_{sc} vs. potential across the space charge layer $\Delta\phi_{sc}$ for an intrinsic semiconductor (theoretical curve)

It should be mentioned here, that the capacity of the space charge layer in an intrinsic semiconductor looks very similar to that of the diffuse Gouy layer in the electrolyte (compare with Eq. 5.8). This is very reasonable because the Gouy layer is also a kind of space charge layer with ions instead of electrons as mobile carriers. C_d was actually derived by the same procedure as given here for C_{sc}. Similarly as in the case of C_H and C_d, the space charge capacity C_{sc} and the Helmholtz capacity C_H can be treated as capacitors circuited in series. We have then

$$\frac{1}{C} = \frac{1}{C_{sc}} + \frac{1}{C_H} \tag{5.30}$$

Accordingly, the space charge capacity can only be measured for $C_{sc} < C_H$. This condition can usually fulfilled with semiconductors of a carrier density smaller than $n_0 = 10^{20}$ cm^{-3}. In addition a thickness of the space charge layer, d_{sc}, can be derived using the relation $d = \varepsilon\varepsilon_0/C$ (valid for a capacitor with fixed plates). Applying this to a depletion layer, one obtains from Eq. (5.27)

$$d_{sc} = 2\,L_{D,eff}\left[\frac{e\Delta\phi_{sc}}{kT} - 1\right]^{1/2} \tag{5.31}$$

The thickness increases with increasing potential $\Delta\phi_{sc}$ across the space charge region. For instance for $\Delta\phi_{sc} \approx 0.5$ V, we have $d_{sc} \approx 10L_{D,eff}$. Taking a semiconductor of a typical doping of $n_0 = 10^{17}$ cm^{-3} one obtains $L_D \approx 10^{-6}$ cm ($\varepsilon = 10$) and accordingly $d_{sc} \approx 10^{-5}$ cm.

It should be mentioned here that the method of surface conductivity can also be applied to studying the space charge layer. The surface conductivity can be derived as follows.

In the space charge region the carrier densities are different from those in the bulk. For an upward band bending, the distance between the Fermi level and conduction band is increased and that between E_F and the valence band is decreased, so that the electron density $n(x)$ is smaller and the hole density $p(x)$ is higher in the space charge region compared with the bulk. The excess of charges over the whole space charge region is defined as [13]

$$\Delta N = \int [n(x) - n_0]\,\mathrm{d}x; \quad \Delta P = \int [p(x) - p_0]\,\mathrm{d}x \tag{5.32}$$

Using Eq. (5.18) this equation can be transformed into

$$\Delta N = L_D n_0 \int_0^{\Delta\phi_{sc}} \left[\frac{\left(\exp\left(-\frac{e\Delta\phi_{sc}}{kT}\right) - 1\right)}{G(\Delta\phi_{sc})}\right] \mathrm{d}(\Delta\phi_{sc}) \tag{5.33a}$$

$$\Delta N = L_D p_0 \int_0^{\Delta\phi_{sc}} \left[\frac{\left(\exp\left(\frac{e\Delta\phi_{sc}}{kT}\right) - 1\right)}{G(\Delta\phi_{sc})}\right] \mathrm{d}(\Delta\phi_{sc}) \tag{5.33b}$$

in which $G(\Delta\phi_{sc})$ is given by Eq. (5.20a). Eqs. (5.33a and b) have been evaluated and tabulated in ref. [14]. The change of the conductivity $\Delta\sigma$ is then given by

$$\Delta\sigma = e(\mu_n \Delta N + \mu_p \Delta P) \tag{5.34}$$

in which μ_n and μ_p are the mobility of electrons and holes, respectively. This method has been successfully applied only in a very few cases (Ge and Si) because of the problem of a parallel conductance through the electrolyte.

5.2.4 Charge Distribution in Surface States

At a semiconductor surface the crystal lattice is interrupted. In consequence surface states within the bandgap may exist at the surface as already discussed in Chapter 2. Charges can be stored in surface states. The occupation by electrons depends on the position of the Fermi level with respect to such a surface state. The charge is given by

$$Q_{ss} = e f N_t \tag{5.35}$$

in which N_t is the density of surface states given in cm^{-2} and f is the Fermi function (compare with Eq. 1.25) as given by

$$f = \frac{1}{1 + \exp\left(\frac{E_F - E_t}{kT}\right)} \tag{5.36}$$

Since the distance between the Fermi level E_F and the energy bands varies with $\Delta\phi_{sc}$ (see Section 5.3) f is also changed. If $E_F = E_t$, the surface state is half-occupied ($f = 0.5$), as shown in Fig. 5.8. Since the charge Q_{ss} depends on the potential across the space charge layer, a differential surface state capacity can be defined by

$$C_{ss} = \frac{dQ_{ss}}{d(\Delta\phi_{sc})} = eN_t \frac{df}{d(\Delta\phi_{sc})} \tag{5.37}$$

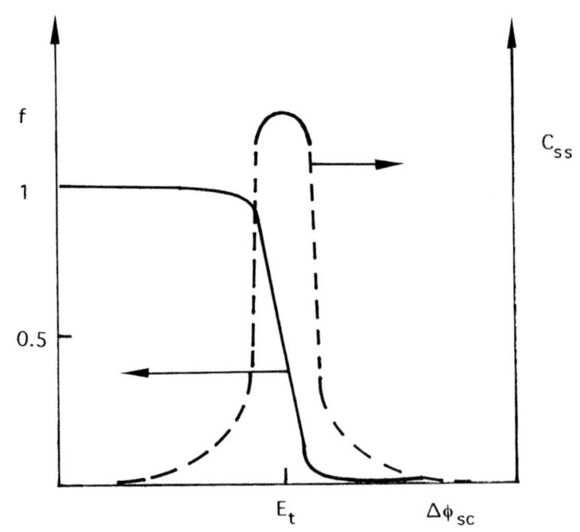

Fig. 5.8 Fermi function f and surface state capacity C_{ss} vs. potential across the space charge layer $\Delta\phi_{sc}$ (theoretical curve)

As can be recognized very easily, $df/(d\Delta\phi_{sc})$ and therefore C_{ss} passes a maximum at $E_F = E_t$ as also shown in Fig. 5.8. Since the charge in the space charge layer and in the surface states is always varied in equal directions C_{ss} must be parallel to C_{sc}.

5.3 Analysis of the Potential Distribution

During the 30 years from 1970, the potential distribution at many semiconductor electrodes has been studied. In the early stage of semiconductor electrochemistry primarily germanium and silicon electrodes were investigated because well-defined single crystals were available. It turned out much later that very important basic information had been obtained, especially with intrinsic germanium electrodes ($n_0 = p_0$), as will be shown below. In the following, Ge, Si and compound electrodes will be treated separately.

According to Fig. 5.4, the potential at a semiconductor electrode interface is composed of a potential across the Helmholtz layer $\Delta\phi_H$ and across the space charge layer $\Delta\phi_{sc}$, provided that the potential across the Gouy layer can be neglected. The electrode potential U_E measured vs. a reference electrode is then given by

$$U_E = \Delta\phi_{sc} + \Delta\phi_H + \text{const} \tag{5.38}$$

The constant contains all potentials at the reference electrode which are not known (see also Chapter 3). In the case of a metal electrode in which no space charge layer exists, it is clear that any variation of the electrode potential leads to a corresponding change of $\Delta\phi_H$. With semiconductor electrodes, however, in principle both potentials $\Delta\phi_{sc}$ and $\Delta\phi_H$ can be varied. This must be analyzed for each material.

5.3.1 Germanium Electrodes

The first capacity measurements with germanium were performed 1957 by Bohnenkamp and Engell [15], who observed a minimum in the capacity when varying the electrode potential. It took another 5 years before the potential distribution was investigated quantitatively at nearly intrinsic Ge electrodes by using capacity [16] and surface conductivity [17] measurements as shown in Figs. 5.9 and 5.10. In both cases a minimum of the C_{sc} and $\Delta\sigma$ vs. U_E curves was found (dots). The theoretical curves were calculated by using Eqs. (5.29) and (5.34) (solid lines) and plotted against $\Delta\phi_{sc}$ (upper scale). The two scales, U_E and $\Delta\phi_{sc}$, were shifted against each other until the best fit between experimental data and theoretical curves were obtained.

Since there was such an excellent agreement between the theoretical and experimental curves, it was concluded that any variation of the electrode potential occurs across the space charge layer, i.e. according to Eq. (5.38)

$$\Delta U_E = \Delta(\Delta\phi_{sc}) \tag{5.39}$$

Therefore, the potential across the Helmholtz layer $\Delta\phi_H$ remains constant. This is a very important result which had already been obtained at a very early stage of research

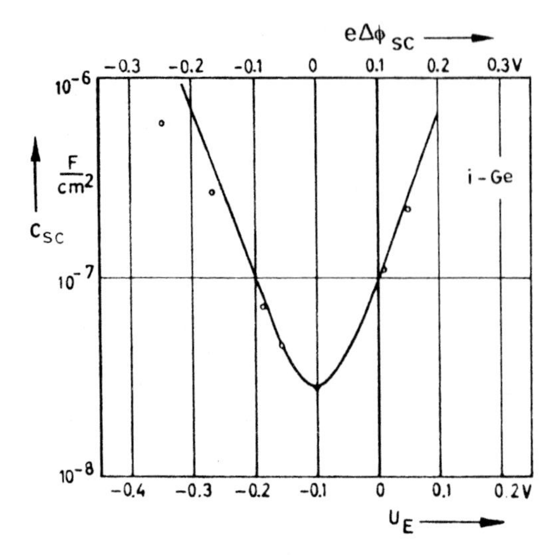

Fig. 5.9 Log C_{sc} vs. electrode potential U_E for an intrinsic germanium electrode in aqueous solutions (see text for fuller description). (After ref. [13])

in semiconductor electrochemistry. The importance of this result will be seen to be even more pronounced in connection with the discussion of compound semiconductors (Section 5.3.3). The minimum of the curves in Figs. 5.9 and 5.10 corresponds then to an electrode potential where the energy bands are flat as far as the surface. The corresponding electrode potential is called the flatband potential U_{fb}.

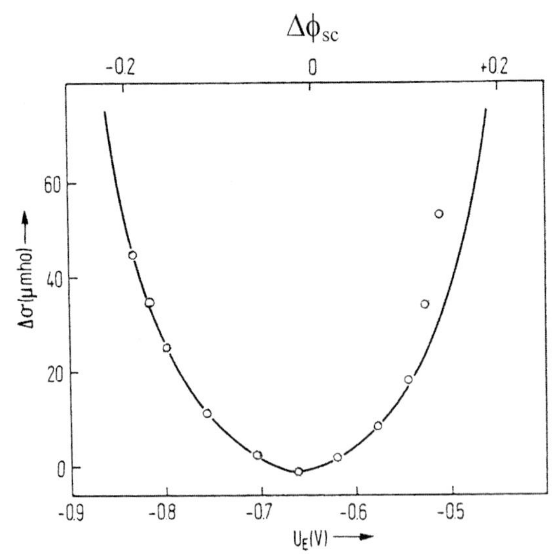

Fig. 5.10 Surface conductivity $\Delta\sigma$ vs. electrode potential U_E for intrinsic germanium electrode in aqueous solutions. (see text for fuller description). (After ref. [17])

It should be mentioned that the theoretical capacity curve could only be fitted to the experimental values the C_{sc} values had been multiplied by a factor of 1.3 which corresponds to a slight linear shift on the logarithmic C_{sc} scale in Fig. 5.9. This has been interpreted as a roughness factor by which the geometric surface area has been corrected.

Another important conclusion has been drawn from the capacity–potential curve in Fig. 5.9. Since the experimental data could be fitted very well to the theoretical C_{sc}–$\Delta\phi_{sc}$ curve it has been concluded that no surface states affect the experimental capacity curve. According to a quantitative evaluation of the experimental results, the density of surface states must be smaller than 10^{10} cm^{-2}. Since this density is very small Brattain and Boddy concluded that a Ge surface which is in contact with H_2O exhibits a „perfect“ surface [16]. This result was very surprising because Ge or Si surfaces produced by cleavage in ultrahigh vacuum exhibit a much higher density of surface states ($\approx 10^{12}$ cm^{-2}). Accordingly, due to the reaction of germanium with H_2O the corresponding germanium hydroxide surface contains an extremely small number of dangling bonds. The electronic states of the Ge–OH surface groups must be located at or very near to the conduction and valence bands so that they are not detectable.

The ideal capacity or surface conductivity curves as presented in Figs. 5.9 and 5.10, have only been obtained after anodic prepolarization of the Ge electrodes and during a fast potential scan from anodic toward cathodic potentials. This result indicates that there are other factors which influence the surface properties of Ge electrodes. Before discussing these phenomena it should be mentioned that the capacity and the surface conductivity curves depend on the pH of the solution. The corresponding minima as determined after anodic prepolarization are given by curve a in Fig. 5.11 as has been measured by several authors [17–19]. The slope of this curve is 64 mV/pH above pH 4. According to this result the electrode potential at which the minimum of the capacity curve occurs ($\Delta\phi_{sc} = 0$) is given by

$$U_{E,\min} = \Delta\phi_H + \text{const} \tag{5.40}$$

This effect has been interpreted by a dissociation of the double layer as given for a (111) surface by [18]

$$\equiv\text{Ge–OH} + \text{OH}^- \Leftrightarrow \equiv\text{Ge–O}^- + \text{H}_2\text{O} \tag{5.41}$$

The equilibrium condition is given by

$$\mu_{\text{Ge–OH}} + RT\ \ln\left(x_{\text{Ge–OH}}\right) = \mu_{\text{Ge–O}^-} + RT\ \ln\left(x_{\text{Ge–O}^-}\right) - F\Delta\phi_H + \mu_H + 2.3\text{pH} \tag{5.42}$$

in which x is the molar ratio and μ_i the chemical potential of the various surface species. For electrostatic reasons $x_{\text{Ge–O}^-}$ must remain small so that $x_{\text{Ge–OH}} \approx 1$. One obtains then from Eq. (5.42)

$$\Delta\phi_H = (RT/F)\text{pH} + \text{const} \tag{5.43}$$

As already mentioned, each of these results was obtained after anodic prepolarization. If the same type of measurements are done after cathodic polarization then the capacity curve and its minimum occur at more cathodic potentials, as illustrated in the upper part of Fig. 5.12. The pH-dependence of this minimum is also given in Fig. 5.11 (curve b). According to this result, the potential across the Helmholtz layer $\Delta\phi_H$ is different

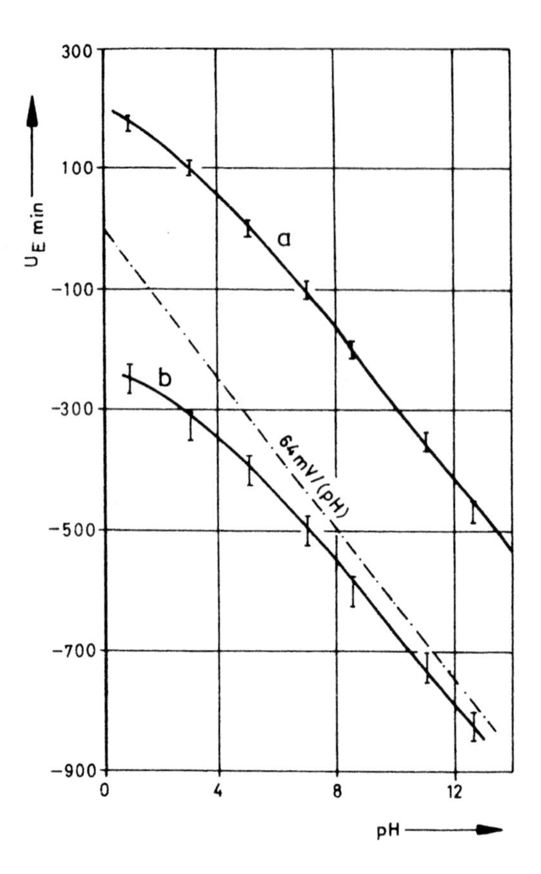

Fig. 5.11 Potentials of capacity minima after anodic (a) and cathodic (b) prepolarization of intrinsic Ge electrodes with dependence on the pH of the solution. (After ref. [18])

from that obtained after anodic prepolarization. Simultaneously, the current–potential curve has been measured with the intrinsic Ge electrode during a cathodic and the corresponding reverse anodic potential sweep (see lower part of Fig. 5.12). Using a certain scan rate (here 0.35 V s^{-1}) the potential scale corresponds also to a time scale. Since current flows only within a certain potential range or within a limited time interval the charge transferred in this process must be limited. An evaluation of the current–time dependence, i.e. an integration over the *j*–*t* peaks, yields a charge corresponding to about one monolayer. According to this result the formation of a hydride layer on the surface has been postulated [20]:

$$\equiv \text{Ge–OH} + 2\text{H}^+ + \text{e}^- \Leftrightarrow \equiv \text{Ge–H} + \text{H}_2\text{O} + \text{p}^+ \tag{5.44}$$

Gerischer et al. predicted that the reduction of the ≡Ge–OH surface occurs by a transfer of an electron via the conduction band and by a simultaneous injection of a hole into the valence band. Details concerning charge transfer processes are given in Chapter 7. A further proof of Eq. (5.44) is given below (see Eqs. 5.46 to 5.48). A pH-dependence was also found in the case of a hydride surface (curve b in Fig. 5.11), which was explained by a dissociation of the double layer as given by [20]

$$\equiv \text{Ge–H} + \text{OH}^- \Leftrightarrow \equiv \text{Ge}^- + \text{H}_2\text{O} \tag{5.45}$$

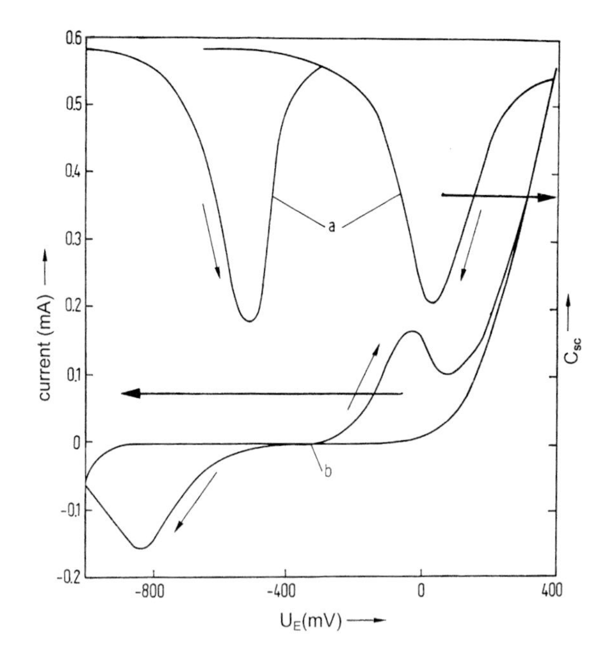

Fig. 5.12 a) Space charge capacity (right scale) of intrinsic Ge vs. electrode potential after anodic and cathodic prepolarization at pH2. b) Current–potential curve. Both measurements were performed at a scan rate of 0.35 V s^{-1}. (After ref. [22])

It is interesting to note that the capacity minima and therefore the potential across the Helmholtz layer become independent of pH below pH 4, as shown by curves a and b in Fig. 5.11. This result is understandable on the basis of the dissociation equilibria given by Eqs. (5.41) and (5.45). However, it may not be concluded from this result that there is no longer a potential drop across the Helmholtz layer at low pH. This becomes clear when comparing the capacity minima for the hydroxide ($U_E \approx +500$ mV) and hydride surfaces ($U_E \approx +30$ mV) at low pH. Since these have different positions, $\Delta\phi_H$ differs by more than 0.5 V for the two cases. This behavior can only be explained by a different electrical moment of the hydroxide and hydride surfaces. Accordingly, the Helmholtz potential at a Ge electrode is caused by a dissociation double layer and a dipole within the surface layer.

Even more insight into the potential and charge distribution at the germanium-electrolyte interface was obtained by experiments with solutions containing H_2O_2. In this case only the capacity curve corresponding to the ≡Ge–OH surface was observed. In addition, a cathodic current was found under stationary polarization which corresponds to the reduction of H_2O_2. Since this current occurred just in that potential range where the reduction of the ≡Ge–OH surface was found, Gerischer et al. postulated a surface radical to be involved as an intermediate in the reduction given by Eq. (5.44) [21]. We have then

$$\equiv \mathrm{Ge-OH} + \mathrm{H}^+ + \mathrm{e}^- \Leftrightarrow \equiv \mathrm{Ge}\bullet + \mathrm{H_2O} \tag{5.46a}$$

$$\equiv \mathrm{Ge}\bullet + \mathrm{H}^+ \Leftrightarrow \equiv \mathrm{Ge-H} + \mathrm{p}^+ \tag{5.46b}$$

In the presence of H_2O_2 the following reactions occur

$$\equiv \mathrm{Ge}\bullet + \mathrm{H}^+ + \mathrm{H_2O_2} + \mathrm{e}^- \rightarrow \equiv \mathrm{Ge{-}OH} + \mathrm{H_2O} \tag{5.47}$$

$$\equiv \mathrm{Ge{-}H} + \mathrm{H_2O_2} \rightarrow \equiv \mathrm{Ge{-}OH} + \mathrm{H_2O} \tag{5.48}$$

In the first reaction (Eq. 5.47), the reduction of H_2O_2 and the oxidation of the radical ≡Ge• involves an electron transfer whereas the other (Eq. 5.48) is a pure chemical process. It is clear from this set of reactions that the hydroxide surface is the only stable component in the presence of H_2O_2. In addition, it explains the occurence of a stationary cathodic reduction current. It should be mentioned here that there are different ways of testing whether a reaction involves electron or hole transfer, as will be discussed in detail in Chapter 7.

Concerning the charge distribution at the interface the formation of a surface radical (dangling bond) is of special interest. This surface species is equivalent to a surface state, and its energy position is expected to occur somewhere within the bandgap of the semiconductor and to be detectable either as an additional capacity (see Eq. 5.37) or by surface recombination measurements (see Section 4.3). Appropriate capacity measurements did not give any indication of an additional capacity which could be related to surface states. On the other hand, a pronounced surface recombination velocity (minimum in the short-circuit current) was detected during the reduction of the ≡Ge–OH surface as shown in Fig. 5.13 [22]. The corresponding experiment was performed as follows. After the electrode had been prepolarized anodically, the first

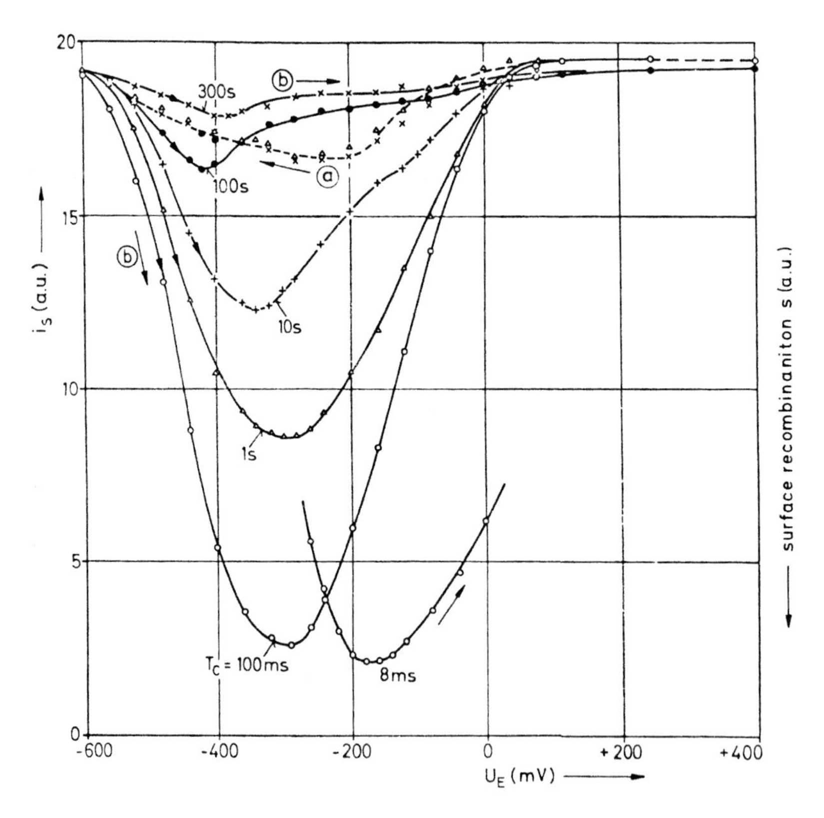

Fig. 5.13 Short circuit current j_s (surface recombination velocity $s \sim j_s$) versus electrode potential of an intrinsic Ge electrode at pH 2. Parameters of the different curves: cathodic prepolarization times at –0.5 V (saturated calomel electrode, SCE); scan rate: 0.5 V s^{-1}. (After ref. [22])

sweep towards cathodic potentials did not show any surface recombination. Stepping, however, quickly to a potential at which ≡Ge• and ≡Ge–H were formed, staying there for a certain time interval and sweeping back toward anodic potentials led to the occurrence of surface recombination (Fig. 5.13).The surface recombination decreased with increasing polarization times and almost disappeared when the formation of ≡Ge–H was completed. In the presence of H_2O_2 in the solution, surface recombination was also detectable under stationary conditions as expected according to the reaction scheme given above. The kinetics of all surface processes have been worked out in detail [22, 23]. A full description would be beyond the scope of this chapter.

It should be mentioned that surface states were also formed on a ≡Ge–OH surface by dipping the electrode into a solution containing a small concentration (10^{-7} M) of Au^{3+}, Ag^{2+} or Cu^{2+} ions, detected as an additional peak in the capacity curve or by surface recombination measurements [16, 24]. A density of surface states in the order of 10^{11} cm^{-2} was determined.

The latter results demonstrate very well the relation between surface chemistry, dangling bonds (radicals) and properties of surface states. With hindsight, it is rather surprising that these results should have been obtained at a somewhat early stage of semiconductor electrochemistry.

5.3.2 Silicon Electrodes

At the early stages of semiconductor electrochemistry, the potential distribution at the silicon–electrolyte interface was also studied using surface conductivity [25] as well as capacity measurements [26]. One example, as measured with an n-type Si electrode, is given in Fig. 5.14. These experiments were performed with slightly doped n- and p-doped Si electrodes, because the resistivity of intrinsic Si would be too high for an electrochemical measurement. In the case of the n-type electrode shown in Fig. 5.14, the theoretical curve could be fitted to the experimental values of the space charge capacity. Accordingly also here, the externally applied voltage occurred only across the space charge layer whereas the potential across the Helmholtz layer remained constant, as already reported for germanium electrodes. On the other hand, the capacity and the surface conductivity curves did not show a minimum, i.e. they did not increase again in the range where inversion is expected. Such an effect occurs if insufficient minority carriers (holes in n-type) are available for building up an inversion layer. The insufficiency of minority carriers at the electrode surface is caused by transfer of minority carriers to a corresponding acceptor, e.g. a redox system, in the electrolyte. Details of charge transfer processes are described in Chapter 7. As a consequence of this phenomenon, the space charge capacity remains determined by the donor and acceptor density in the semiconductor, even in the range of inversion. In other words, instead of an inversion layer we have only a depletion layer over the whole potential range. As derived in Section 5.2.3, in this range the relation between space capacity C_{sc} and potential across the space charge region $\Delta\phi_{sc}$ can be approximated by the Mott–Schottky equation as given by Eq. (5.27). A corresponding re-plot of the normal capacity curve is also given in Fig. 5.14 (right scale).This phenomenon has been observed with all doped semiconductor electrodes, as will be discussed in more detail in Section 5.3.3.

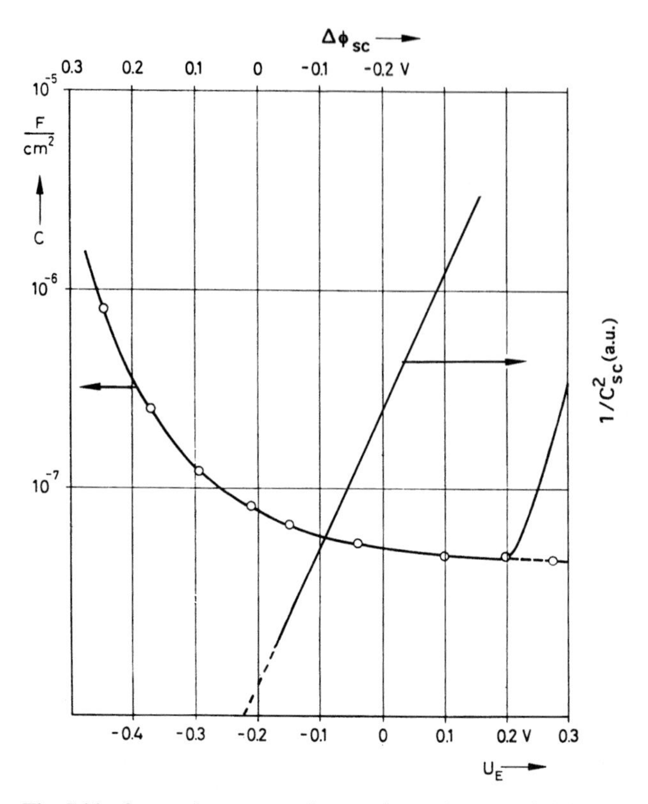

Fig. 5.14 Space charge capacity vs. electrode potential for an n type silicon electrode (1 Ω cm material) in 10 M HF; open circles: experimental values; solid line theoretical curve. a) Linear plot of C_{sc}; b) Mott–Schottky type of plot. (After ref. [26])

Concerning the potential distribution at the silicon–electrolyte interface, it should be further mentioned that here also an ≡Si–OH surface is assumed in aqueous solutions. Investigations of the potential distribution, however, are rather difficult because of the oxide formation. Very reliable measurements were possible in hydrofluoric acid in which any oxide formation can be avoided [26]. Also in non-aqueous electrolytes such as H_2O-free methanol, linear Mott–Schottky curves with the proper slope have been obtained [27].

5.3.3 Compound Semiconductor Electrodes

The potential distribution at the interface has been studied for many compound semiconductors. Usually a straight line has been obtained when plotting $1/C_{sc}^2$ vs. U_E according to Eq. (5.27), as shown, for example, in Fig. 5.15 for an n-type CdS electrode in an aqueous electrolyte [28]. In addition, the flatband potential U_{fb} ($\Delta\phi_{sc} = 0$), has been determined by extrapolating the straight line to $1/C_{sc}^2 \rightarrow 0$. Such a linear dependence found experimentally, does not necessarily mean that the externally applied voltage occurs entirely across the space charge layer. This can only be decided by com-

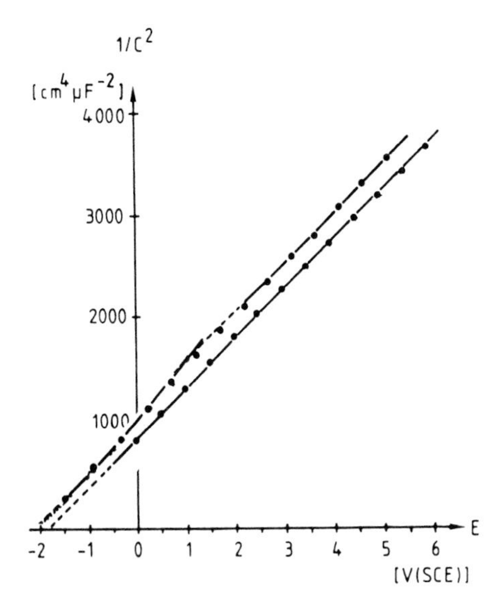

Fig. 5.15 Mott–Schottky plot of C_{sc} vs. U_E for a CdS electrode at pH5: upper curve, after etching in concentrated HCl for 30 s; lower curve, after etching in 20 % HCl for 3 s and cathodic prepolarization in an oxygen-saturated solution. (After ref. [28])

paring the slope of the linear dependence with the theoretical value which can be calculated if the donor or acceptor density in the semiconductor is known with sufficient accuracy (see Eqs. 5.27 and 5.28). This has been tested for several compound semiconductors and it has been shown that the externally applied voltage occurred across the space charge layer. There still remains some uncertainty because of the roughness factor introduced in investigations of germanium electrodes (Section 5.3.1).

In addition it should be mentioned that many published Mott–Schottky curves measured within a rather small potential range (around 0.5 V) look like straight lines. A closer inspection over a much larger range has frequently shown, however, that these are bent. Under these conditions the slope of the Mott–Schottky curves often exhibited a frequency dispersion. The deviation from linearity may have different origins. One reason is that frequently the equivalent circuit has not been carefully investigated, i.e. the measuring frequency was too low. Such a test is especially important when measuring in a potential range where the Faraday current increases (see also Section 4.4.1). Another reason could be that the electrode surface is not clean. This is illustrated for CdS (E_g = 2.5 eV) which tends to be covered by sulfur produced by etching or anodic reaction (Fig. 5.15). The CdS electrode could be cleaned by cathodic prepolarization in an O_2-saturated solution or by working in a polysulfide solution which dissolves the sulfur on the surface [28]. After such cleaning, Mott–Schottky curves were found which were linear over more than 6 V. In some cases, for example TiO_2, the deviation is due to inhomogeneous dopong after hydrogen treatment. There are various other factors which influence the Mott–Schottky curves which cannot be treated here in detail. Most other effects are due to surface reactions.

The flatband potential of many semiconductor electrodes depends on the pH, as already described for germanium electrodes. This indicates that many semiconductor surfaces exhibit a strong interaction with water. Some semiconductors, such as layered transition metal compounds and especially their basal planes, show a weaker inter-

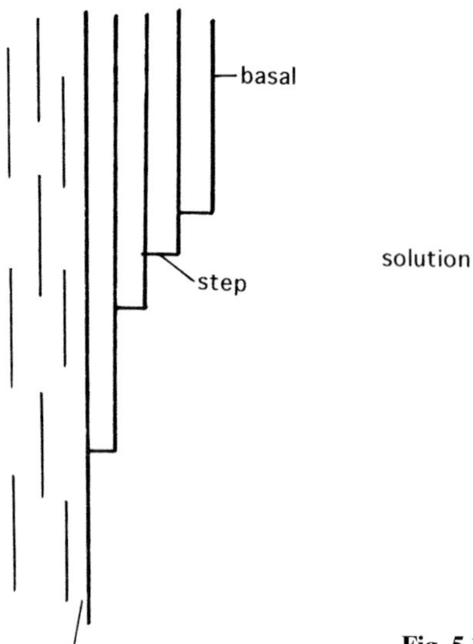

Fig. 5.16 Schematic presentation of a step at the surface of a layered compound such as WSe_2.

action. This became obvious in the case of WSe_2 electrodes with which no pH-dependence was found [29, 30]. This has been explained by the layer structure, i.e. the metal atom which could form a hydroxide, is shielded from the solution by Se atoms (Fig. 5.1d). Only if many steps from one layer to the other are present (Fig. 5.16), can the metal at each step contact the liquid. Accordingly, a pH-dependence of the flatband potential occurred after steps were created by scratching the surface [29]. This result was confirmed by photoelectron spectroscopic studies. Adsorption of H_2O on van der Waals surfaces produced by cleavage in ultrahigh vacuum, was only found with samples cooled to liquid nitrogen temperatures. When these samples were heated up to room temperature, H_2O was completely desorbed [31].

In the case of GaAs a change of the potential across the Helmholtz layer was observed upon anodic and cathodic prepolarization, which was interpreted in terms of hydroxyl and hydride surface layers, as for Ge (see Section 5.3.1) A linear Mott-Schottky dependence for an n-GaAs electrode was only found at sufficiently high scan rates after anodic or cathodic prepolarization as shown in Fig. 5.17 [40]. It is worth mentioning that all reliable capacity measurements could be interpreted in terms of space charge capacities, i.e. additional capacities due to surface states were not found.

5.3.4 Flatband Potential and Position of Energy Bands at the Interface

Concerning the potential distribution, comparisons of the Mott–Schottky curves and the flatband potentials as obtained with n- and p-type electrodes of the same semiconductor are of special interest. One example is GaP which has a relatively large bandgap (2.3 eV). The Mott–Schottky plots of the n- and p-type electrodes are given in Fig. 5.18 [32]. Since their slopes agree very well with those predicted upon the doping of the

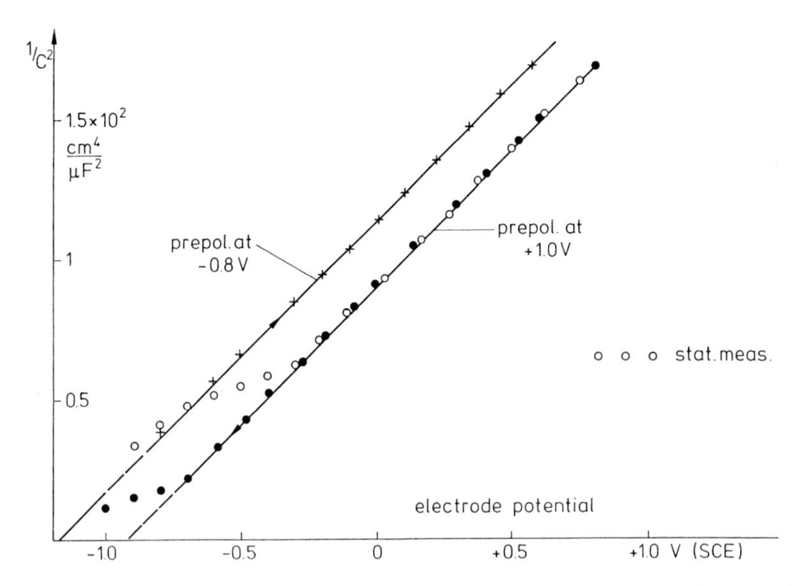

Fig. 5.17 Mott–Schottky plot of the space charge capacity vs. electrode potential for n-GaAs in aqueous solutions under stationary conditions and after different prepolarizations; scan rate: $0.2\ V\ s^{-1}$. (After ref. [40])

material, any variation of the electrode potential must occur across the space charge layer ($\Delta U_E = \Delta(\Delta\phi_{sc})$). The extrapolations of the two linear dependencies yield a flatband potentials of –0.9 V for the n-type electrode and +1.25 V for the p-type electrode. At first sight it may be surprising that the flatband potentials are extremely different and one may be tempted to conclude that $\Delta\phi_H$ is different for the n- and p-type elec-

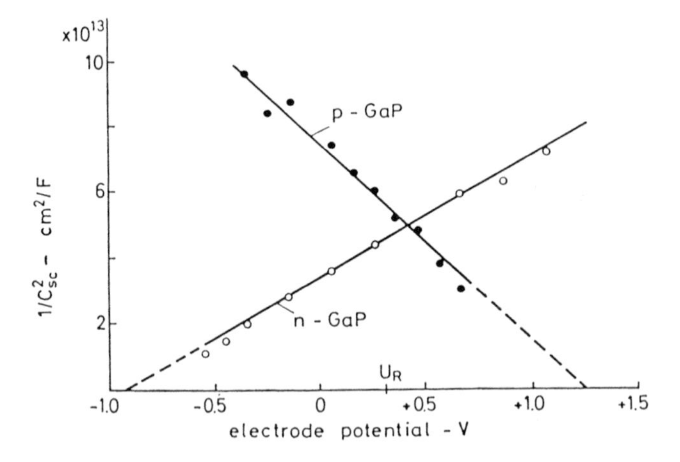

Fig. 5.18 Mott–Schottky plots of the space charge capacity vs. electrode potential for n- and p-type GaP electrodes in 0.1 M H_2SO_4. (After ref. [32])

trodes. On the contrary, this result is just a proof that $\Delta\phi_H$ is identical for n- and p-type materials, as can be derived from a complete potential and energy diagram, as given in Fig. 5.19. In the upper part of this figure, the energy schemes of an n-and p-type electrode in contact with an aqueous solution of a redox system, at equilibrium are shown. Under these conditions the Fermi level is constant throughout the whole system, i.e. $E_F = E_{F,redox}$ (compare also with Section 3.2.4). The electrode potential (eU_E in an energy diagram) measured at the rear ohmic contact with respect to a reference electrode, is here the energy difference between the Fermi level of the electrode and the corresponding energy level of the reference electrode. It is clear from Fig. 5.19 that the electrode potential is identical for the n- and p-type electrodes at equilibrium. According to the scale introduced in Section 3.2.4, the Fermi level of the semiconductor electrode is moved upward when the potential is made more cathodic, and downward when the potential becomes positive. Assuming that the energy bands at the surface of both electrodes have the same energy position, then the energy bands of the n-electrode are bent upward and of the p-type downward, at equilibrium. According to the experimental results presented in Fig. 5.18, any variation of the electrode potential occurs across the space charge layer which leads to an equivalent change of the band bending ($\Delta(eU_E) = \Delta(e\Delta\phi_{sc})$ in the energy diagram). As is quite obvious from Fig. 5.19, the energy bands at the surface remain pinned during such a potential change.

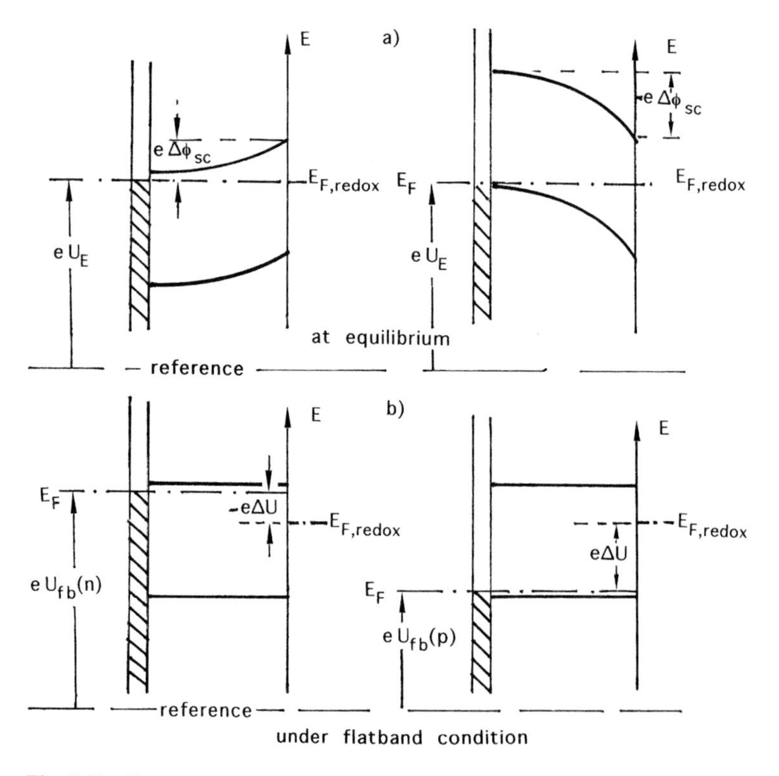

Fig. 5.19 Energy scheme of the n- (left) and p-type (right) semiconductor–liquid interface: a) at equilibrium; b) under flatband conditions

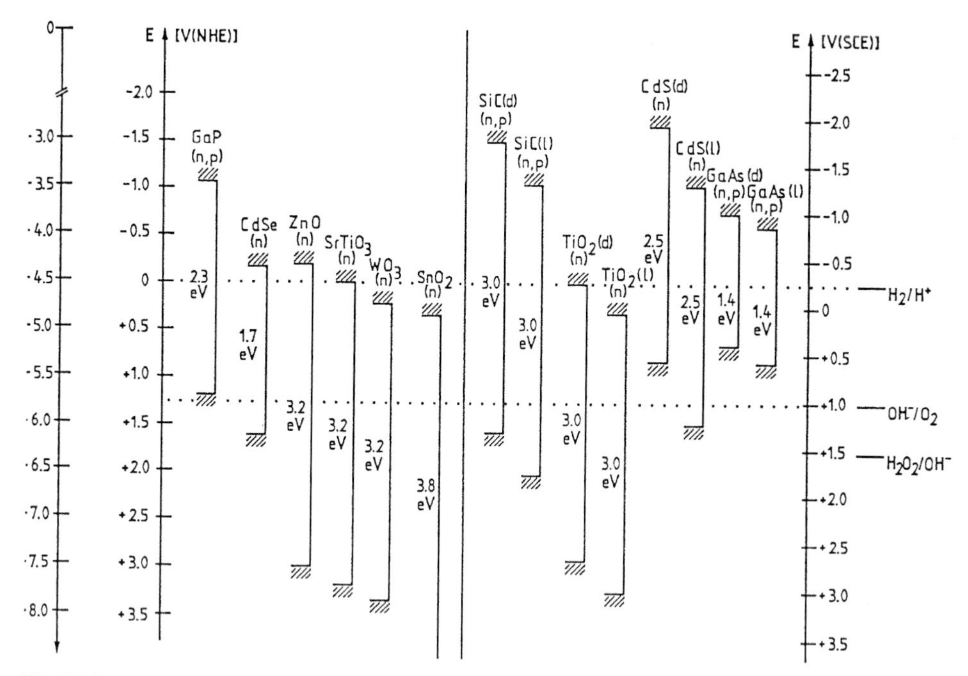

Fig. 5.20 Position of energy bands at the surface of various semiconductors in aqueous solutions at pH 0: (d), dark; (I), illuminated. (After ref. [45])

Accordingly, it can be concluded that pinning of bands occurs if the potential across the Helmholtz layer remains constant ($\Delta(\Delta\phi_H) = 0$).

It is further clear from Fig. 5.19 that the n-electrode has to be polarized cathodically with respect to the equilibrium potential, and the p-electrode anodically, in order to reach the corresponding flatband situation (see lower part of Fig. 5.19), provided that the positions of the energy bands at the surface are the same for the two types of electrodes. Keeping in mind that the electrode potential refers to the Fermi level of the electrode, then the difference of flatband potentials corresponds exactly to the difference of the two Fermi levels. Since the Fermi level in the bulk of a semiconductor with the usual doping ($\geq 10^{17}$ cm^{-3}) is rather close to the corresponding band, the difference in the flatband potentials approximates the bandgap of the semiconductor as found with GaP.

This result has been confirmed by investigations of various other semiconductors which are available as n- as well as p-doped material, such as GaAs, InP and SiC. As previously mentioned, the flatband potential of many systems depends on the pH of the electrolyte. An increase of pH always leads to a cathodic shift of U_{fb} and a corresponding change of $\Delta\phi_H$. Accordingly, the energy bands of the semiconductor at the surface are always shifted upward by an increase of pH.

Another important result is that the flatband potentials and therefore the position of the energy bands at the semiconductor surface contacting an aqueous electrolyte, are usually independent of any redox system added to the solution. Hence, the interaction between semiconductor and H_2O determines the Helmholtz layer and the position of the energy bands. Therefore, it is reasonable to characterize semiconductor electrodes by their positioning of energy bands at the surface for a given pH. A selection is shown in Fig. 5.20.

It has been emphasized by Bard et al. that there may be exceptions to the model derived above, insofar as Fermi level pinning by surface states may occur in a similar fashion to that at semiconductor–metal junctions [33]. Such an effect would lead to an unpinning of bands at the interface. There are some examples in the literature, such as FeS_2 in aqueous solutions [34, 35] and Si in methanol [36] for which an unpinning of bands has been reported. In some cases, such as TiO_2, experimental values of flatband potentials scatter considerably. This is mostly due to changes in surface chemistry and doping profiles.

The situation is quite different for non-aqueous electrolytes. Recently it has been shown by capacity measurements that the flatband potential, and consequently the position of energy bands, of GaAs electrodes in acetonitrile or methanol solutions depends strongly on the redox couple added to the electrolyte [37]. Actually the flatbands and therefore the positions of the energy bands vary by more than 1 eV, if the standard potential of various metallocenes as redox couple were changed by the same value as shown in Fig. 5.21. Obviously here the interaction of the semiconductor with the redox system is much stronger than with the solvent. This result was interpreted as Fermi level pinning by surface states, the latter being located about 0.3 to 0.4 eV above the valence band [37]. More recent studies have led to another interpretation. According to these investigations, metallocenes are adsorbed on the GaAs surface and the interaction between GaAs and metallocenes varies from one system to another [38]. For instance, the flatband potential is shifted toward cathodic potentials upon addition of the reduced species of cobaltocene ($Co(Cp)_2^0$) to the acetonitrile electrolyte

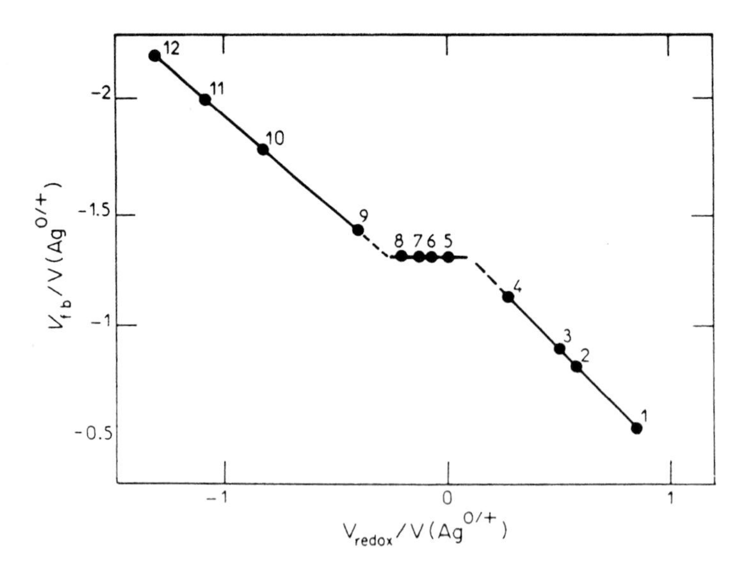

Fig. 5.21 Flatband potential of n-GaAs in acetonitrile versus standard potentials of various metallocenes redox systems. (After ref. [37])

whereas the oxidized species ($Co(Cp)_2^+$) has only a little effect. In the case of ferrocene, the opposite effect was observed, i.e. here the oxidized species ($Fe(Cp)_2^+$) causes a large anodic shift of U_{fb} [39]. These differing properties of the two redox systems can be related to differences in the electron configuration within the two molecules.

5.3.5 Unpinning of Energy Bands during Illumination

In 1980, in other words, only at a rather advanced state of semiconductor electrochemistry, it was reported for the first time [43, 44] that Mott–Schottky curves and therefore the flatband potentials are shifted upon illumination of the semiconductor electrode, toward cathodic potentials with p-type electrodes and toward anodic with n-type electrodes, as illustrated for a p-GaAs electrode in an aqueous electrolyte in Fig. 5.22. Meanwhile, this effect has been found with almost all semiconductors studied so far. Mostly, shifts of few hundred millivolts have been found [45]. We have here an unpinning of bands upon illumination, and the shift of U_{fb} must be interpreted by an equivalent movement of the energy bands, as illustrated for a p-type semiconductor in Fig. 5.23. The origins of this effect can be manifold. In most cases it is explained by trapping of minority carriers in surface states, a process which competes with the transfer of minority carriers in the electrolyte. In the case of a p-GaAs electrode immersed into an acid electrolyte, electrons excited into the conduction band are either transferred to protons or captured in surface states, as shown in the center part of Fig. 5.23. Since the hole density is small at the surface for potentials negative of U_{fb}, the recombination rate of the trapped electrons with holes is very low. Provided that a sufficient number of surface states is available, a considerable charge can be stored, leading finally to a change of the potential distribution and an upward shift of the bands. The stored charge ΔQ_{ss} can be calculated from the corresponding change of $\Delta\phi_H$ by using the relation (see also Eq. 5.6)

$$\Delta Q_{ss} = C_H \left[U_{fb}(\text{light}) - U_{fb}(\text{dark})\right] = C_H \left[\Delta(\Delta\phi_H)\right]_{h\nu} \tag{5.49}$$

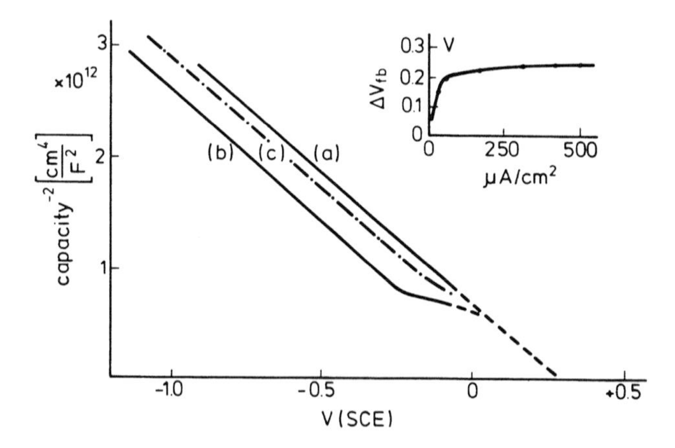

Fig. 5.22 Mott–Schottky plot of the space charge capacity vs. electrode potential for p-GaAs at pH 1, in the dark and under illumination. Insert: U_{fb} vs. limiting photocurrent. (After ref. [43])

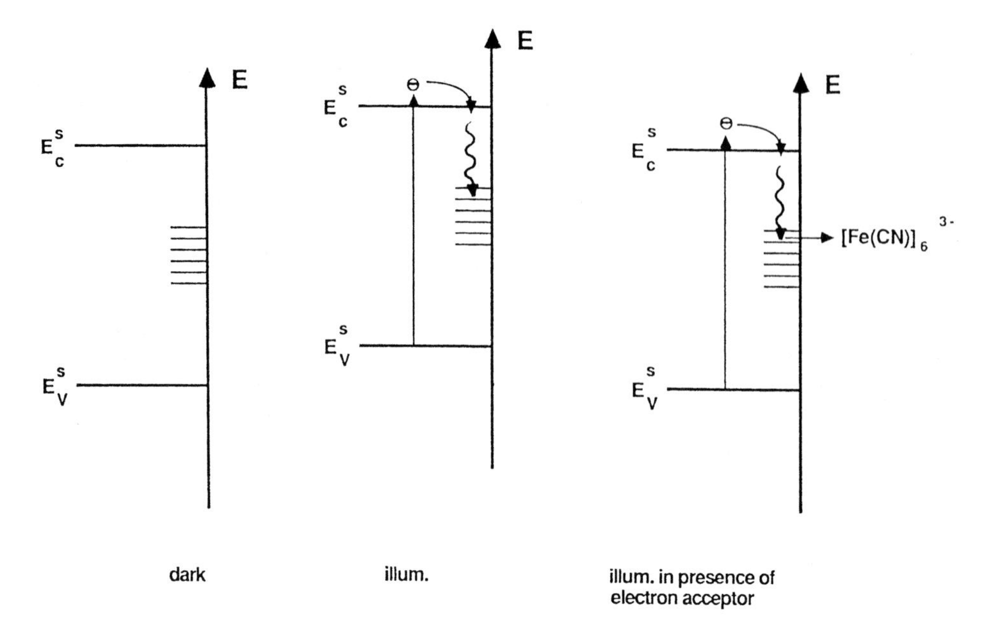

Fig. 5.23 Position of energy bands at the surface of p GaAs in the dark and under illumination. (After ref. [44])

If all surface states are occupied by electrons their density is given by $N_t = \Delta Q_s/e$ (Eq. 5.35). Assuming a Helmholtz capacity of $C_H \approx 10^{-5}$ F cm^{-2} one obtains N_t in the order of about 10^{13} cm^{-2} for a shift of $\Delta U_{fb} = 0.2$ V. It is interesting to note that the shift of the flatband potential occurs mostly at very low light intensities, and it saturates at higher light intensities because then all surface states are filled [45]. In Fig. 5.20 the shift of energy bands is indicated for some semiconductors.

The accumulation of minority carriers and the resulting shift of energy bands can be avoided if a suitable redox system is added to the electrolyte. For instance, according to capacity measurements performed with p-GaAs electrodes, the energy bands are shifted back to their original values upon addition of $[Fe(CN)_6]^{3-}$ as an electron acceptor. It is believed that the electrons captured by surface states, are transferred from these states to the electron acceptor in the solution, as illustrated on the right side of Fig. 5.23. The same type of shift of U_{fb} occurs in the dark if the minority carriers are injected from a redox system in the solution, which is consistent with the model. This has been found with n-GaAs and n-WSe_2 using Ce^{4+} as the hole injection agent [40, 46].

As mentioned previously, the phenomenon of unpinning of bands upon light excitation has been found with many semiconductor electrodes. It is not primarily necessary, however, to make the trapping of minority carriers in surface states responsible for the band edge movement. In the first step, minority carriers produced by light excitation are drawn towards the surface because of the electric field across the space charge region. In the case of a p-type electrode the energy bands are bent downwards so that the electrons are moved toward the surface, as already illustrated in Fig. 5.23. Considering a p-type electrode in contact with a rather inert non-aqueous electrolyte (containing no electron acceptor), then electrons excited into the conduction band cannot be transferred across the interface and must accumulate at the surface. This leads to tremendous changes in the potential distribution.

There are some semiconductors (typically n-WSe_2, n-$MoSe_2$, n-RuS_2) with which very large anodic shifts of the flatband potential were found when they were immersed into an aqueous electrolyte. As already mentioned, WSe_2 and $MoSe_2$ are layered compounds, the basal planes of which show very low interaction with H_2O because the metal is shielded from the solution (see Section 5.3.4). This has consequences for the anodic decomposition reaction (see Section 8.1.4) because it occurs primarily at steps. If their density is small then the kinetics of the dissolution should be slow. In the case of n-WSe_2 electrodes, an anodic shift of the Mott–Schottky plot by 0.6 V has been observed upon illumination. Since the anodic dissolution process requires holes, the shift has been interpreted as hole accumulation because of slow kinetics of the hole transfer [46]. Here again any shift of U_{fb} could be avoided by adding a proper hole acceptor, such as $[Fe(phen)_3]^{2+}$, to the electrolyte. Details of the charge transfer process are given in Section 8.1.4.

RuS_2 electrodes present rather extreme examples, with which a shift of the Mott-Schottky plots of up to 2 V have been observed by Kühne and Tributsch [47]. Here the shift of the bands is even considerably larger than the bandgap ($E_g = 1.25$ eV). It cannot be interpreted on the basis of a simple surface state model, because an estimation of the corresponding density of states by using Eq. (5.49) would lead to $N_t > 10^{14}$ cm^{-2}; this is an unreasonable number because it would correspond to more than 10 % of the surface atoms. More information can be obtained by comparing the photocurrent–potential dependence and the Mott–Schottky curves, as measured by the same authors, in the presence of a redox system in the solution under illumination [47] (Fig. 5.24). According to this figure, the Mott–Schottky plots under illumination occur at different potentials, depending on the redox system added to the supporting electrolyte, whereas in the dark the Mott–Schottky curve occurs at much more negative potentials, independently from the redox system. All oxidation processes occur via the valence band, i.e. a hole is transferred to a corresponding hole acceptor in the solution. Since RuS_2 is an n-type semiconductor, a corresponding current can only be produced by light excitation as indicated in the insert of Fig. 5.24b (see also Chapter 7). The highest shift upon illumination was found in H_2SO_4 without any redox system. Since RuS_2 is very stable against corrosion, the holes are here consumed for O_2 formation. According to thermodynamic requirements, a hole transfer is only feasible if the valence band occurs below the standard potential of the H_2O/O_2 couple. Since this requirement is not fulfilled in the dark, the holes produced by light excitation accumulate at the surface which leads to a downward shift of the bands until the valence band finally occurs below $E°(H_2O/O_2)$. Only then does a hole transfer becomes possible as measured by the anodic photocurrent (Fig. 5.24b). Interestingly, the photocurrent onset occurs at less anodic potentials in the presence of halide ions, such as I^-, Br^-, or Cl^-, in the electrolyte, because the standard potentials are less positive. Simultaneously, the shift of the Mott–Schottky curves upon illumination is also smaller (Fig. 5.24a). In all cases, the flatband potential occurs close to the onset of the photocurrent. In the case of RuS_2 electrodes even this model is still not sufficiently defined for the following reason.

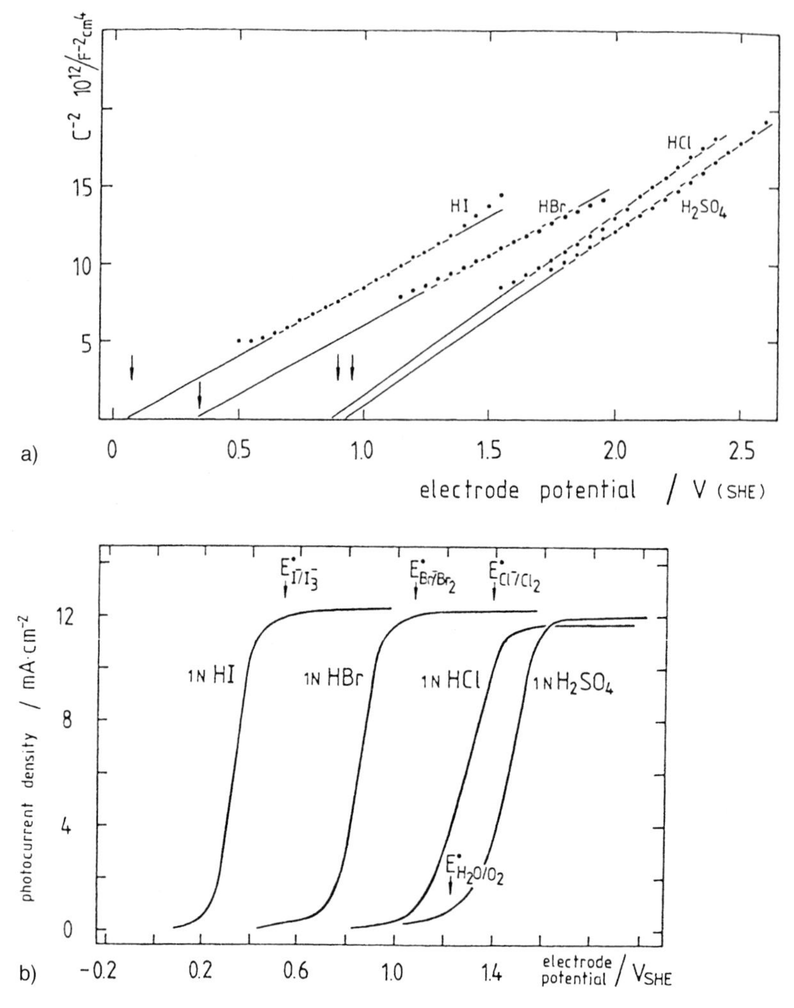

Fig. 5.24 Mott–Schottky plot (a) and photocurrent (b) vs. electrode potential for n-RuS_2 in aqueous solutions. (After ref. [47])

According to a surface spectroscopic analysis (XPS), one or two monolayers of RuO_2 are formed during anodic polarization [48]. This result indicates that the formation of the oxide prevents RuS_2 from any anodic dissolution. Since RuO_2 is a metal-like conductor, a kind of semiconductor–metal Schottky junction exists on the surface as illustrated in Fig. 5.25. Assuming a low overvoltage for the oxidation of halides at the RuO_2 layer, the Fermi level in RuO_2 should be pinned close to the redox potential of the corresponding halide. A comparison of the redox potential with the flatband potentials during illumination yields finally the energy diagram as given in Fig. 5.25. This diagram leads to a more or less constant barrier at the RuS_2–RuO_2 interface, as is typical for semiconductor–metal junctions. According to this model, the origin of the

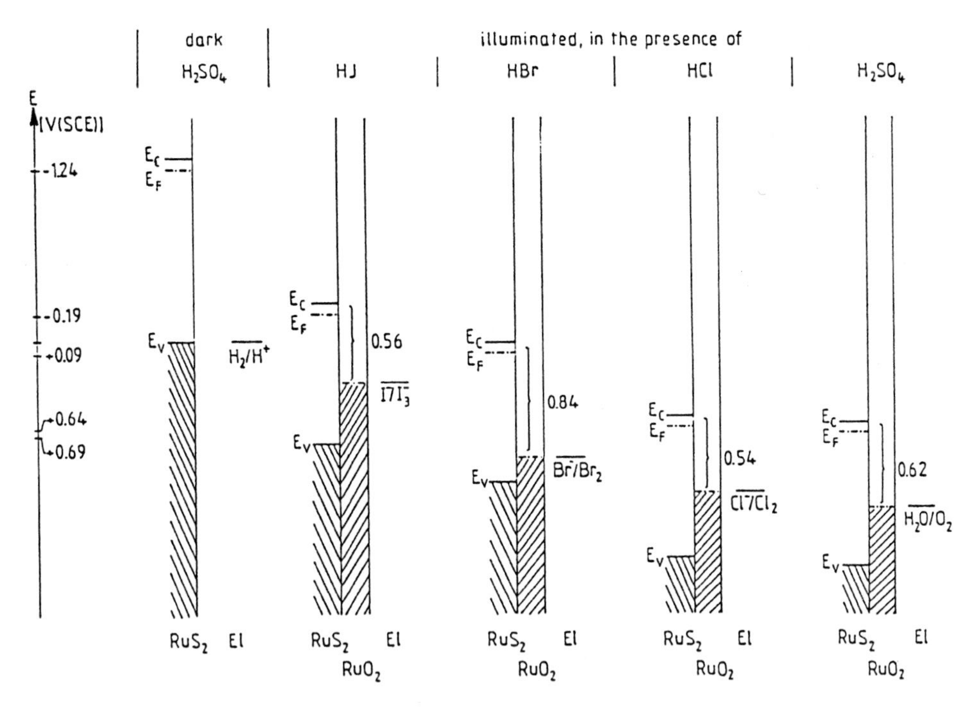

Fig. 5.25 Positions of energy bands at the surface of n-RuS_2. (After ref. [50])

flatband shift upon illumination is a complete pinning of the quasi-Fermi level of holes to that in the RuO_2 layer.

The same type of Fermi level pinning also occurs also when the surface of InP is reduced by cathodic polarization leading to a thin In layer, or when another metal, such as Pt, is deposited [49].

6 Electron Transfer Theories

In the early 1960s various scientists, including Marcus [1, 2], Hush [3] and Gerischer [6, 7], started to develop modern theories of electron transfer between molecules in homogeneous solutions and at electrodes. The Russian school, Levich [5], Dogonadze [4], Kuznetzov [4], have added quantum mechanical aspects to the originally more classical approaches. In all these theories the interaction between the reactants and the products on the one hand and the polar solvent on the other hand is considered. Since Marcus' theory is applicable for reactions in homogeneous solutions as well as at electrodes it is the most widely used theory. Gerischer's model is generally applied in electrochemistry. Therefore we concentrate at first on these theories. Recently some new treatments have been published which will be discussed later.

6.1 The Theory of Marcus

6.1.1 Electron Transfer in Homogeneous Solutions

Marcus has developed his primarily classical theory for electron transfer reactions in homogeneous solutions. In the simplest type of reaction a single electron is transferred without breaking or forming bonds. A bimolecular exchange reaction is given by

$$D + A \rightarrow D^{+} + A^{-} \tag{6.1}$$

Here an electron is transferred from a donor molecule D to an acceptor molecule A. In addition it is useful to define self-exchange reactions of the components as given by

$$A + A^{-} \rightarrow A^{-} + A \tag{6.2a}$$

$$D + D^{+} \rightarrow D^{+} + D \tag{6.2b}$$

In the latter case (Eq. 6.2) the reaction is isoergonic, i.e. the free energy of the reaction is zero ($\Delta G = 0$). The elementary electron transfer step occurs when the reactants are sufficiently close to each other. This is still a rather vague description for which more specific details will be given later on. Since an electron can tunnel over such a short distance, the reaction would be expected at first sight to be very fast. However, electron transfer reactions in condensed media involve considerable reorganization or rearrangement in the surrounding medium. Using for instance water as a solvent, the ions involved in the reaction are highly solvated, the water dipoles in their vicinity being oriented and polarized by the field of the ion. In the course of electron transfer the interaction between the reacting molecules and the solvent changes. Thus although a self-exchange reaction is isoergonic ($\Delta G^{0} = 0$), the reaction is an activated and not a resonance process. Accordingly, the reaction pathway can now be described by (see Eq. 1):

$$\begin{array}{ccccccccccc} D+A & \Leftrightarrow & (D,A) & \Leftrightarrow & (D,A)^{\#} & \Leftrightarrow & (D^{+},A^{-}) & \Leftrightarrow & (D^{+},A^{-}) & \Leftrightarrow & D^{+}+A^{-} \\ & & \Uparrow & & \Uparrow & & \Uparrow & & \Uparrow & & \\ & & R_0 & & R^{\#} & & P^{\#} & & P_0 & & \end{array} \tag{6.3}$$

in which R_0 and P_0 represent an encounter complex with D, A, D^+ and A^- in their equilibrium configurations, respectively, whereas $R^{\#}$ and $P^{\#}$ are the activated complexes.

The activated complex $R^{\#}$ is a non-equilibrium state, a small portion of which is formed by the fluctuation of the solvent environment. Fluctuations are essentially changes in the redox system configuration due to thermal motion. In the activated state the isoergonic electron transfer occurs to form $P^{\#}$, the activated product complex which has the same nuclear configuration. $P^{\#}$ relaxes then to the equilibrium configuration P_0.

It should be emphasized here that the electron transfer in the activated state is a very fast process which occurs within a time interval of about 10^{-15} s. The relaxation times for the solvent and the reacting nuclei are much longer, typically 10^{-11} to 10^{-13} s for vibrational motion and 10^{-9} to 10^{-11} s for rotational motion. Accordingly, it is a reasonable approximation that the positions of the nuclei are unchanged in the course of the electron transfer. This condition is called the Franck–Condon principle. It is well known from studies of absorption and emission of light by molecules.

According to this transition state concept the corresponding electron transfer rate constant k_{ET} is given by

$$k_{et} = \kappa\, \nu\, \exp\left(\frac{\Delta G^{\#}}{kT}\right) \tag{6.4}$$

where $\Delta G^{\#}$ is the Gibbs energy of activation and ν the frequency of nuclear motion through the transition state ($\nu \approx 10^{12}$–10^{13} s^{-1}), whereas κ is a transmission coefficient ranging between 0 and 1. It is customary to illustrate the energetics of electron transfer reactions by so-called energy surface diagrams in which the Gibbs energy, free energy or potential energy of the system is illustrated as a function of nuclear coordinates q_i. In Marcus' theory the Gibbs energy profile of the initial system (reactants, R) and the final system (product, P) are represented by two intersecting parabolas of the same curvatures as illustrated in Fig. 6.1. Since the energy curves are given here for equilibrium ($\Delta G = 0$) their minima occur at the same energy. The activation energy $\Delta G^{\#}$ occurs as the difference between the crossing point of the two curves and their minima. It should be emphasized that a parabolic energy surface means that the fluctuation of the nuclei follows that of a harmonic oscillator. Fig. 6.1 is certainly very simplified; since the energy surfaces for the position of all nuclei have to be considered, one obtains a high dimensional set of curves.

The profile itself is strongly determined by the reorganization energy λ which is by definition the energy of the product with respect to its equilibrium state when its solvent coordinate is still the same as that of the reactant state. If the curvatures of the reactant and the product parabolas are identical, the reorganization energy can also be defined as the work required to distort the reactant (D,A) from its equilibrium coordinate q_R^0 (Fig. 6.1) to the equilibrium coordinate q_P^0 of the product without any electron transfer (Fig. 6.1).

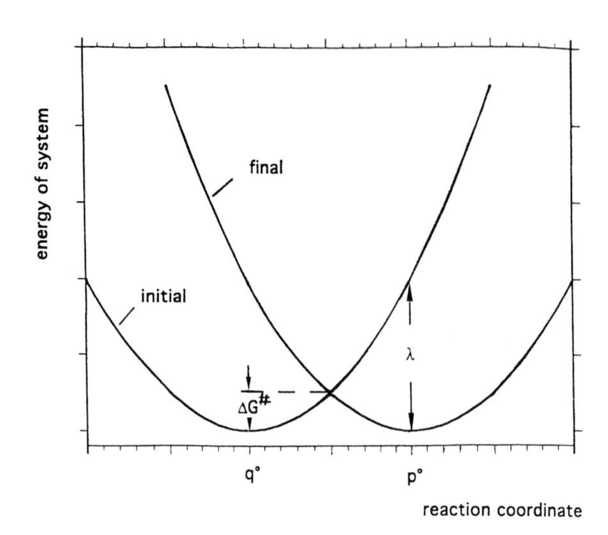

Fig. 6.1 Free energy profile along the reaction coordinate q for a non-adiabatic process; $\Delta G^0 = 0$

The activation energy $\Delta G^{\#}$ changes, of course, when the Gibbs energy of a reaction is not zero, as illustrated for $\Delta G > 0$ in Fig. 6.2 and for a series of cases of $\Delta G < 0$ in Figs. 6.3a–c. $\Delta G^{\#}$ can be related to the reorientation energy λ as follows.

Assuming that the fluctuation of the environment of the initial and final system oscillates like a harmonic oscillator, then the enthalpies are given by (compare with Fig. 6.3)

$$G_R - G_R^0 = \gamma \left(q - q_R^0\right)^2 \tag{6.5}$$

for the initial system and

$$G_R - G_R^0 = \gamma \left(q - q_R^0\right)^2 + G^0 \tag{6.6}$$

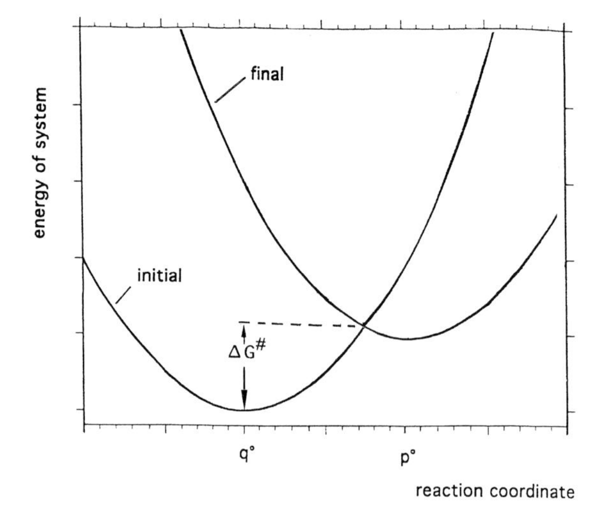

Fig. 6.2 Free energy profile along the reaction coordinate q for a non-adiabatic process; $\Delta G^0 \neq 0$

for the final system. In both equations γ is a constant which does not need to be specified further. At the intersection of both curves ($q = q^{\#}$) the energies of the initial and final system are equal, i.e.

$$G_{\mathrm{R}} = G_{\mathrm{P}}; \quad q = q^{\#} \tag{6.7}$$

The reorganization energy is given by

$$\lambda = G_{\mathrm{R}}^{p} - G_{\mathrm{R}}^{0} = \gamma\left(q_{\mathrm{p}}^{0} - q_{\mathrm{p}}^{0}\right)^{2} = G_{\mathrm{p}}^{\mathrm{R}} - G_{\mathrm{p}}^{0} \tag{6.8}$$

Inserting Eqs. (6.5) and (6.6) into (6.7), $q^{\#}$ can be calculated. The activation energy $\Delta G^{\#}$ can be expressed in terms of Eq. (6.5):

$$\Delta G^{\#} = \gamma\left(q^{*} - q_{\mathrm{R}}^{0}\right)^{2} \tag{6.9}$$

Inserting the value for $q^{\#}$ and using Eq. (6.8) one obtains

$$\Delta G^{\#} = \frac{\left(\Delta G^{0} + \lambda\right)^{2}}{4\,\lambda} \tag{6.10}$$

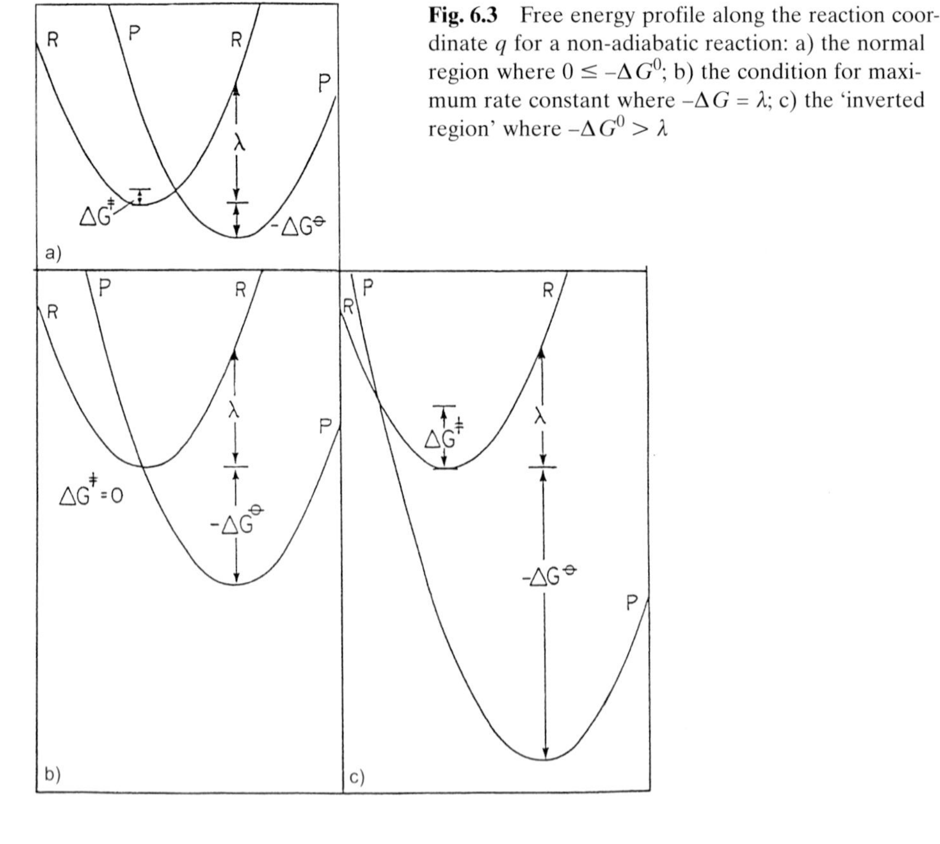

Fig. 6.3 Free energy profile along the reaction coordinate q for a non-adiabatic reaction: a) the normal region where $0 \leq -\Delta G^{0}$; b) the condition for maximum rate constant where $-\Delta G = \lambda$; c) the 'inverted region' where $-\Delta G^{0} > \lambda$

Accordingly, in the case of a harmonic fluctuation the activation energy can be simply expressed in terms of ΔG^0 and λ. Combining Eqs. (6.4) and (6.10) one obtains as a rate constant

$$k_{et} = \kappa\, \nu\, \exp\left[-\frac{(\Delta G^0 + \lambda)^2}{4kT\lambda}\right] \tag{6.11}$$

This model predicts for endergonic reactions that the activation energy $\Delta G^{\#}$ increases whereas the rate constant k_{et} decreases (compare also with Fig. 6.2). On the other hand, for weak exergonic reactions, i.e. negative ΔG values, $\Delta G^{\#}$ becomes smaller (Fig. 6.3a) and k_{et} increases. At $-\Delta G^0 = \lambda$ the reaction becomes activationless ($\Delta G^{\#} = 0$, see Fig. 6.3b) and the rate constant reaches its maximum value of $k_{et} = \kappa\nu$. *Any further increase of* $-\Delta G^0$ leads, surprisingly, again to a new increase of $\Delta G^{\#}$ and a corresponding decrease of k_{et}. As illustrated in Fig. 6.3c the crossing point of the two energy surfaces occurs now on the left side of the reacting system. The physical explanation for the increase of the activation energy is that the environment of the reactants must be distorted before an electron transfer can occur. This range is the so-called 'Marcus inverted region (see also Section 7.3.6).

6.1.2 The Reorganization Energy

The reorganization energy, λ, is usually treated as the sum of two contributions, an inner (λ_{in}) and an outer (λ_{out}) sphere reorganization energy, i.e.

$$\lambda_{in} + \lambda_{out} = \lambda \tag{6.12}$$

Marcus has considered the vibration of the ion–solvent bond within the first solvation shell and treated it as an harmonic oscillator. He obtained

$$\lambda_{in} = \sum_j \frac{f_j^r f_j^p}{f_j^r + f_j^p} (\Delta q_i)^2 \tag{6.13}$$

in which f_j^r and f_j^p are the jth ion–solvent bond, and

$$\Delta q_i = q_j^r - q_j^p \tag{6.14}$$

where q_j^r and q_j^p are the equilibrium values of the bond coordinates of the ion in its reactant and product state, respectively.

The outer sphere reorganization energy λ has been derived by Marcus [2] as well as by Levich et al. [5] by using a continuum model for describing the solvent. In a polar medium the solvent molecules (dipoles) are oriented around an ion (Fig. 6.4) and this structure will be changed when the charge of the ion changes upon electron transfer. This leads locally to a corresponding change of polarization of the liquid. The total polarization is related to the static dielectric constant ε_s by:

$$P_{tot} = \frac{(\varepsilon_s - 1)}{4\pi\,\varepsilon_s} D \tag{6.15}$$

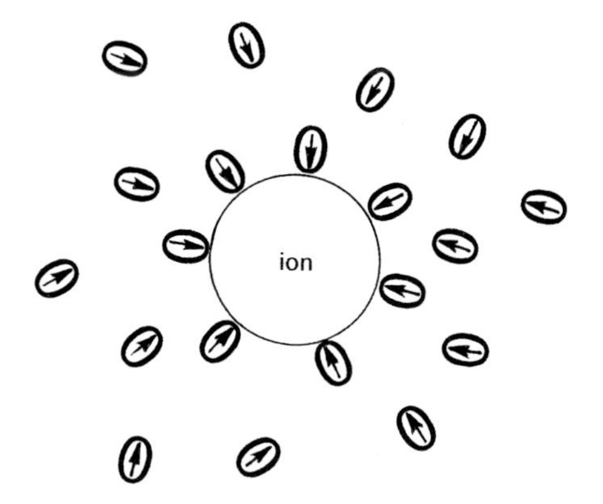

Fig. 6.4 Redox ion surrounded by inner-sphere H_2O dipoles and outer-sphere dipoles

in which D is the electric induction. Such a polarization consists of various contributions such as orientation polarization arising from the permanent dipole moment and from the distortion of the solvent nuclei. Changes of these polarizations occur at frequencies of the order of 10^{11} Hz. Much faster is the distortion of electrons within the solvent molecule. In the latter case the static dielectric constant has to be replaced by the optical dielectric constant $\varepsilon_{\mathrm{opt}} = n^2$ (n = refractive index) and the fast polarization is given by

$$P_{\mathrm{fast}} = \frac{(n^2 - 1)}{4\pi\, n^2} D \tag{6.16a}$$

The slow component due to orientation polarization is then given by

$$P_{\mathrm{slow}} = P_{\mathrm{tot}} - P_{\mathrm{fast}} = \left(\frac{1}{n^2} - \frac{1}{\varepsilon_{\mathrm{s}}}\right)\frac{D}{4\pi} \tag{6.16b}$$

The energy involved upon any change of polarization is given by

$$\mathrm{E} = \int\left(\int D\,\mathrm{d}P\right)\mathrm{d}V = \int\left(\int P\,\mathrm{d}D\right)\mathrm{d}V \tag{6.17}$$

With respect to the electron transfer, we are interested in the rather slow orientation change of the solvent molecules. The corresponding energy involved in such a process can be obtained by performing a thought experiment as follows. In the first step the electric field is switched on very slowly so that all processes can follow. In the second step the field is switched off very rapidly so that only the distortion of electrons can follow. The outer sphere reorientation energy is thus given by

$$\lambda_{\mathrm{out}} = E_{\mathrm{slow}} = \int\left[\int_0^{D_0} \frac{\varepsilon_{\mathrm{s}} - 1}{4\pi\,\varepsilon_{\mathrm{s}}} D\,\mathrm{d}D - \int_{D_0}^{D} \frac{n^2 - 1}{4\pi\, n^2} D\,\mathrm{d}D\right]\mathrm{d}V \tag{6.18}$$

Assuming that the molecules are spherical in shape, the integration of this equation leads to

$$\lambda_{\mathrm{out}} = \frac{(\Delta e)^2}{4\pi\,\varepsilon_0}\left[\frac{1}{2\,r_{\mathrm{D}}} + \frac{1}{2\,r_{\mathrm{A}}} - \frac{1}{r_{\mathrm{DA}}}\right]\left(\frac{1}{n^2} - \frac{1}{\varepsilon_{\mathrm{s}}}\right) \tag{6.19}$$

in which r_D, r_A are the radii of the donor and acceptor molecule, respectively, r_{DA} the distance between the reacting molecules (center to center) and ε_0 the permittivity in free space. This equation is similar to that of the solvation energy as derived by Born (see Chapter 3).

It is assumed here that only one charge is transferred in such a reaction because the simultaneous transfer of two electrons is very unlikely. Considerable values of λ_{out} can be reached, which can be estimated easily from Eq. (6.19). Considering for instance the electron transfer between two equal molecules or ions with a fixed inner solvation shell (e.g. $Fe^{2+/3+}(H_20)_6$) then we have $r_A = r_D$. Assuming further that the transfer occurs between molecules which are in direct contact ($r_{DA} = 2r_D$) and using H_2O as a liquid ($\varepsilon_s = 80$; $n = 1.3$) then one obtains for $r_{DA} = 7$ Å an outer sphere reorganization energy of $\lambda_{out} \approx 1.2$ eV. According to this estimate fairly large values are obtained in polar solvents. Experimental data are of the same order of magnitude (Section 7.5). It should be pointed out that the reorganization energy is of the same order of magnitude as the bandgap of semiconductors, a result which will be of importance later on. In non-polar solvents, for which $\varepsilon_{0pt} \approx \varepsilon_s$, $\lambda_{out} \rightarrow 0$, i.e. then only λ_{in} remains.

6.1.3 Adiabatic and Non-adiabatic Reactions

Until now it has been assumed that there is no additional interaction between the reacting species, i.e. the parabola of the reactant and the product in Figs. 6.1 to 6.3 are unperturbed at their intersection point. Since, however, the molecules are rather close to each other when electron transfer occurs, some electronic interaction is expected. Such an electronic coupling leads to a splitting of the energy surfaces at the crossing point, as illustrated in Fig. 6.5. The electronic coupling energy V_{RP} between reactant and product surfaces is defined by

$$V_{RP} = \left\langle \Psi_R^0 \mid H \mid \Psi_P^0 \right\rangle \tag{6.20}$$

in which $\psi_R{}^0$ and $\psi_P{}^0$ are the electronic wave functions of the equilibrium reactant and product states, whereas H is the total electronic Hamiltonian for the system. If V_{RP} is large then the energy surfaces are well separated in the crossing region, and the reaction pathway occurs along the lower energy surface, as indicated by the arrow in Fig. 6.5. When these conditions are met the electron transfer process is called an adiabatic reaction. In this case the transmission factor κ in the rate equation (Eq. 6.11) is approximately unity.

In the other case, when the interaction V_{RP} is very small the reaction is said to be non- adiabatic. The energy surfaces are perturbed only slightly as illustrated in Fig. 6.5b. The reactant system remains mostly on the reactant surface also when passing the crossing point and returns to its equilibrium state (see large arrow in Fig. 6.5b) without much electron transfer. Thus, $\kappa \ll 1$. According to these definitions, the Marcus model describes primarily adiabatic reactions.

The question arises: above which interaction energy must a reaction be considered to be adiabatic? This is difficult to answer, especially for electrode reactions, because it depends on the distance of the reacting species during the electron transfer. In the case of reactions in homogeneous solutions Newton and Sutin [8] have estimated for

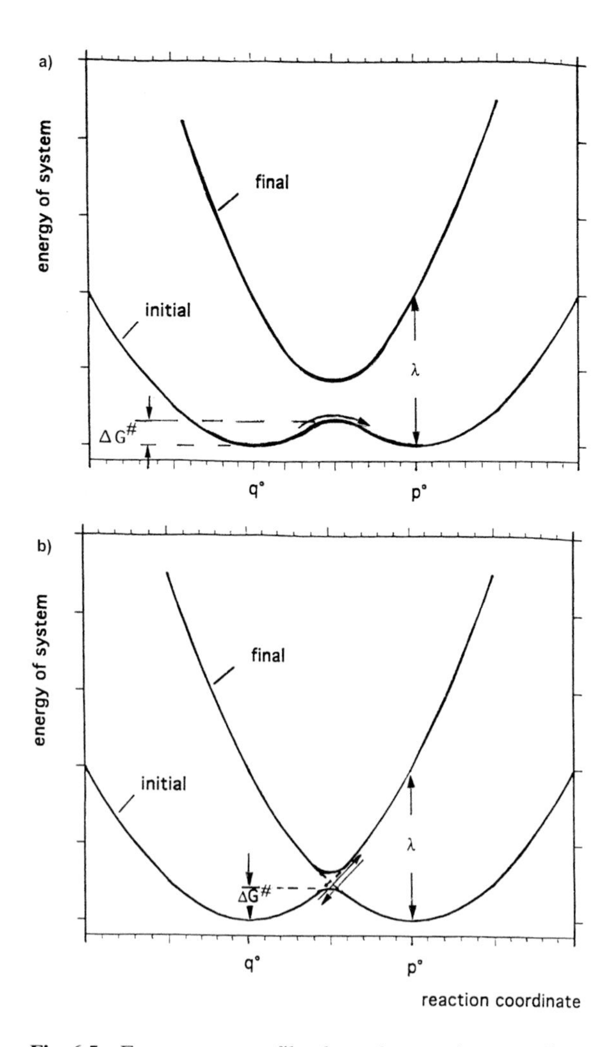

Fig. 6.5 Free energy profile along the reaction coordinate q for an adiabatic reaction (a) and a non-adiabatic reaction (b); $\Delta G^0 = 0$

typical transition-metal redox reactions that $V_{RP} \approx 0.025$ eV is a reasonable limit above which a reaction must be considered to be adiabatic. This problem will be discussed again later in connection with some quantum mechanical models for electron transfer.

6.1.4 Electron Transfer Processes at Electrodes

The basic reaction between two molecules has been given in Eq. (6.1). The corresponding redox reaction at an electrode is given by

$$\mathrm{Ox} + \mathrm{e}^- \xrightarrow{k^{El}} \mathrm{Red} \tag{6.21}$$

In terms of the Marcus theory the rate constant for the electron transfer is given by

$$k_{\mathrm{et}}^{\mathrm{El}} = \kappa\, Z^{\mathrm{El}} \exp\left(-\frac{(\Delta G^0 + \lambda)^2}{4kT\,\lambda}\right) \tag{6.22}$$

in which Z^{El} is the frequency of the reaction coordinate. The charge transfer occurs in the solid via occupied or empty states at energy E in the electrode, depending on whether we have a reduction or an oxidation of the redox system. The enthalpy ΔG^0 of such a reaction is then given by

$$\Delta G^0 = E - E_{\mathrm{redox}} = e(U_{\mathrm{E}} - U_{\mathrm{redox}}) \tag{6.23}$$

in which U_{E} and U_{redox}^0 are the electrode potential and the standard redox potential, respectively. Inserting Eq. (6.23) into (6.22) we have

$$k_{\mathrm{et}}^{\mathrm{El}} = \kappa\, Z^{\mathrm{El}} \exp\left(-\frac{(E - eU_{\mathrm{redox}}^0 + \lambda)^2}{4kT\,\lambda}\right) \tag{6.24}$$

In the case of a metal electrode the electron transfer occurs around the Fermi level ($E = E_{\mathrm{F}}$). Here the rate constant $k_{\mathrm{et}}^{\mathrm{El}}$ can be varied over several orders of magnitude by applying an overvoltage of $\eta = (E_{\mathrm{F}}/e) - U_{\mathrm{redox}}$. Using the energy profile presentation, an application of an external voltage simply means a shift of the reactant energy surface R against P. At equilibrium the Fermi level of the electrode and the redox potential are equal, i.e. $E_{\mathrm{F}} = E_{\mathrm{redox}}^0$ and $\Delta G^0 = 0$. The situation is quite different for semiconductor electrodes insofar as the charge transfer occurs via one of the energy bands, namely via the conduction band (with an energy of E_{c} at its lower edge) or the valence band (with an energy of E_{v} at its upper edge). Accordingly, we have to replace E by E_{c} or E_{v} in Eq. (6.24). Since the positions of the energy bands are usually fixed at the surface, i.e. they remained unchanged upon varying the electrode potential (for details see Chapter 5), the energy difference $E - eU_{\mathrm{redox}}^0$ remains constant. This has an important consequence, namely that the rate constant remains constant for reactions at semiconductor electrodes. Equilibrium is here achieved by a corresponding change of the carrier density at the surface, so that E_{F} becomes equal to E_{redox}^0.

The reorganization energy can be treated in a similar way. The inner sphere component is only half of that for a self-exchange reaction because only one ion is involved, i.e.

$$\lambda_{\mathrm{in}}(\text{electrode}) = 0.5\,\lambda_{\mathrm{in}}(\text{homogneous solution}) \tag{6.25}$$

In the case of the outer sphere component, λ_{out}, it has to be realized that only one molecule needs to be considered, either a donor or an acceptor molecule, i.e. only the$1/2r_{\mathrm{D}}$ or the $1/r_{\mathrm{A}}$ term has to be considered in Eq. (6.19). It must also be taken into account that the reacting ion forms an image charge in the metal electrode having the same distance from the surface as the ion. For metal electrodes this leads to

$$\lambda_{\mathrm{out}}(m) = \frac{(\Delta \mathrm{e})^2}{4\pi\,\varepsilon_0}\left[\frac{1}{2r_{\mathrm{m}}} - \frac{1}{4d}\right]\left(\frac{1}{n^2} - \frac{1}{\varepsilon_{\mathrm{s}}}\right) \tag{6.26}$$

in which r_M is the radius of the molecule in solution and d is the distance of the ion from metal surface. Assuming $r_M = d$, then we have

$$\lambda_{out}(\text{metalelectrode}) = 0.5\,\lambda_{out}(\text{homogeneous solution}) \tag{6.27}$$

The question arises, however: which λ_{out} can be used for reactions at a semiconductor electrode? In this context a more general equation is of interest which has recently been derived by Marcus by using a dielectric continuum model [9]. He considered two adjacent dielectrics having charges in both phases, e.g. ions in two immiscible liquids. We omit here the complete derivation because the basic physical picture for the reorganization of the liquid has already been presented above. The final result as derived by Marcus [9] is given by:

$$\begin{aligned}\lambda_{out} &= \frac{(\Delta e)^2}{8\pi\,\varepsilon_0\,r}\left(\frac{1}{\varepsilon_{opt,1}} - \frac{1}{\varepsilon_{s,1}}\right) \\ &\quad - \frac{(\Delta e)^2}{8\pi\,\varepsilon_0\,R}\left(\frac{1-\varepsilon_{opt,1}/\varepsilon_{opt,2}}{1+\varepsilon_{opt,1}/\varepsilon_{opt,2}}\,\frac{1}{\varepsilon_{opt,1}} - \frac{1-\varepsilon_{s,1}/\varepsilon_{s,2}}{1+\varepsilon_{s,1}/\varepsilon_{s,2}}\cdot\frac{1}{\varepsilon_{s,1}}\right)\end{aligned} \tag{6.28}$$

in which the subscripts 1 and 2 refer to the two phases. It is useful to see what this equation reduces to when one phase is a classical metal conductor. In this limit both ε terms in phase 2 are replaced by infinity and it can be immediately recognized that Eq. (6.28) is reduced to Eq. (6.26) for $R = d$. In the case of a semiconductor, at least for crystals of relatively low doping, the full equation has to be applied. The static dielectric constant of many semiconductors are in the range of $\varepsilon_{s,2} = 5$ to 20, i.e. they are usually much smaller than that of H_2O. Since the optical dielectric constant is much smaller than the static value, the second term in the second bracket of Eq. (6.28) can be neglected. The optical dielectric constants, $\varepsilon_{op,2}$, range typically between 4 and 16. Taking values such as $\varepsilon_{s,2} = 16$, $\varepsilon_{op,2} = 9$ ($n = 3$) for the semiconductor and $\varepsilon_{s,1} = 80$, $\varepsilon_{op,1} = 1.7$ ($n = 1.3$) for water one obtains ($R \approx 2r$) : $\lambda_{out} = 0.4$ eV assuming $r = 7$ Å. Using the same data for calculating for the same reaction at a metal electrode then Eq. (6.26) yields $\lambda_{out} = 0.3$ eV, i.e. about the same value. Considerable differences occur, however, if $\varepsilon_{op,2}$ approaches $\varepsilon_{op,1}$. Then the first term in the second bracket of Eq. (6.28) becomes very small and λ_{out} increases for a semiconductor–liquid interface and may be even twice as large as for metal electrodes. At the same time, Smith and Koval derived an even more detailed model [11]. According to the latter authors, the physical reason for the effects described above, is that in the metal–liquid case the image charge tends to reduce the effect of the approaching charge on the solvent polarization, whereas in a semiconductor–liquid system it concentrates the effect of charge in the liquid and hence increases the amount of reorganization required for the reaction [9, 10].

In all cases, i.e. for electron transfer reactions in homogeneous solutions as well as for electrochemical processes, $\Delta G^{\#0} = \lambda/4$ at equilibrium ($\Delta G^0 = 0$). Comparing, under these conditions, the rates for an electron transfer between molecules in homogeneous solutions with that of molecule at a metal electrode, and assuming $\kappa \approx 1$, then the corresponding rate constants can be related by a simple equation. Using Eqs. (6.11), (6.22) and (6.27) one obtains

$$k_{et}/Z = \left(k_{et}^{El}/Z^{El}\right)^{1/2} \tag{6.29}$$

It should emphasized again that this relation is usually not valid for reactions at semiconductor electrodes because the reorganization energy may be larger than 0.5λ (homogeneous solutions).

6.2 The Gerischer Model

Gerischer approached the problem of charge transfer at electrodes in a completely different way. In 1960 he developed a model in which the charge transfer is considered in terms of electronic energies in the solid and of energy levels in solution, the latter being associated with ions [6, 7]. This was a very ingenious approach especially for semiconductor electrodes because electrons can only be transferred via the conduction or valence band. However, the description of energy levels in solution is more complex than is familiar from solid state physics because of the effect of the polar solvent surrounding the ions. For instance in the case of water, the H_2O dipoles can move and rotate which is not usually found in solids. Before electron transfer rates are discussed, energy levels in the liquid are first introduced in the next section. As will be shown below, the Gerischer model is only applicable for weak interactions between the redox system and the electrode, i.e. mainly for non-adiabatic processes.

6.2.1 Energy States in Solution

Considering a simple redox reaction such as

$$\mathrm{Ox} + \mathrm{e}^- \Leftrightarrow \mathrm{Red} \tag{6.30}$$

then the Nernst equation (see Chapter 5) can be written in terms of electrochemical potentials by

$$\overline{\mu}_{\mathrm{e,redox}} = \mu^0_{\mathrm{e,redox}} + kT \ln\left(\frac{c_{\mathrm{ox}}}{c_{\mathrm{red}}}\right) \tag{6.31}$$

in which $\overline{\mu}_{\mathrm{e,redox}}$ is the electrochemical potential of electrons in the redox system which are dissolved in a liquid such as water, and c_{ox} and c_{red} are the concentrations of the Ox and Red species. As already described in Chapter 3, the electrochemical potential $\overline{\mu}_{\mathrm{e,redox}}$ is equivalent to a Fermi level of a redox system, $E_{\mathrm{F,\,redox}}$, provided that the same reference level is used for the solid electrode and the redox system [6]. We have then

$$E_{\mathrm{F,redox}} = \overline{\mu}_{\mathrm{e,redox}} \tag{6.32}$$

Usually the corresponding redox potential is given on a conventional scale using the normal hydrogen electrode (NHE) or a saturated calomel electrode (SCE) as a reference electrode, as already described in Section 3.2.3. Different scales are given in Chapter 5. Theoretically, however, one can also use the vacuum level as a reference, as is common in solids.

In the case of a typical redox couple, such as $M^{(z+1)}.aq/M^{z+}.aq$, dissolved e.g. in water, the $M^{(z+1)}.aq$ ions represent unoccupied electron energy levels and the $M^{z+}.aq$ ions the

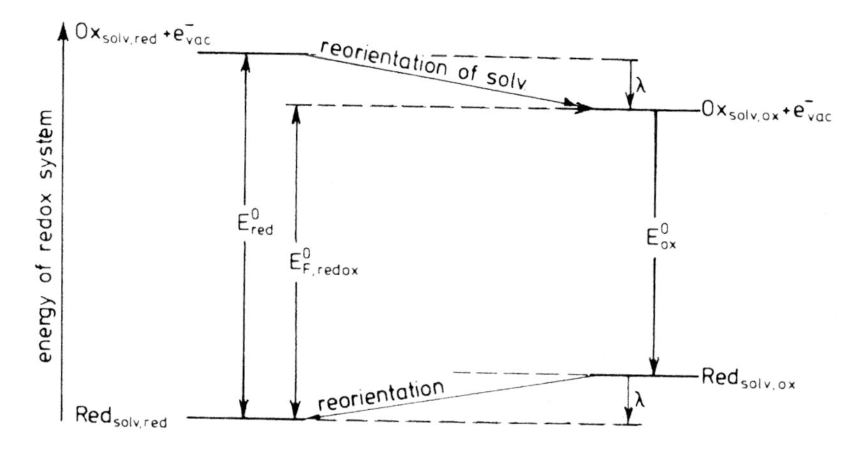

Fig. 6.6 Energy of a redox system in its oxidized and reduced states. The energy differences E^0_{red}, E^0_{ox} and $E_{F,\,\text{redox}}$ are electron energies. (After ref. [13])

occupied states. These ions are surrounded by a specific solvation shell, the latter being different for the reduced and oxidized species because the interaction with the solvent is different. In consequence, the energy levels involved in an electron transfer differ from the thermodynamic value $E_{\text{F, redox}}$ as can be shown by the following reaction cycle (Fig. 6.6).

The total energy of a redox system in its reduced state occurs at equilibrium by $\text{Red}_{(\text{solv,red})}$. The subscript (solv,red) indicates that the reduced species is surrounded by a solvation shell typical for this ion. Starting from this state, the electronic energy E^0_{red} required to transfer an electron from the $\text{Red}_{(\text{solv, red})}$ state into the vacuum leading to the formation of the Ox species. Since, however, this electron transfer into the vacuum is expected to be very fast compared with the reorganization of the solvation shell and the solvent dipoles (Frank–Condon principle), one ends up at first with an Ox species still surrounded by a solvation shell typical for the red species as indicated by $\text{Ox}_{(\text{solv, red})}$ in Fig. 6.6. After the electron transfer step the solvent dipoles reorganize themselves until the ox species reaches its equilibrium state $\text{Ox}_{(\text{solv,ox})}$. The energy involved in the relaxation process is the reorganization energy λ which has already been introduced in the section on the Marcus theory (see Section 4.1). The reverse process, the electron capture by the Ox species, proceeds in a similar way. In a first step the electronic energy E^0_{ox} is gained leading to $\text{Red}_{(\text{solv,ox})}$ followed by a reorganization of the solvent dipoles, as illustrated in Fig. 6.6. It is clear from Fig. 6.6 that electronic energies occur as energy differences in this diagram, i.e. also the electrochemical potential of the electrons or the Fermi level of the redox system $E_{\text{F,redox}}$. Plotting now only the electronic energies in a pure electron energy diagram, using the vacuum level as a reference (absolute scale), one obtains an energy diagram as given in Fig. 6.7a. This energy diagram is very similar to that of solids (compare with Chapter 1). It should be mentioned that the electronic energy E^0_{red} is actually an ionization energy I^0, whereas E^0_{ox} corresponds to electron affinity A^0. We can further derive from Fig. 4.12 the relation

$$E^0_{\text{red}} - \lambda_{\text{red}} = E^0_{\text{ox}} + \lambda_{\text{ox}} = E^0_{\text{F,redox}} \tag{6.33}$$

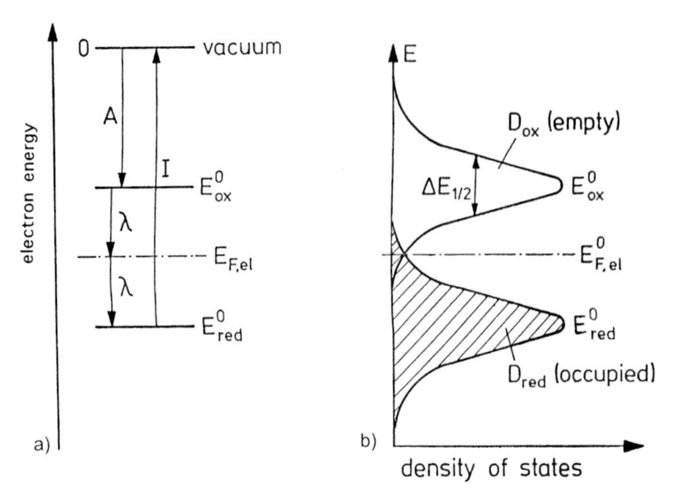

Fig. 6.7 Electron energies of a redox system (differences in Fig. 6.6); a) E^0_{red} occupied states, E^0_{ox} empty states; b) the corresponding distribution functions, D. (After ref. [13])

The energy diagram is only a simplified version of the Gerischer model. More insight into this theory can be gained when the motion and rotation of the solvent molecules are included. Therefore it is useful to consider the free energy of the system with regard to dependence on a reaction coordinate as illustrated in Fig. 6.8. The lower curve represents the Red species whereas the upper corresponds to the Ox species and the electron at infinity. One can easily recognize that the energy terms given in Fig. 6.6 are included in Fig. 6.8. The reaction coordinate represents here something like the average distance of the solvent molecules. Assuming an harmonic oscillation for the fluctuation of the solvent molecules, then the two energy curves in Fig. 6.8 have a parabolic shape around the minimum, i.e.

$$E(\mathrm{Red}_{\mathrm{solv,red}}) = \gamma(\rho - \rho_{0,\mathrm{red}})^2 \tag{6.34a}$$

$$E(\mathrm{Ox}_{\mathrm{solv,ox}}) = \gamma(\rho - \rho_{0,\mathrm{ox}})^2 \tag{6.34b}$$

in which γ is a force constant whereas $\varrho_{0,\mathrm{red}}$ and $\varrho_{0,\mathrm{ox}}$ represent values of the reaction coordinate at which the two energy curves have their minima.

For simplicity we have assumed that the shapes of the two curves are equal (only one γ value in both cases). In consequence the reorganization values are equal, i.e. $\lambda_{red} = \lambda_{ox} = \lambda$. The fluctuation of the solvent molecules leads finally to a broadening of the electronic levels as derived below.

The distribution function of solvation states is given by

$$W_{\mathrm{ox}}(\rho) = W^0 \exp\left[-\frac{E(\mathrm{Ox}_{\mathrm{solv,ox}}) - E^0\left(\mathrm{Ox}_{\mathrm{solv,ox}}\right)}{kT}\right] \tag{6.35}$$

$$W_{\mathrm{red}}(\rho) = W^0 \exp\left[-\frac{E(\mathrm{Red}_{\mathrm{solv,red}}) - E^0\left(\mathrm{Red}_{\mathrm{solv,red}}\right)}{kT}\right] \tag{6.36}$$

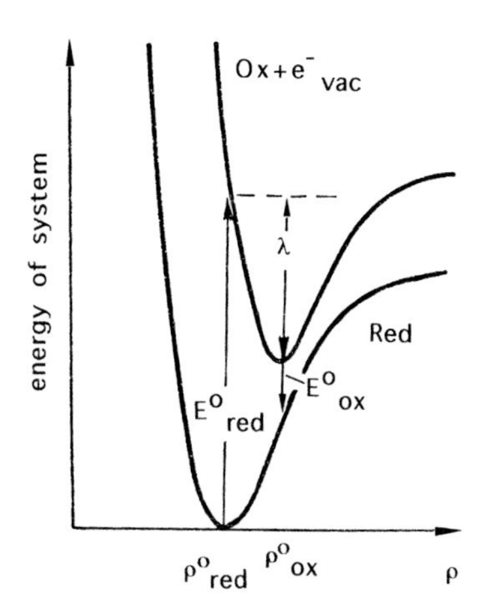

Fig. 6.8 Energy of a redox system vs. the average distance ϱ of the solvent molecules. (After ref. [6])

in which $E^0(\mathrm{Red}_{\mathrm{solv,red}})$ and $E^0(\mathrm{Ox}_{\mathrm{solv,ox}})$ represent the minima in the two free energy curves in Fig. 6.8. These equations can be converted into distribution functions $W_{\mathrm{ox}}(E)$ and $W_{\mathrm{red}}(E)$, which depend only on electronic energies, by inserting Eqs. (6.33) and (6.34) into Eqs. (6.35) and (6.36). After some mathematical rearrangement one obtains:

$$W_{\mathrm{ox}}(E) = W^0 \exp\left[-\frac{(E - E^0_{\mathrm{F,redox}} + \lambda)^2}{4kT\lambda}\right] \tag{6.37}$$

$$W_{\mathrm{red}}(E) = W^0 \exp\left[-\frac{(E - E^0_{\mathrm{F,redox}} - \lambda)^2}{4kT\lambda}\right] \tag{6.38}$$

where the pre-exponential factor is a normalizing constant in order to make the integrated probability unity ($\int W(E)\,\mathrm{d}E = 1$). It is given by

$$W^0 = (4kT\lambda)^{-1/2} \tag{6.39}$$

It should be mentioned here that only one single electronic state of the redox system is considered in this model. W_{ox} and W_{red} are then only the probabilities of finding the empty and occupied electronic state, respectively, at the energy E. The variation of the electronic state is only due to the interaction of the redox couple with the solvent.

The gaussian types of functions (Eqs. 6.37 and 6.38) yield the distribution of electronic levels of the redox system. W_{red} describes the fluctuation and therefore the distribution of the occupied electronic level and W_{ox} the distribution of the empty level. The density of electronic states is proportional to the concentration of the reduced (c_{red}) and oxidized species (c_{ox}) of the redox system. Accordingly, the total distribution is given by:

$$D_{\mathrm{ox}}(E) = c_{\mathrm{ox}} W_{\mathrm{ox}}(E); \quad D_{\mathrm{red}}(E) = c_{\mathrm{red}} W_{\mathrm{red}}(E) \tag{6.40}$$

The corresponding distributions are illustrated in Fig. 6.7b for equal concentrations ($c_{\mathrm{ox}} = c_{\mathrm{red}}$). In this case, the distributions of the reduced and oxidized species are equal at the standard electrochemical potential of the redox couple (standard Fermi level $E^0_{\mathrm{F,redox}}$).

The model can be further tested by varying the concentration of one of the species as illustrated in Fig. 6.9. In this case, the Fermi level of the redox system is shifted according to the Nernst equation. One can easily prove by using Eqs. (6.37), (6.38) and (640) that in this case D_{ox} and D_{red} are equal at $E = E_{\mathrm{F,redox}}$.

The half-width of the distribution curve is given by

$$\Delta E_{1/2} = 0{,}53\lambda^{1/2}\,\mathrm{eV} \tag{6.41}$$

Since the reorganization energies are typically in the order of 1 eV, the width can reach values of about 0.5 eV. These are values which are found for bandwidths in solids. However, such a comparison should be handled with care. In the case of solids, the energy levels are fixed, whereas the energy levels of a redox system are based on the fluctuation of the solvent molecules. It should be emphasized that occupied and empty energy states of a redox system have different energy positions because the interaction of the ox and red species with the solvent is different, whereas in solids the position of energy states usually remains constant when the occupation of a level is changed. Accordingly also, no optical excitation of electrons from an occupied into an empty level is possible [13].

Various scientists consider the time-fluctuating energy levels (Fig. 6.7) as bands of energy levels. Such a description is very convenient, especially for semiconductor-liquid interfaces, but must be used with caution. As Morrison has already pointed out in his book [12], these „bands“ arise from the fluctuation of the solvent and they have different properties from the fixed bands in solids. There is an essential difference in concept between, on the one hand, electron–phonon interactions causing a fluctuation of electronic energy in a static distribution of levels, and, on the other hand, ion-phonon interactions causing a fluctuation of the energy levels themselves. For instance, it is not possible to have an optical transition between the occupied and unoccupied levels.

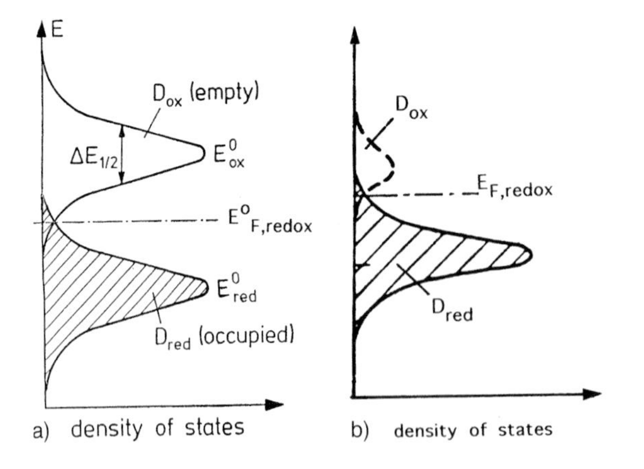

Fig. 6.9 Electron energies of a redox system vs. density of states: a) $c_{\mathrm{ox}} = c_{\mathrm{red}}$; b) $c_{\mathrm{ox}} \ll c_{\mathrm{red}}$

6.2.2 Electron Transfer

In the Gerischer model, an electron transfer occurs from an occupied state in the metal or the semiconductor to an empty state in the redox system, as illustrated in Fig. 6.10. The reverse process occurs then from an occupied state in the redox system to an empty state of the solid (not shown). The electron transfer takes place at a certain and constant energy as indicated by arrows in Fig. 6.10. This means that the electron transfer is faster than any rearrangement of the solvent molecules, i.e. the Frank–Condon principle is valid. In this approach, the rate of an electron transfer depends on the density of energy states on both sides of the interface. For instance, in the case of an electron transfer from the electrode to the redox system the rate is given by

$$\text{Transfer rate} \Rightarrow \kappa Z \int f(E)\rho(E)W_{\text{ox}}(E)\,\text{d}E \tag{6.42}$$

in which κ is again the transmission coefficient as derived in Section 6.1.3; $\varrho(E)$ the distribution of energy state in the electrode; $f(E)$ the Fermi function in the solid (see Eq. 1.25), whereas $W_{\text{ox}}(E)$ is the distribution function of the empty states as defined in Eqs. (6.37) and (6.38). In Eq. (6.42) one has to integrate over all possible electron energies. Concerning Eq. (6.42), one has to differentiate between an electron transfer via the conduction and via the valence band. Taking the electron transfer from the conduction band as an example, the corresponding cathodic current can be determined from Eq. (6.42) by using Eq. (6.37):

$$j_{\text{c}}^{-} = \frac{e\,Z\,c_{\text{ox}}}{(\pi\,kT\,\lambda)^{1/2}}\kappa \int_{E_{\text{c}}}^{\infty} n_{\text{s}} \exp\left[-\frac{(E - E^{0}_{\text{F,redox}} + \lambda)^2}{4\,kT\,\lambda}\right] \text{d}E \tag{6.43}$$

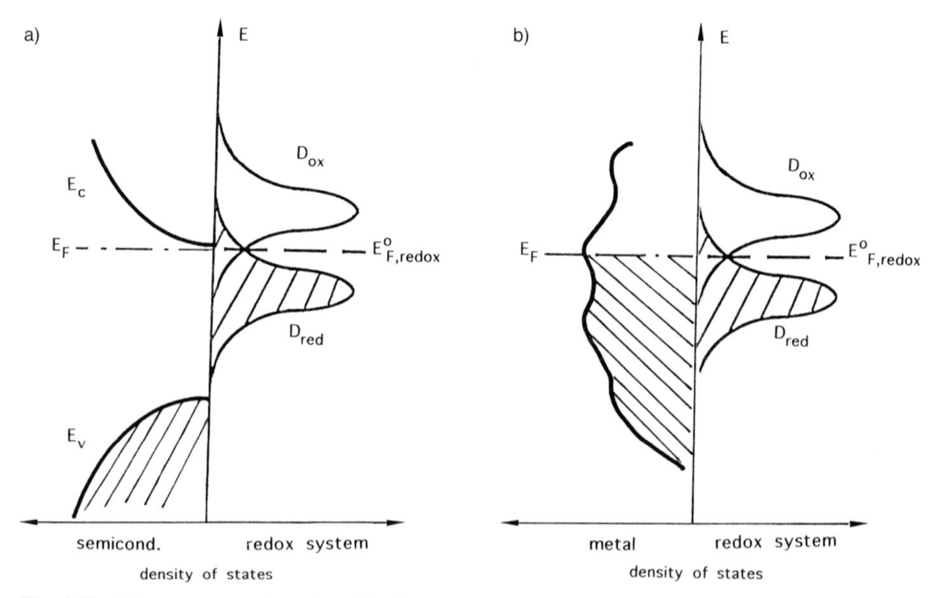

Fig. 6.10 Electron energies of solid electrodes in contact with a redox system vs. density of states: a) semiconductor electrode; b) metal electrode

Here the term $(\pi kT\lambda)^{-1/2}$ is a normalization factor and the density of occupied states in the conduction band is equal to the density of free electrons in the conduction band, i.e. $n_s = f(E)\varrho(E)$, where n_s is actually the electron density at the surface. The integral can only be solved by making some approximations as follows. Since the distribution function W_{ox} varies exponentially with E^2 and since , in most cases, the overlap between energy states on both sides of the interface is mostly limited to a rather small energy range (see e.g. Fig. 6.10), the electron transfer is assumed to occur mainly within 1 kT at the edge of the conduction band. Using this approximation, the integral in Eq. (6.43) can be replaced by inserting $dE = 1\ kT$ and $E = E_c$. One then obtains

$$j_c^- = e\,Z\,c_{ox}\left(\frac{kT}{\pi\lambda}\right)^{1/2}\kappa\,n_s\,\exp\left[-\frac{(E_c - E^0_{F,redox} + \lambda)^2}{4kT\lambda}\right] \tag{6.44}$$

The reverse process, namely an electron transfer from the redox system to the semiconductor, is in general also described by Eq. (6.42), when W_{ox} is replaced by W_{red} and inserting the density of unoccupied states in the conduction band, i.e. $(1 - f(E))\varrho(E)$. Since at usual doping only few energy states are occupied by electrons, the density of empty states is equal to N_c (N_c = density of states at the lower edge of the conduction band). Using the same approximations as above, we have

$$j_c^+ = e\,Z\,c_{red}\left(\frac{kT}{\pi\lambda}\right)^{1/2}\kappa\,N_c\,\exp\left[-\frac{(E_c - E^0_{F,redox} - \lambda)^2}{4kT\lambda}\right] \tag{6.45}$$

Similar results are obtained for a valence band process (see Section 7.3.2).

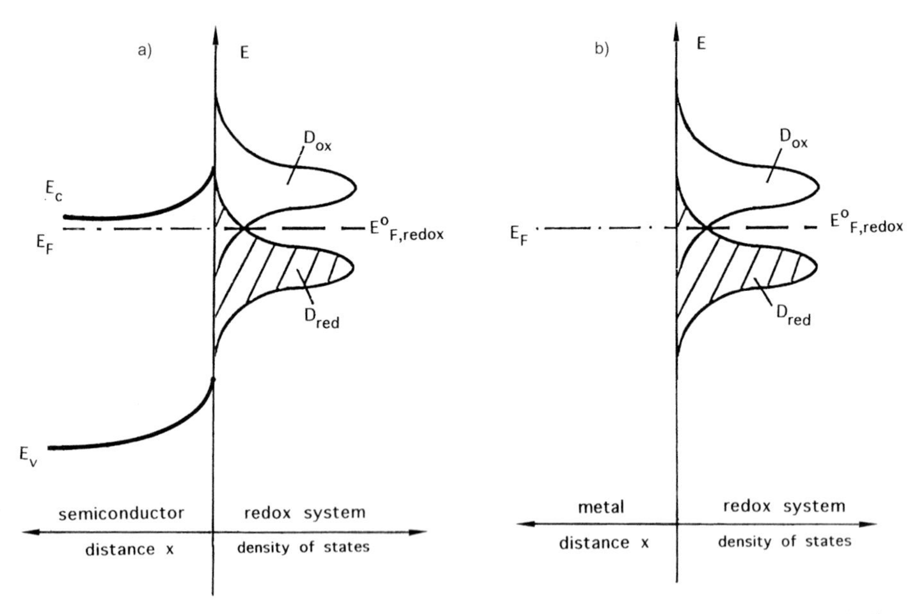

Fig. 6.11 Electron energies of solid electrodes and redox system (compare with Fig. 6.10)

The essential term is the exponential function of W_{ox} (here a gaussian distribution function of the empty energy states in the redox system), which is mathematically the same as in the original Marcus theory. Accordingly, the rate equation is almost identical for both theories although the basic models are conceptually different. The reason is that a harmonic oscillator type of fluctuation of the solvent molecules is assumed in both cases.

The Gerischer description of electron transfer, in terms of electron exchange between empty and occupied energy states on both sides of the interface, has been applied by many researchers in semiconductor electrochemistry because from such an energy scheme one can predict very easily which energy band of the semiconductor electrode is involved in the charge transfer process. It should be emphasized again that this model is only valid for weak interactions; which has not always been appreciated by various scientists. Besides the energy diagram in terms of densities of states at both sides of the interface, a modified energy scheme is used more frequently, where the lower edge of the conduction band and the upper edge of the valence band of the semiconductor are plotted versus distance from the surface, as illustrated in Fig. 6.11.

6.3 Quantum Mechanical Treatments of Electron Transfer Processes

Since the early work of Levich and Dogonadze [4, 5] various modifications of these investigations and a number of new approaches have been published. During the late 1990s the problem of electron transfer has been studied in more detail, leading, however, to very complex results. The solution of corresponding equations became possible because of the great progress in computer simulations within a reasonable period of time. Most of these papers are rather difficult to read for non-theoreticians and especially for students who are new in this field. Therefore, some general approaches, usually applied in all quantum mechanical derivations, will be given in the following introduction.

6.3.1 Introductory Comments

a) Quantum Mechanical Analysis of a Two-level System

Adiabatic and non-adiabatic reactions have already been defined in Section 6.1.3. In both cases an electronic coupling between the reactants or between the initial and final state occurs, leading to a splitting of the energy surfaces as qualitatively illustrated in Fig. 6.5. This is a typical problem of a two-level system which is quite common in physics. A full quantum mechanical treatment for weak as well as strong interactions in a single model was not really possible. Therefore different approaches have been used for strong and weak interactions. In the case of weak coupling, the problem is usually solved by using a first-order perturbation theory as briefly shown below. When the intensity of the perturbation is sufficiently weak, it can be proved that its effect on the two states can be calculated, to a first approximation, by ignoring all the other energy levels of the system.

Considering a system of the two eigenstates $|\varphi_1\rangle$ and $|\varphi_2\rangle$ of the Hamiltonian H_0 whose eigenvalues are E_1 and E_2, respectively, then the Schrödinger equation is given by (see e.g. ref. [14])

$$H_0 \mid \varphi_1\rangle = E_1 \mid \varphi_1\rangle \tag{6.46a}$$

$$H_0 \mid \varphi_2\rangle = E_2 \mid \varphi_2\rangle \tag{6.46b}$$

Introducing a small perturbation, initially neglected in H_0, the Hamiltonian becomes:

$$H = H_0 + V \tag{6.47}$$

This perturbed system is then described by a new set of eigenstates and eigenvalues of H denoted by $|\psi_+\rangle$, $|\psi_-\rangle$, and E_+, E_-, respectively. We have then

$$H \mid \psi_+\rangle = E_+ \mid \psi_+\rangle \tag{6.48a}$$

$$H \mid \psi_-\rangle = E_- \mid \psi_-\rangle \tag{6.48b}$$

Usually H_0 is called the unperturbed Hamiltonian and V the perturbation or coupling. In addition it is assumed that V is time-independent. In the absence of coupling, E_1 and E_2 are the possible energies of the system, and the states $|\varphi_1\rangle$ and $|\varphi_2\rangle$ are stationary states, i.e. if the system is placed in one of these states, it remains there indefinitely. The energies have now to be evaluated after having introduced the coupling V.

In the $\{\varphi_1\rangle, |\varphi_2\rangle\}$ basis, the matrix representation is written as:

$$(H) = \begin{vmatrix} E_1 & V \\ V & E_2 \end{vmatrix} \tag{6.49}$$

After diagonalization of this matrix (see any book on quantum mechanics, e.g. ref. [14]), i.e.

$$\begin{vmatrix} E_1 - E & V \\ V & E_2 - E \end{vmatrix} = 0 \tag{6.50}$$

we find the eigenvalues of the perturbed system as given by:

$$E_+ = \tfrac{1}{2}(E_1 + E_2) + \tfrac{1}{2}\sqrt{(E_1 - E_2)^2 + 4|V|^2} \tag{6.51a}$$

$$E_- = \tfrac{1}{2}(E_1 + E_2) - \tfrac{1}{2}\sqrt{(E_1 - E_2)^2 + 4|V|^2} \tag{6.51b}$$

Studying the effect of coupling V on the energies E_+ and E_- in terms of the unperturbed energy values E_1 and E_2 it is useful to introduce the parameters:

$$E_2 = E_1 + \Delta E \tag{6.52}$$

According to this definition, ΔE is the difference between the two energy states. This kind of definition will play an important role in the description of the solid–liquid interface. Substituting Eqs. (6.52) into (6.51) one obtains:

$$E_+ = E_1 + \frac{\Delta E}{2} + \tfrac{1}{2}\sqrt{(\Delta E)^2 + 4|V|^2} \tag{6.53a}$$

$$E_- = E_2 + \frac{\Delta E}{2} - \tfrac{1}{2}\sqrt{(\Delta E)^2 + 4|V|^2} \tag{6.53b}$$

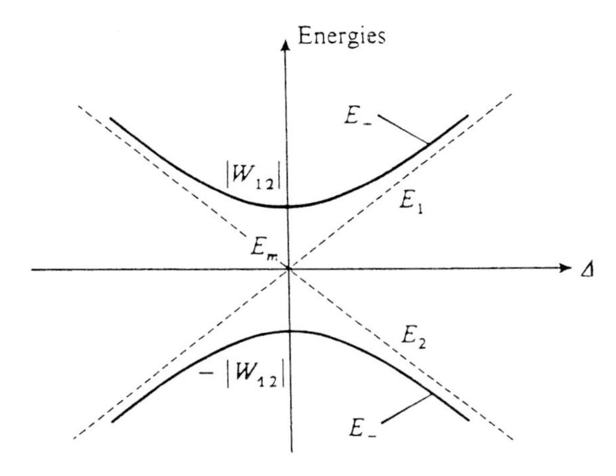

Fig. 6.12 Variation of the energies E_+ and E_- with respect to the energy difference $\Delta = (E_1 - E_2)/2$. In the absence of coupling the levels cross at the origin (dashed straight lines); in the presence of coupling the curves giving E_+ and E_- in terms of Δ are branches of a hyperbola (solid lines)

According to these equations the variation of E_+ and E_- with respect to ΔE is very simple as illustrated for the four energies E_1, E_2, E_+ and E_- in Fig. 6.12. When ΔE is small compared with $|V|$, then the minimum separation of the branches is $2|V|$. In the absence of coupling, the energies E_1 and E_2 of the two levels cross at $\Delta E = 0$ ($E_1 = E_2$). In this range the coupling has its largest effect and we have according to Eqs. (6.53a) and (6.53b):

$$\begin{aligned} E_+ &= E_1 + |V| \\ E_- &= E_1 - |V| \end{aligned} \tag{6.54}$$

On the other hand, if $\Delta E \gg |V|$ then the roots in Eq. (6.53) can be expanded for a limited power if the coupling is weak, i.e. $|\Delta E| \gg |V|$, and one obtains:

$$E_+ = E_\mathrm{m} + \frac{\Delta E}{2} + \frac{\Delta E}{2}\left(1 + 2\left|\frac{V}{\Delta E}\right|^2 + \ldots + \ldots\right) \tag{6.55a}$$

$$E_- = E_\mathrm{m} + \frac{\Delta E}{2} - \frac{\Delta E}{2}\left(1 + 2\left|\frac{V}{\Delta E}\right|^2 + \ldots + \ldots\right) \tag{6.55b}$$

Accordingly, the effect of coupling is much more important if the two unperturbed levels have the same energy. This is exactly the situation in the energy surface diagram for non-adiabatic reactions (Fig. 6.1) where the two energy surfaces cross. The effect of perturbation is then first-order, as given by Eq. (6.54), while it is of second-order when $|\Delta E| \gg |V|$ (Eq. 6.55).

Using such a perturbation theory, the Schrödinger equation of the corresponding system (see Eq. 6.48) has to be fulfilled. In the case discussed above, we can distinguish again between two cases. At the center of the hyperbola, when $E_1 = E_2$ ($\Delta E = 0$), the wave vectors can be written as:

$$|\psi_+\rangle = \frac{1}{\sqrt{2}}\left[e^{-\frac{i\varphi}{2}}|\varphi_1\rangle + e^{i\varphi/2}|\varphi_2\rangle\right] \tag{6.56a}$$

$$|\psi_-\rangle = \frac{1}{\sqrt{2}}\left[-e^{-\frac{i\varphi}{2}}|\varphi_1\rangle + e^{i\varphi/2}|\varphi_2\rangle\right] \tag{6.56b}$$

and it can be easily seen that this sum also fulfils the Schrödinger equation. Near the asymptotes, i.e. for $\Delta E \gg |V|$, we have, to first-order in $|V|/\Delta E$:

$$|\psi_+\rangle = e^{-i\varphi/2}\left[|\varphi_1\rangle + e^{i\varphi}\frac{|V|}{2\Delta E}|\varphi_2\rangle + \ldots + \ldots\right] \tag{6.57a}$$

$$|\psi_-\rangle = e^{-i\varphi/2}\left[|\varphi_2\rangle + e^{i\varphi}\frac{|V|}{2\Delta E}|\varphi_1\rangle + \ldots + \ldots\right] \tag{6.57b}$$

Accordingly, the perturbed states differ only very slightly from the unperturbed states for weak coupling. Within a global phase factor $e^{-i\varphi/2}$, $|\psi_+\rangle$ is equal to the state $|\varphi_1\rangle$, slightly changed by a small contribution from state $|\varphi_2\rangle$. On the other hand, for strong coupling ($\Delta E \ll |V|$), Eq. (4.39) indicates that the states $|\psi_+\rangle$ and $|\psi_-\rangle$ are very different from the states $|\varphi_1\rangle$ and $|\varphi_2\rangle$.

In various theoretical derivations of electron transfer the validity of approximations to the Schrödinger equation is frequently not proved or it is hidden and therefore difficult for non-theoreticians to recognize.

b) Golden Rules for Transition Rates

So far a description of only the static properties of the quantum system has been given. In order to derive the probability for a transition from one stationary state to another (e.g. from R_0 to P_0 in Fig. 6.5), we have to consider the dynamics of the quantum system. For solving such a problem, time-dependent wave functions and the time-dependent Schrödinger equation have to be used. It is possible to obtain an exact solution of the quantum mechanical equation of motion of the system, when the perturbation causing a transition varies harmonically in time. However, a static perturbation, that is present for only a finite amount of time (e.g. from $t = 0$ to $t = t'$) depends implicitly on time and, therefore, can also induce transitions between stationary states of the unperturbed system. Since a complete derivation is not of any help in understanding the next chapters we omit it here and refer the reader to other textbooks. According to a corresponding derivation for a two-level system the transition probability P from state i to state f is given by

$$P = \frac{2\pi}{\hbar}\overline{|V|^2}\,\rho(E_f)t \tag{6.58}$$

in which $\overline{|V|^2}$ is the value of the absolute square of the perturbation matrix element averaged over the final states, whereas $\varrho(E_f)$ represents the density of final states assuming that many states are available around E_f. The transition rate from the initial to the final state may be regarded as the transition probability per unit time which is given by

$$k_{tr} = \frac{P}{t} = \frac{2\pi}{\hbar}\overline{|V|^2}\,\rho(E_f) \tag{6.59}$$

This equation can be used for solving many problems. Its only restriction is that the perturbation is weak. Because of the general applicability of this equation it is named „Fermi's Golden Rule“ of quantum mechanics. This rule has been applied In many quantum mechanical derivations of the electron transfer rate without its being particularly mentioned. One can identify it immediately by the appearance of the $|V|^2$ term in a relevant equation. We return to this point in the following sections.

6.3.2 Non-adiabatic Reactions

a) The Levich–Dogonatze Treatment

Quantum mechanical approaches for describing electron transfer processes were first applied by Levich [4] and Dogonadze, and later also in conjunction with Kuznetsov [5]. They assumed the overlap of the electronic orbitals of the two reactants to be so weak that perturbation theory, briefly introduced in the previous section, could be used to calculate the transfer rate for reactions in homogeneous solutions or at electrodes. The polar solvent was here described by using the continuum theory. The most important step is the calculation of the Hamiltonians of the system. In general terms the latter are given for an electron transfer between two ions in solution by

$$H_{\text{total}} = H_{\text{e}} + H_{\text{ion-solv}} + V_{\text{elec-solv}} \tag{6.60}$$

where H_{e} is the Hamiltonian of the transferring electron in the field of two ions, $H_{\text{ion-solv}}$ is the Hamiltonian of the solvent in the presence of ions and $V_{\text{elec–solv}}$ describes the interaction of the electron with the solvent.

Since the ions are relatively heavy, they are considered to be stationary during electron transfer, i.e. the Hamiltonian for the electron H_{e} can be written in terms of the kinetic energy of the electron and its interaction with the ions as

$$H_{\text{e}} = \frac{\hbar}{2m_{\text{e}}}\frac{\partial^2}{\partial r^2} + \frac{(z_1\, e_0)^2}{r} + \frac{(z_2\, e_0)^2}{|r - R|} \tag{6.61}$$

in which z_1 and z_2 correspond to the valencies of ion A and B, respectively. The electronic coordinate $\vec{r}$ presents the position of the electron with respect to the ion and $\vec{R}$ represents the distance between the two ions (Fig. 6.13).

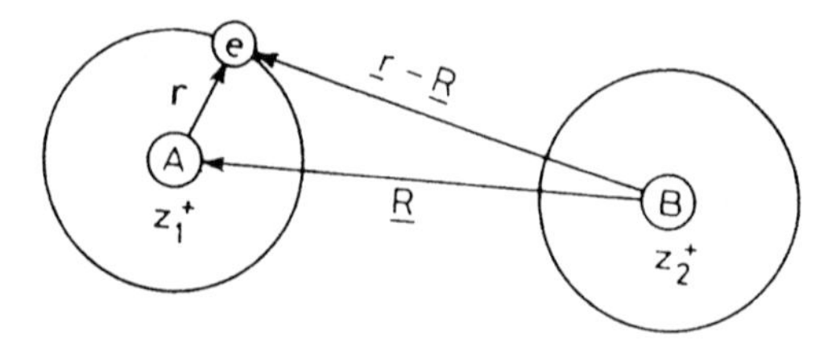

Fig. 6.13 The coordinate representation and the separation of two reactant ions (see text)

The Hamiltonian of the solvent $H_{\text{ion-solv}}$ depends on two parts, one concerning the solvent itself, H_{solv}, and another term which specifies the role of the ion, H_{ion}, (see below), so that we have

$$H_{\text{ion-solv}} = H_{\text{ion}} + H_{\text{solv}} \tag{6.62}$$

The Hamiltonian H_{solv} has been derived on the assumption that the solvent fluctuation occurs again as an harmonic motion. The harmonic of a single harmonic oscillator is given by

$$\tfrac{1}{2}k_s r^2 + \frac{p^2}{2m} \tag{6.63}$$

in which k_s is the force constant, r the displacement of the solvent molecule, m its mass and p the momentum operator as defined by

$$p = -i\hbar \frac{\partial}{\partial r} \tag{6.64}$$

The first term corresponds to the kinetic and the second to the potential energy. Since the frequency of the motion is given by $\omega_0 = (k/m)^{1/2}$ the Hamiltonian is given by

$$H_{solv} = \frac{\hbar^2}{2m}\frac{\partial^2}{\partial r^2} + \tfrac{1}{2}\omega_0\, mr^2 \tag{6.65}$$

In the case of solvent molecules or dipoles it is useful to introduce a solvent coordinate such as

$$q^2 = \hbar\omega_0\, mr^2 \tag{6.66}$$

One obtains then

$$H_{solv} = \tfrac{1}{2}\sum_{\nu} \hbar\,\omega_0 \left(q_\nu^2 + \frac{\partial^2}{\partial q_\nu^2}\right) \tag{6.67}$$

where the subscript v refers to the mode which is involved. The description of the Hamiltonian in terms of the dimensionless „solvent coordinate“ q_v, is very useful for such a complex movement (see ref. [15]).

As already mentioned, Levich et al. described the solvent by a continuum model, i.e. in terms of induction D and polarization P. We have already applied this model to the derivation of the reorganization energy (Section 6.1.2). This is a model similar to that in polar crystals in which we consider the interaction of a charge with the liquid dipoles that constitute the liquid. These dipoles interact, after having received the interaction energy from the ions, back upon the ion and consequently change the energy of the ion itself. The energy which is involved here [5, 15] is given by

$$H_{ion} = \int \left(\int_0^D (D\mathrm{d}P)\right)\mathrm{d}V + \frac{(\varepsilon_{opt} - 1)}{4\pi\,\varepsilon_{opt}} \int \left(\int_0^{D_i} (D\,\mathrm{d}D)\right)\mathrm{d}V \tag{6.68}$$

Finally we have to find the interaction energy $V_{elec\text{-}solv}$ between the electron and the solvent. This again is determined by the induction and polarization induced by the electron. One obtains

$$V_{elec\text{-}solv} = \int \vec{D}_e \vec{P}\mathrm{d}V \tag{6.69}$$

The wave function describing the total system, which consists of the solvent, the two ions and the electron, can be found by solving the Schrödinger equation using the first-order perturbation theory. The energy values of the unperturbed system can be approximated by adding up the following terms (using only one mode):

$$\begin{aligned} E(q) &= E_e + E_{\text{ion-solv}} \\ &= \frac{(z_1 e)^2}{\vec{r}} + \frac{(z_2 e)^2}{(r - \vec{R})} + \frac{(\varepsilon_{\text{opt}} - 1)}{8\pi \, \varepsilon_{\text{opt}}} \int D_i^2 \, \mathrm{d}V + \gamma \left(q - q^0\right)^2 \\ &= E_{\min} + \gamma \left(q - q^0\right)^2 \end{aligned} \tag{6.70}$$

In this equation only the last term, introduced already in Eq. (6.70), depends on the reaction coordinate. The other terms actually determine the free energy of the minimum of the $E(q)$ curves. We have to deal with two energy profiles, one for the initial state, $E_i(q)$, and another for the final state, $E_f(q)$. Their minima occur at different q values, as discussed previously in Section 6.1.1. and already illustrated in Fig. 6.5. According to the rules of the perturbation theory we now have to replace E_1 and E_2 in the matrix element (Eq. 6.33) by E_i and E_f, respectively. Since there is only a small splitting of the energy levels near the transition point (Fig. 6.5), E_i and E_f also correspond to E_+ and E_- outside the crossing point. The same basic equation was obtained by Levich et al., using Fermi's Golden Rule. Omitting the complete derivation by Levich et al., the probability for a transition from the initial to the final state can be expressed as

$$P_{if} = \frac{2\pi}{\hbar} |V|^2 \, A \, v \left| \int |\phi_f(q) \, \phi_i(q) \, \mathrm{d}\tau| \right|^2 \delta(E_f - E_i) \tag{6.71}$$

in which Av is a statistical averaging (of the thermal motion) over the initial states. The Dirac δ function describes the energy conservation law and corresponds to radiationless transfer The δ function in Eq. (6.71) is zero for $E_i \neq E_f$, thus transfer occurs only between states having the same energy. The Dirac δ function in general can be expressed as

$$\delta(E_f - E_i) = \frac{1}{2\pi \, h} \int_{-\infty}^{+\infty} \exp[i(E_i - E_f)t/h] \, \mathrm{d}t \tag{6.72}$$

The complete derivation of Eq. (6.71) is omitted here because there is no point in proving again the validity of Fermi's Golden Rule (see previous section). For details see ref. [15].

On the basis of the q-dependence of the free energy as given by Eq. (6.70) and Fig. 6.5 and by using Eq. (6.71), the final equation for the probability of electron transfer is then given by

$$P_{if} = \frac{|V|^2}{h} \left(\frac{\pi}{kT\,\lambda}\right)^{1/2} \exp\left[-\frac{(\Delta G_0 + \lambda)^2}{4kT\,\lambda}\right] \tag{6.73}$$

in which $(p/kT\lambda)^{1/2}$ is a normalization factor. According to this equation, the rather qualitative transmission factor κ originally introduced by Marcus (see Section 6.1), is now replaced by a quantity which contains mainly the interaction energy $|V|$. Since the perturbation theory is only applicable for weak interactions, Eq. (6.73) is valid for a range where $\kappa \ll 1$. This is still a rather vague statement and we will come to this problem again in the next subsection.

Although this model was first developed for transfer processes in homogeneous solutions, it can also be applied for electrochemical reactions. In this case some of the Hamiltonians are somewhat different. In addition the interaction between the electrode and the redox system has to be derived. However, as long as only small coupling is assumed, there will be no essential change in Eq. (6.73). In other words the same equation can be used for electron transfer processes between a metal electrode and a redox system.

b) Other Treatments

In more modern theories the Hamiltonian of the solvent coordinates is not described in terms of a dimensionless reaction coordinate q_v as defined in Eq. (6.67) but is modeled by

$$H_{solv} = p_s^2/2m_s + \tfrac{1}{2}k_s(\Delta E)^2 \tag{6.74}$$

in which p is the momentum operator of the solvent as already defined in Eq. (6.64), m_s is the solvent mass and k_s is again a force constant. The second term is the potential energy, i.e. the free energy of the solvent coordinates expressed in terms of an energy difference ΔE, the latter being explained below. This Hamiltonian actually describes the interaction between the redox species and the solvent. The energy of interaction considered here corresponds to the electrostatic interaction of the total charge of the redox species with all solvent molecules. Accordingly, we are considering here mainly the Coulomb energy of interaction of the oxidized species with the fixed solvent configuration, $E_{coul}(ox)$, and of the reduced species with the same fixed solvent configuration, $E_{coul}(red)$. The difference of these Coulomb energies is basically ΔE, or a more complete approximation is given by

$$\Delta E \approx E_{coul}(ox) + E_{solv} - E_{coul}(red) + E_{solv} \tag{6.75}$$

in which E_s is the energy of interaction of all the solvent dipoles with each other. The first term in brackets gives the total energy for the system with oxidized species in the system and the term in the second bracket the energy with the reduced species in it. Since E_s cancels out we have

$$\Delta E \approx E_{coul}(ox) - E_{coul}(red) \tag{6.76}$$

If this total energy difference is zero ($\Delta E = 0$) then the transition state is reached and an electron transfer can occur under conservation of energy.

It should be emphasized again that only the coulomb forces were considered in the above, which corresponds to a long-range type of interaction (classical approach). Of course each molecule has also an internal energy, the energy associated with its electron, and there are short-range interactions between the molecules due to quantum effects. These short-range interactions are extremely important in determining the configuration of the redox molecule and the solvent. The short-range interactions give for instance water its familiar shape and all redox species and all molecules in general. The sum of the long-range Coulombic and short-range electronic interactions give the true total energy of the system (the electronic and nuclear energy) which finally determines the true ΔE. Therefore, Eq. (6.76) is an approximation. Only for very polar solvents is the true ΔE well approximated by this equation.

The main advantage of using ΔE instead of the very vague reaction coordinate q used so far, is the fact that ΔE is a well-defined microscopic quantity which can be introduced in quantitative calculations. It has been of widespread use in homogeneous electron transfer models and has also been used in electrochemical models [16]. More physical insight in the quantity ΔE will be given Section 6.3.3b. Let us describe here only the free energy curves in terms of ΔE for a weak interaction.

Assuming symmetric free energy curves of equal and parabolic curvature for the reacting and product diabatic curves (i.e. E_1 and E_2 in Eq. (6.49) vary with $(\Delta E)^2$), the initial (E_i) and final (E_f) state can be expressed by

$$E_i = \tfrac{1}{2}k_s(\Delta E + \Delta E_0)^2; \quad E_f = \tfrac{1}{2}k_s(\Delta E - \Delta E_0)^2 \tag{6.77}$$

where $\pm\Delta E_0$ represents the displacement of the minima along the ΔE axis, as indicated in Fig. 6.14, assuming that the minima occur at the same energy (here at $E = 0$). In this case we have equilibrium and the activation energies are equal for the forward and back reactions. The crossing point of both curves occurs at $\Delta E = 0$. The energies $E_i(\Delta E)$ and $E_f(\Delta E)$ are linked by Eq. (6.52). The reorganization energy is defined again in the usual way, i.e. under these circumstances $E_i = \lambda$ for $\Delta E = +\Delta E_0$. Inserting these data into Eq. (6.75) the reorganization energy λ is determined by

$$\lambda = 1/(2k_s) \tag{6.78}$$

For a non-equilibrium case, the two curves are shifted vertically against each other. The vertical energy difference of the two minima is approximately $\lambda - \Delta E_0$ as can easily be derived from the above equations. Accordingly if there is no shift, $\Delta E_0 = \lambda$. Inserting Eqs. (6.52) and (6.77) into (6.51), one can easily calculate from the latter equation the complete energy profile for a given coupling energy $|V|$. This is illustrated by the solid curves in Fig. 6.14 for $\lambda = 1$ eV and $|V| = 0.1$ eV.

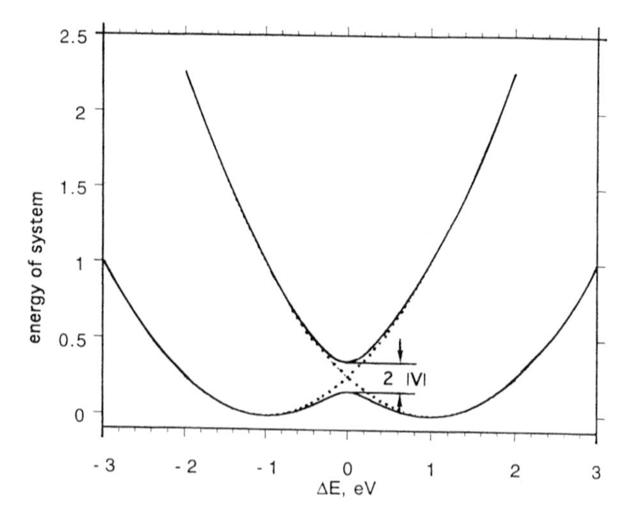

Fig. 6.14 Free energy of a system vs. the energy difference ΔE (see text) under adiabatic conditions

This derivation shows that the description of the free energy in terms of ΔE is a very convenient formalism. In order to describe a real system including the minimum energies quantitatively, however, the complete Hamiltonians of the solvent and the electron have to be derived, as in the previous case. One example will be given in Section 6.2.3.

This kind of calculation leads to an equation for the rate constant which is essentially identical to that given by Eq. (6.59), i.e. the pre-exponential factor is mainly determined by the coupling energy $|V|$. In other cases the pre-exponential factor is simply expressed in terms of a transmission factor κ (see e.g. Eq. (6.4). The latter has actually been derived much earlier by Landau [17] and Zener [18], although within another context. These authors were interested in the dissociation of diatomic molecules and described the system by two potential energy curves (corresponding to two electron terms) vs. the diatomic distance. These curves were also expected to cross similarly as in our case. Landau and Zener have determined the transition probability by using a quantum mechanical method. Applying their treatment one can derive the transition probability as follows.

As already shown in Section 6.3.1, the energy levels $E_i(q)$ and $E_f(q)$ of the reacting system split into two new levels E_a and E_b upon weak interaction. The corresponding wave functions ψ_a and ψ_b are given by Eq. (6.57). Far from the crossing point the interaction is negligible and then $E_a \rightarrow E_i$ on the left side of the transition point and $E_a \rightarrow E_f$ on the right side of the transition point as illustrated in Fig. 6.14. The stationary wave function of the complete system is given by

$$\psi(\Delta E) = a\psi_a(\Delta E) + b\psi_b(\Delta E) \tag{6.79}$$

In order to calculate the transition probability we have to start with the time-dependent wavefunction which is given in accordance with Eq. (6.79)

$$\psi(\Delta E, t) = a(t)\,\psi_a(q)\,\exp\left[\frac{i\,E_a\,t}{\hbar}\right] + b(t)\,\psi_b(q)\,\exp\left[\frac{i\,E_b\,t}{\hbar}\right] \tag{6.80}$$

which satisfies the time-dependent Schrödinger equation

$$-i\hbar\frac{\partial\psi}{\partial t} = H\psi \tag{6.81}$$

If the equation is solved with the boundary condition $a = 1$, $b = 0$ as $t \rightarrow -\infty$, then $|b(+\infty)|^2$ gives the probability that the molecule enters the state υ_b, representing a transition from curve a to curve b, as the system passes through the activation state at the coordinate $q = q^*$. Similarly, $|a(+\infty)|^2 = 1 - |b(+\infty)|^2$ is the probability that the system remains on curve a. Since we are interested particularly in this case, we can express the transfer probability by

$$P = |a(+\infty)|^2 = 1 - |b(+\infty)|^2 \tag{6.82}$$

The factors $a(t)$ and $b(t)$ can be derived from Eq. (6.80) by using Eq. (6.81). Zener finally obtained:

$$P = 1 - \exp\left[-\left(\frac{2\pi\,V}{h\,v\,S_a - S_b}\right)\right] \tag{6.83}$$

Here $|S_a - S_b|$ is the difference in slope of the two undisturbed energy profiles and v is the velocity at which the system passes the crossing point. At equilibrium we have $S_a = -S_b$ and the latter quantities can be derived by using Eqs. (6.5 to 6.8). In the case of v, usually the Boltzmann averaged velocity is taken, as given by

$$v = \frac{(2\,kT)^{1/2}}{\pi\, m'} \tag{6.84}$$

in which m' is the effective mass of the system. Determining S_a and S_b as described above, inserting the corresponding values into Eq. (6.83) and using Eq. (6.84), one obtains

$$P_0 = 1 - \exp(-2\pi\Gamma) \tag{6.85}$$

in which

$$2\pi\,\Gamma = \frac{|V|^2\,\pi^{3/2}}{h\,\nu_{\mathrm{eff}}\,(kT\,\lambda)^{1/2}} \tag{6.86}$$

In the general case the system is transformed by multiple passages, to give the overall transition probability [18]

$$\kappa_{\mathrm{LZ}} = P_0 + (1 - P_0)P_0(1 - P_0) + \ldots\ldots) = 2P_0\,/\,(1 + P_0) \tag{6.87}$$

which is usually called the Landau–Zener transmission coefficient. It is clear from Eq. (6.83) that the exponent depends strongly on the interaction energy $|V|$. Since the whole Zener–Landau derivation is only valid for small interactions, Eq. (6.83) can be expanded into a series for $2\pi\Gamma \ll 1$. We have then for extremely small interactions

$$\kappa_{\mathrm{LZ}} \rightarrow 2\pi\,\Gamma \tag{6.88}$$

The rate equation is given by

$$k = \kappa_{\mathrm{LZ}}\frac{\omega}{2\pi}\left[\exp\left(-\frac{(\Delta G^0 + \lambda)^2}{4\,kT\,\lambda}\right)\right] \tag{6.89}$$

in which κ_{LZ} can be inserted from Eq. (6.87) or (6.88).

6.3.3 Adiabatic Reactions

a) Metal Electrodes

In the Marcus classical theory of electron transfer, adiabatic reactions are classified according to a strong coupling between the initial and final state leading to a large splitting of energy states (see Fig. 6.5) as described already in Section 6.1.3. In this case the transmission factor approaches $\kappa \rightarrow 1$. This problem certainly can not be solved by applying the perturbation theory because the coupling is too large. Around the late 1980s, the adiabatic electron transfer process between a redox system and a metal electrode was approached theoretically by Schmickler [19], and has been tackled more recently by Smith and Hynes [16].

Schmickler [19] was the first to introduce the so-called Anderson–Newns model to problems of electron transfer at metal electrodes. In the latter model Anderson [20] and also Muscat and Newns [21] have considered the chemisorption of a single atom on a metal surface in vacuum. Chemisorption implies a reasonably large energy required to remove the adsorbate from the surface, namely in the order of 1 eV. Thus, a strong coupling exists between adsorbate and metal. Anderson and Newns approached the problem by using a Hartree–Fock approximation. According to this quantum mechanical method, the wavefunction of the whole system is described by the product of the wavefunctions of the atomic orbitals. It has been shown, for instance, that the energy levels within the valence band of a solid (tight binding case) are well described by using the Hartree–Fock approximation (for details see e.g. ref. [22]). Anderson made a further simplification insofar as he used a one-electron Fock Hamiltonian with spin. Even then the solution of the problem is rather complicated. Finally, the density of electronic states in the adatom has been obtained as given by

$$\rho_{\mathrm{a}} = \frac{\Delta}{\pi\left[(\varepsilon - \varepsilon_{\mathrm{a}} - \Lambda)^2 + \Delta^2\right]} \tag{6.90}$$

in which $\Delta(\varepsilon)$ is the width (broadening) of the energy states in the adatom and $\Lambda(\varepsilon)$ their shift; ε_{a} and ε are the electronic energies before and after contacting the solid. $\Delta(\varepsilon)$ and $\Lambda(\varepsilon)$ are given by the following two equations:

$$\Delta(\varepsilon) = \pi \sum_{\kappa} |V_{\mathrm{ak}}|^2\, \delta(\varepsilon - \varepsilon_{\mathrm{k}}) \tag{6.91}$$

$$\Lambda(\varepsilon) = P \sum_{\kappa} \frac{|V_{\mathrm{ak}}|^2}{\varepsilon - \varepsilon_{\mathrm{k}}} \tag{6.92}$$

in which $|V_{ak}|$ represents the corresponding coupling energy and P the Cauchy principal value whereas ε_{k} is the electronic energy of the k state in the solid. All energies refer to the vacuum level.

A full quantum mechanical derivation of Eqs. (6.90) to (6.92) would be beyond the scope of this chapter. The best derivation is given in a review article by Muscat and Newns [21]. The physical model resulting from this treatment is illustrated in Fig. 6.15. The broadening is mainly due to the overlap of one orbital in the adatom with all the k orbitals in the metal. The energy scheme given in Fig. 6.15 is still oversimplified because the energy shift $\Lambda(\varepsilon)$ of the adatom is a complicated function of ε. The energy shift can only be taken to be constant if the width of the valence band of the metal is much larger than the half-width of the distribution of energy states in the adatom [21]. This situation is actually shown in Fig. 6.15. The latter approximation is not applicable for semiconductor electrodes where two energy bands have to be taken into account.

As already mentioned, Schmickler considered the adiabatic electron transfer between a redox center and a metal electrode and derived free energy profiles of the system as follows.

In this system, the redox center at a fixed distance from the electrode surface is characterized by an orbital $|\varphi>$. In the case of an isolated redox center the energy of the orbital is ε. The electronic Hamiltonian is then simply

$$H_{\mathrm{el,r}} = \varepsilon n \tag{6.93}$$

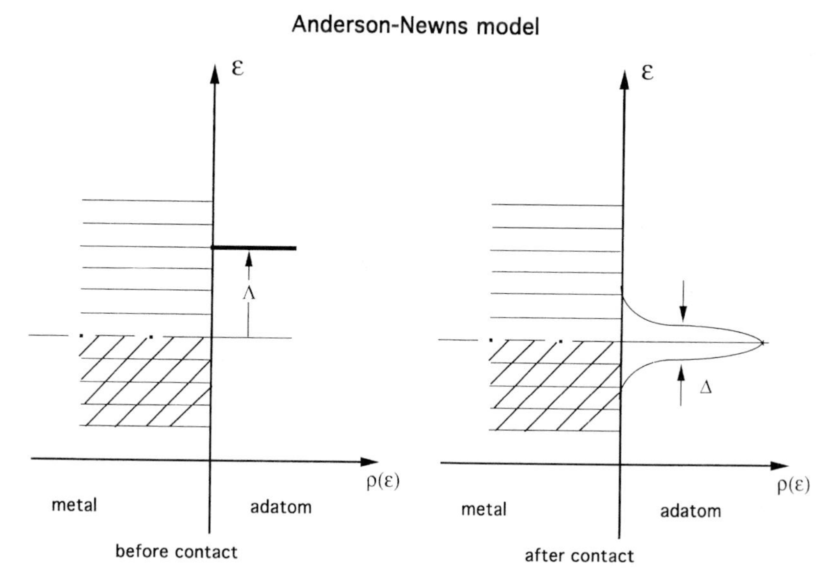

Fig. 6.15 Energy scheme for a metal and an adatom before (left) and after contact (right), derived according to the theoretical calculations given in refs. [20] and [21]

in which n is the operator (occupation number operator) indicating whether the electron is present or not. The metal electrode is described by a collection of spinless, quasi-free electrons, with the Hamiltonian

$$H_{\mathrm{el,m}} = \sum_k (\varepsilon_{\mathrm{k}} n_{\mathrm{k}} + V_{\mathrm{k}} c_{\mathrm{a}}^{\ddagger} c_{\mathrm{k}} + V_{\mathrm{k}}^{\ddagger} c_{\mathrm{k}}^{*} c_{\mathrm{a}}) \tag{6.94}$$

Here k is the momentum of the quasi-free electrons, whose single-particle energies ε_{k} include the effect of electron–electron repulsion renormalization and n_k is the occupation number operator for state $|\mathrm{k}\rangle$. In Eq. (6.94) the electronic coupling between electrode and redox center is included which is governed by the matrix element V_k between states $|k>$ and $|\varphi>$; $c_a{}^{\ddagger}$ and $c_k{}^{\ddagger}$ are the creation operators for the states in the redox system and the metal, respectively, whereas c_a and c_k are the corresponding annihilation operators. Creation and annihilation means that an orbital k in the metal becomes occupied by an electron or emptied, respectively; here in the presence of the electronic coupling.

The Hamiltonian of the solvent has already been derived in Section 6.3.2. and is given again (compare with Eq. 6.67) by:

$$H_{\mathrm{solv}} = \tfrac{1}{2} \sum_{\nu} \hbar\omega_{\nu} (p_{\nu}^2 + q_{\nu}^2) \tag{6.95}$$

in which p_{v} *and* q_{v} are again dimensionless moments and reaction coordinates; ω is the fluctuation frequency and v refers to vibrational mode. It is assumed that only modes are involved where $\hbar$-$\omega < kT$. Since the momentum is independent of q it can also be neglected. Schmickler assumed the interaction between the solvent and the redox couple to be linear, i.e.

$$H_{\text{int}} = (z - n)\left[\sum_{v} \hbar\omega_v\, g_v\, q_v\right] \tag{6.96}$$

where g_v is a kind of force constant (coupling constant); z is the charge number of the redox system in its oxidized state, and n indicates whether the corresponding orbital in the redox system is occupied ($n = 1$) or empty ($n = 0$). The factor $(z - n)$ ensures the proper sign in Eq. (6.96). The assumption of a linear interaction is rather arbitrary and has not been commented upon by Schmickler. We will return to this problem later.

The sum of Eqs. (6.93) to (6.96) leads to the usual Hamiltonian H for an electron transfer reaction, i.e.

$$H = H_{\text{el,r}} + H_{\text{el,m}} + H_{\text{solv}} + H_{\text{int}} \tag{6.97}$$

whereas the total free energy is given by

$$F = E_{\text{electr}} + E_{\text{solv}} + E_{\text{int}} \tag{6.98}$$

The electrical term consists of both, the redox system and the electrode. Using the Anderson–Newns model one can replace E_{electr} by the electronic energy ε multiplied by the distribution of energy states $\varrho(\varepsilon)$, given by Eq. (6.90). Eq. (6.99) then becomes:

$$E = \int_{-\infty}^{0} \varepsilon\, \rho(\varepsilon)\, \mathrm{d}\varepsilon + \tfrac{1}{2}\sum_{v} h\omega_v\, q_v^2 + z\sum_{v} h\omega_v\, g_v\, q_v \tag{6.99}$$

Here it is integrated over all energy states taking the Fermi level as a reference point.

As already mentioned before, Λ represents in Eq. (6.90) the energy shift of the electron level of an adsorbed species due to the strong interaction with the metal. In the case of a redox couple which is dissolved in a liquid, an additional shift is caused by the interaction between the redox center and the solvent, which Schmickler defined as

$$\Lambda_{\text{s}} = -\left[\sum_{v} \hbar\omega_v\, g_v\, q_v\right] \tag{6.100}$$

Using further the definition

$$\hat{\varepsilon}(q_v) = \varepsilon_{\text{a}} + \Lambda_{\text{s}} \tag{6.101}$$

and assuming that the interaction with the solvent dominates over that with the solid ($\Lambda \ll \Lambda_{\text{s}}$), then we have

$$\rho(\varepsilon) = \frac{\Delta}{\pi\left[\left(\varepsilon - \hat{\varepsilon}(q)\right)^2 + \Delta^2\right]} \tag{6.102}$$

We are now mainly interested in deriving corresponding free energy profiles of the complete system. This can be achieved by the following procedure.

It is of special interest to find the configurations for which the electronic energy has its minimum. According to the Hellman–Feynman theorem (quantum mechanical rule) this can be obtained from

$$\left\langle \frac{\partial H}{\partial q_v} \right\rangle = 0 \tag{6.103}$$

Applying this to Eq. (6.97), after having inserted Eq. (6.101), one obtains

$$q_v^0 = -(z - \langle n \rangle) g_v; \quad \varepsilon^0 = \hat{\varepsilon} + 2(z - \langle n \rangle)\lambda \tag{6.104}$$

in which λ is the reorganization energy of the redox system and $<n>$ the occupation probability. The latter is defined as

$$\langle n(q_v) \rangle = \frac{1}{2\pi} \int_{-\infty}^{+\infty} f(\omega)\, \rho(\omega)\, d\omega \tag{6.105}$$

in which $f(\omega)$ is the Fermi–Dirac distribution function (see Chapter 1) and $\varrho(\varsigma)$ the density of electronic states in the redox system as given by Eq. (6.90). The solution of Eq. (6.105) for the stationary state is given by

$$\langle n \rangle_0 = \frac{1}{\pi} \operatorname{arc\,cot} \frac{\varepsilon + 2(z - \langle n \rangle_0\, \lambda)}{\Delta} \tag{6.106}$$

Schmickler finally derived the potential energy $F[<n>]$ of the system from Eq. (6.99) by using Eqs. (6.106) and (6.104). (More specifically, Schmickler actually determined the energy difference between two configurations in order to avoid any divergence of the integral; for details see ref. [19].) Fig. 6.16 shows the corresponding potential energies vs. the reaction coordinate. They were calculated for different Δ values, the latter being an important parameter in the Anderson–Newns model describing the interactions of an adatom on the electrode surface as given by Eq. (6.91). In the case of small

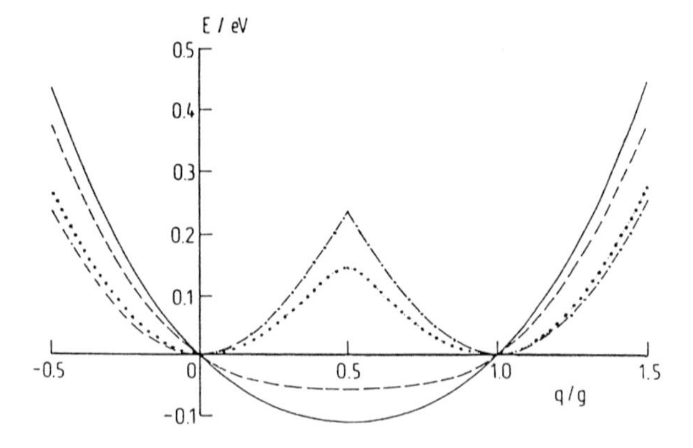

Fig. 6.16 Potential energy curves for a one-dimensional model for a metal–liquid (redox system) interface for $\lambda = 1.0$ eV: $\Delta = 0.01$ eV (– · – · –); $\Delta = 0.1$ eV (·····); $\Delta = 2\lambda/\pi$ (– – –); $\Delta = 1$ eV (——) (see text). (After ref. [19]).

Δ values, two well-defined minima occur, corresponding to the reduced and oxidized state (Fig. 6.16). This curve is in principle identical to the lower curve in Fig. 6.5), whereas the upper curve is not shown in Fig. 6.16. The minima are separated by an energy barrier of a height which is close to $\lambda/4$, i.e. a value which is obtained for the non-adiabatic case in the Marcus theory. With increasing energy width, the energy barrier is lowered; it disappears at $\Delta = 2\lambda/\pi$, and for even higher Δ values only one minimum occurs. Accordingly, in the case of very strong interaction, the electron transfer between a redox system and a metal electrode occurs without any activation. Applying an external voltage to the system leads to a variation of the potential curves similar to that found with traditional theories.

Smith and Hynes used the same approach in principle for adiabatic reactions at metal electrodes [16]. They described the solvent in terms of the single collective variable ΔE, rather than by a collection of harmonic oscillators as already introduced in Section 6.3.2.

b) Semiconductor Electrodes

B. Smith et al. extended the basic Anderson–Newns model introduced in the previous section to electron transfer reactions at semiconductor–liquid interfaces, related them to molecular orbital theory, and addressed certain inherent energy dependencies in them [23]. These authors also performed for the first time electronic structure calculations coupled to molecular dynamics simulations, i.e. they carried out „first principle" molecular dynamic calculations. Their principal approach is as follows.

Similarly to other approaches, Smith et al. first derived the Hamiltonians of the system. The Hamiltonian associated with the solvent nuclei is expressed in terms of ΔE instead of a reaction coordinate q. Using Eq. (6.74) (introduced already in Section 6.3.2b) we have

$$H_{\mathrm{solv}} = p_{\mathrm{s}}^2/(2m_{\mathrm{s}}) + \tfrac{1}{2}k_{\mathrm{s}}(\Delta E)^2 \tag{6.107}$$

The solvent mass m_{s} and the force constant k_{s} are related to the solvent frequency ω_{s} by

$$k_{\mathrm{B}}T/m_{\mathrm{s}} = \langle (\Delta\dot{E})^2 \rangle; \quad k_{\mathrm{s}}/m_{\mathrm{s}} = \omega_{\mathrm{s}}^2 \tag{6.108}$$

The electronic Hamiltonian is given by

$$H_{\mathrm{el}} = n_{\mathrm{a}\sigma}(\Delta E + \varepsilon_{\mathrm{a}\sigma}) + \sum_{K}\left[\varepsilon_{\mathrm{k}} n_{\mathrm{k}\sigma} + \left(V_{\mathrm{ak}} c_{\mathrm{a}\sigma}^{\#} c_{\mathrm{k}\sigma} + V_{\mathrm{ka}} c_{\mathrm{k}\sigma}^{\#} c_{\mathrm{a}\sigma}\right)\right] \tag{6.109}$$

in which $n_{a\sigma}$ *and* $n_{k\sigma}$ represent the operator indicating whether the electron in the redox system and in the semiconductor, respectively is present or not. The other quantities have already been defined in connection with Eq. (6.95). In principle, this Hamiltonian corresponds to the sum of the Hamiltonians ($H_{\mathrm{el,r}} + H_{\mathrm{el,m}}$) given by Eqs. (6.93) to (6.95) as derived by Schmickler (see previous section). The first term describes the redox system and the second the semiconductor. However, there are some essential differences compared with Schmickler's derivation, firstly that the electron spin is included, denoted by the subscript σ.

Concerning the first term corresponding to $H_{\mathrm{el,r}}$ in Schmickler's derivation, it may be surprising that ΔE occurs here as an additional term to $\varepsilon_{\mathrm{a}\sigma}$ ($\varepsilon_{eff} = \varepsilon_{a\sigma} + \Delta E$). However, the problem becomes clear if it is remembered that ΔE essentially is the Coulomb part of the energy difference between the reduced and oxidized species resulting from the long-range interaction of the atomic charges of the redox molecule and of the solvent, as discussed already in Section 6.3.2b. On the other hand, the internal electronic energy of the redox molecule is accounted for by $\varepsilon_{a\sigma}$. Accordingly, $\varepsilon_{a\sigma}$ is basically the ionization energy of the redox species in a vacuum. It is well known from solid state physics that the minimum energy necessary for an electron to escape is affected by the applied electric field and the electron's own image potential which occur as additive quantities to the ionization potential. In the case of a redox system, we have an analogous situation where an electric field is set up by the interaction between redox species and solvent. The ΔE term describes the electric field potential caused by the solvent because the major component is just given by

$$\Delta E = \sum_i \frac{q_{\rm red}\,\delta_{{\rm solv}(i)}}{r_{\rm i}} - \sum_i \frac{q_{\rm ox}\,\delta_{\rm solv\,(i)}}{r_{\rm i}} \tag{6.110}$$

The first term in this equation describes the Coulomb interaction of the total charge of the reduced species ($q_{\rm red}$) with the partial charge on each solvent atom ($\delta_{{\rm solv},i}$). The second term is the same interaction for the oxidized species and r_i is the distance between the center of the redox species and each solvent molecule. If only one electron is transferred we obtain approximately

$$\Delta E = \sum_i \frac{(q_{\rm red} - q_{\rm ox})\,\delta_{\rm solv\,(i)}}{r_{\rm i}} = \sum_i \frac{(-1)\,\delta_{\rm solv\,(i)}}{r_{\rm i}} \tag{6.111}$$

Thus, ΔE basically adds the Coulomb potential produced by the solvent to the redox species vacuum ionization energy $\varepsilon_{a\sigma}$ *and* $\varepsilon_{eff} = (\varepsilon_{a\sigma} + \Delta E)$ which is roughly the ionization energy of the redox species in the presence of the solvent. Schmickler described the interaction by a separate Hamiltonian (Eq. (6.96)) which is linear with respect to the reaction coordinate. In Smith's model the interaction is already described in $\varepsilon_{\rm eff}$ which also depends linearly on ΔE.

In an adiabatic regime where nuclear motion is effectively decoupled from electronic motion (due to their differences in time scale) the full Hamiltonian could be broken into two parts, namely the electronic part and the nuclear part. Accordingly, the electrons can be described by a wavefunction which by itself obeys the Schrödinger equation. Considering the ground state we have then

$$H_{\rm el}(\Delta E)\,|\psi_0(\Delta E)\rangle = E_0(\Delta E)\,|\psi_0(\Delta E)\rangle \tag{6.112}$$

in which E_0 is the electronic ground state energy. It is assumed that the nuclei can be treated classically so that their energy is just a function of the reaction parameter ΔE and the momentum, i.e. the Schrödinger equation is not needed.

Accordingly, Eqs. (6.107) and (6.109), the total Hamiltonian ($H_{\rm el} + H_{\rm solv}$), can be expressed by

$$\begin{aligned} H &= P_{\rm s}^2/(2m_{\rm s}) + \tfrac{1}{2}k_{\rm s}(\Delta E)^2 + E_0(\Delta E) \\ &= P_{\rm s}^2/(2m_{\rm s}) + F(\Delta E) \end{aligned} \tag{6.113}$$

where $F(\Delta E)$ is the total free energy of the system

$$F(\Delta E) = \tfrac{1}{2}k_{\rm s}(\Delta E)^2 + E_0(\Delta E) \tag{6.114}$$

The ground state energy E_0 was derived as follows. The ground state occupation number of the redox species orbital is given by Eq. (6.105); written in terms of ΔE this leads to

$$\langle n(\Delta E)\rangle = \int f(\varepsilon)\rho(\varepsilon, \Delta E)\,\mathrm{d}\varepsilon \tag{6.115}$$

Using the Hellman–Feynman theorem, the ground state electronic energy for the system is then

$$E_0(\Delta E) = \int \langle n(\Delta E)\rangle\,\mathrm{d}(\Delta E) \tag{6.116}$$

Concerning the density of state function $\varrho(\varepsilon,\Delta E)$ Eq. (6.90) resulting from the Newns–Anderson model it must be emphasized, that in the case of semiconductor electrodes the broadening $\Delta(\varepsilon)$ and the shift $\Lambda(\varepsilon)$ of orbitals of the redox species cannot be neglected or taken to be independent of energy. This makes, of course, an evalution of E_0 and therefore of the free energy profile very difficult. Smith et al. applied the method of molecular dynamics (MD) simulations [16, 23]. This method can be used to calculate the change in solvent and redox species configuration with time, and also to calculate the change of energy with time. Since these authors included quantum mechanical calculations at each time step in the MD simulations, they could determine potentials, the electronic structure of the complete system (i.e. of the semiconductor also), and also the flow of electrons between the semiconductor and the redox species. A detailed description of these „first principle" calculations would be beyond the scope of this chapter. They have calculated the free energy profile of a few systems from their three-dimensional MD simulation. One example, that of InP electrode-water–$Fe(H_2O)_6^{2+/3+}$ is given in Fig. 6.17 [24]. This profile was derived by holding the redox molecules fixed at about 5 Å from the electrode, but allowing them to move transversely to the surface.

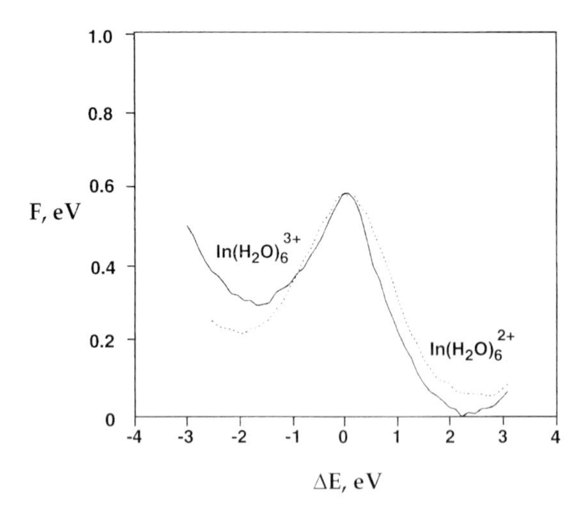

Fig. 6.17 Free energy surfaces for the InP–liquid interface calculated from the molecular orbital molecular dynamics (MD). The solid curve is for the case of fixed semiconductor atoms; the dashed curve is calculated by including the dynamics of the semiconductor atoms. (After ref. [24])

Using the transition state theory the rate constant is then given by

$$k_{\mathrm{TST}} = (\omega_{\mathrm{R}}/2\pi)\,\exp(-\Delta F^{\ddagger}/k_{\mathrm{B}}T) \tag{6.117}$$

in which ς_{R} is the reactant well frequency. The latter is given by

$$\omega_{\mathrm{R}} = (m_{\mathrm{s}})^{1/2}(\partial^2 F/\partial(\Delta E)^2)^{1/2} \quad \text{at} \quad \Delta E = \Delta E_{\mathrm{R}}$$

i.e. it is proportional to the square root of the second derivative of the free energy profile at the well of the reactant. The activation energy is obviously the barrier height in the free energy profile which can be calculated from the MD simulation. These rate constants will be discussed in more detail in the next section.

6.4 The Problem of Deriving Rate Constants

There are various techniques for measuring rate constants experimentally. In the case of reactions in homogeneous solutions, the flux is determined by fast analytical tools whereas for electrochemical reactions interfacial currents are measured. Considering at first a simple electron transfer between a donor and an acceptor in homogeneous solutions, such as

$$\mathrm{D} + \mathrm{A} \Leftrightarrow \mathrm{D}^+ + \mathrm{A}^-$$

then the rate is given by

$$\frac{\mathrm{d}c_{\mathrm{D}}}{\mathrm{d}t} = k c_{\mathrm{D}} c_{\mathrm{A}} \tag{6.118}$$

in which c_{D} and c_{A} are the donor and acceptor concentrations, respectively. In order to avoid dimension problems the concentrations will be given here in molecules per cm^3. Accordingly, the second order rate constant k is given in units of $(\mathrm{cm}^3\ \mathrm{s})^{-1}$. It must be emphasized that the rate constant k_{et} derived theoretically (see e.g. Eq. 6.97) has a completely different unit, namely s^{-1} because of the pre-exponential factor $\omega/2\pi$. The reason is that k_{et} in Eq. (6.97) describes just a single elementary process, whereas in an experiment the possibility of electron transfer from one molecule to many other acceptors has to be considered and we have to integrate over the whole space. Therefore, k_{et} is a local rate constant. This is illustrated in Fig. 6.18.

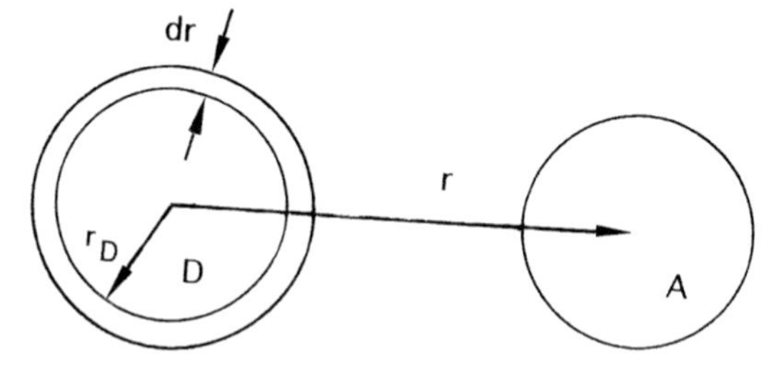

Fig. 6.18 Geometric considerations for the calculation of real rate constants

The rate constant k can be related to k_{et} by

$$k(\mathrm{cm}^3\mathrm{s})^{-1} = \int k_{\mathrm{et}} F_{\mathrm{sphere}}\,\mathrm{d}r \tag{6.119}$$

in which F_{sphere} is the surface of the donor sphere. This relation has to be considered if any prediction for rate constants is made.

A similar problem arises for electrochemical reactions. Again considering here the reaction

$$Ox + e^- \Rightarrow Red$$

the interfacial current at a metal electrode is given by

$$j = ekc_{ox}$$

or the electron flux by

$$\text{flux} \Rightarrow j/e = kc_{ox} \tag{6.120}$$

in which c_{ox} is the concentration of the ox species of the redox system again in units of cm^{-3}. Since the current is given in units of ($A\ cm^{-2}$) and the corresponding flux in units of ($cm^{-2}\ s^{-1}$), the first-order rate constant k is given in ($cm\ s^{-1}$). The latter can be related to the local rate constant k_{et}, defined by Eq. (6.89) or (6.117), by

$$k(\text{cm}\,\text{s}^{-1}) = 1/c_{ox}^{0} \int k_{et}(z)\, c_{ox}(z)\, dz \tag{6.121}$$

in which c_{ox}^{0} is the concentration in the bulk of the solution. The z-axis refers to a direction perpendicular to the electrode surface. The distance dependence of $k_{et}(z)$ refers to the possibility that an electron transfer to an acceptor molecule is not only restricted to adsorbed molecules but may occur also with molecules which are located at a certain distance from the electrode surface. Since the electronic coupling matrix element goes to zero roughly exponentially with distance [25], the local rate constant $k_{et}(z)$ goes to zero as z goes to infinity, so that the integral converges also for a molecule concentration which is constant throughout the system. It is clear from Eq. (4.104) that the main difficulty for a quantitative prediction of the rate constant k is the calculation of the distance dependence of $k_{et}(z)$.

The problem becomes even more severe when using a semiconductor electrode. Since the electron density at the semiconductor surface n_s is variable and small compared with that at metal electrodes, the electron flux is defined as

$$\text{flux} \Rightarrow kc_{ox}n_s \tag{6.122}$$

Again using the unit cm^{-3} for c_{ox} and n_s, then the rate constant k has the unit $cm^4\ s^{-1}$ in the above equation. Inserting the local rate constant, we have here

$$k\text{cm}^4\,\text{s}^{-1} = c_0^{-1} n_s^{-1} \int_x \int_y \int_z \int_{z'} k_{et} c_{ox}(z')\, n_s(x, y, z)\, dx\, dy\, dz\, dz' \tag{6.123}$$

where x, y, z correspond to the Cartesian coordinates in the half-plane of the semiconductor and z' is a Cartesian coordinate in the liquid half-space describing the perpendicular distance to surface. The calculation of this rate constant is rather difficult because the interaction of the redox species with the semiconductor depends on the distance of these two components. There may be an adiabatic electron transfer from the semiconductor to the nearest redox molecule but a non-adiabatic one for a molecule which is at a larger distance from the surface. Accordingly, free energy profiles and the activation energy $\Delta F^{\ddagger}$ must be calculated for various distances in order to obtain values of the local rate constant k_{et}.

6.5 Comparison of Theories

In the previous sections the main models and theories have been presented. In all models the fluctuation of the solvent molecules is described by a harmonic oscillation leading to a gaussian type of exponential function in the corresponding rate equations. The weak interaction between the solid electrode and redox system (still non-adiabatic) and the strong interaction (adiabatic) have been described by quantum mechanical approaches. These interactions lead to a splitting of the free energy profiles near the crossing point of the reactant and of the product. It is still an open question how large the energy splitting must be for an adiabatic process (transmission factor $\kappa = 1$). Schmickler concluded from his model that an energy splitting of 0.08 eV should be sufficient if a reorganization of inner sphere modes is required, and for outer sphere reactions he assumed values even in the order of 10^{-3} eV, i.e. that relatively little interaction is needed to make a reaction adiabatic.

The Gerischer model is conceptually different in that electron energy states of the redox system are introduced. In this model it is simply the Coulombic interaction between the solvent and the redox species that changes the energy of the system and creates a distribution of energies, i.e. the solvent fluctuation leads to a broadening of these states. This model was derived on the basis of a one-molecular orbital of the redox couple. Interestingly, Smith and Nozik also obtained energy states of the redox system from their MD simulations. In their case, however, many molecular orbitals have been included. One example is the couple $Fe(H_2O)_6^{2+/3+}$. These authors calculated by the MD method the distribution of s, p and d orbitals of redox molecules

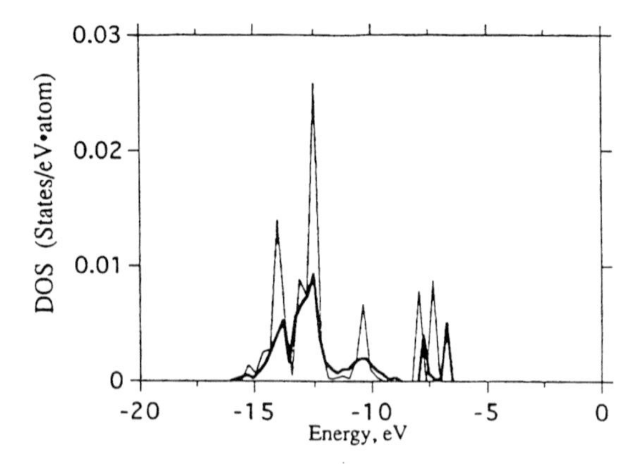

Fig. 6.19 Distribution of energy states of $Fe(H_2O)_6^{2+/3+}$ near an InP electrode surface (heavy curves show DOS with water present). The figure shows the distribution of d orbitals of the redox molecules which are located within a distance of 4 Å from a (100) InP surface. (After ref. [24])

which were located within a distance of 4 Å from a (100) face of InP. The corresponding density of states (DOS) for the d orbitals of the reduced species, $Fe(H_2O)_6^{2+}$, is given in Fig. 6.19. The heavy curve shows the DOS as calculated with solvent (H_2O) present, i.e. with 40 H_2O molecules closest to Fe and the lighter curve represents the results obtained without the additional H_2O molecules. A broadening of the electronic levels occurs due to the Anderson–Newns model, and on the other hand a broadening and shift occurs because of the Coulombic interaction between the redox species and the solvent. Similar results have been obtained for the s and p orbitals, which are not shown here because only the d states are involved in the electron transfer.

If such a multiorbital system were to be used in the Gerischer model then the corresponding distribution of W_{ox} and W_{red} or D_{ox} and D_{red} would look more complicated.

7 Charge Transfer Processes at the Semiconductor–Liquid Interface

Electron transfer reactions at metal electrodes had been studied long before investigations of processes at semiconductor electrodes were started. They were even studied long before Marcus published his model of electron transfer processes. Early in this century, kinetic models on electron transfer processes had already been developed, which are still used for analyzing experimental data obtained with metal electrodes. Since the corresponding descriptions of the electrochemical kinetics and the application of various techniques are also of importance in semiconductor electrochemistry, the essential results obtained with metal electrodes will be briefly presented in the first section.

7.1 Charge Transfer Processes at Metal Electrodes

7.1.1 Kinetics of Electron Transfer at the Metal–Liquid Interface

In this section we only consider electron transfer processes between a redox couple dissolved in the electrolyte and an inert metal electrode such as platinum. Here an inert electrode means that we work in a potential range where essentially no other electrochemical reactions take place. Considering a single electron transfer step as given by

$$\mathrm{Ox} + \mathrm{e}^- = \mathrm{Red} \tag{7.1}$$

the anodic current j^+ (oxidation of the redox system) and cathodic current j^- (reduction of the redox system) are given by

$$j^+ = ek^+c_{\mathrm{red}} \tag{7.2a}$$
$$j^- = ek^-c_{\mathrm{ox}} \tag{7.2b}$$

and the total current density by

$$j = j^+ - j^- \tag{7.2c}$$

in which c_{red} and c_{ox} are the concentrations of the reduced and oxidized species, respectively. In Eqs. (7.2.a) and (7.2.b) we have current densities in units A cm^{-2}. If we take the concentrations in terms of number of molecules or ions per cm^{-3} and the charge e in As, then the rate constants k^+ and k^- have dimensions of cm s^{-1}. In traditional electrochemistry usually the Faraday constant F instead of the elementary charge e is used. Then the concentrations should be given in mol l^{-1} which finally leads to the same dimension for k. We prefer the first dimensions because it is then easier to compare redox processes at metal and semiconductor electrodes. It should be men-

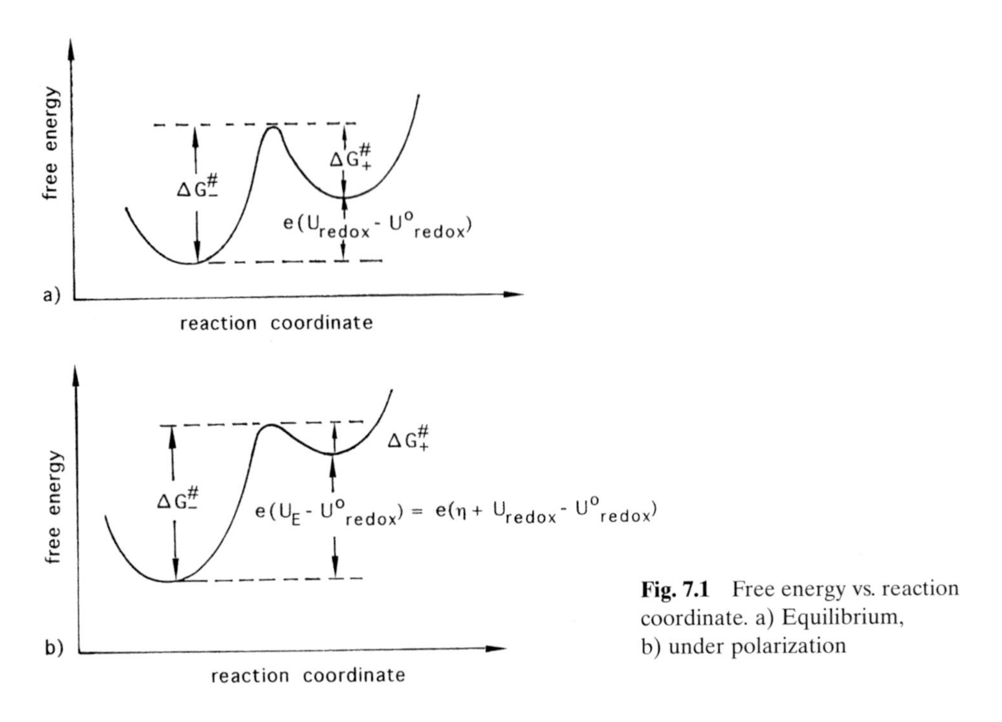

Fig. 7.1 Free energy vs. reaction coordinate. a) Equilibrium, b) under polarization

tioned that the rate constants defined here, are related to the local rate constant k_{et} (dimension s^{-1}) by Eq. (6.121). The local rate constant has been calculated by using various theories depending on weak or strong interactions (see Chapter 6).

As already mentioned in the introduction to this chapter, kinetic models describing electron transfer processes at metal electrodes, had been used for a long time before Marcus developed his theory. At a fairly early stage a transition state model was applied; the rate constants were described in terms of an activation energy so that we have

$$k^{+} = k' \exp\left(-\frac{\Delta G_{+}^{\#}}{kT}\right) \tag{7.3a}$$

$$k^{-} = k' \exp\left(-\frac{\Delta G_{-}^{\#}}{kT}\right) \tag{7.3b}$$

where the $\Delta G^{\#}$ terms are the free energies of activation. The pre-exponential factor will be discussed later.

Considering now the equilibrium case, the two partial currents given in Eq. (7.2) must be equal ($j^{+} = j^{-}$). Substituting Eq. (7.3) into (7.2), one obtains after rearranging the equation

$$\ln\frac{c_{ox}}{c_{red}} = \ln\frac{k_0^{+}}{k_0^{-}} = \frac{\Delta G_{-}^{\#} - G_{+}^{\#}}{kT} \tag{7.4}$$

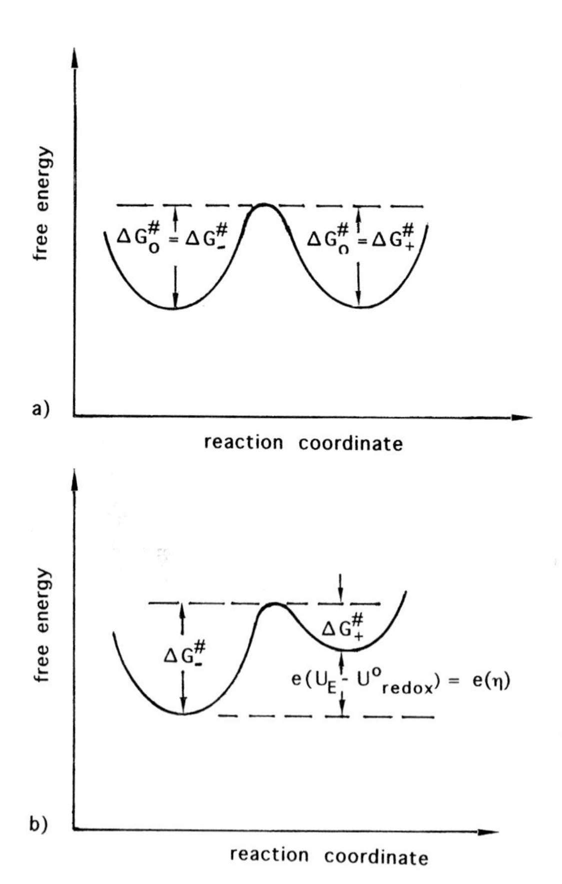

Fig. 7.2 Free energy vs. reaction coordinate under standard conditions. a) Equilibrium, b) under polarization

where k_0^+ and k_0^- are the rate constants at zero current. In the case of equilibrium the concentrations of the redox species are also determined by the Nernst equation as given by

$$\frac{e(U_{\text{redox}} - U^0_{\text{redox}})}{kT} = \ln \frac{c_{\text{ox}}}{c_{\text{red}}} \tag{7.5}$$

where U^0_{redox} is the standard redox potential ($c_{\text{ox}} = c_{\text{red}}$) and U_{redox} the actual redox potential for a given concentration ratio $c_{\text{ox}}/c_{\text{red}} \neq 1$. Combining Eqs. (7.4) and (7.5) one obtains

$$\Delta G^{\#}_{+} - \Delta G^{\#}_{-} = e(U_{\text{redox}} - U^0_{\text{redox}}) \tag{7.6}$$

The transition state is shown in Fig. 7.1 in terms of a free energy–reaction coordinate diagram. It is useful to also consider the same type of diagram under standard conditions ($c_{\text{ox}} = c_{\text{red}}$) as shown in Fig. 7.2. Here the free energy minima of the Ox and Red species occur at the same level because $\Delta G^{\#}_{+} - \Delta G^{\#}_{-}$ in Eq. (7.6) becomes zero ($\Delta G = 0$).

The rate constants defined by Eq. (7.3) and consequently the anodic and cathodic currents can be changed by varying the electrode potential via an external voltage to the cell. Considering again at first the standard case (see Fig. 7.2), the barrier height

$\Delta G_+^\#$ for an anodic reaction is decreased and $\Delta G_-^\#$ increased by applying a positive voltage with respect to the equilibrium condition (Fig. 7.2b). According to this model, the potential dependence of the interfacial current is caused by the potential dependence of the rate constants. As will be shown later, in this aspect metal electrodes behave completely different from semiconductor electrodes. It also becomes clear from Figs. 7.1 and 7.2 that a variation of the concentration ratio leads to the same effect as that caused by an application of a voltage to the cell. This is reasonable because an increase of c_{red} speeds up the reaction rate, i.e. the barrier height must be smaller as shown in Fig. 7.1a.

At first sight one might expect from Fig. 7.2 that an application of an overvoltage $\eta = U_E - U^0{}_{redox}$ leads to an identical decrease of the barrier height. The consequence would be that the other barrier height, $\Delta G_-^\#$, and consequently the cathodic current would remain constant. Since such a model was not acceptable, a linear relation between electrode potential or overvoltage and the activation energy was assumed as given by

$$\Delta G_-^\# = \Delta G_0^\# + \alpha e(U_E - U_{redox}^0) \tag{7.7a}$$

$$\Delta G_+^\# = \Delta G_0^{\#-} (1-\alpha)\, e\, (U_E - U_{redox}^0) \tag{7.7b}$$

in which $\Delta G_0^\#$ is the activation energy under standard conditions ($c_{ox} = c_{red}$). By introduction of the transfer factors α and $(1 - \alpha)$ it has been assured that both activation energies are varied upon application of an external voltage. In the case of $\alpha = 0.5$, the overvoltage varies both activation energies by the same amount. The α factor is further discussed below. In addition, it is useful to define a rate constant k_0 for the equilibrium case under standard conditions ($U_{redox} = U_{redox}^0$) which is given by

$$k_0 = k' \exp\left(-\frac{\Delta G_0^\#}{kT}\right) \tag{7.8}$$

There is only one k_0 value for reactions in both directions because the barrier heights are equal at equilibrium under standard conditions. It is also useful to derive the rate constants k_0^+ and k_0^- for any other redox potential at equilibrium ($U_E = U_{redox}$) which can be obtained from Eqs. (7.3a) and (7.3b) after inserting Eqs. (7.8), (7.7a) and (7.7b). One obtains

$$k_0^+ = k_0 \exp\left(\frac{(1-\alpha)e(U_{redox} - U_{redox}^0)}{kT}\right) \tag{7.9a}$$

$$k_0^+ = k_0 \exp\left(\frac{(1-\alpha)e(U_{redox} - U_{redox}^0)}{kT}\right) \tag{7.9a}$$

The rate constants under polarization are obtained from Eq. (7.3) by using Eqs. (7.7) and (7.9). We have then

$$k^+ = k_0^+ \exp\left(\frac{(1-\alpha)e(U_E - U_{redox})}{kT}\right) \tag{7.10a}$$

$$k^- = k_0^- \exp\left(-\frac{\alpha e(U_E - U_{redox})}{kT}\right) \tag{7.10b}$$

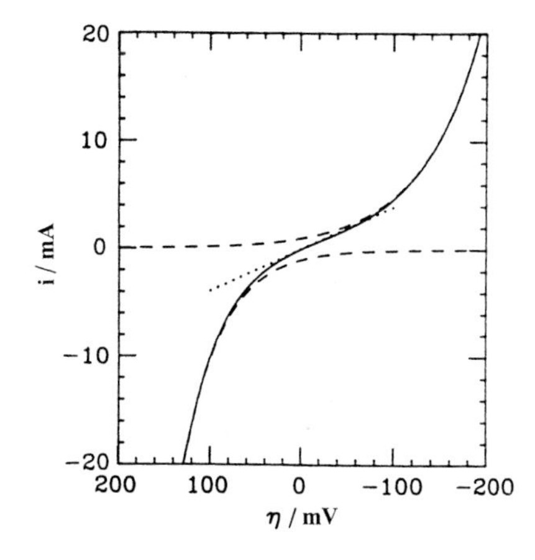

Fig. 7.3 Current–potential curve according to the Butler–Volmer equation for $\alpha = 0.5$. Dashed lines represent the partial currents. (After ref. [2])

The term $U_E - U_{redox} = \eta$ is the applied overvoltage. The current–potential dependence can be obtained from Eq. (7.2) after substitution of k_0^+ and k_0^- by Eqs. (7.10a) and (7.10b). One obtains

$$j = j_0 \left[\exp\left(\frac{(1-\alpha)e\eta}{kT} \right) - \exp\left(-\frac{\alpha e\eta}{kT} \right) \right] \tag{7.11}$$

In traditional electrochemistry this equation is known as the Butler–Volmer relation in which j_0 is the exchange current. The latter is equal to the anodic and the cathodic partial currents at equilibrium ($\eta = 0$) and are given by

$$\begin{aligned} j_0 &= k_0 c_{ox} \exp\left(-\frac{\alpha e(U_{redox} - U^0_{redox})}{kT} \right) \\ &= k_0 c_{red} \exp\left(\frac{(1-\alpha)e(U_{redox} - U^0_{redox})}{kT} \right) \end{aligned} \tag{7.12}$$

A theoretical current–potential curve (j/j_0 vs. η) is given in Fig. 7.3 for $\eta = 0.5$. It should be emphasized here that Eq. (7.11) is only valid in this simple form if the current is really kinetically controlled, i.e. if diffusion of the redox species toward the electrode surface is sufficiently fast. According to the Butler–Volmer equation (Eq. 7.11) the current increases exponentially with potential in both directions. In this aspect charge transfer processes at metal electrodes differ completely from those at semiconductors. When the overpotential is sufficiently large, $e\eta/kT \gg 1$, one of the exponential terms in Eq. (7.11) can be neglected compared to the other. In this case we have either

$$\ln j = \ln j^+ = \ln j_0 + \frac{(1-\alpha)e\eta}{kT} \tag{7.13a}$$

$$\ln j = \ln j^- = \ln j_0 - \frac{\alpha e\eta}{kT} \tag{7.13b}$$

This is the so-called Tafel equation (Tafel was the first person to find empirically a linear relation between log j and η. Meanwhile many redox reactions at metal electrodes

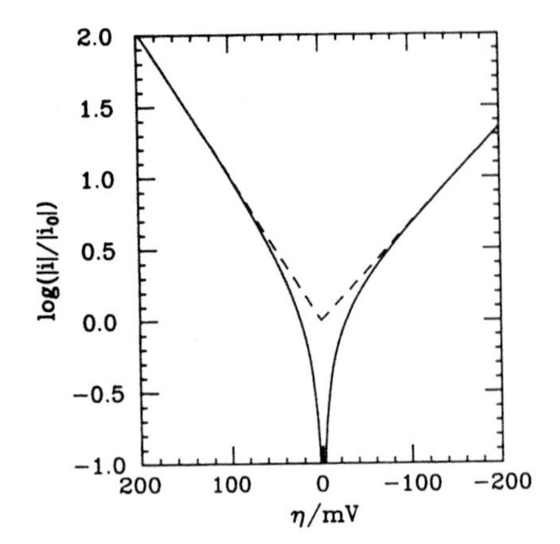

Fig. 7.4 Tafel plot of current–overvoltage. (After ref. [2])

have been investigated (see e.g. ref. [1]). One example is shown in Fig. 7.4. It should be mentioned that the relevant experiments are performed with electrolytes containing a sufficiently high concentration of a conducting salt so that the externally applied potential occurs only across the Helmholtz double layer. Usually α values between 0.4 and 0.6 have been found. An extrapolation of the linear portions of the log j–η curves to $\eta = 0$ yields the exchange current j_0 which is concentration-dependent. Usually j_0 values are given for standard conditions, i.e. for $c_{ox} = c_{red} = 1$ mole l^{-1}. For further details on the evaluation of more complex systems the reader is referred to refs. [1, 2].

As already mentioned above, the derivation of the Butler–Volmer equation, especially the introduction of the transfer factor α, is mostly based on an empirical approach. On the other hand, the model of a transition state (Figs. 7.1 and 7.2) looks similar to the free energy profile derived for adiabatic reactions, i.e. for processes where a strong interaction between electrode and redox species exists (compare with Section 6.3.3). However, it should also be possible to apply the basic Marcus theory (Section 6.1) or the quantum mechanical theory for weak interactions (see Section 6.3.2) to the derivation of a current–potential. According to these models the activation energy is given by (see Eq. 6.10)

$$\Delta G^{\#}_{+} = \frac{e(U_E - U_{redox}) - \lambda^2}{4kT\lambda} \tag{7.14a}$$

$$\Delta G^{\#}_{-} = \frac{e(U_E - U_{redox}) + \lambda^2}{4kT\lambda} \tag{7.14b}$$

Restricting the derivation to an anodic process, the current is given by

$$j^{+} = ek_0^{+} c_{red} \exp\left[-\frac{(e\eta - \lambda)^2}{4kT\lambda}\right] \tag{7.15}$$

When $\eta = 0$

$$j^{+} \rightarrow j_0 = ek_0^{+} c_{red} \exp\left[\frac{\lambda}{4kT}\right] \tag{7.16}$$

It is clear from Eq. (7.15) that the theoretical models derived in Chapter 6, do not yield a simple exponential j–η relation. Only in the case of large reorganization energies ($\lambda \gg e\eta$) can one neglect the η^2 term in the numerator of the exponential function and one obtains a Tafel type of relation:

$$j^+ = j_0 \exp\left[\frac{e\eta}{2kT}\right] \tag{7.17}$$

Interestingly, this approximation yields an exponential term with a transition factor of $\alpha = ½$ which would correspond to a slope for the Tafel equation of 120 mV per current decade. There are certainly many redox systems with which this slope has nearly been verified ($\alpha \approx 0.4$–0.5). On the other hand, one would expect a deviation from a linear log j–η plot for redox couples with a reorganization energy of $\lambda < 1$ eV. This is, however, difficult to prove because small λ values lead to large exchange currents (Eq. 7.16). Since then the electron transfer process is very fast the current becomes easily diffusion-controlled (see Section 7.1.2). The absence of clear evidence for a curved Tafel slope has raised criticism of the electron transfer theories as described in Chapter 6 (see ref. [3]). However, experimental proofs for the electron transfer theories were provided by measurements with electrodes coated with an insulating film which decreased the reaction rate by several orders of magnitude. In this case the current could be measured at high overpotentials without having problems with transport limitations [4, 5]. Miller et al. investigated various outer-sphere redox reactions at gold electrodes coated with hydroxythiol layers (thickness 20 Å). In all cases they found the expected curvature. Further details are given in Section 7.36.

The redox process at metal electrodes described above, should also be briefly discussed in terms of the Gerischer model (see Section 6.2). Assuming equal concentrations for the reduced and oxidized species of the redox system then the energetics of the metal liquid interface are given in Fig. 7.5 for equilibrium, cathodic and anodic polarization. The anodic and cathodic currents are then given by (see Eq. 6.42):

$$j^+ = e\bar{k}c_{\text{red}} \int_{E_F}^{\infty} (1 - f(E))\, \rho(E) \exp\left[-\frac{(E - E_{F,\text{redox}} - \lambda)^2}{4kT\lambda}\right] \mathrm{d}E \tag{7.18a}$$

$$j^- = e\bar{k}c_{\text{ox}} \int_{-\infty}^{E_F} f(E)\rho(E) \exp\left[-\frac{(E - E_{F,\text{redox}} + \lambda)^2}{4kT\lambda}\right] \mathrm{d}E \tag{7.18b}$$

Here $\varrho(E)$ is the distribution of energy states in the metal whereas $f(E)$ is the Fermi distribution function as given by Eq. (1.25), i.e. $f(E)\varrho(E)$ is the number of occupied and $(1–f)\varrho(E)$ the number of empty states in the metal. The exponential terms correspond to the distribution functions of the empty and occupied states of the redox system as illustrated in Fig. 7.5. All terms describing the interaction between electrode and redox system and other factors such as a normalization are summarized in the pre-exponential factor k^- which will not be discussed here.

Since the externally applied voltage occurs only across the Helmholtz layer at the metal electrolyte interface, the energy levels on both sides of the interface are shifted against each other as illustrated in Fig. 7.5. Upon cathodic polarization, an electron transfer occurs from the occupied states in the metal where the latter overlap with the

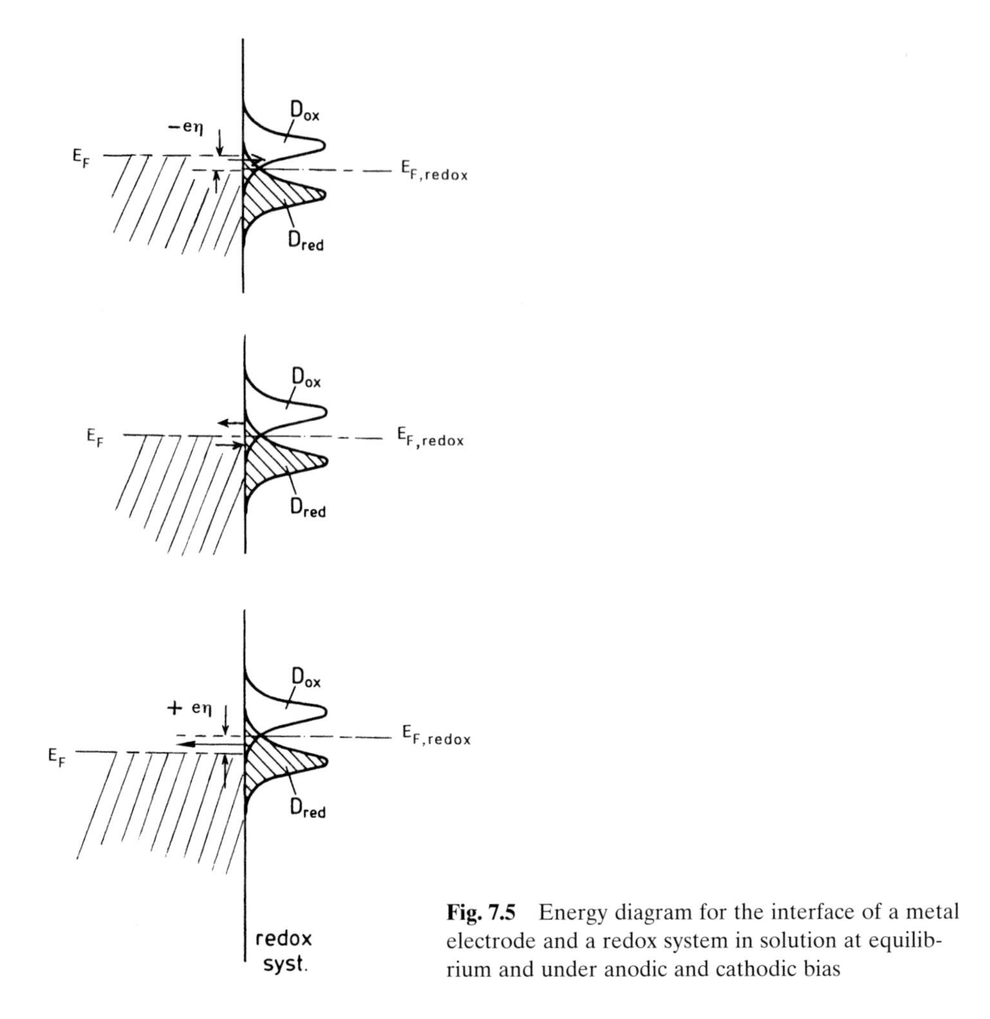

Fig. 7.5 Energy diagram for the interface of a metal electrode and a redox system in solution at equilibrium and under anodic and cathodic bias

empty states, D_{ox}, of the redox system (Fig. 7.5b). The reverse process occurs from the occupied states, D_{red}, of the redox system into the empty states of the metal. Accordingly, one has to integrate Eq. (7.18) over a relatively large energy range. An analytical solution of this equation is, however, not possible. Since the distribution functions vary exponentially with E^2 and since in most cases the overlapping between corresponding states on both sides of the interface is limited to a rather small energy range the electron transfer occurs mainly within 1 kT around the Fermi level. Hence, it is a reasonable approximation to replace the integrals by inserting $dE = 1$ kT and $E = E_F$. One then obtains equations which are identical to Eqs. (7.15a) and (7.15b). This result is not surprising because both models (Marcus and Gerischer) contain the same exponential function which is based on the assumption that the solvent molecules fluctuate like an harmonic oscillator. For details see Chapter 6.

Finally it should be mentioned that a kinetically controlled process is called an *irreversible reaction.*

7.1.2 Diffusion-controlled Processes

If the kinetics of electron transfer at an electrode are very fast then the interfacial current becomes controlled or even limited by the diffusion of the redox species toward the electrode. In this case the concentration of the reacting species at the electrode surface is decreased with respect to its bulk concentration and will be zero for large polarizations. Then the current reaches a limiting value which can only be increased by disturbing the solution or by rotating the electrode. Quantitative information can be obtained by solving the diffusion equation. Since we have two diffusing species, one diffusing toward the electrode and the other, namely the product, away from the surface, in fact two diffusion equations have to be solved. They are given by

$$\frac{\partial c_{\mathrm{ox}}}{\partial t} = D_{\mathrm{ox}} \frac{\partial^2 c_{\mathrm{ox}}}{\partial z^2} \tag{7.19a}$$

$$\frac{\partial c_{\mathrm{red}}}{\partial t} = D_{\mathrm{red}} \frac{\partial^2 c_{\mathrm{red}}}{\partial z^2} \tag{7.19b}$$

in which D_{ox} and D_{red} are the diffusion coefficients of the Ox and Red form, respectively. In the above equations it is assumed that the solution is not disturbed and that the diffusion of molecules occurs only perpendicular to a planar electrode surface along the z-axis. It is further assumed that the initial concentration of the Red species is constant and equal to the bulk concentration c_{red}^0 whereas the concentration of the Ox species is zero. Then the two equations can be solved by using the method of Laplace transformation. Omitting here the detailed derivation, the solutions of Eqs. (7.19a) and (7.19b) are

$$c_{\mathrm{red}}(c, t) = c_{\mathrm{red}}^0 \frac{\left(\dfrac{D_{\mathrm{red}}}{D_{\mathrm{ox}}}\right)^{1/2} \dfrac{c_{\mathrm{red}}^{\mathrm{s}}}{c_{\mathrm{ox}}^{\mathrm{s}}} + \mathrm{erf}(z/2D_{\mathrm{red}}^{1/2} t^{1/2})}{1 + \left(\dfrac{D_{\mathrm{red}}}{D_{\mathrm{ox}}}\right)^{1/2} \dfrac{c_{\mathrm{red}}^{\mathrm{s}}}{c_{\mathrm{ox}}^{\mathrm{s}}}} \tag{7.20a}$$

$$c_{\mathrm{ox}}(z, t) = c_{\mathrm{ox}}^0 \frac{\dfrac{D_{\mathrm{red}}}{D_{\mathrm{ox}}} \left[1 - \mathrm{erf}(z/2D_{\mathrm{ox}}^{1/2} t^{1/2})\right]}{1 + \left(\dfrac{D_{\mathrm{red}}}{D_{\mathrm{ox}}}\right)^{1/2} \dfrac{c_{\mathrm{red}}^{\mathrm{s}}}{c_{\mathrm{ox}}^{\mathrm{s}}}} \tag{7.20b}$$

in which erf() is the error function whereas $c_{\mathrm{red}}^{\mathrm{s}}$ and $c_{\mathrm{ox}}^{\mathrm{s}}$ are the concentrations of the Red and the Ox species, respectively. For details see for instance refs. [2, 6].

We are mainly interested in the surface concentrations at $z = 0$. Since $\mathrm{erf}(0) = 0$ the surface concentrations $c_{\mathrm{red}}^{\mathrm{s}}$ and $c_{\mathrm{ox}}^{\mathrm{s}}$ can be obtained from Eq. (7.20):

$$c_{\mathrm{red}}^{\mathrm{s}} = c_{\mathrm{red}}^0 \frac{\left(\dfrac{D_{\mathrm{red}}}{D_{\mathrm{ox}}}\right)^{\frac{1}{2}} \dfrac{c_{\mathrm{red}}^{\mathrm{s}}}{c_{\mathrm{ox}}^{\mathrm{s}}}}{1 + \left(\dfrac{D_{\mathrm{red}}}{D_{\mathrm{ox}}}\right)^{\frac{1}{2}} \dfrac{c_{\mathrm{red}}^{\mathrm{s}}}{c_{\mathrm{ox}}^{\mathrm{s}}}} \tag{7.21a}$$

$$c_{\mathrm{ox}}^{\mathrm{s}} = c_{\mathrm{ox}}^0 \frac{\left(\dfrac{D_{\mathrm{red}}}{D_{\mathrm{ox}}}\right)^{\frac{1}{2}} \dfrac{c_{\mathrm{red}}^{\mathrm{s}}}{c_{\mathrm{ox}}^{\mathrm{s}}}}{1 + \left(\dfrac{D_{\mathrm{red}}}{D_{\mathrm{ox}}}\right)^{\frac{1}{2}} \dfrac{c_{\mathrm{red}}^{\mathrm{s}}}{c_{\mathrm{ox}}^{\mathrm{s}}}} \tag{7.21b}$$

These equations look rather complex. Since the electron transfer itself was assumed to be very fast so that the Ox and Red species are always in equilibrium at the electrode surface, the ratio of the surface concentrations is given by the Nernst equation. Accordingly we have

$$\frac{c_{\mathrm{ox}}^{\mathrm{s}}}{c_{\mathrm{red}}^{\mathrm{s}}} = \exp\left[\frac{e(U_{\mathrm{E}} - U_{\mathrm{redox}}^{0})}{kT}\right] \tag{7.22}$$

Therefore, Eq. (7.21) actually relates the surface concentrations to the electrode potential. It is also useful to define the thickness of a depleted layer, L_{diff}, as the distance over which a linear concentration gradient would produce the same flux at the electrode surface as calculated from an exact solution to the diffusion equation. The flux is given by Fick's first law:

$$\text{flux} = D_{\mathrm{red}} \frac{\partial c_{\mathrm{red}}(x, t)}{\partial z} \tag{7.23}$$

where c_{red} is given by Eq. (7.20a). Differentiating Eq. (7.20a) one obtains

$$\text{flux} = c_{\mathrm{red}} \frac{(D_{\mathrm{red}}/\pi t)^{1/2}}{1 + \left(\dfrac{D_{\mathrm{red}}}{D_{\mathrm{ox}}}\right)^{1/2} \dfrac{c_{\mathrm{red}}^{\mathrm{s}}}{c_{\mathrm{ox}}^{\mathrm{s}}}} \tag{7.24}$$

(The differentiation of an error function yields an exponential function.) The flux through a linear concentration gradient can also be expressed by

$$\text{flux} = D_{\mathrm{red}} \frac{c_{\mathrm{red}}^{0} - c_{\mathrm{red}}^{\mathrm{s}}}{L_{\mathrm{diff}}} \tag{7.25}$$

Substituting $c_{\mathrm{red}}^{\mathrm{s}}$ from Eq. (7.21a), we have

$$\text{flux} = c_{\mathrm{red}}^{0} \frac{D_{\mathrm{red}}/L_{\mathrm{diff}}}{1 + \left(\dfrac{D_{\mathrm{red}}}{D_{\mathrm{ox}}}\right)^{1/2} \dfrac{c_{\mathrm{red}}^{\mathrm{s}}}{c_{\mathrm{ox}}^{\mathrm{s}}}} \tag{7.26}$$

Comparing this flux with that obtained from the exact solution (Eq. (7.24) for $z = 0$, one obtains

$$L_{\mathrm{diff}} = (\pi D_{\mathrm{red}} t)^{1/2} \tag{7.27}$$

One may also interpret L_{diff} as the diffusion length of the redox molecules. In principle this relation is identical to the diffusion length of minority carriers in semiconductor crystals ($L = (D\tau)^{1/2}$, see Eq. 2.26). The only difference is that the lifetime τ of the minority carriers in a semiconductor is a material constant, whereas L_{diff} depends on the time. In Fig. 7.6 the concentration profile in terms of L_{diff} is illustrated.

As already mentioned before, the diffusion of the redox species can be enhanced by disturbing the solution. The most well-defined mass transport is obtained by using a rotating disc electrode as described in Section 4.2.3). As derived at first by Levich, the

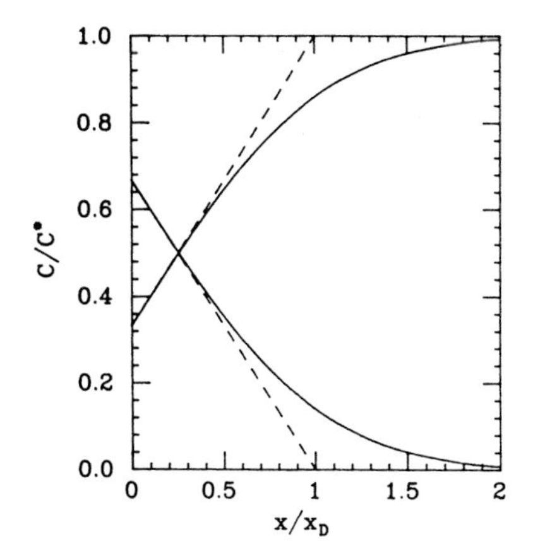

Fig. 7.6 Concentration profile for diffusion to a planar electrode (x_D corresponds to L_{diff} in the text). (After ref. [2])

thickness of the diffusion layer increases with the rotation speed ω of the electrode disc. The quantitative relation is given by

$$L_{\text{diff}} = 1.61 D^{1/3} \omega^{-1/2} \nu^{1/6} \tag{7.28}$$

in which ν is the kinematic viscosity of the solution. We have omitted here the complete derivation, and the reader is referred to the relevant literature [7–9]. Concerning the dimensions of ω, it must be mentioned that the latter is an angular speed in radians per second (2π times the rotation speed in hertz) in Eq. (7.28). Since ω is usually given in rotations per minute (rpm), one should bear in mind that 1 radiant corresponds to about 10 rpm. Taking typical values of D and ν ($D = 10^{-5}$ cm^2 s^{-1} and $\nu = 10^{-2}$cm^2 s^{-1}) and a rotation speed of 500 rpm, then one obtains $L_{diff} \approx 10^{-4}$ cm.

More interesting is the current due to the flux of molecules toward the surface. Considering again only the diffusion of the Red species toward the electrode the flux is given by Eq. (7.24). The net current at $z = 0$ is then given by

$$j^+_{\text{diff}}(0, t) = ec^0_{\text{red}} \frac{(D_{\text{red}}/\pi t)^{1/2}}{1 + \left(\dfrac{D_{\text{red}}}{D_{\text{ox}}}\right)^{1/2} \dfrac{c^s_{\text{red}}}{c^s_{\text{ox}}}} \tag{7.29}$$

This equation is only valid for $t > 0$. When the electrode potential is large and positive, i.e. $U_E \gg U^0_{redox}$, then c^s_{red} becomes very small according to Eq. (7.22). In this case, each Red molecule that arrives at the surface is reduced. Any further potential increase does not make the current any larger because it is limited by the rate of diffusion of Red to the electrode. Setting $c^s_{red} = 0$ one obtains from Eq. (7.25) as a limiting current

$$j^+_{\text{lim}} = ec^0_{\text{red}} (D_{\text{red}}/\pi t)^{1/2} \tag{7.30a}$$

$$j^+_{\text{lim}} = ec^0_{\text{red}} D_{\text{red}}/L_{\text{diff}} \tag{7.30b}$$

and substituting πt by using Eq. (7.27) yields

This limiting current increases linearly with the bulk concentration and decreases with the square root of time. Accordingly, this current should decrease by a factor 10 within a time interval from 1 to 100 s. In most experiments such a variation in current is not observed because of some convection of the liquid.

The potential dependence of the diffusion current is clearly seen if the time in Eq. (7.29) is eliminated by dividing j^{+}_{diff} by j^{+}_{lim}, the latter being given by Eq. (7.30), and by substituting $c^{s}_{\text{red}}/c^{0}_{\text{ox}}$ by using Eq. (7.22). We have then

$$U_{\text{E}} = U^{0}_{\text{redox}} + \frac{kT}{e}\ln\left(\frac{D_{\text{ox}}}{D_{\text{red}}}\right) + \frac{kT}{e}\ln\left(\frac{j^{+}_{\text{lim}}}{j^{+}_{\text{diff}}} - 1\right) \tag{7.31}$$

It should be remembered that this equation is only valid for experiments where j^{+}_{diff} and j^{+}_{lim} are measured at the same time. It is further useful to introduce a so-called half-wave potential $U_{1/2}$ where $j^{+}_{\text{diff}} = \frac{1}{2}j^{+}_{\text{lim}}$. As can be easily derived from Eq. (7.31), it is given by

$$U_{1/2} = U^{0}_{\text{redox}} + \frac{kT}{e}\ln\left(\frac{D_{\text{ox}}}{D_{\text{red}}}\right) \tag{7.32}$$

Eq. (7.31) can then be written as

$$U_{\text{E}} = U_{1/2} + \frac{kT}{e}\ln\left(\frac{j^{+}_{\text{lim}}}{j^{+}_{\text{diff}}} - 1\right) \tag{7.33}$$

This equation was derived at first by Heyrovsky and Ilkovic [10] and is usually called the Heyrovsky–Ilkovic equation. In most cases the diffusion coefficients of the Ox and Red species are not very different, so that $U_{1/2}$ is essentially the standard redox potential. A theoretical current–potential curve in terms of $j^{+}_{\text{diff}}/j^{+}_{\text{lim}}$ vs. $U_{\text{E}} - U_{1/2}$ is shown in Fig. 7.7. A semilogarithmic plot would yield a straight line (see Section 7.1.3). It should be mentioned here that Eq. (7.32) can also applied to majority carrier processes at semiconductor electrodes (see e.g. Section 7.3.4).

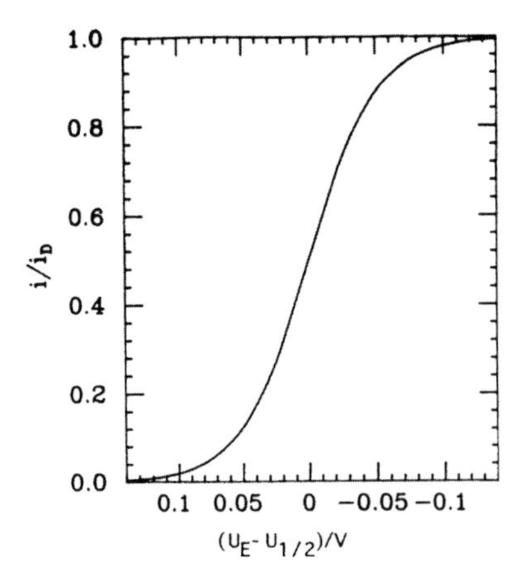

Fig. 7.7 Current–potential curve for a reversible reaction and diffusion limitation

The limiting diffusion current, of course, can also be increased by rotating the electrode. Using now Eq. (7.28) instead of (7.27) in the derivation of j^{+}_{lim} then one obtains

$$j^{+}_{\text{lim}} = 0.62ec^{0}_{\text{red}}(D_{\text{red}})^{2/3}\,\nu^{-1/6}\,\omega^{1/2} \tag{7.34}$$

In this case j^{+}_{lim} is constant; it is independent of time.

Finally it should be mentioned that a diffusion-controlled process is called a *reversible reaction*.

7.1.3 Investigations of Redox Reactions by Linear Sweep Voltametry

Current–potential curves are usually measured with metal and semiconductor electrodes by scanning the electrode potential over a certain potential range. Accordingly, the potential scale corresponds also to a tim scale. The potential scan leads typically to a current peak and at higher potentials the current levels off into the diffusion-limited current j^{+} as shown in Fig. 7.8. The peak occurs because during the first time interval sufficient redox molecules are available. The same type of peak occurs when scanning back in the reverse direction (Fig. 7.8). This kind of behavior is expected fore diffusion-controlled as well as for kinetically controlled reactions at metal electrodes, and also at semiconductor electrodes as long as majority carriers are involved in the charge transfer process (see also Section 7.3.4).

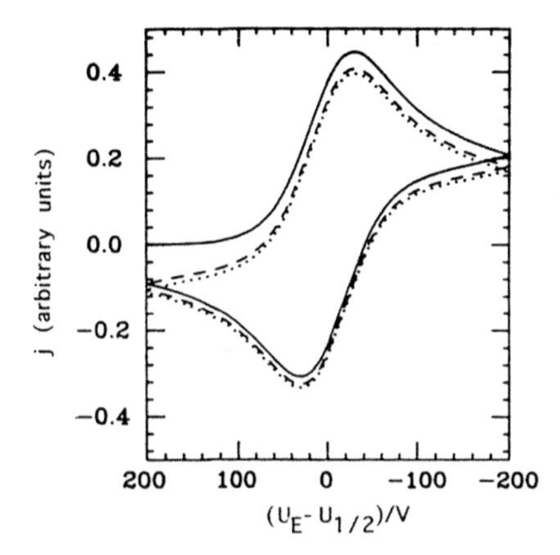

Fig. 7.8 Cyclic voltammogram for various potential scans

In the case of a diffusion-controlled reaction a current–potential curve can be evaluated quantitatively. The diffusion equation has to be solved again by using time-dependent boundary conditions. The mathematics, however, are very complicated and cannot be shown here. They end up with an integral equation which has to be solved numerically [11]. The peak current, j_{p}, for a diffusion-controlled process (reversible reaction) is found to be

$$j_{\text{p}} = 0.446e(ve/kT)^{1/2}\,D^{1/2}_{\text{red}}c^{0}_{\text{red}} \tag{7.35}$$

where v is the potential scan rate in V s^{-1}. The current peak is a little displaced from $U_{1/2}$ and we have

$$U_p = U_{1/2} \pm 1,11kT/e \tag{7.36}$$

in which the positive sign is valid for the anodic and the minus sign for the cathodic peak. According to this equation the shift amounts to 28.5 mV and the separation of the two peaks to 57 mV in a cyclic scan. The easiest way of getting $U_{1/2}$ and U^0_{redox} is given by $U_{1/2} = 1/2\ (U_{p,a} + U_{p,c})$. It should be emphasized that the peak current increases with the square root of the scan rate; its position on the potential scale, however, is independent of the scan rate, provided that the electron transfer is diffusion-controlled.

In the case of a kinetically controlled reaction (irreversible process), the situation is different. Here current peaks also occur because the current finally becomes diffusion-limited at large polarization. However, the position of the current peak is shifted to higher potentials when the scan rate is increased.

7.1.4 Criteria for Reversible and Irreversible Reactions

If an electrode reaction is not entirely controlled by the kinetics but also by diffusion we can express the overvoltage η by two terms, namely

$$\eta = \eta_{kin} + \eta_{diff} \tag{7.37}$$

with

$$\eta_{kin} = 1/e(E_F - E^s_{F,redox}) = (U_E - U^s_{redox}) \tag{7.38}$$

$$\eta_{diff} = 1/e(E^s_{F,redox} - E_{F,redox}) = (U^s_{redox} - U_{redox}) \tag{7.39}$$

These energies and potentials are illustrated in an energy diagram as given in Fig. 7.9. In the non-equilibrium case the anodic current can be derived from Eqs. (7.2c), (7.10) and (7.25) and one obtains

$$\frac{1}{j} = \frac{1}{j_{kin}} + \frac{1}{j_{diff}} \tag{7.40}$$

in which j_{kin} is the kinetically controlled current as given for an oxidation by

$$j_{kin} = ek^+c^s_{red} \tag{7.41}$$

which differs from Eq. (7.2a) only insofar as we have here the surface concentration of the reduced species.

In the case of slow reaction kinetics, quite high overvoltages are required to obtain a measurable current. Then j_{kin} is determined by Eq. (7.11) and j_{diff} by j^+_{lim} (Eq. (7.30b). We have then

$$\ln\left(\frac{j^+_{lim}}{j} - 1\right) = \frac{e(1-\alpha)}{kT}(U_{1/2} - U_E) \tag{7.42}$$

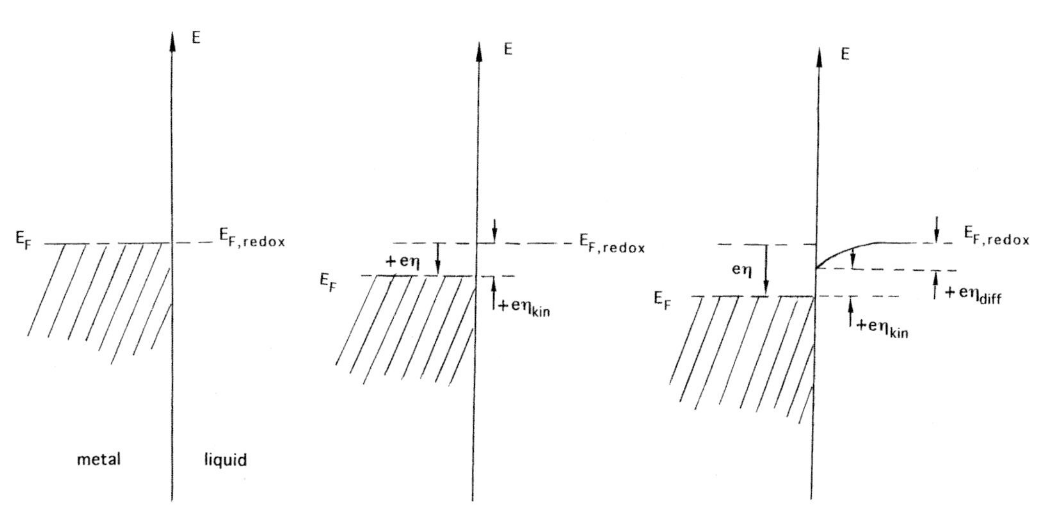

Fig. 7.9 Representation of overvoltages, $e\eta_{kin}$ and $e\eta_{diff}$, in an energy diagram

with

$$U_{1/2} = U^0_{redox} + \frac{kT}{e} \ln\left(\frac{D_{red}}{L_{diff} k_0}\right) \tag{7.43}$$

This equation looks similar to Eq. (7.33) which was derived for the diffusion-controlled case. Eq. (7.42) differs from Eq. (7.33), however, insofar as here the half-wave potential depends on L_{diff} and therefore on the rotation speed of the electrode. On the other hand, if the current is only diffusion-controlled, Eq. (7.33) determines the current–potential curve. In this case $U_{1/2}$ is independent of L_{diff} and therefore also independent of the rotation speed (Eq. 7.32).

According to these differences with respect to $U_{1/2}$, an investigation of the rotation dependence yields the best proof either for a kinetically or a diffusion-controlled reaction. This is also true for majority carrier processes at a semiconductor electrode. In the case of a metal electrode, one may be tempted to distinguish between kinetically and diffusion-controlled processes via the slope of $\ln[(j^+_{lim}/j) - 1]$ vs. U_E because the factor $(1 - \alpha)$ occurs in the equation for the kinetically controlled current (Eq. 7.42) and not in the other (Eq. 7.33). This method can lead to misinterpretations, however [2]. In the case of semiconductors, the latter method would even be useless because then $\alpha = 0$ (see Section 7.3.4).

7.2 Qualitative Description of Current–Potential Curves at Semiconductor Electrodes

In principle any electron transfer at a semiconductor–liquid interface can only occur via the conduction or valence band. Whether then a corresponding current is possible depends on various factors, such as the position of the energy bands and the occupation of the energy states in the bands by electrons. This basic behavior is already obvious from current–potential curves as measured with semiconductor electrodes in aqueous electrolytes without any redox system. Typical examples are the current–potential curves as obtained with n- and p-type GaAs electrodes in H_2SO_4 (Fig. 7.10). Here the cathodic process corresponds to the reduction of protons, i.e. H_2 formation, whereas in the anodic range, the electrode is dissolved. These are processes which also occur at less noble metal electrodes. Oxygen formation takes place only at stable oxide electrodes as will be discussed separately in Section 11.1.2

It is interesting to see that the current–potential curves for n- and p-type electrodes look very different. For instance, the cathodic current due to the formation of H_2, rises steeply with increasing cathodic potential at the n-type electrode whereas a very small current occurs at the p electrode (Fig. 7.10). This result is a clear indication that the electrons required for the reduction of protons are transferred from the conduction band to the protons. This conclusion is supported by the result that the cathodic current at the p-type electrode is enhanced by light excitation. In the latter case electrons are excited from the valence band into the conduction band from where the electrons are transferred to the protons.

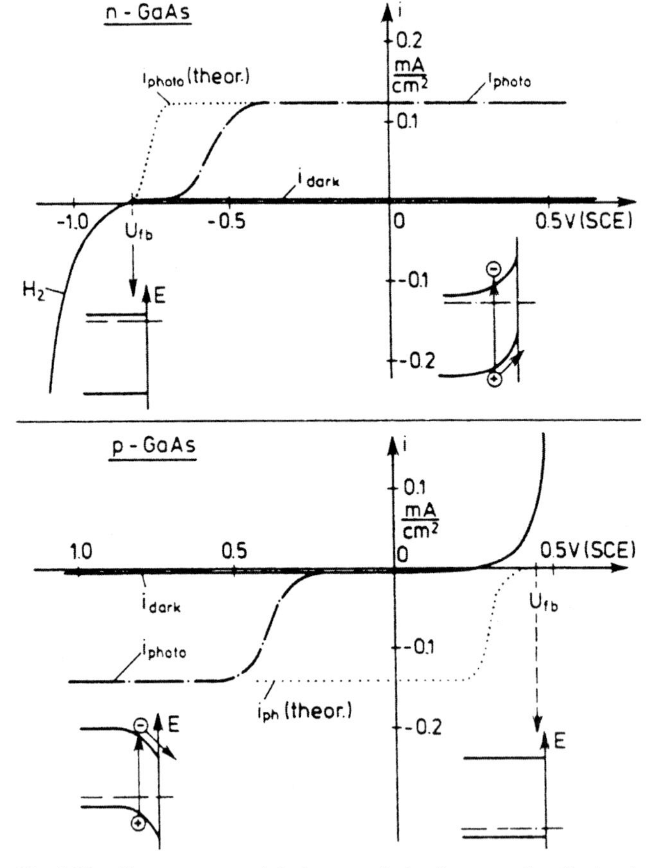

Fig. 7.10 Current–potential characteristics for n- and p-GaAs electrodes in 0.1 M H_2SO_4

The same type of arguments prove that the anodic decomposition reaction occurs via the valence band. Here we see that the corresponding anodic current at p-GaAs increases steeply with increasing anodic polarization whereas a very small anodic current is found with n-type electrodes. The latter could be increased by light excitation. Accordingly, holes from the valence band are required for the anodic decomposition of the semiconductor. It should be emphasized here that not holes but electrons are actually transferred across the interface, but an injection of electrons into the valence band is only possible if holes are present at the semiconductor surface.

These are typical phenomena which are found in principle with all semiconductor electrodes. These rules are also valid for redox processes. However, redox reactions may occur either via the conduction or the valence band whereas the anodic decomposition occurs always via the valence band and the H_2 formation always via the conduction band. Frequently, investigations of redox processes are limited in aqueous solutions because of interference with reactions of H_2O at the electrode. This can only be avoided by using non-aqueous electrolytes. More details concerning charge transfer reactions between a semiconductor electrode and a redox system will be given in the following section.

Many of the basic processes have also been treated in review articles and some books [12–20].

7.3 One-step Redox Reactions

7.3.1 The Energetics of Charge Transfer Processes

As already mentioned in the previous section, any electron transfer across the semiconductor–liquid interface occurs via the energy bands. There may also be an electron transfer via surface states at the interface; the electrons or holes, however, must finally be transported via one of the energy bands. This is possible by capturing an electron from the conduction band or a hole from the valence band in the surface states. In the present section the basic rules for the charge transfer will be given. in particular, physical factors which determine whether an electron transfer occurs via the conduction or the valence band, will be derived. For illustration, the Gerischer model will be used here because it best shows the energetic conditions.

As described in Section 5.3, impedance measurements with various semiconductor electrodes in aqueous solutions have shown that the positions of energy bands at the surface are usually pinned. This observation was made with electrodes of different dopings, i.e. n- and p-type electrodes of the same material exhibit the same band positions at the surface (see also Fig. 5.11). Accordingly, the band positions can be given on the usual elctrochemical scale, i.e. vs. a normal hydrogen electrode (NHE) or saturated calomel electrode (SCE) as a reference electrode. The energy bands also usually remain pinned after the addition of a redox system to an aqueous electrolyte. Accordingly, the interaction of the semiconductor with water is stronger than with the redox system. The situation is different in non-aqueous solutions. Here considerable shifts of the band positions at the surface have been observed upon addition of a redox system [21]. Considering at first only systems with aqueous solutions, then one can easily find

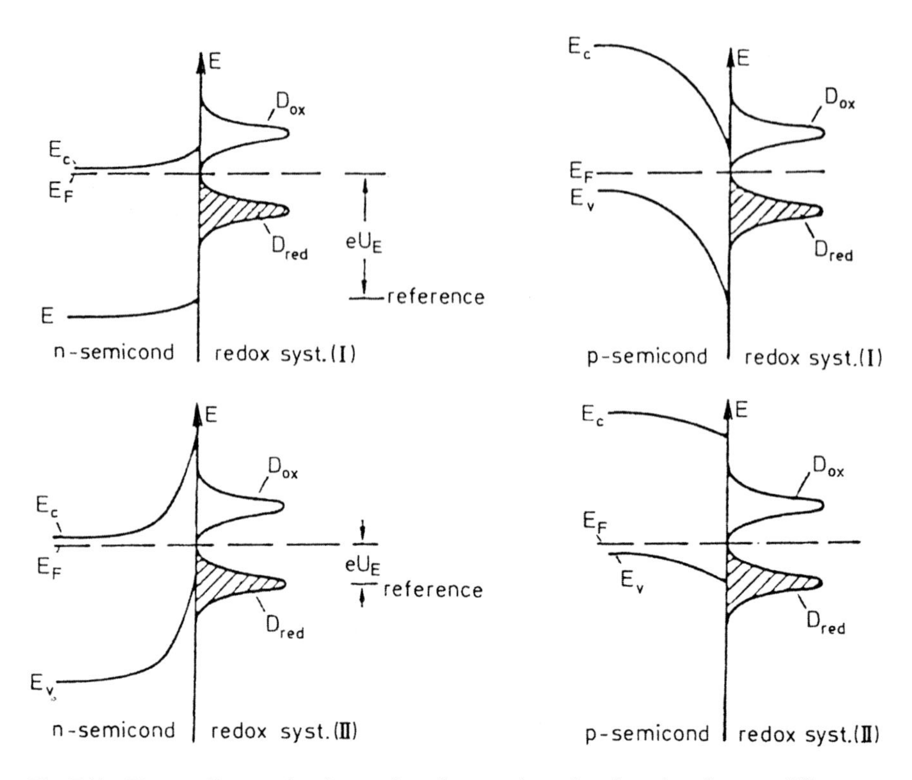

Fig. 7.11 Energy diagram for the semiconductor–electrolyte interface for two different standard redox potentials (syst. I and II); left side: n-type semiconductor; right side: p-type semiconductor

out the relative positions of the energy levels of the semiconductor and a redox system from Fig. 5.20, in which the standard potentials of some redox systems are given besides the positions of bands of semiconductors. Choosing, for instance, a redox couple with a rather negative standard potential, the Fermi level of a redox system will occur fairly close to the conduction band of a semiconductor, as illustrated in Fig. 7.11a, on the left side for an n-type and on the right side for a p-type semiconductor electrode at equilibrium. On the other hand, choosing a redox couple of very positive standard potential, then the energy levels of the redox system occur relatively close to the valence band (Fig. 711b). In the first case where the energy levels of the redox system occurs close to the conduction band, an electron transfer via the conduction band is expected. In the other case an electron transfer via the valence band should occur because of the same kind of reasoning. This implies that any charge transfer occurs horizontally, i.e. that the electron does not loose any energy during the transfer process. The latter condition is required in all theoretical models, It means that the electron transfer is much faster than any reorganization of the solvation shell or of the solvent dipoles (Frank–Condon principle).

According to measurements of the space charge capacity dependent on the electrode potential (Mott–Schottky measurements, see Section 5.3), any variation of the electrode potential leads usually only to a corresponding change of the potential, $\Delta\phi_{sc}$,

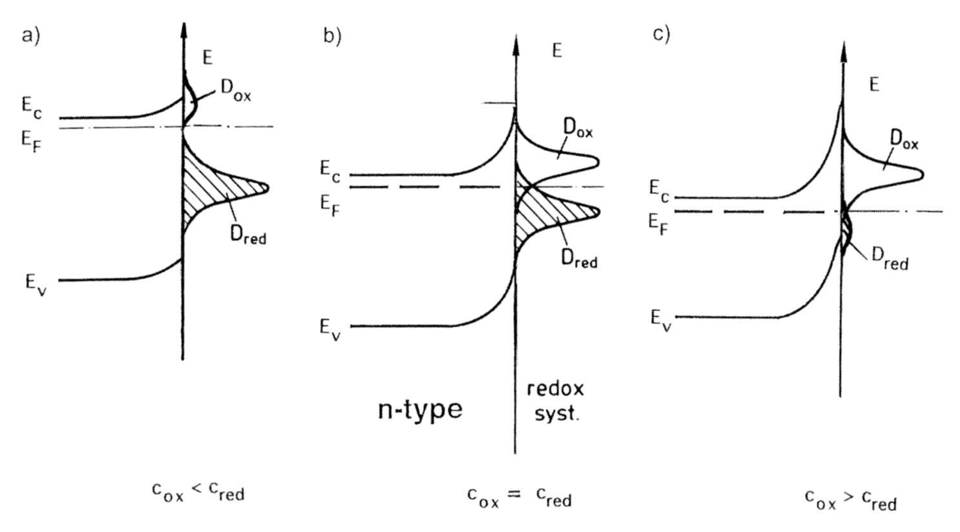

Fig. 7.12 Energy diagram for the semiconductor–electrolyte interface at equilibrium for different concentrations

across the space charge layer and therefore of the bending of the band (see Fig. 5.15). Hence, the potential across the Helmholtz double layer, $\Delta\phi_H$, remains constant. As already discussed in Chapter 5, this result is caused by the strong interaction between electrode and water. In consequence of this result, the position of the energy levels of the redox system remains unchanged with respect to the energy bands of the semiconductor. This is a very important consequence insofar as the analysis of the charge transfer kinetics is then much simpler. Fortunately, the latter conclusion is also valid for systems with non-aqueous electrolytes. Although here the interaction between electrode and redox system mainly determines the relative position of energy states on both sides of the interface, any potential variation still occurs across the space charge region and the relative position of energy states remains constant.

A variation of the concentration of one of the redox species leads to a change of the redox potential and of the Fermi energy $E_{F,redox}$ as given by the Nernst equation. Reducing, for instance, the concentration of the reduced species (D_{red} is decreased), then the redox potential becomes more positive with respect to the standard potential and $E_{F,redox}$ moves downward as shown in Fig. 7.12c. The opposite effect occurs if D_{ox} is lowered (Fig. 7.12a). It is important to note that in all cases the relative positions of energy states remain unchanged. In all other models it means that the activation energy is independent of the relative concentrations of the redox system and also independent of the externally applied voltage.

7.3.2 Quantitative Derivation of Current–Potential Curves

Since the charge transfer across a semiconductor–electrolyte interface can only occur via the conduction or valence band, the processes via the two bands have to be treated separately. In general anodic and cathodic currents are given by equations similar to

Eqs. (7.18a) and (7.18b), but now using boundary conditions which are specific for semiconductors. Since the integral in Eq. (7.18) cannot be solved analytically, we assumed in the case of metal electrodes that the electron transfer occurs mainly around the Fermi level. As proved in Section 7.1, this is a satisfactory approximation. Using an equivalent approach for charge transfer processes at semiconductor electrodes, the anodic current corresponding to an electron transfer from the occupied states of the redox system to empty states of the conduction band, is given by [22]

$$j_c^+ = ek_0\,(1 - f(E_c))\,\rho(E = E_c)c_{red}\,\exp\left[-\frac{(E_c^s - E_{F,redox}^0 - \lambda)^2}{4kT\lambda}\right] \tag{7.44}$$

in which the product of c_{red} and of the exponential function corresponds to the density of occupied states of the redox system at the energy of the lower edge of the conduction band at the surface ($E = E_c^s$). The density of energy states in the semiconductor at the lower edge of the conduction band (i.e. the number of states within an energy interval of 1 kT) is $\varrho(E = E_c) = N_c$ (see Chapter 1). The occupation factor $f(E)$ given by the Fermi distribution function, is very low even if the Fermi level is close to the conduction band. For instance, assuming the Fermi level to be located 4 kT (= 0.1 eV) below the conduction band, $f \approx 1\,\%$ at $E = E_c$. Accordingly, most of the energy states in the conduction band are empty ($(1 - f) \approx 1$), i.e. $(1 - f)\varrho = N_c$. We have then

$$j_c^+ = ek_0N_cc_{red}\,\exp\left[-\frac{(E_c^s - E_{F,redox}^0 - \lambda)^2}{4kT\lambda}\right] \tag{7.45}$$

Since this equation contains only constant parameters for a given system, the anodic current j_c^+ is independent of the electrode potential. Its absolute value depends essentially on the energy terms in the exponent.

The reverse current, j_c^-, which corresponds to an electron transfer from the conduction band to the empty states of the redox system, is given by

$$j_c^- = ek_0f(E_c)\rho(E = E_c)c_{ox}\,\exp\left[-\frac{(E_c^s - E_{F,redox}^0 + \lambda)^2}{4kT\lambda}\right] \tag{7.46}$$

The product of c_{ox} and the exponential term corresponds to the density of empty states of the redox system. In this case $f(E)\varrho(E)$ at $E = E_c^s$ is the density of occupied states at the bottom of the conduction band at the surface, i.e. a number which is equal to the density of free electrons n_s at the surface. Eq. (7.46) turns then into

$$j_c^- = ek_0n_sc_{ox}\,\exp\left[-\frac{(E_c^s - E_{F,redox}^0 + \lambda)^2}{4kT\lambda}\right] \tag{7.47}$$

in which n_s is related to the bulk electron density n_0 by the Boltzmann distribution function (compare with Eq. 2.44)

$$n_s = n_0\,\exp\left(-\frac{e\Delta\phi_{sc}}{kT}\right) \tag{7.48}$$

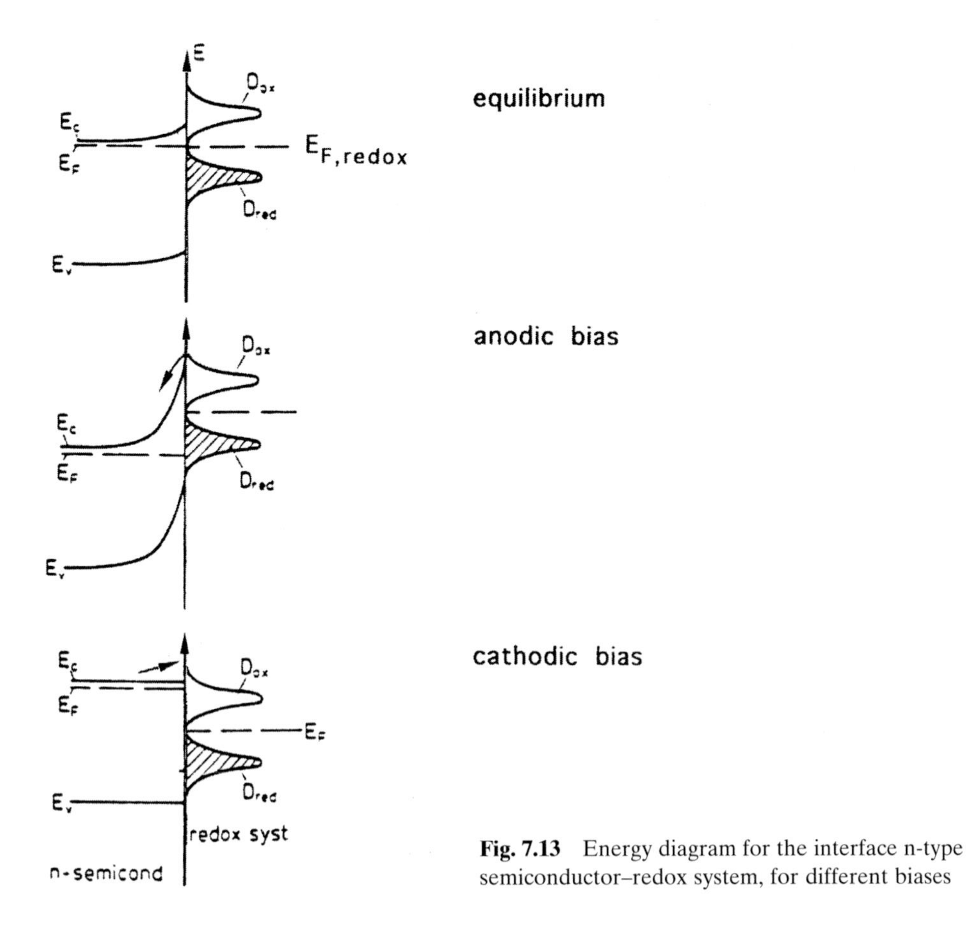

Fig. 7.13 Energy diagram for the interface n-type semiconductor–redox system, for different biases

in which ϕ_{sc} is the potential across the space charge region as already defined in Chapter 5. In Section 5.3 it has been shown that any variation of the electrode potential usually occurs entirely across the space region ($\Delta U_E = \Delta(\Delta\phi_{sc})$). Accordingly, j_c^- becomes potential-dependent via n_s. These two cases are illustrated in Fig. 7.13. The same procedure can be used to derive a current corresponding to an electron transfer via the valence band. One obtains

$$j_v^+ = ek_0 p_s c_{red} \exp\left[-\frac{(E_c^s - E_{F,redox}^0 - \lambda)^2}{4kT\lambda}\right] \tag{7.49}$$

$$j_v^- = ek_0 N_v c_{red} \exp\left[-\frac{(E_c^s - E_{F,redox}^0 + \lambda)^2}{4kT\lambda}\right] \tag{7.50}$$

Here N_v is the density of energy states at the upper edge of the valence band and E_v^s occurs as the energy of the valence band in the exponential term. In the case of a valence band process the cathodic current is constant (Eq. 7.50) whereas the anodic current depends on the hole density at the surface. The latter is given by

$$p_s = p_0 \exp\left(\frac{e\Delta\phi_{sc}}{kT}\right) \tag{7.51}$$

The anodic current is expected to increase exponentially with the electrode potential provided that the condition $\Delta U_E = \Delta(\Delta\phi_{sc})$ is again fulfilled. In such an anodic reaction in which electrons are transferred into the valence band, holes must be available at the surface. Frequently, scientists then argue in terms of hole transfer. This is only a rather lax description and has no real physical basis for the process at the boundary of two different phases.

In the preceding derivations we have used the Gerischer model, i.e. we have described the currents in terms of a charge transfer between occupied states on one side of the interface and empty states on the other. In principle one obtains the same equations when using one of the other theories described in Chapter 6. The reason is that in all theories the same exponential term occurs which originates from the assumption that the fluctuation of the solvent molecules or dipoles is assumed to behave like an harmonic oscillator. Quantitatively speaking, the use of the different theories mainly leads to different pre-exponentials [19]. Since the exponential terms are independent of potential, it is useful to include them in the rate constant [19]. The conduction band processes can then be described by

$$j_c^+ = ek_c^+ N_c c_{red} \tag{7.52}$$

in which

$$k_c^+ = k_{c,max}^+ W_{red} = k_{c,max}^+ \exp\left[-\frac{(E_c - E_{F,redox}^0 - \lambda)^2}{4kT\lambda}\right] \tag{7.53}$$

The cathodic current via the conduction band is given by

$$j_c^- = -ek_c^- n_s c_{ox} \tag{7.54}$$

with

$$k_c^- = k_{c,max}^- W_{ox} = k_{c,max}^- \exp\left[-\frac{(E_c - E_{F,redox}^0 + \lambda)^2}{4kT\lambda}\right] \tag{7.55}$$

The valence processes are then given by (see Eqs. 7.49 and 7.50)

$$j_v^+ = ek_v^+ p_s c_{red} \tag{7.56}$$

in which

$$k_v^+ = k_{v,max}^+ W_{red} = k_{v,max}^+ \exp\left[-\frac{(E_v - E_{F,redox}^0 - \lambda)^2}{4kT\lambda}\right] \tag{7.57}$$

and

$$j_v^- = -ek_v^- N_V c_{ox} \tag{7.58}$$

with

$$k_v^- = k_{v,max}^- W_{ox} = k_{v,max}^- \exp\left[-\frac{\left(E_v - E_{F,redox}^0 + \lambda\right)^2}{4kT\lambda}\right] \tag{7.59}$$

The currents given by Eqs. (7.52), (7.54), (7.56) and (7.58) can also be expressed in terms of the equilibrium conditions. The latter is characterized by

$$j_c^+ = j_c^- = j_c^0 \quad \text{and} \quad j_v^+ = j_v^- = j_v^0 \tag{7.60}$$

This condition must also be fulfilled if both bands are involved in one redox reaction at a semiconductor electrode. In addition we have $n_s = n_s^0$ and $p_s = p_s^0$. In the case of a conduction band process one obtains then, by using Eqs. (7.52) and (7.54), and with $\phi_{sc} - \phi_s{}^0 = \eta$

$$j_c = -j_c^0\left[\frac{n_s}{n_s^0} - 1\right] = -j_c^0\left[\exp\left(-\frac{e\eta}{kT}\right) - 1\right] \tag{7.61a}$$

and for a valence band process using Eqs. (7.56) and (7.58)

$$j_v = j_v^0\left[\frac{p_s}{p_s^0} - 1\right] = j_v^0\left[\exp\left(-\frac{e\eta}{kT}\right) - 1\right] \tag{7.61b}$$

The anodic and cathodic currents are given in a semilogarithmic plot vs. overvoltage in Fig. 7.14. It should be emphasized that the Tafel curves have a slope of 60 mV/decade which corresponds to a transition factor of $\alpha = 1$, whereas with metal electrodes an α value of about 0.5 is usually found (see Section 7.1). Semiconductor and metal electrodes behave differently because any overvoltage occurs across the space charge region of the semiconductor, whereas it leads to change of the Helmholtz potential in the case of metal electrodes.

The absolute values of the four maximum rate constants, $k_{1,max}^{j}$, depend on the theory applied here. The rate constant k_1^j is a second-order rate constant with a dimension of $cm^4\ s^{-1}$, provided that the concentration of the carrier density and of the redox system are given in units of cm^{-3}, and k_1^j is related to the local rate constant k_{et} [s^{-1}] by

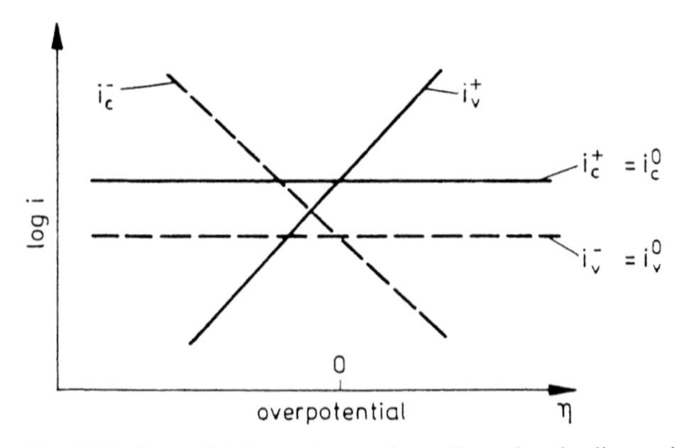

Fig. 7.14 Potential dependence of anodic and cathodic partial currents

Eq. (6.123). As shown in Chapter 6, the pre-exponential factor in Eq. (6.123) depends primarily on the interaction between electrode and redox system (adiabatic or non-adiabatic reaction) and also on the characteristics of the specific model. Experimental values of $k_{\rm i}^{\rm i}$ will be given in Section 7.3.4.

So far we have not taken into account the influence of doping of the semiconductor (n- or p-type). Considering for instance an electron transfer from the redox system into the conduction band, this process leads to a constant current. This electron transfer is in principle possible with n- as well as with p-type electrodes. Since the electrons transferred into an n-type electrode are majority carriers they can easily be transported to the rear contact which leads to a constant anodic current. In the case of a p-type electrode, the injected electrons are minority carriers. These processes are more complex and will be treated, together with light-induced reaction, in the following Section (7.3.3). On the other hand, a cathodic current via the conduction band of an n-type electrode rises exponentially with increasing negative overpotential because sufficient electrons are available. Accordingly, only majority carriers are involved in a conduction band process at an n-type electrode. Provided that such a reaction is entirely kinetically controlled then it is quantitatively described by Eq. (7.61a). A theoretical $j_{\rm c}^{-}$–η curve is illustrated in Fig. 7.14 for various $j_{\rm c}^{0}$ values. Using a p-type electrode, the cathodic current would be limited to a very low level because the electron density in the conduction band is extremely small. This current, however, can be enhanced by light excitation as will be described in the following section.

Concerning valence band processes, one observes the opposite effect. Here the anodic current at a p-type electrode rises with increasing anodic overvoltage because sufficient holes are available. At an n-type electrode, only a small current occurs which can be enhanced again by excitation. In the case of a cathodic current at a p-type electrode, the holes injected into the valence band, are easily transported to the rear contact. Accordingly, valence band processes at p-type electrodes are majority carrier reactions. The kinetics of this process are determined by Eq. (7.61b) and the corresponding theoretical $j_{\rm v}$–η curves are given in Fig. 7.14 for various $j_{\rm v}^{0}$ values.

7.3.3 Light-induced Processes

Light excitation is of interest if minority carriers are involved in the charge transfer. The excitation and recombination phenomena are treated in Section 1.6. The absorption of a photon always leads to the formation of one electron–hole pair. In doped semiconductors, the increase of majority carriers is negligibly small compared with the number of carriers present due to the doping. On the other hand, the density of the minority carriers is usually higher, by orders of magnitude, than that in the dark. The minority carriers generated by light within the space charge region are driven toward the interface by the electric field so that they can be transferred across the interface. Carriers produced in the bulk, i.e. outside the space charge region ($z > d_{\rm sc}$), can only diffuse toward the surface or to the space charge region, as illustrated in Fig. 7.15 for an n-type electrode. The distance, however, across which electron–hole pairs diffuse, is limited because holes (minority carriers in the n-type electrode) also recombine. They can only diffuse approximately over a distance as given by the diffusion length $L_{\rm p}$. Accordingly, only holes created within an approximate range of $d_{\rm sc} + L_{\rm p}$, will reach the

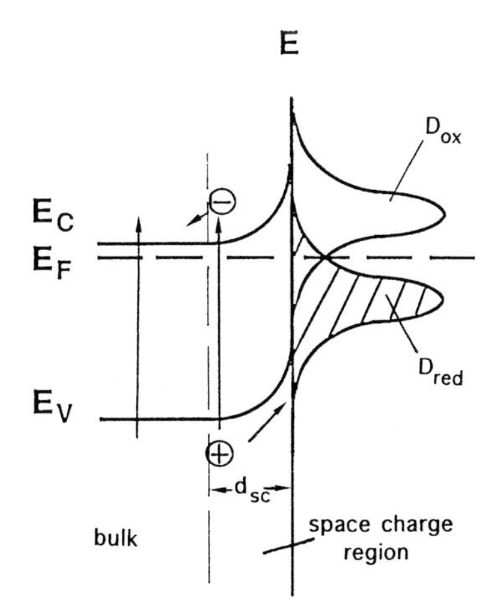

Fig. 7.15 Excitation of electron–hole pairs by light and separation of charge carriers by the field across the space charge layer

interface. Considering at first an n-type semiconductor, the interfacial current under illumination can be derived as follows by using Reichman's method [19, 23].

To derive the photocurrent density, the carrier densities at the interface have to be determined. The hole density can be obtained by solving the diffusion equation for minority carriers produced in the neutral region (the bulk). The diffusion equation for holes is given by (compare also with Section 2.2.3)

$$D\frac{d^2p}{dz^2} - \frac{p - p_0}{\tau} + I\alpha \exp(-\alpha z) = 0 \tag{7.62}$$

in which D is the diffusion coefficient; τ the lifetime of the minority carriers; α the absorption coefficient in the semiconductor as already defined in Chapter 1; p_0 is the equilibrium hole density, and I_0 is the monochromatic photon flux incident on the semiconductor surface at $z = 0$. The first term describes the diffusion, the second the recombination and the third the generation of minority carriers. One obtains from the above equation

$$j_{\mathrm{diff}} = eD\mathrm{grad}p\big|_{z=0} \tag{7.63}$$

and the hole current density at the edge of the depletion layer ($z = d_{sc}$) is obtained by using the boundary conditions $p = p_0$ at $z = \infty$ and $p = p_d$ at $z = d_{sc}$:

$$j_{\mathrm{diff}} = -j_0\left[\frac{p_d}{p_0} - 1\right] + \frac{eI_0\alpha L_p}{1 + \alpha L_p}\exp(-\alpha z) \tag{7.64}$$

with

$$j_0 = \frac{eDn_i^2}{N_D L_p} \tag{7.65}$$

In the last equation we have used the relation $L_p = (D\tau)^{1/2}$ (Eq. 2.16) and the equilibrium condition $n_0 p_0 = n_i^2$. Excitation of holes within the space charge region leads to the current

$$j_{sc} = eI_0\left[1 - \exp(-\alpha d_{sc})\right] \tag{7.66}$$

Here it has been assumed that there is no recombination within the space charge region. Accordingly, all holes generated within the space region are transferred. The two components, j_{sc} and j_{diff} add up to the total hole current

$$j_v = j_{diff} + j_{sc} \tag{7.67}$$

This current must be equal to the hole current expressed in terms of surface hole densities (Eq. 7.61b). This yields a relation from which the ratio p_s/p_s^0 can be obtained. Substituting it into Eq. (7.61b) leads to

$$j_v = \frac{j_g - j_0 \exp\left(-\frac{e\eta}{kT}\right)}{1 + j_0/j_v^0 \exp\left(-\frac{e\eta}{kT}\right)} \tag{7.68}$$

in which the generation current is given by

$$j_g = j_0 + eI_0\left[1 - \frac{\exp(-\alpha d_{sc})}{1 + \alpha L_p}\right] \tag{7.69}$$

i.e. we have

$$j_g = j_0 + i_{ph} \tag{7.70}$$

It should be mentioned here that the second term in Eq. (7.69) has already been derived by Gärtner [24] for the photocurrent of a semiconductor–metal junction under reverse bias, assuming $p_d = 0$. In the derivation of Reichmann [23], however, p_d was obtained in a manner consistent with the interface boundary conditions. Although the derivation of Reichmann is much more general, most scientists applied only the

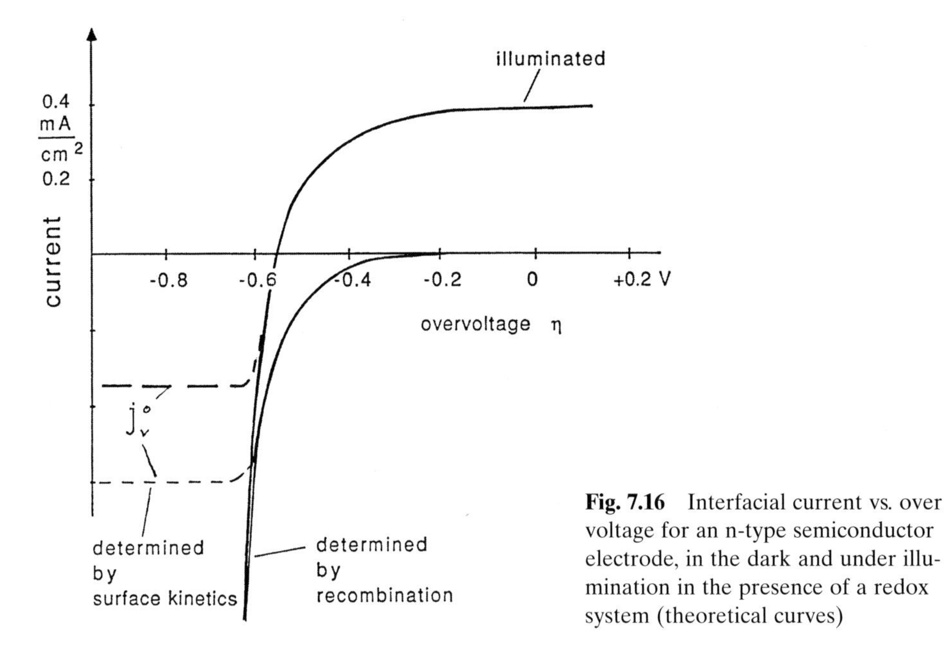

Fig. 7.16 Interfacial current vs. over voltage for an n-type semiconductor electrode, in the dark and under illumination in the presence of a redox system (theoretical curves)

Gärtner equation [25]. Later on Wilson extended the Gärtner model by including recombination via surface states [26].

The Reichmann derivation is of special interest because the final current equation (Eq. 7.68) describes the complete valence band process at an n-type electrode for anodic as well as for cathodic polarization. Eq. (7.68) looks rather complex because it contains two saturation currents, j_0 and j_v^0. The general issue of this equation becomes clearer when the dark current is also considered. Setting $I_0 = 0$ one obtains from Eq. (7.68)

$$j_v(\mathrm{dark}) = -\frac{j_0\left[\exp\left(-\frac{e\eta}{kT}\right) - 1\right]}{1 + \frac{j_0}{j_v^0}\exp\left(-\frac{e\eta}{kT}\right)} \tag{7.71}$$

A theoretical current–potential curve as calculated from Eqs. (7.68) and (7.71) is displayed in Fig. 7.16. At anodic polarization (positive η values), the dark current will become $j_v(\mathrm{dark}) \rightarrow j_0$ and the total current under illumination $j_v \rightarrow j_g$. The cathodic behavior depends on the ratio of j_0/j_v^0. Referring to Eqs. (7.60), (7.58) and (7.65), these two currents represent the generation/recombination rate of minority carriers in the bulk of the semiconductor (j_0) and the rate of hole transfer at the interface (j_v^0). As can be seen from Eq. (7.65), j_0 depends entirely on the properties of the semiconductor. On the other hand, j_v^0 is controlled by the kinetics of the actual hole transfer at the interface (Eq. 7.58). Accordingly, the ratio j_0/j_v^0 controls whether the generation/recombination process or the surface kinetics are rate-determining. More insight can be obtained by considering several extreme cases. Assuming, for instance, that the kinetics, i.e. the hole injection, is relatively slow ($j_v^0 \ll j_0$), then one obtains for large negative polarization, from Eq. (7.71)

$$j_v(\mathrm{dark}) \rightarrow j_v^0 \tag{7.72}$$

as indicated by the dotted curve in Fig. 7.16. This kinetically controlled current depends linearly on c_{ox}. On the other hand, if the recombination controls the current ($j_v^0 \gg j_0$), then one obtains from Eqs. (7.71) and (7.68)

$$j_v = -j_0\left[\exp\left(-\frac{e\eta}{kT}\right) - 1\right] + j_{ph} \tag{7.73}$$

This relation is identical to that derived for a pure solid state device which is determined by minority carrier transfer and recombination, such as a pn junction (see Section 2.3) or semiconductor–metal contact (see Section 2.2.3). The corresponding current–potential curves in the dark and under illumination are given by the solid lines in Fig. 7.16. Taking the complete Eq. (7.71), there may be a certain potential range where the recombination current determines the process until the current levels off to a constant j_v^0. For very large j_v^0 values, the cathodic current can ultimately be diffusion-limited, which can be checked experimentally by using a rotating electrode.

A similar scheme can easily be derived for the electron transfer at p-type electrodes, which will not be given here. In conclusion it can be stated that the forward current in minority carrier processes is usually governed by the recombination. In the case of holes, these recombine with electrons which finally carry the current. However, frequently it is difficult to decide whether a cathodic current which rises exponentially with the overvoltage, is really due to injection and recombination of minorities. Since the distribution of the energy states in redox systems is relatively wide due to large reorganization energies (in the order 1 eV), it can happen that the reduction of the redox system (cathodic current) is caused by an electron transfer via the conduction band whereas the oxidation occurs via the valence band. This can be checked by also measuring the cathodic current at a p-type electrode at which holes are majority carriers. In this case, the cathodic current would be controlled by the kinetics of the hole transfer as given by Eq. (7.61b). If, however, the reduction is a conduction band process, then the cathodic current would be very small in the dark, but could be increased by light excitation.

7.3.4 Majority Carrier Reactions

Dark current–potential curves representing a majority carrier transfer to a redox system have been measured by many research groups. Mostly cathodic currents at n-type electrodes have been studied rather than anodic currents at p-type semiconductors. This is because anodic hole consumption from p-type electrodes usually results in corrosion of the material. At least it is difficult to find a redox system where the oxidation of the redox couple competes sufficiently quickly with the corrosion.

Although many current–potential curves have been measured, most of them were not evaluated with respect to the models derived in Chapter 6. From the early stages of semiconductor electrochemistry there was only one report, published by Morrison [27], in which corresponding processes at n-type ZnO were quantitatively investigated. The cathodic reduction of various redox systems at different pH values of the solution were measured. In addition the flatband potential was determined by measurements of the space charge capacity. These showed that the flatband potential and hence the position of the energy bands were pH-dependent as expected for an oxide semiconductor (Chapter 5). Since they did not change upon addition of a redox system it was clear that there was no further interaction between ZnO and redox system. In order to compare the results obtained with different redox systems the currents were plotted versus the potential across the space charge region (Fig. 7.17).

According to these results, the log j^- vs. $\Delta\phi_{sc}$ plots are linear and have a slope of nearly 60 mV/decade of current, as expected from the theory (see Eq. 7.61a). Morrison also evaluated the log j^- vs. $\ll \Delta\phi_{sc}$ plots in terms of rate constants. Taking one example, the reduction of $[Fe(CN)_6]^{3-}$ at pH 3.8, $k_c^- = 1.3 \times 10^{-17}$ cm^4 s^{-1} was obtained. The maximum rate constant $k_{c,max}^-$, as defined by Eq. (7.55), can only be calculated if the reorganization energy λ is known besides the energy difference $E_c^s - E_{F,redox}^0$. The standard redox potential of the $[Fe(CN)_6]^{3-/4-}$ couple is $E_{F,redox}^0 = +0.2$ eV and $E_c^s = -0.2$ eV as calculated from Mott–Schottky measurements. Two different λ values, namely 0.5 and 0.75 eV, were reported for $[Fe(CN)_6]^{3-/4-}$ [14, 28]. The corresponding maximum rate constants are then $k_{c,max}^- = 1.6 \times 10^{-17}$ and 6.8×10^{-17} cm^4 s^{-1}, respectively. The rate constants k_c^- and $k_{c,max}^-$ do not differ very much because the conduction band is rather close to the maximum of the density of empty states of the redox system.

The question arises concerning which value of the anodic reverse j_c^+ can then be expected for the same redox system at ZnO. Assuming the same maximum rate con-

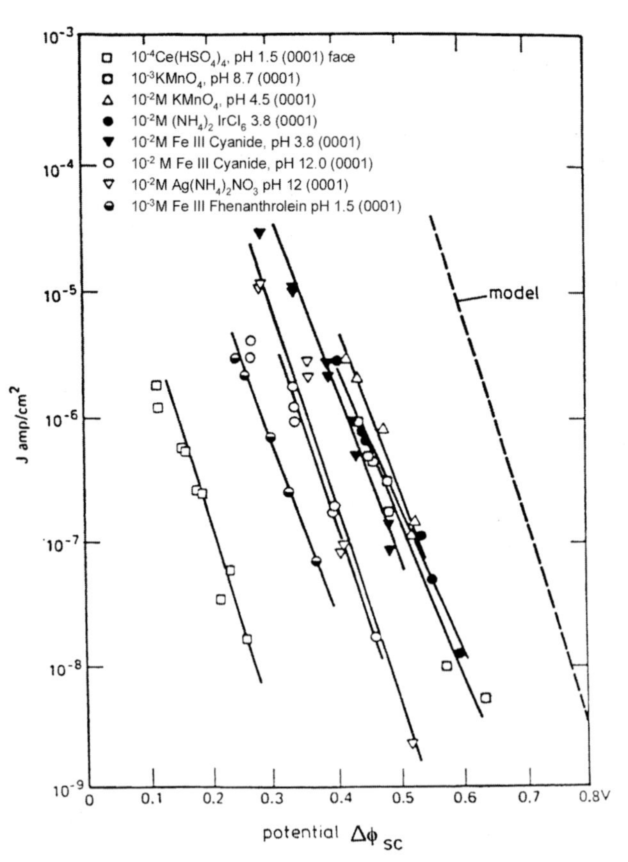

Fig. 7.17 Variation of cathodic current with the potential across the space charge layer for n-ZnO. (After ref. [27])

stant and the same λ values as for the reduction of the $[Fe(CN)_6]^{3-/4-}$ system, then one obtains according to Eq. (7.53), $k_c^+ = 1.5 \times 10^{-24}$ cm^4 s^{-1}, of course for both λ values. This is a very low value because the occupied states of the redox system are mainly distributed below the conduction band. The anodic current which is expected to be independent of the band bending, is given by Eq. (7.52). Assuming $N_c \approx 10^{19}$ cm^{-3} and $c_{red} = 6 \times 10^{18}$ cm^{-3} (corresponding to 10^{-2} M) then one obtains $j_c^+ \approx 10^{-4}$ A cm^{-2}, i.e. a current density which should be measurable. Morrison tried to measure it; unfortunately, however, the currents were not reproducible.

During the 1960s and early 1970s some research groups studied majority carrier reactions with few semiconductors or redox systems, without much success, however. Most current–potential curves were not properly evaluated, or the log j–U_E curves exhibited the wrong slope (sometimes of more than 100 mV/decade), or they were not even linear. Only in the case of the anodic dissolution of silicon, germanium and GaAs was the theoretically required slope of around 60 mV/decade found (see Chapter 8). In some cases, it has been speculated that the deviation from an ideal slope is due to a charge transfer via surface states or due to Fermi level pinning, as already discussed by Morrison in his book [15]. Obviously, ZnO was a very suitable semiconductor for quantitative investigations, perhaps because of its large bandgap (E_g = 3.1 eV). On the other hand, corresponding studies with TiO_2 which has also a large bandgap (E_g = 3 eV) failed completely [29], probably because the surface and the region below the surface is easily changed by diffusion of hydrogen into the electrode during cathodic polarization and of OH during anodic polarization. Later, scientists became mainly interested in photoeffects at the semiconductor–liquid interface because of the possible application in photoelectrochemical solar cells (see Chapter 11). Unfortunately, nobody realized at that time that the evaluation of the forward dark currents was an essential task for the improvement of solar cells.

Surprisingly it was around 25 years before a few research groups realized that the problem of majority carrier transfer was not really solved. Several majority carrier processes at Si, GaAs and InP electrodes have been investigated in the late 1990s which yielded interesting quantitative results, as will be described below. Many of these investigations were performed in non-aqueous solutions. This is advantageous because any interference with the anodic dissolution and with H_2 formation could be avoided.

Redox systems such as ferrocene ($Fc^{0/1+}$) and cobaltocene ($CoCp_2^{0/1+}$) and their derivatives are usually assumed to be suitable non-adsorbing outer sphere redox couples for use in non-aqueous solutions such as methanol or acetonitrile. One example is

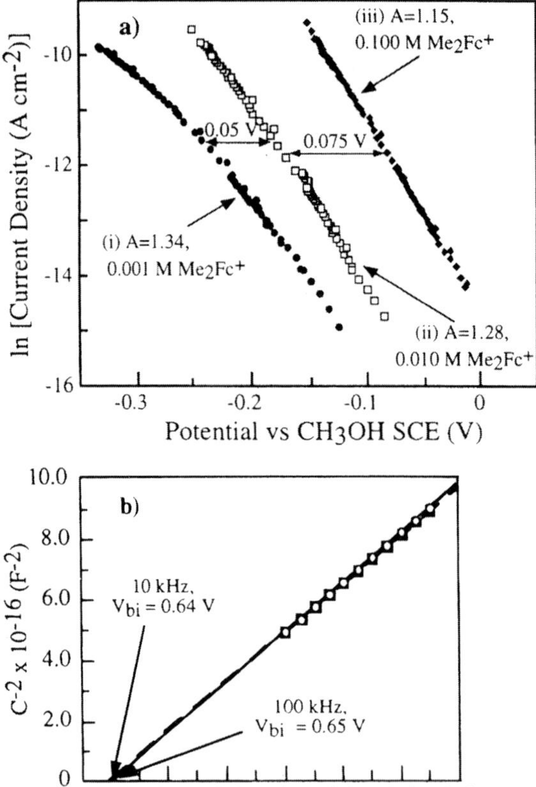

Fig. 7.18 Current–potential and Mott–Schottky plot for an n-type InP electrode in methanol in the presence of methyl ferrocene $(Me_2Fc)^+$: a) ln j vs. U_E; b) Mott–Schottky plot. (After ref. [30])

the cathodic reduction of methyl ferrocenium (Me_2Fc^{1+}) at n-InP in H_2O-free methanol. The log $j_c^- - U_E$ curve and the corresponding Mott–Schottky plot are given in Fig. 7.18 as published by Lewis and co-workers [30]. The current–potential curves were measured at different concentrations of the redox system. The shift of the curves in Fig. 7.18 by 60 mV/decade of concentration indicates that the current increases linearly with concentration. These curves also exhibit the correct slope of nearly 60 mV/decade of current as expected according to Eqs. (7.47) and (7.48). The rate constant determined from these measurements was, on average, $k_c^- = 9.5 \times 10^{-17}$ cm^4 s^{-1}. According to the capacity measurements, $E_c^s - E_{F,redox}^0 = -0.6$ eV. Using this value and $\lambda = 0.8$ eV, the authors obtained, by using Eq. (7.55): $k_{c,max}^- = 1.9 \times 10^{-16}$ cm^4 s^{-1}. Since k_c^- is very close to $k_{c,max}^-$, the conduction band of InP has nearly an optimal position with respect to the energy levels of the redox system. The same authors have tested other redox systems at n-InP and found similar $k_{c,max}^-$ values [30].

In addition, Lewis and co-workers have investigated the electron transfer from the conduction band of n-Si electrodes to viologen in H_2O free CH_3OH [31]. Here also the currents vary linearly with the concentration of the corresponding redox system and the current–voltage curves have a slope close to 60 mV/decade. The reported rate constants were in the range of $k^-_c = 10^{-17}$ to 10^{-16} cm$^{4=}$s^{-1} depending on the redox system. The maximum rate constant calculated according to Eq. (7.55), was around $k_{c,max}^- \approx$ 10^{-16} cm^4 s^{-1}.

Similar investigations were performed with cobaltocene at n-GaAs ($E_g = 1.4$ eV) and n-$GaInP_2$ ($E_g = 1.85$ eV) electrodes, and at GaAs capped with a thin 80-Å layer of $GaInP_2$ (GIP/GaAs) in acetonitrile containing tetrabutylammonium hexafluorophosphate ($TBAPF_{6)}$ as a conducting salt [32]. In the case of $GaInP_2$ (GIP) and of GIP/GaAs electrodes, the current–potential curves also showed a slope of 60 mV/decade. The currents varied linearly with the concentration of $CoCp_2^+$ (variation of c_{ox} over nearly four orders of magnitude). These measurements yielded rate constants of $k_c^- = 1 \times 10^{-16}$ cm^{4}s^{-1} for bulk n-GIP and 2×10^{-17} cm^{4}s^{-1} for n-GIP/GaAs, values which were independent of the concentration of the redox system. Assuming a reorganization energy of $\lambda = 0.6$ eV for $Co(Cp)_2^{0/1+}$, the maximum rate constant, $k_{c,max}^-$, was estimated to be in the order of 10^{-16} to 10^{-15}.

In contrast to the former cases, the reduction of $Co(Cp)_2^{1+}$ at GaAs electrodes was found to be an extremely fast reaction [32]. At low $Co(Cp)_2^{1+}$ concentrations ($\ll 10^{-3}$ M) the current depended linearly on the $Co(Cp)_2^{1+}$ concentration. However, above 1×10^{-3} M the current became nearly independent of the concentration [32]. The corresponding current–potential curve and the Mott–Schottky plot as measured at $c_{ox} = 1.3$ mM, are given in Fig. 7.19. The space charge capacity was determined by measuring the complete impedance spectrum as described in Chapter 4. This was necessary for also determining C_{sc} in the potential range where the current increased. The evaluation of the impedance spectra yielded the charge transfer resistance R_{ct} and the potential $\Delta\phi_{sc}$ across the space charge layer. The current and R_{ct} are plotted versus $\Delta\phi_{sc}$ in Fig. 7.20. According to this figure, an ideal slope of 60mV/decade has been found for both quantities, log j_c^- and R_{sc}. The same authors have shown that the log j_c^- – $\Delta\phi_{sc}$ curve can be fitted very well by the thermionic emission model. This model has successfully been applied to the interpretation of the current–voltage behavior of metal–semiconductor junctions (Schottky junctions) as described in Section 2.2.2. In this model the number of carriers moving towards the surface are calculated and it is

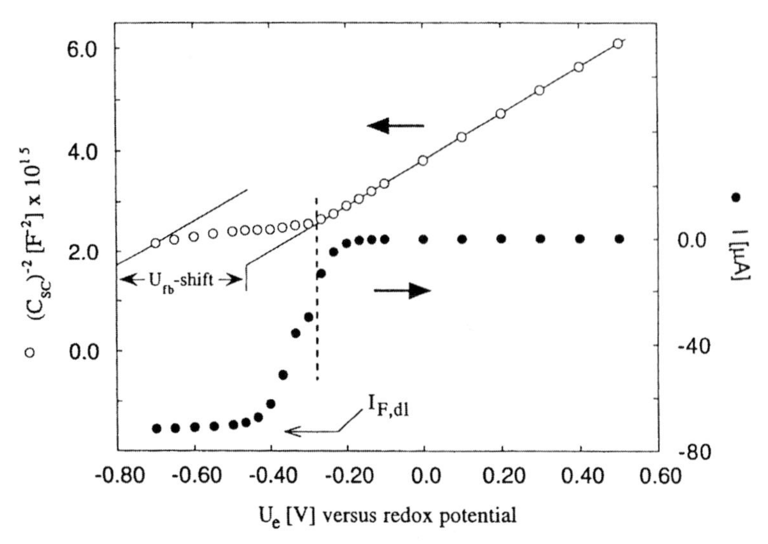

Fig. 7.19 Mott–Schottky plot of the capacitance vs. electrode potential and current–potential curve for n-GaAs in acetonitrile in the presence of 1.3 M $[Co(C_p)_2^+]$ and 0.9 M $Co(C_p)_2^0$. (After ref. [33])

assumed that all carriers which have finally reached the surface are transferred across the interface. Corresponding theoretical curves of j_c^- versus ϕ_{sc} and R_{ct} $(= \partial j_c^- / \partial_{sc})$ versus $\Delta\phi_{sc}$ as calculated by using Eq. (2.14), are given by the solid lines in Fig. 7.20. Accordingly, the transfer rates have reached extremely high values so that the current is finally determined by the transport of electrons through the space charge region. Such high currents are only understandable if the adsorption of the redox species is involved. This has actually been proved by a detailed analysis of the impedance spectrum and by quantitatively measuring the amount of adsorbed molecules using the method of quartz crystal microbalance (QCM) [33]. It was found that the molecules are physically adsorbed and the number of adsorbed molecules could be described by a Langmuir isotherm. The composition of the adsorbed layer (ratio of the oxidized and reduced species of cobaltocene) determines the potential across the Helmholtz layer and therefore the flatband potential of n-GaAs. During cathodic polarization, the number of $Co(Cp)_2^{1+}$ molecules in the layer is reduced leading to a shift of the flatband potential (Fig. 7.19).

Using the kinetic model, it is possible to calculate a lower limit for the rate constant from the charge transfer resistance, the latter being given by [33]

$$R_{ct} = \left(\frac{\partial U_E}{\partial j}\right) = \frac{kT}{e^2 A N_{ad,max}} \frac{1}{k_c^- n_0 \Theta_{ox}} \exp\left(\frac{e\Delta\phi_{sc}}{kT}\right) \tag{7.74}$$

where A is the surface area of the electrode; N_{ad}^{max} is the maximum number of adsorption sites and Θ_{ox} the degree of coverage of $Co(Cp)_2^{1+}$. A value of $k_c^- \geq 1.2 \times 10^{-7}$ cm^3 s^{-1} was obtained using the experimental data given in Fig. 7.19. The rate constant has another unit (cm^3 s^{-1} instead of cm^4 s^{-1}) and therefore it is not easy to compare it with the other rate constant discussed above. The reason is that the number of surface sites,

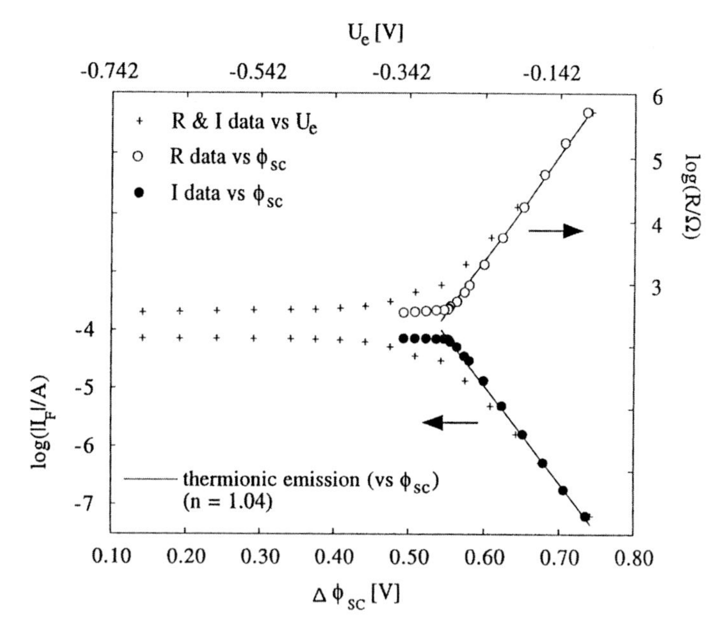

Fig. 7.20 Faraday current and charge transfer resistance vs. the potential across the space charge potential; data taken from Fig. 7.19. (After ref. [33])

N^{ad}_{max}, is expressed as an areal density and given in cm^{-2}. If one takes the thickness of the layer containing the adsorbed $Co(Cp)_2^+$ to be 8 Å, then the effective value of the second-order rate constant using volumetric units for the acceptor is $k_{et} \geq 1.2 \times 10^{-7} \times 8 \times 10^{-8}$ which is $\cong 1 \times 10^{-14}$ $cm^4\,s^{-1}$. The value of 8 Å, which includes the sum of the molecule thickness and the distance from the surface, was selected for this calculation because theoretical calculations have shown that the adsorption energy passes a maximum at that distance [33]. The second-order rate constant can also be converted into an electron transfer velocity (S_{et} in cm s^{-1}) or into a cross-section (σ_{et}) by using the following relationships: $S_{et} = k_c^- \, N_{ad,ox} \approx 5 \times 10^6$ cm s^{-1}, and $\sigma_{et} = S_{et}(v_{th} N_{ad,ox})^{-1} = 1.2 \times 10^{-14}$ cm^2 where $v_{th} = 10^7$ cm $^{s-1}$ is the thermal velocity. Both of these numbers indicate ultrafast electron transfer dynamics. To provide such a high rate for the electron transfer from the conduction band of GaAs electrode to the adsorbed $Co(Cp)_2{}^{1+}$ molecules, the resulting Red species must be oxidized at an equivalent rate. It was shown that a self-exchange reaction in which an electron is exchanged between an adsorbed Red species and an Ox species in the solution, fulfilled this condition [33].

Rate constants for electron transfer were recently also determined at thin As-capped GaAs quantum wells (thickness 50 Å) by measuring the photoluminescence decay in the presence of ferrocenium ions as electron acceptors [34]. In this case, rate constants were obtained which are about one order of magnitude smaller than those reported by Meier et al. and no adsorption was found. These results indicate that the composition of the GaAs surface plays an important role.

Concerning aqueous solutions, two other systems have recently been reported with which majority carrier processes were studied quantitatively. One is the oxidation of Cu^{1+} at GaAs in solutions of 6 M HCl. Since the standard potential of the couple is

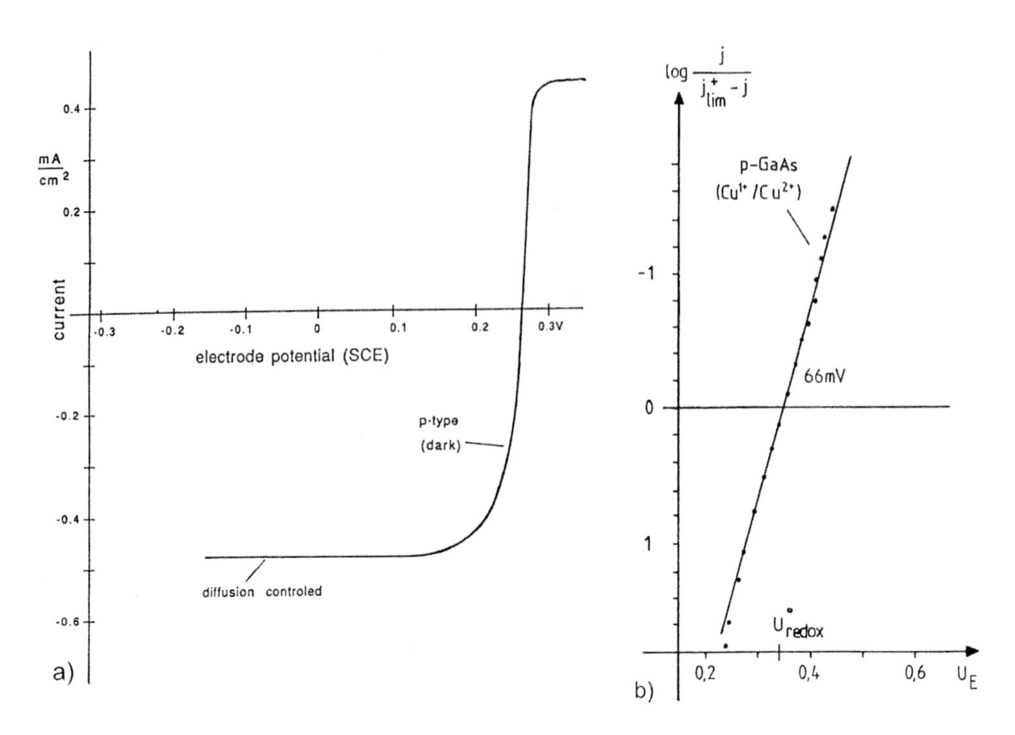

Fig. 7.21 Current–potential dependence for p-GaAs in 3 M HCL in the presence of 10^{-2}M $Cu^{1+/2+}$: a) linear current scale; b) logarithmic current scale. (After ref. [35])

U^0_{redox} = +0.33 V vs. SCE and the valence band is located at around E_V = +0.5 eV, the oxidation of the redox system is expected to occur via the valence band as also proved experimentally. Using here a p-type electrode, the oxidation is a majority carrier reaction. Since the energy difference between valence band and standard redox potential is rather small, the current and accordingly the transfer rates are very large [35]. The corresponding current–potential curve is given in Fig. 7.21a. Plotting the current as log $[j/(j_{lim} - j)]$ vs. U_E (Heyrovsky–Ilkovic plot, see Section 7.1.2) yields a straight line of an ideal slope of 60 mV/decade (Fig. 7.21b). Since the half-wave potential ($U_{1/2}$ = +0.336 V) occurs at the standard redox potential and was found to be independent of the rotation speed of the electrode (Eq. 7.33), Reineke et al. concluded that the oxidation of Cu^{1+} is a diffusion-controlled process. Accordingly, the 60-mV slope is due to the variation of the Cu^{1+} and the Cu^{2+} concentrations at the electrode surface (Nernst law) as discussed in Section 7.1 and is not determined by the kinetics of the charge transfer. The evaluation of the kinetic parameters was rather difficult because of the high rates. The authors reported that the rate constant could only be determined from the potential dependence of the charge transfer resistance (R_{ct}), the latter being obtained from impedance measurements. The impedance spectra were characterized by two half-circles, in which the high frequency half-circle is governed by the space charge capacity C_{sc} and the charge transfer resistance R_{ct}, whereas the other is determined by the Warburg impedance. The rate constants were evaluated by using Eq. (4.20). The transfer rate showed the required potential dependence although the current range was limited

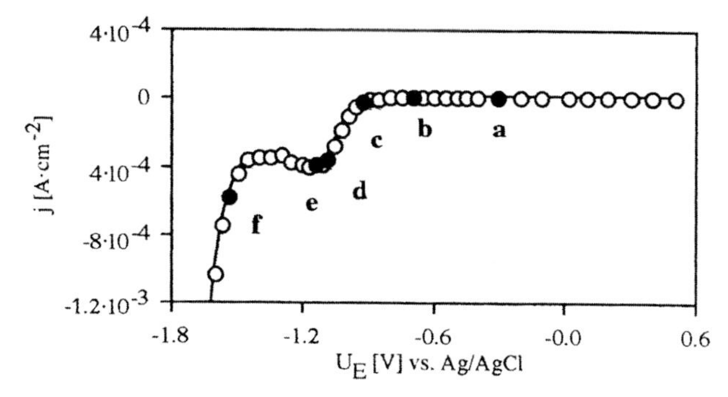

Fig. 7.22 Current–potential curve for n-GaAs in aqueous solution at pH 3 (H_2 formation). (After ref. [36])

because of the interference with the anodic dissolution of GaAs. According to this evaluation a rate constant of $k_v^+ = 5 \times 10^{-19}$ cm^4 s^{-1} was obtained. A value of $k_{v,max}^+$ has also been estimated by using Eq. (7.57). Unfortunately, the reorganization energy for $Cu^{1+/2+}$ was not known. By comparison with other ions such as $Fe^{2+/3+}$ it seemed reasonable to assume a value in the range between 1 and 1.5 eV. The authors estimated a maximum rate constant of around $k_{v,max}^+ = 10^{-16}$ cm^4 s^{-1}.

According to Fig. 7.21a, the reverse reaction, i.e. the reduction of the Cu^{2+}–ions, also occurs via the valence band and the corresponding cathodic current is diffusion-limited. The kinetic data, evaluated from the impedance spectra, yielded here the product of the rate constant and the density of states in the valence band, $k_v^- N_v = 10^{-2}$ cm s^{-1}. Using Eq. (7.58) one obtains $j_v^- = 1$ mA cm^{-2} for a Cu^{2+} concentration of 0.7×10^{-3} M. This kinetically controlled current is considerably larger than the experimentally determined diffusion-limited current which is in agreement with the model.

Besides these classical redox reactions, the formation of H_2, i.e. the reduction of protons has also been studied with n-GaAs. As already mentioned in Section 7.2, this is a conduction band process. The corresponding majority carrier reaction with n-GaAs was recently investigated by Uhlendorf et al. [36]. The current–potential curve as measured at pH 3, is characterized by a current increase around –0.8 V, followed by a range where the current is saturated, and further current increase above around –1.5 V (Fig. 7.22). The first current increase corresponds to the reduction of protons which becomes diffusion-limited; the second increase is due to the reduction of H_2O. From the impedance spectra measured under stationary conditions over the whole potential range, the space charge capacity, C_{sc}, was determined, the results of which are presented as C_{sc}^{-2} vs. the electrode potential in Fig. 7.23a. For comparison the current–potential curve in a semilogarithmic plot is given in the same figure. At relatively anodic potentials (range 1) the Mott–Schottky plot shows a fairly good straight line and the slope is determined by the doping of the electrode. An extrapolation to $C_{sc}^{-2} \rightarrow 0$ yields a flatband potential of about –1.0 V. In range 2, characterized by a small cathodic current, the C_{sc}^{-2} values are almost constant. It was concluded that the potential across the space charge layer $\Delta\phi_{sc}$ remains nearly constant in this range, i.e. the flatband potential is shifted toward cathodic potentials and the bands become unpinned at the surface. This shift had been already found earlier and was interpreted as a change of the Helmholtz double layer, possibly due to a change of a hydroxyl to

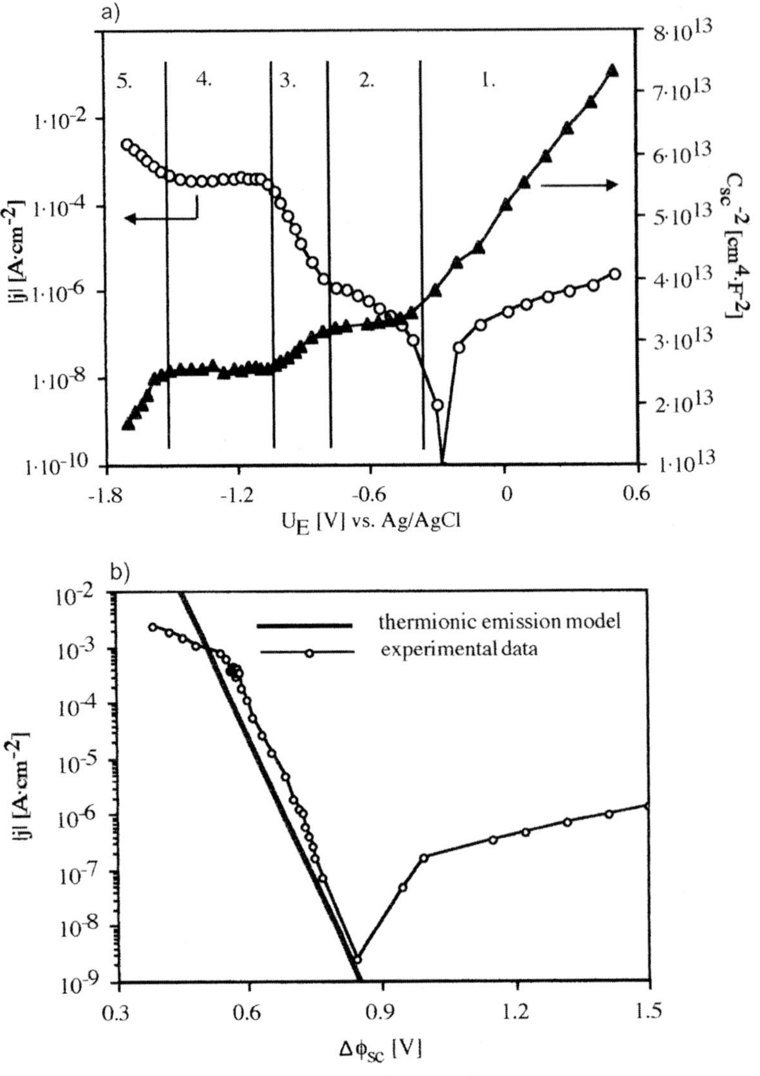

Fig. 7.23 a) Current–potential curve and $1/C_{sc}^2$ vs. electrode potential for n-GaAs at pH 3 (the same experimental conditions as in Fig. 7.22). b) Log j vs. $\Delta\phi_{sc}$; data taken from Fig. 7.22. (After ref. [36])

a hydride surface [37]. In range 3, C_{sc}^{-2} decreases again which means ϕ_H is constant. Obviously, the hydride layer is completed here. This has been proved by measuring C_{sc}^{-2} during a fast scan back towards anodic potentials (see dashed line in Fig. 7.23a). In range 4, where the current is diffusion-limited, C_{sc}^{-2} again varies very little with potential. Here the pH of the solution rises near the surface because the concentration of protons decreases which leads to a further upward shift of the bands at the surface. Finally, C_{sc}^{-2} varies with potential again in the usual manner in the range where the electrons are transferred to H_2O.

Although the variation of the potential distribution is rather complex, the space charge capacity could be related to the potential across the space charge layer, $\Delta\phi_{sc}$, for each electrode potential. This evaluation allowed a plot of log j_c^- vs. $\Delta\phi_{sc}$ as shown in Fig. 7.23b. Interestingly, this plot is linear and exhibits an ideal slope of 60 mV/decade. Moreover, it has been shown that in this case also the electron transfer can be described by the thermionic emission model, similarly to the findings for the reduction of cobaltocene at n-GaAs in acetonitrile. Accordingly, the transfer rate must be very high. This again is difficult to interpret on the basis of a simple electron transfer. Possibly adsorption of protons is involved, similarly to the reduction of $Co(Cp)_2^{1+}$ in acetonitrile [36].

There are a few other semiconductors and other redox systems where the charge transfer processes have been analyzed quantitatively but where the current–potential curves did not fulfil the basic requirements of the theory such as linearity and a slope of 60 mV/decade. Examples are layered compounds such as, WSe_2 or $MoSe_2$. When the basal planar surfaces are exposed to the electrolyte and a crystal with a low density of steps is used, then only a weak interaction between such a semiconductor and the solvent or the redox system is expected because the metal (W in WSe_2) is screened from the surface. From this point of view, one would expect ideal properties when using a suitable outer sphere redox couple. Appropriate investigations performed with n-WSe_2 in acetonitrile using ferrocene ($Me_2Fc^{0/1+}$), were not successful, insofar as the slope of the log j_c^- vs. U_E curves exhibited a slope of more than 90 mV per decade which cannot be explained by the model [38]. Bard and co-workers suggested that the steps at the crystal surface are responsible for this deviation from ideal behavior [39]. Bard et al re-investigated the problem by using the SCEM technique (see Section 4.2.4) which made it possible to study the current–potential behavior within small surface areas. They studied the oxidation of various redox systems at p-type WSe_2 electrodes in aqueous solutions. In only one case, namely the reduction of $Ru(NH_3)_3^{3+}$, was a slope of 60 mV/decade found. However, they did not find any influence on the slope from steps on the crystal surface, either with $Ru(NH_3)_3^{3+}$ or with any other redox system [39]. A slope of 60 mV/decade has also been reported for the oxidation of $Fe(phen)_3^{2+/3+}$ in aqueous electrolytes p-WSe_2 (valence band process). In this case, however, the reaction was diffusion-controlled, similarly to the case of $Cu^{1+/2+}$ at p-GaAs (see above) [48].

According to the experimental results, the experimentally determined rate constants, scatter over a considerable range. In some cases, the maximum rate constants ($k_{c,max}$ and $k_{v,max}$) were also determined, using Eqs. (7.53), (7.55), (7.57) and (7.59), if data about the reorganization energy were available. These data also scatter over several orders of magnitude. In this context, it should be mentioned that models have been developed by which absolute values of rate constants can be derived. Lewis [40] has derived a model on the basis of Marcus' theory on electron transfer between different molecules, the latter being dissolved in two immiscible liquids (see Section 6.1.4). In the derivation of Lewis, one liquid is replaced by the semiconductor and the electrons in the solid correspond to the molecules in the former liquid. According to this model, Lewis has determined a maximum rate constant of $k^{j}_{1,max} \approx 10^{-17}$–$10^{-16}$ cm^4 s^{-1}[40]. Smith et al. have shown that Lewis's model leads to a maximum value which is 2–3 orders of magnitude higher [41, 42]. Probably a more rigorous method needs to be applied to obtain a reliable theoretical value. Since the situation is not clear yet, the problem of theoretical rate constants will not be discussed further here.

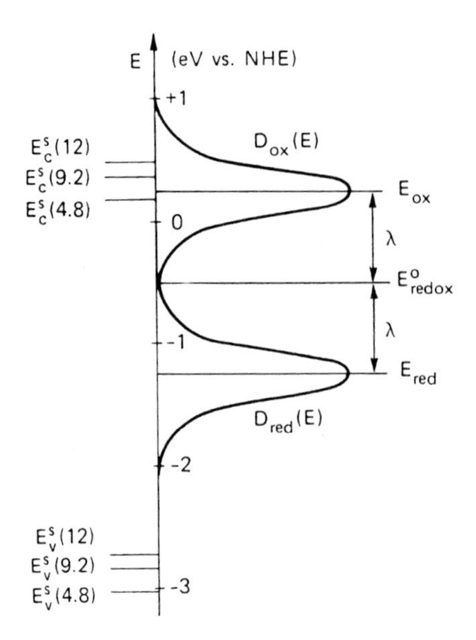

Fig. 7.24 Energy scheme for n-ZnO in contact with an electrolyte containing equal concentrations of $[Fe(CN)_6]^{-3}$ and$[Fe(CN)_6]^{-4}$ at different pH values

The ideal behavior of cells presented here, is also of significance in view of prior reports according to which Fermi level pinning instead of band pinning occurs, as found with Schottky junctions (see Section 2.2) [43, 44]. It is surprising that it took more than 25 years until reliable results on majority carrier processes were published. Many issues, such as the influence of surface chemistry on the rate constants, are still open.

Finally it should be emphasized that the energy difference $E_c - E^0_{F,redox}$ was smaller than λ in all the systems presented here. There was no system where $E_c - E^0_{F,redox} > \lambda$ which would correspond to Marcus' so-called inverted region (see Section 6.1). In the latter range the rate constant should decrease again as found in molecular donor–acceptor complexes. In this context some relevant investigations were carried out by Van den Berghe et al., who studied the electron capture by $[Fe(CN)_6]^{3-}$ at ZnO electrodes at different pH values [45]. They shifted the energy bands upwards upon increasing the pH as indicated in Fig. 7.24. The highest cathodic current was found at pH 5, the lowest at pH 12. The authors analyzed the data in terms of Eq. (7.59) and concluded that the Gaussian of the empty states of the redox system is as indicated in Fig. 7.24. The best fit was obtained with $\lambda = 0.75$ eV. Such an evaluation implies, however, that $k^-_{c,max}$ is not changed by varying the pH. Nevertheless this is the only quantitative result indicating a decrease of rate constant in the inverted region.

7.3.5 Minority Carrier Reactions

a) Transfer from the Semiconductor to the Solution Under Illumination

If the charge carriers to be transferred across the interface are minority carriers, then the corresponding current is very low because the minority carriers are mainly produced by thermal generation. The current is given by Eq. (7.65), i.e. it is determined

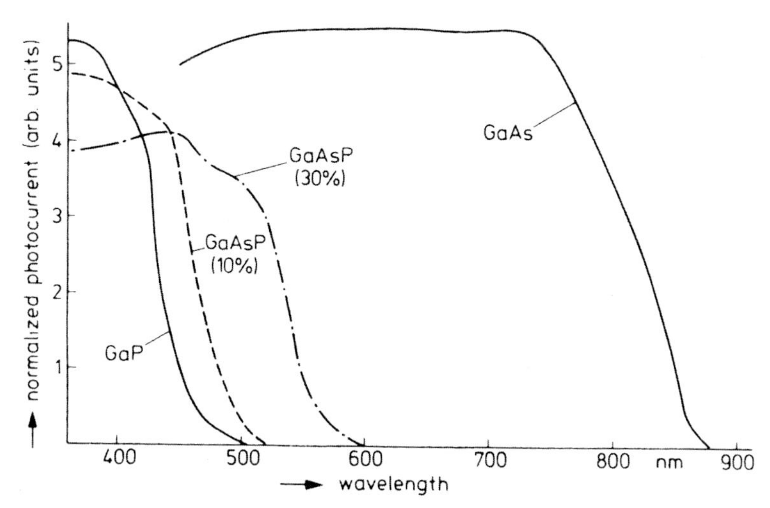

Fig. 7.25 Excitation spectra of photocurrents at various semiconductor electrodes in 0.1 M H_2SO_4. (After ref. [18])

by quantities which actually determine the bulk recombination. Since recombination and generation are equal at equilibrium and since the generation does not increase with increasing potential, the generation is also determined by Eq. (7.65) over the whole potential range. Taking typical values for D, n_i^2, p or n and L, then one has a current density in the order of $j_0 \approx 10^{-12}$ A cm^{-2} for Si, i.e. a current which is not really detectable. Usually one measures higher currents which are determined by side effects. As frequently mentioned, these small dark currents can be considerably enhanced by light excitation (see e.g. Fig. 7.10). A corresponding photocurrent is observed for photon energies which are larger than the bandgap. Excitation spectra for a selection of semiconductors normalized to equal light intensity, are given in Fig. 7.25. The photocurrents rise with increasing absorption coefficients (compare with absorption spectra given in Fig. 1.8) and reach a nearly constant value at higher photon energies. The limiting value is obtained if all excited minority carriers reach the surface where they are consumed in the reaction process. Accordingly, the quantum efficiency is unity in the potential range where the photocurrent saturates. In this range the photocurrent is proportional to the light intensity. In the case of semiconductors with an indirect bandgap, for instance GaP, the photocurrent is not detectable in the range of indirect bandgap absorption and starts to rise at photon energies which correspond to the direct bandgap transition. Since the absorption for an indirect transition is low the penetration of light is large, so that the minority carriers recombine before they reach the surface.

Concerning the potential dependence of the interfacial current under illumination, it is frequently useful to measure it in the presence of only one species of the redox couple, the reduced species for an anodic and the oxidized species for a cathodic reaction. Taking n-WSe_2 as an example, then the current–potential curve under illumination, as measured in an aqueous solution free from any redox system, is presented in Fig. 7.26. The cathodic dark current which occurs cathodic of the flatband potential, is due to H_2 formation (conduction band process). The anodic photocurrent which starts

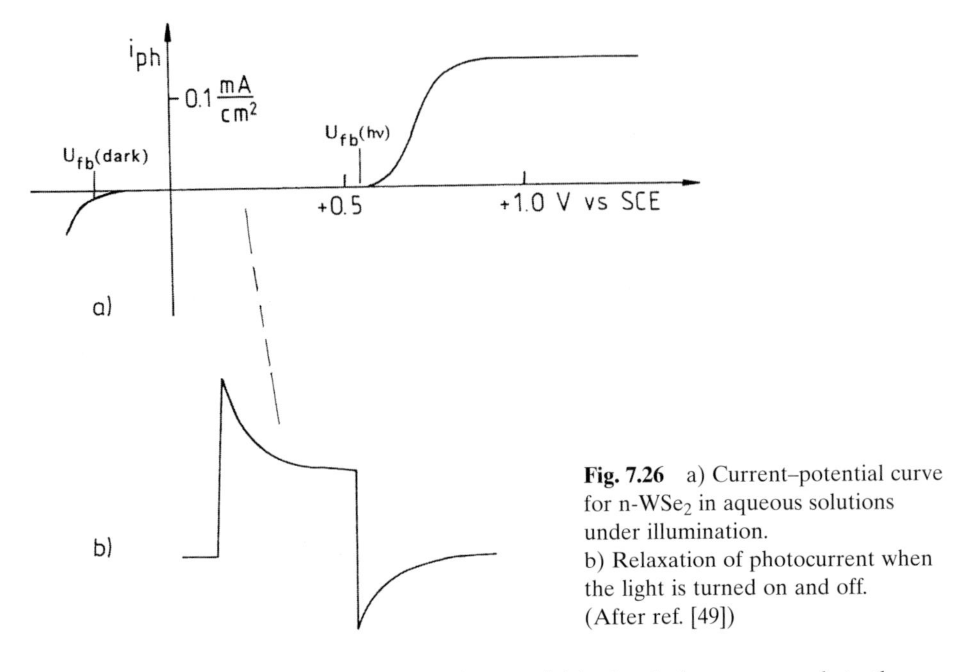

Fig. 7.26 a) Current–potential curve for n-WSe_2 in aqueous solutions under illumination.
b) Relaxation of photocurrent when the light is turned on and off. (After ref. [49])

to rise about 0.6 V anodic of the flatband potential in the dark, corresponds to the corrosion of the semiconductor. Theoretically, however, the photocurrent should start around the flatband potential because the energy bands of an n-type semiconductor are bent upward for potentials which are anodic from the flatband potential, and the corresponding electrical field across the space charge layer drives the holes toward the surface as illustrated in Fig. 7.27. Since a relatively large overvoltage is required for the photocurrent at n-WSe_2, it must be concluded that the electron–hole pairs created by light excitation, are lost by recombination within WSe_2 or at its surface as realized by other authors [46, 47]. The question arises, however, what is the origin for a surface recombination, a high recombination velocity via surface states or a slow kinetics for a hole transfer? In the case of WSe_2, it has been shown that the Mott–Schottky plots are shifted by about 0.6 V upon illumination of the semiconductor [49] as already discussed in Chapter 5. This has caused a shift of the flatband potential to U_{fb}(light) = +0.4 V, i.e. very close to the onset of the photocurrent. (U_{fb}(dark) and U_{fb}(light) are indicated in Fig. 7.26). This result indicates that the WSe_2 surface is positively charged

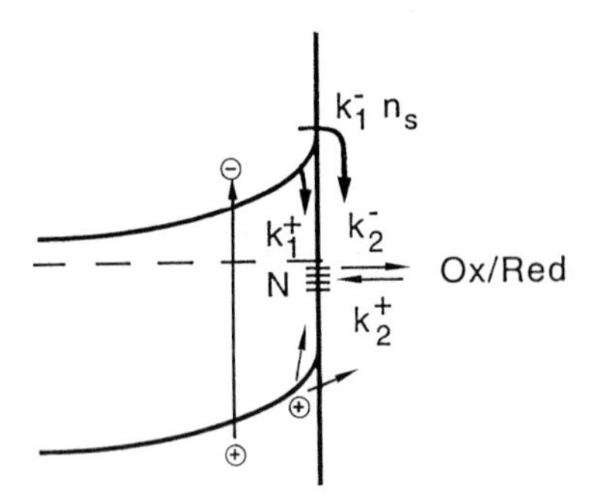

Fig. 7.27 Excitation, relaxation and electron transfer during illumination

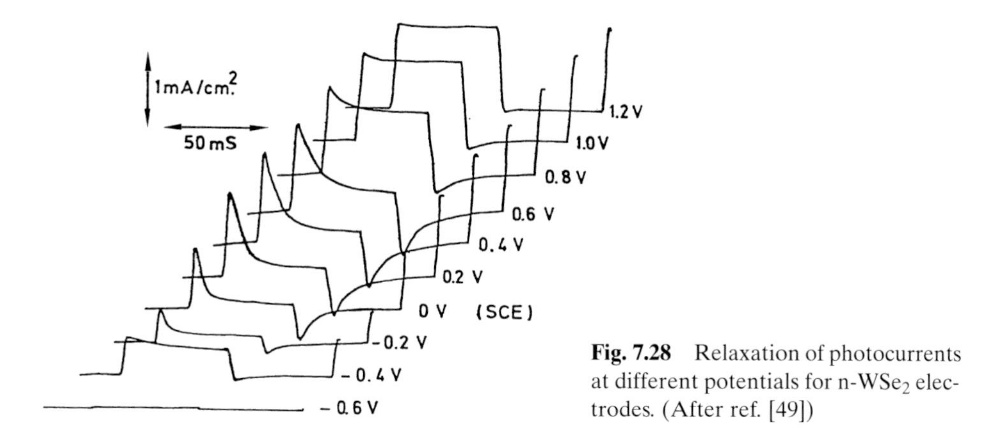

Fig. 7.28 Relaxation of photocurrents at different potentials for n-WSe_2 electrodes. (After ref. [49])

by hole capture which leads to a downward shift of the energy bands. Measurements with chopped light supported this interpretation. In the whole potential range between U_{fb}(dark) and U_{fb}(light), photocurrent transients were found as shown in Fig. 7.28 which have the typical shape for charging a capacitor. Upon addition of a suitable redox system, the standard potential of which occurs rather close to the valence band, such as $Fe(phen)_3^{2+}$ (U^0_{redox} = +0.85 V vs.SCE), the shift of the Mott–Schottky curve was avoided and the photocurrent onset occurred near U_{fb}(dark) as also shown in Fig. 7.26. Simultaneously, the transients disappeared because a constant photocurrent was found in the corresponding potential range.

According to these experimental results the primary effect here is an accumulation of holes at the surface because the anodic dissolution seems to be a very slow reaction. Working at a potential between U_{fb}(dark) and U_{fb}(light) the bands become flattened due to their shift. Since then the majority carrier density is increased near and at the surface, the recombination rate increases (the recombination rate is proportional to n and p, see Section 1.6). Accordingly, the high recombination is a consequence of the bandedge shift at the surface. In the case of WSe_2 it is not primarily due to an especially high surface recombination velocity compared with the hole transfer rate. This aspect has been overlooked in the publications referred to above (e.g. in ref. [47]).

Certainly there are also cases where surface recombination does play a dominant role. Examples are cathodic processes at p-GaAs electrodes. Here also a considerable overvoltage for the photocurrent onset has been found (≈0.5 V) as shown in Fig. 7.29 [50]. Identical photocurrent–potential dependencies were obtained with solutions containing redox systems such as Eu^{3+} (E^0_{redox} = –0.7 V (SCE)), Cr^{3+} (E^0_{redox} = –0.67 V) and without any redox system (H^+ reduction). Since the conduction band was located at –0.9 eV the reduction of the redox couples were expected to occur via the conduction band. In addition a cathodic shift of energy bands by 0.25 V was found upon illumination. Accordingly, the overvoltage of 0.5 V could only be partly due to a shift of energy bands and surface recombination dominates over all other steps. Surface recombination was measured by using the thin slice method, i.e. having a p–n junction on the rear of the electrode (see Section 4.3). According to Eq. (4.5) the surface recombination velocity S is inversely proportional to the short circuit current j_{sc} provided that $S \gg D/L$. Appropriate measurements performed with light of sufficiently small penetration depth, have yielded val-

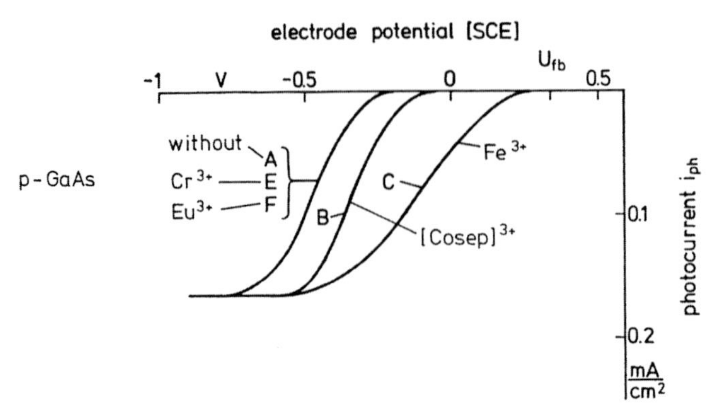

Fig. 7.29 Cathodic photocurrent vs. electrode potential for p-GaAs in the presence of various redox system in 10^{-2}M M_2SO_4

ues of $S \approx 10^7$ cm s^{-1} in the potential range where no photocurrent occurred [50]. These values are also found with GaAs samples in air. The whole process has been described quantitatively using a kinetic model as follows (see also Fig. 7.30).

$$\frac{dn_s}{dt} = g - k_n n_s N_t (1 - f) - k_c^{-} c_{ox} n_s \tag{7.75}$$

$$N_t \frac{df}{dt} = k_n N_t (1 - f) n_s - k_p N_t f p_s - k_s^{-} N_t f c_{ox} \tag{7.76}$$

in which g is the generation rate for electrons, N_t the density of surface states in units of cm^{-2} and f the fraction of centers occupied by electrons. The different rate constants are given in Fig. 7.30. It should be mentioned that the terms $S_{n0} = \sigma_n v_{th} N_t$ and $S_{p0} = \sigma_p v_{th} N_t$ (σ_n, σ_p capture cross-sections, v_{th} thermal velocity) are used in the semiconductor terminology where $\sigma_n v_{th}$ and $\sigma_p v_{th}$ correspond to k_n and k_p, respectively. The Faraday flux via the conduction band and the surface states are given by

$$j_c^{-} = k_c^{-} c_{ox} n_s \tag{7.77}$$

$$j_s^{-} = k_s^{-} c_{ox} N_t f \tag{7.78}$$

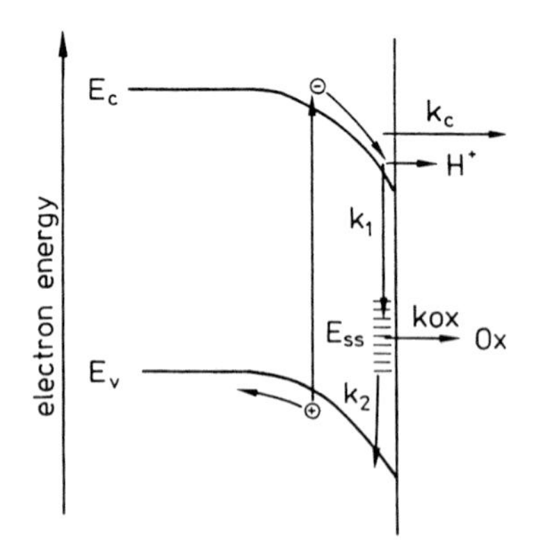

Fig. 7.30 Electron transfer at a p-GaAs electrode as illustrated by an energy diagram. (After ref. [50])

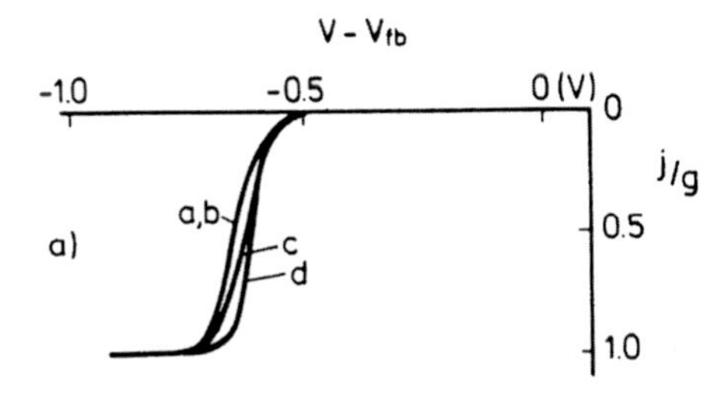

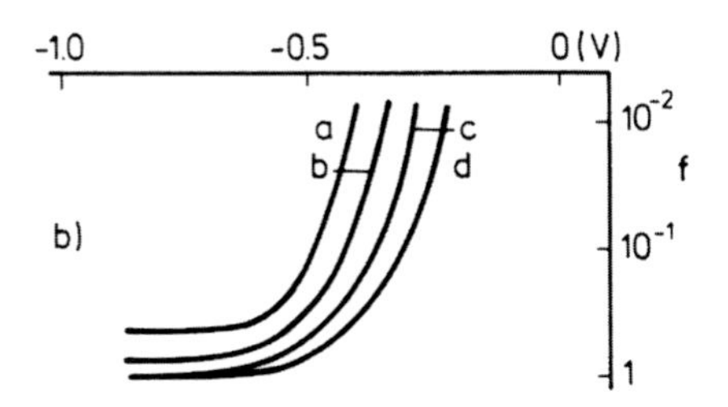

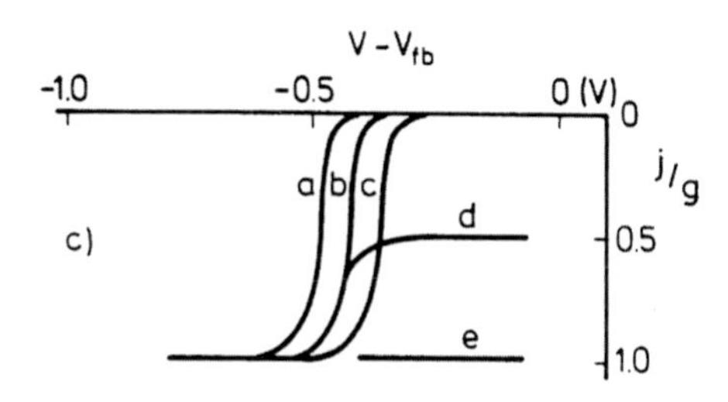

Fig. 7.31 Results calculated for simultaneous electron transfer from the conduction band and surface states to an oxidizing agent in solution, for electron transfer processes from the conduction band (rate constant $k_c^- c = 10^3$ cm s^{-1}) and from the surface states (k_s); a) and c) current normalized to equal intensities; b) Fermi function describing the occupation of surface states; c) as a) but with $k_s = 0$ (compare with Fig. 7.30. The parameters are different light intensities varying from $g = 10^{14}$ (a), to 10^{17} cm^{-2} (d). (After ref. [50])

The shift of the Mott–Schottky curve was explained by a trapping of electrons in surface states which leads to a change of the potential across the Helmholtz double layer by $\Delta(\Delta\phi_H)$. We have then according to Eq. (5.49)

$$\Delta(\Delta\phi_H) = \Delta Q/C_H = eN_t f/C_H \tag{7.79}$$

The rate constants are also affected by the shift of bands. Instead of Eq. (7.55) we have now

$$k_c^- = k_{c,\max}^- \exp\left[-\frac{\left(E_c - E_{F,\text{redox}}^0 - e(\Delta\phi_H) + \lambda\right)^2}{4kT\lambda}\right] \tag{7.80}$$

and for an electron transfer via the surface states (Fig. 7.30)

$$k_s^- = k_{s,\max}^- \exp\left[-\frac{\left(E_t - E_{F,\text{redox}}^0 - e(\Delta\phi_H) + \lambda\right)^2}{4kT\lambda}\right] \tag{7.81}$$

The total current normalized to equal light intensities, which is given by

$$j/g = (j_c^- + j_s^-)/g \tag{7.82}$$

can be calculated by numerical methods using Eqs. (7.75) to (7.81). Two examples are given in Fig. 7.31. It is important to note that for $k_s^- = 0$, the onset of the normalized photocurrent becomes dependent on the light intensity (Fig. 7.31c) which was not found experimentally. The intensity dependence obviously disappears for reactions via surface states (Fig. 7.31a) because an increase of the intensity fills the surface states more rapidly but simultaneously speeds up the rate of electron transfer from these states to the Ox species of the redox system. No attempt was made to fit the experimental photocurrent–potential curve quantitatively to the theoretical model because the values of too many critical parameters, such as the reorganization energy, density of surface states and electron trapping rate constants, are not known with sufficient accuracy. The only aim here was to show that an electron transfer from the surface states is essential.

The same observations have been made with p-GaAs electrodes in non-aqueous electrolytes such as acetonitrile. For instance, in the presence of cobaltocenium ions ($CoCp_2^+$) an overvoltage of around 0.5 V has also been found. Interestingly, the surface recombination could be reduced to negligible levels after a 30-Å layer of $GaInP_2$ had been deposited on p-GaAs [51]. In this case the flatband potential was not affected by light excitation and the onset of photocurrent occurred close to U_{fb}. In the range of the photocurrent increase it became dependent on the concentration of the redox system. The latter data are difficult to evaluate by means of the kinetic model introduced above, because a shift of the flatband potential must occur under potentiostatic conditions if no redox system as an electron acceptor is present (H_2 formation is not possible in acetonitrile). The photoelectrochemical results were proved by measurements of the photoluminescence lifetime. Here also a much longer lifetimewas observed with a p-type electrode capped with a 30- or 50-Å $GaInP_2$ layer. According to these results a very low surface recombination velocity, such $S = 200\,\text{cm s}^{-1}$, and a rate constant for the electron transfer from the conduction band to $CoCp_2^+$ of $k_c^- = 4 \times 10^{-16}\,\text{cm}^4\,\text{s}^{-1}$ has been obtained. This value actually agrees with that obtained with an n-type GaAs electrode also capped with $GaInP_2$, in the dark (compare with Section 7.3.4).

The same effects occur with most other semiconductors but were not further analyzed. As already discussed in Chapter 5, the proof of the existence of surface states is rather difficult because spectroscopic methods are not sufficiently sensitive. Another way of reducing the influence of surface states is to catalyze a charge transfer by depositing a metal or metal oxide on the semiconductor surface. This problem will be treated separately in Section 7.9.

b) Minority Carrier Injection into the Semiconductor

All light-induced processes and the resulting photocurrents are related to the transfer of minority carriers from the semiconductor to the electrolyte. The reverse process, the injection of minority carriers from a redox system into the semiconductor, is only possible with redox systems, where the standard potential is very close to the corresponding energy band of the semiconductor as illustrated in Fig. 7.32. This could be accomplished with redox systems of a fairly positive standard potential such as $Ce^{3+/4+}$

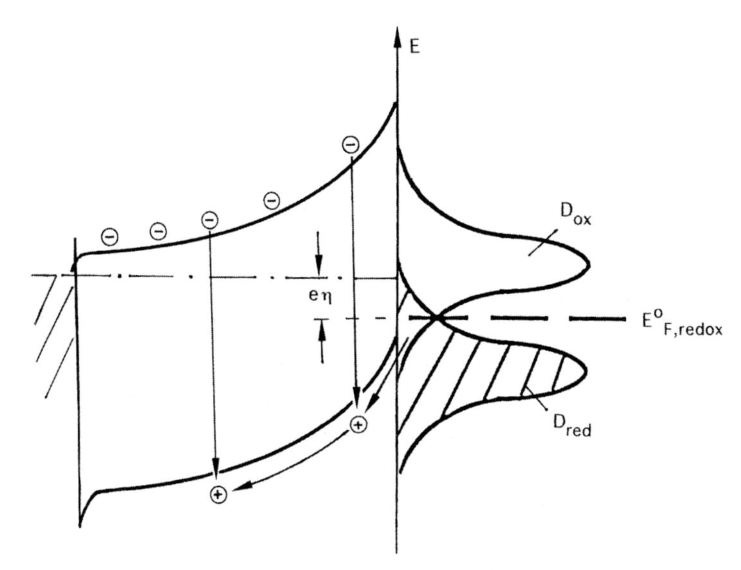

Fig. 7.32 Injection of minority carriers (holes) from a redox system into the valence of an n-type semiconductor electrode and their subsequent recombination upon anodic bias

(U^0_{redox} = +1.36 V vs. SCE), at n-type semiconductors such as WSe_2, Ge, Si, GaAs, InP and GaP but not with most large bandgap semiconductors such as TiO_2. In the latter case the valence band is far too low. Injection of holes was also possible with $[Fe(CN)_6]^{3-/4-}$ (U^0_{redox} = +0.2 V vs. SCE) in alkaline solutions at Ge and GaAs [22]. Here, the corresponding cathodic currents reach high values so that they are finally diffusion-limited. This is important insofar as it is very difficult to decide whether this current is really due to a hole injection into the valence band or due to electron transfer via the conduction band. At first sight it may be useful to see whether a corresponding cathodic current occurs at a p-type electrode in the dark. However, this is not necessarily a sufficient proof for a valence process at the n-type electrode because it can happen that a redox reaction occurs via the valence band at the p-electrode and via the conduction band at the n-electrode because there are sufficient electrons available at the surface of the latter. Only if the dark current at the p-electrode is very large, i.e. diffusion-limited, can one be fairly sure that the reaction also occurs via the valence band when an n-type electrode is used. In Fig. 7.33 a typical current–potential curve is shown as measured with n-GaAs in HCl using $Cu^{1+/2+}$ (U^0_{redox} = +0.33 V vs.SCE) as a redox couple. Since the dark current at p-GaAs is diffusion-limited (Fig. 7.21a) it can be concluded that the cathodic dark current as measured at the n-GaAs electrode is really due to an injection of holes into the valence band. We selected this example here because it seems to be the only one which is more quantitatively investigated [35]. The current–potential curve given in Fig. 7.33 can be analyzed as follows.

First of all one can recognize that the current is diffusion-limited at very negative electrode potentials because the saturation current depends on the rotation speed of the rotating GaAs electrode. In addition, the current increase occurs with a considerable overvoltage with respect to the standard redox potential of the $Cu^{1+/2+}$ couple (Fig. 7.33). Measurements with a ring-disc assembly (the technique described in Sec-

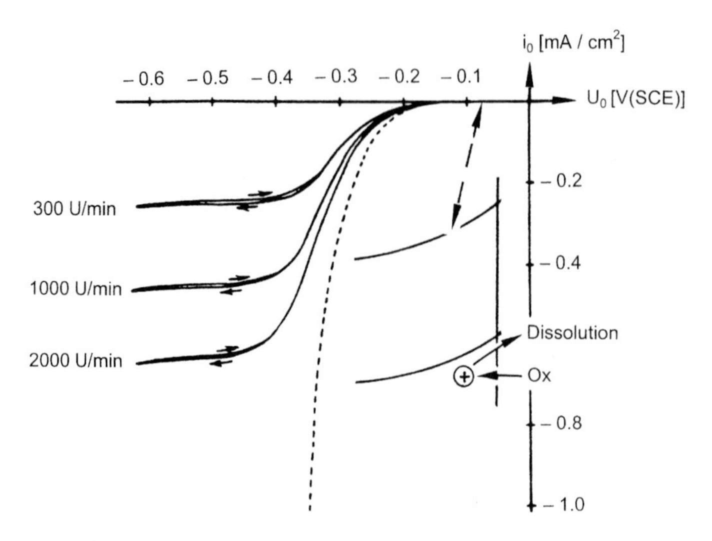

Fig. 7.33 Electrochemical reduction of Cu^{2+} ions at a rotating n-GaAs electrode in 6 M HCL at different rotation velocities. (After ref. [35])

tion 4.2.3) have shown that the oxidized species (Cu^{2+}) is also reduced in the range where the cathodic current is nearly zero. This phenomenon can be explained by an injection of holes, the latter being immediately consumed for the anodic dissolution of GaAs. This corresponds to two currents (anodic and cathodic) which cancel to zero as illustrated in the insert of Fig. 7.33. The latter observations have also been made with several other redox systems (e.g. $Ce^{3+/4+}$, $[Fe(CN)_6]^{3-/4-}$) and semiconductors (e.g. Ge, GaAs, InP, GaP). We will return to this problem in Section 7.4. In order to check whether the current is diffusion-controlled it was measured at different rotation speeds ω. Using Eq. (7.31) we have after rearrangement

$$k_{\mathrm{diff}}^{-} = k_{\mathrm{lim}}^{-} \exp\left[-\frac{e\left(U_{\mathrm{E}} - U_{1/2}\right)}{kT}\right] \tag{7.83}$$

where j_{lim}^{-} is proportional to $\omega^{1/2}$ (see Eqs. 7.30b and 7.28). Since $1/j = 1/j_{\mathrm{kin}} + 1/j_{\mathrm{diff}}$ (Eq. 7.40) a plot of $1/j$ versus $\omega^{-1/2}$ should yield straight lines. These were indeed obtained (Fig. 7.34) when using the data of Fig. 7.33. The kinetically controlled current, j_{kin}, was obtained by extrapolation of theses lines to $\omega^{-1/2} \rightarrow 0$. A semilogarithmic plot of j_{kin} versus U_{E} yielded a straight line with a slope of 92 mV/decade. Using the same procedure for analyzing the current–potential behavior of the p-type GaAs electrode in Cu^{2+} solutions (majority carrier process) linear $1/j - \omega^{-1/2}$ plots have also been obtained. An extrapolation to $\omega^{-1/2} \rightarrow 0$ yielded, however, intercepts around $1/j = 0$, i.e. j_{kin} was extremely large and could not be evaluated. Accordingly, the redox reaction at p-GaAs is entirely controlled by diffusion as already proved in Section 7.35 (compare with Fig. 7.21b). At first sight, these results seem to be inconsistent. However, in the case of the n-type electrode obviously the injection current is governed by the recombination of electrons with the injected holes. Accordingly, the rate constant for the hole injection itself is also very high for n-type, but here the

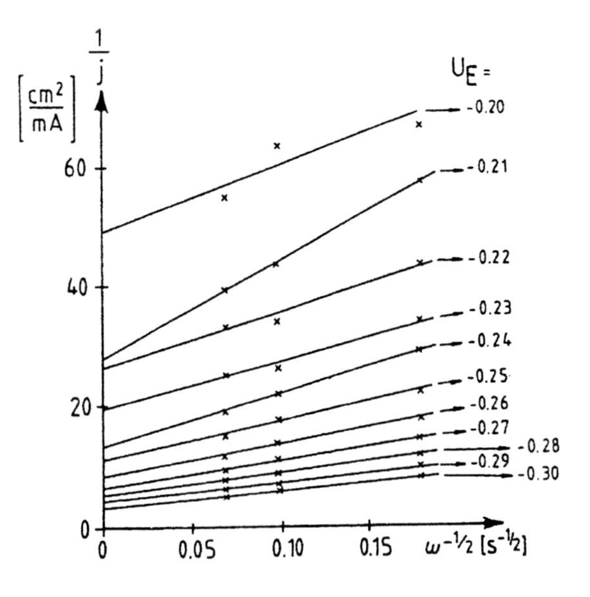

Fig. 7.34 Reciprocal values of cathodic currents vs. $\omega^{-1/2}$ at n-GaAs electrodes; data taken from Fig. 7.33. (After ref. [35])

rate-determining step is the recombination process which explains the lower j_{kin} values obtained from Fig. 7.34.

As mentioned above, the slope of the log j_{kin}–potential dependence given in Fig. 7.35, has a slope of 92 mV/decade. According to the Shockley–Read recombination model (see Section 1.6) which is valid for a recombination in the bulk of a semiconductor, a slope of 60 mV/decade would be expected. In addition, recombination can occur within the space charge layer of a semiconductor which leads to a slope of 120 mV/decade [52] (see also Section 2.3). With most p–n junctions which are minority carrier devices, a slope between 60 and 120 mV has been found. Therefore, the slope of 92 mV

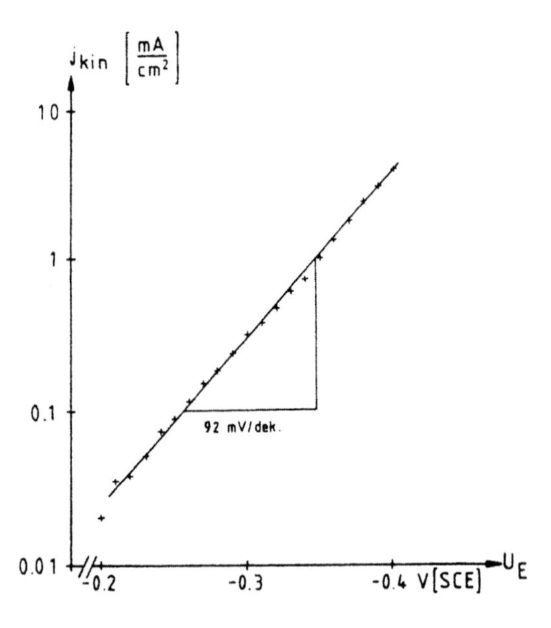

Fig. 7.35 Kinetic cathodic current vs. electrode potential at n-GaAs electrodes; the kinetic currents were obtained by extrapolating the curves in Fig. 7.34 to $\omega^{-1/2} \rightarrow 0$. (After ref. [35])

found with n-GaAs was explained by assuming recombination in the bulk and in the space charge layer [54].

The complete current–potential relation under illumination has already been derived in Section 7.3.3 (Eq. 7.68). In this case it was assumed that the cathodic dark current is only due to the injection of holes into the valence band of an n-type electrode. It was further shown that the current–potential relation could be simplified if the recombination is the rate-determining step (Eq. 7.73). The pre-exponential factor in Eq. (7.73), j_0, mainly depends on material parameters such as diffusion constant and length of minority carriers as given by Eq. (7.65). For instance, the recombination is fast if the diffusion length is short, which leads to high j_0 values and thereby to large cathodic dark currents (Eq. 7.73). As already mentioned, there are many cases where the photocurrent is due to a hole transfer to occupied states of the redox system but the dark current corresponds to an electron transfer from the conduction band to the empty states of the redox system. In this case the current–potential dependence for an n-type electrode has in principle the same shape

$$j = -j_0\left[\exp\left(-\frac{e\eta}{kT}\right)\right] + j_{ph} \tag{7.84}$$

in which η is the overvoltage (compare also with Eq. 7.61a). The physical difference compared with the former case, occurs only in the pre-exponential factor. Here j_0 depends on kinetic parameters such as rate constant and majority carrier density at the electrode surface (Eqs. 7.60 and 7.54). The dark current is zero at $\eta = 0$, i.e. at the redox potential ($U_E = U_{redox}$). Under illumination, however, the total current is zero if the cathodic dark current and the anodic photocurrent are equal ($j = 0$). According to Eq. (7.84), the corresponding η_{ph} is then given by

$$\eta_{ph} = -kT/e \ln\left[\left(j_{ph}/j_0\right) + 1\right] \tag{7.85}$$

Accordingly, η_{ph} depends mainly on the exchange current j_0 for a given photocurrent (Fig. 7.36). This is an interesting feature with regard to photoelectrochemical solar cells (see Chapter 11).

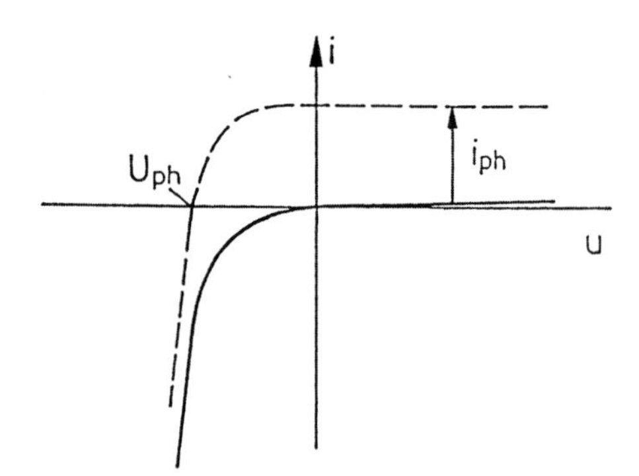

Fig. 7.36 Current–potential curve for an n-type semiconductor electrode in the dark and under illumination

7.3.6 Electron Transfer in the 'Inverted Region'

According to Marcus' theory, the rate constant for an electron transfer between to molecules in solution increases with increasing negative values of the free energy difference (ΔG^0) (see Eq. 6.10). The highest rate constant is obtained for $\Delta G^0 = \lambda$ and it is expected to decrease again for $\Delta G^0 > \lambda$. This has never been verified for an intermolecular electron transfer between single donor and acceptor molecules in solution, i.e. the rate constant remained at its maximum value if $\Delta G^0 > \lambda$. This is due to the diffusion limitation of the electron transfer reaction. On the other hand, if the donor and acceptor molecules were coupled via a hydrocarbon link, then the rate constant passes a maximum and decreases again when varying ΔG^0 (intramolecular process).

In principle, the rate constant for an electrochemical charge transfer process should also pass a maximum if ΔG^0 is increased. Applying this model to an electron transfer from a metal electrode to a redox system in the electrolyte, the corresponding current should increase exponentially with the overvoltage only as long as $e\eta \ll \lambda$. At higher potentials a curved $\log j - \eta$ dependence would be expected. As already mentioned in Section 7.1.1, this was not found with bare metal electrodes because it is not possible to separate transport from kinetic effects. A breakthrough was recently achieved by C.J. Miller et al. [4, 53] who used Au electrodes, the surface of which was modified by a monolayer of ω-hydroxy thiol of a thickness of 25 Å. When the Gerischer model is applied, the conditions for the normal and inverted region are schematically illustrated in Fig. 7.37 for an electron transfer from the metal to the empty states of a redox system. The tunneling of electrons through the ω-hydroxyl monolayer is the dominant mechanism for the electron transfer, i.e. there is no potential drop across this layer. The kinetics of the electron transfer are slowed so far down by this layer that transport limitations can be avoided. According to this scheme, the application of a negative potential to the metal leads to an upward shift of the Fermi level in the metal (compare B and C with A). This causes an increasing overlap of the occupied states in the metal with the empty states of the redox system, and the energy range in which tunneling

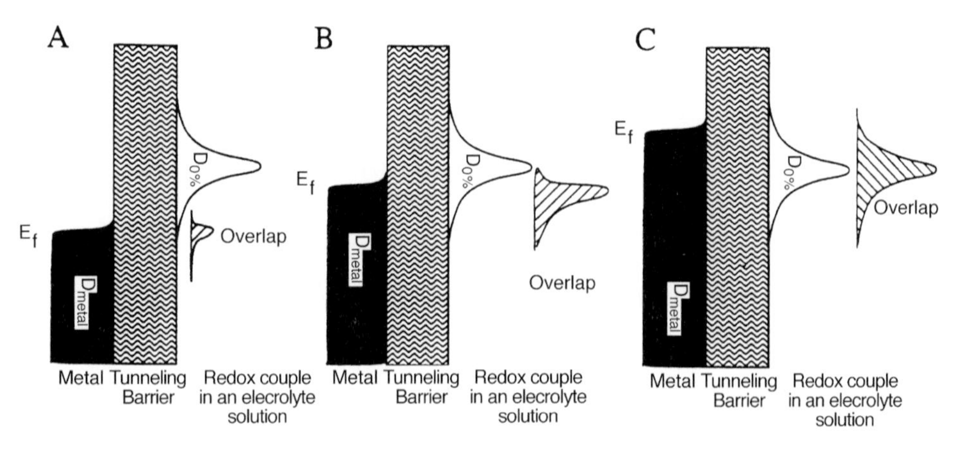

Fig. 7.37 A series of density of electronic state diagrams for a metal electrode coated with a tunneling barrier in contact with an electrolyte solution containing an electron acceptor at different anodic potentials (compare with Fig. 7.9). (After ref. [53])

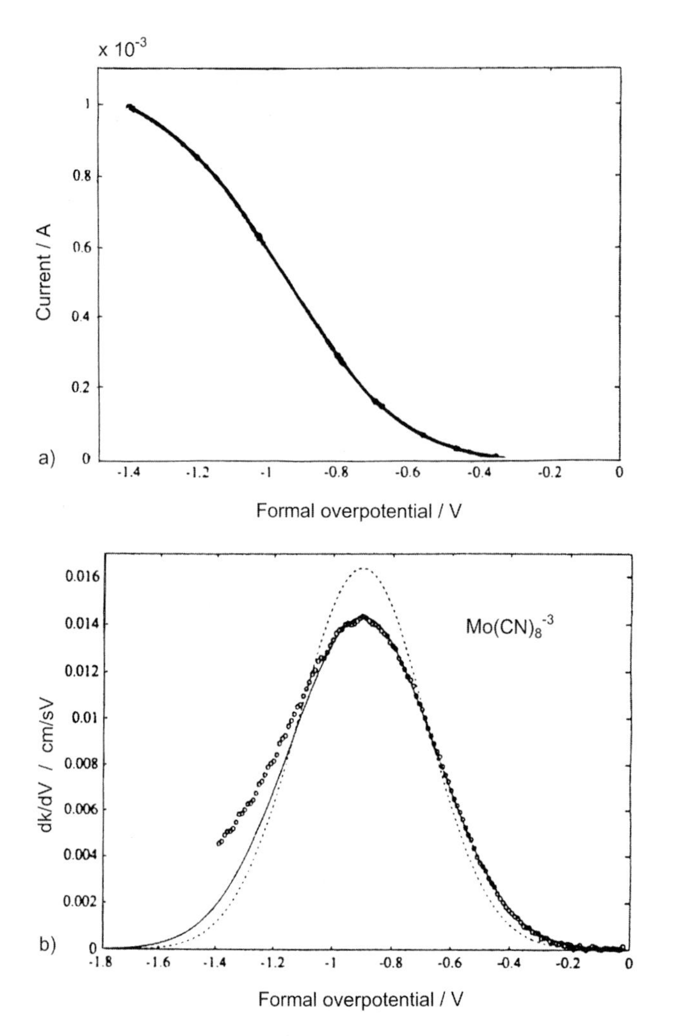

Fig. 7.38 a) Single reduction wave, corrected for diffusion, for a 10 mM $Mo(CN)_8^{3-}$ solution at an Au electrode derivatized with a tetradecanethiol monolayer. b) Density of electronic states for $Mo(CN)_8^{3-}$ calculated as the derivative of the heterogeneous electron transfer rate constant measured at the insulating layer on Au (Figs. 7.36 and 7.37). Solid curve, best fit to the experimental results; dashed curve, the Gaussian distribution predicted by the Marcus theory. (After ref. [53)])

occurs becomes larger and larger. One example is the reduction of $[Mo(CN)_8]^{3-}$. The corresponding current– potential curve shows clearly the saturation of the current as expected from the model (Fig. 7.38a). A quantitative analysis yielded a gaussian shape for W_{ox}, as given in Fig. 7.38b [53]. From the peak position a value of the reorganization energy was obtained ($\lambda = 0.9$ eV). This result proves very nicely the validity of the Marcus/Gerischer model or, more precisely, the validity of the harmonic oscillator picture derived in Chapter 6. Further information on the evaluation of λ values of other redox couples is given in Section 7.5.

In the case of semiconductor electrodes, it is impossible to obtain the same information because the energy bands are fixed at the surface and any potential variation occurs only across the space charge layer. Here the maximum rate constant is expected if the peak of the distribution curve occurs at the lower edge of the conduction band of an n-type semiconductor. Therefore, the experimental results obtained with the modified metal electrodes, are of great importance for the quantitative analysis of rate constants from current–potential curves measured with semiconductor electrodes (see e.g. Section 7.3.4).

7.4 The Quasi-Fermi Level Concept

7.4.1 Basic Model

When an external voltage is applied to a semiconductor electrode then it occurs as the difference between the electrochemical potential (Fermi level of the redox system) and the Fermi level in the semiconductor electrode, as previously illustrated for an n-type semiconductor in Fig. 7.13. More precisely, the latter is the Fermi level of the majority carriers. It is also sufficient to consider only this Fermi level if majority carriers are transferred across the interface. The difference between the Fermi levels on each side of the interface is, thermodynamically speaking, the driving force for the corresponding reaction. If no other reaction occurs then the minority carriers are still in equilibrium with the majority carriers, i.e. the quasi-Fermi levels of holes and electrons remain equal ($E_{F,n} = E_{F,p}$) even during current flow. The Fermi level is constant within the solid including the space charge region as proved by measurements of the space charge capacity (Mott–Schottky plot).

The situation is quite different if minority carriers are involved. Then electrons and holes are not in equilibrium and their quasi-Fermi levels become different. In the case of an n-type semiconductor, $E_{F,p}$ can be located above or below $E_{F,n}$, depending on the minority carrier process, i.e. on whether minority carriers are extracted from or injected into the semiconductor. However, quasi-Fermi levels have been qualitatively used in the theory of non-equilibrium processes in solid state devices, such as the excitation and recombination of electrons and holes (see Section 1.6), and also for the descriptions of charge transfer processes in p–n junctions (see Section 2.3). In this section a quantitative analysis of reactions at n- and p-type electrodes in terms of quasi-Fermi levels will be derived [19, 54].

In the present derivation we take a valence band process as an example. According to the quasi-Fermi level concept, it is assumed that the same reaction with identical rates, i.e. equal currents, takes place at an n- and p-type semiconductor electrode if the densities of holes, p_s, at the surface – or equivalently the quasi-Fermi levels, $E_{F,p}$ – are equal at the surfaces of both types of electrodes, as illustrated for an illuminated n-electrode and a p-electrode in the dark in Fig. 7.39. Since holes are majority carriers in a p-type semiconductor, the position of the quasi-Fermi level $E_{F,p}{}^{s}$ is identical to the electrode potential U_E (right side of Fig. 7.39) and therefore directly measurable with respect to a reference electrode. The density of the surface hole density, p_s, can be easily calculated, from Eq. (1.58) provided that the positions of the energy bands at the

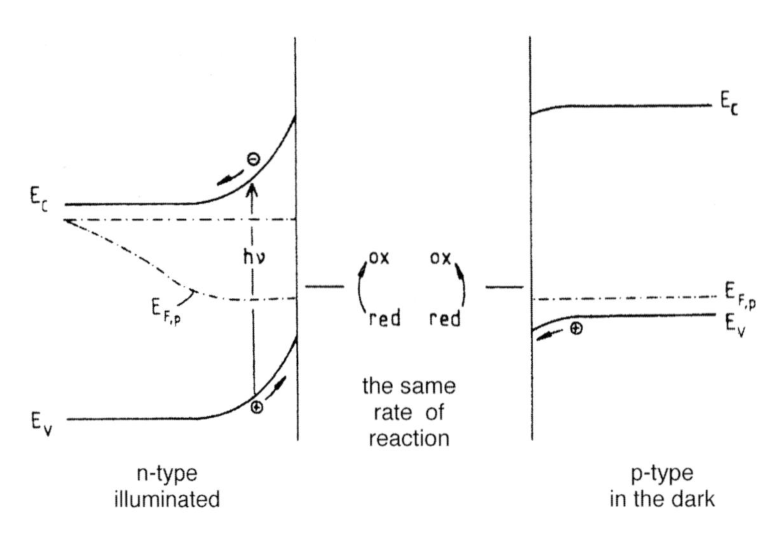

Fig. 7.39 Principle of the comparability of reactions at n- and p-type electrodes

surface are known. The measurement of a current–potential curve at the p-electrode also yields automatically the relation between current and quasi-Fermi level of holes because $E^{s}_{F,p} = eU_E$ at the p electrode in the entire potential range. The basic concept implies that the position of the quasi-Fermi level of holes, $E^{s}_{F,p}$, at the surface of an n-type semiconductor and the corresponding hole density p_s can be derived for a given photocurrent, because the same relationship between current and $E^{s}_{F,p}$ holds. However, since $E^{s}_{F,p}$ is the quasi-Fermi level of the minority carriers it is not equal to the electrode potential (see left side of Fig. 7.39).

This model is applicable if three conditions are fulfilled:
(1) At equilibrium, the conduction and valence band edges at the surface of the n- and p-type electrode have the same position.
(2) All reactions at the electrode can be described as a function of the surface hole density.
(3) The holes at the surface of the p-type electrode are nearly in equilibrium with those in the bulk, i.e. the Fermi level of the majority carriers is constant within the electrode.

Equivalent conditions can be expressed for a conduction band process. In this case electrons are the majority carriers.

This model has been proved experimentally by studying the competition of the anodic decomposition reaction and the oxidation of Cu^{1+} at p-GaAs in the dark and at n-GaAs under illumination [54]. This is a suitable redox system because reduction and oxidation occur via the the valence band and because the anodic oxidation of Cu^{1+} proceeds independently from the corrosion. Accordingly, the ratio of the oxidation current (oxidation of the redox system), j_{ox}, and the total current, j, is independent of the total current at p-GaAs as proved by using the rotating ring disc assembly (see also Section 8.5). The same result has been obtained with n-GaAs under illumination, and an identical j_{ox}/j ratio was found [54]. Other authors also have observed that the reactions at n- and p-type electrodes are comparable [55–57].The advantage of the model presented here is that the quasi-Fermi level of the majority carriers (holes in p-type) can be determined because it is identical to the electrode potential. Accordingly, the current–potential curve measured with the p-GaAs electrode yields directly the relation between $E^{s}_{F,p}$ and the hole current (dark current at p-GaAs). Since the same relation holds for processes at an n-GaAs electrode, the position of the quasi-Fermi level of holes (minority carriers) at the surface of the n-type electrode is then also known for a given current (dark or photocurrent).

It should be emphasized that an anodic current (valence band process) at the p-type electrode as well as at the corresponding n-type electrode occurs only if $E^{s}_{F,p}$ is located below the redox potential, as illustrated in Fig. 7.39. This is required in the dark and during light excitation. When an n-type electrode is polarized anodically, then usually a very small, but constant anodic current is observed in the dark. During this polarization, the band bending is increased and the quasi-Fermi level of electrons (majority carriers) is moved downward. The quasi-Fermi level of holes, however, remains at the same position (very slightly below the redox potential), because the current remains constant.

More insight into the quasi-Fermi level concept and its application has been obtained by studying the charge transfer processes between GaAs and the $Cu^{1+/2+}$ redox system in more detail [54]. The basic steps have already been discussed in Section 7.3.4. In Fig. 7.40, curve (a) represents the current–potential curve as obtained

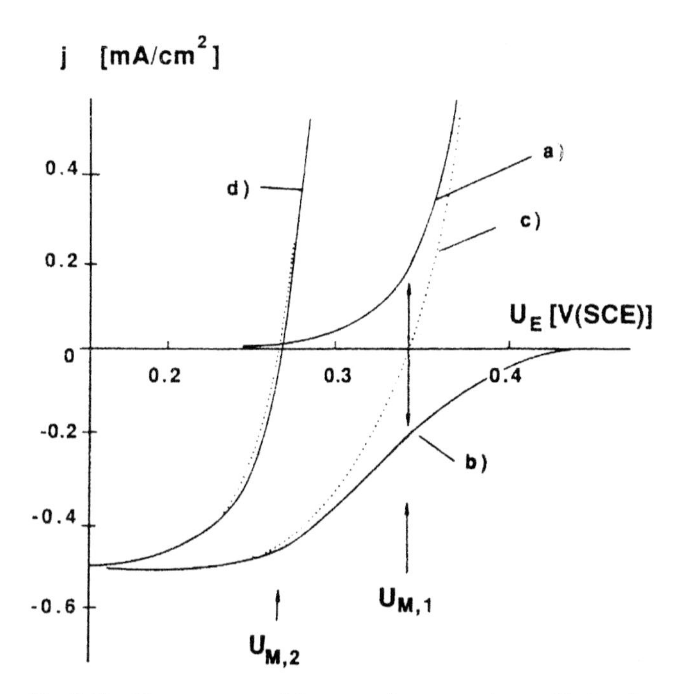

Fig. 7.40 Current–potential curves for a rotating p GaAs electrode in 6 M HCL: a) anodic decomposition current; b) partial current of Cu^{2+} reduction (0.7 mM); c) total current (dotted line); d) total current upon addition of Cu^{1+} ions (50 mM). (After ref. [54])

with p-GaAs in an electrolyte without any redox system. The anodic current corresponds to the decomposition of GaAs. After addition of Cu^{2+} as an oxidizing agent (hole injector), a corresponding cathodic current is visible (curve b). At the mixed potential, $U_{M,1} = 0.34$ V ($j = 0$), the two currents are equal. The position of $E^{S}_{F,p}$ in the p-GaAs electrode at this potential ($eU_{M,1} = E^{S}_{F,p}$) is illustrated in Fig. 7.41a. The same experiment has also been performed with n–GaAs. Here, the total dark current is nearly zero over a large potential range (>0.6 V), as shown in Fig. 7.42, i.e. $eU_{M,1} = E^{S}_{F,p} = +0.34$ eV, as illustrated for two potentials in Fig. 7.41b and c. In the range where the total current is zero, the latter is composed of two partial currents (dashed curves in Fig. 7.42). The cathodic partial current, j_{red}, corresponding to the reduction of Cu^{2+} at n-GaAs, was determined by using a rotating ring disc electrode (Pt ring, n-GaAs disc). The Pt ring electrode was polarized anodic of the redox potential so that the Cu^{1+} ions formed at the GaAs disc were re-oxidized. The cathodic disc current j_{red} was calculated from the corresponding ring current (see Section 4.3). The value of j_{red} was the same as was found with p-GaAs at $U_E = U_{M,1}$ ($j = 0$), provided that equal concentrations of Cu^{2+} ions were used. After addition of Cu^{1+} ions, the current–potential measurement yielded a mixed potential of $U_{M,2} = 0.26$ V (curve (d) in Fig. 7.40). In this case the anodic current is dominated by the oxidation of Cu^{1+} and in the range around $U_{M,2}$ the partial current due to the decomposition of GaAs is much smaller than around $U_{M,1}$. The measurements with the ring disc assembly (Pt ring, n-GaAs disc) yielded again the same partial current as found with the p-electrode. The positions of the quasi-Fermi levels for p- and n-type GaAs electrodes are given in Fig. 7.43, which is similar to Fig. 7.41.

This example shows very nicely how the quasi-Fermi level concept operates. There are various other results given in the literature which can be interpreted on the basis of this concept. In particular, the investigations of the reactions of $[Fe(CN)_6]^{3-/4-}$ at GaAs and the etching behavior confirm this model [58, 59].

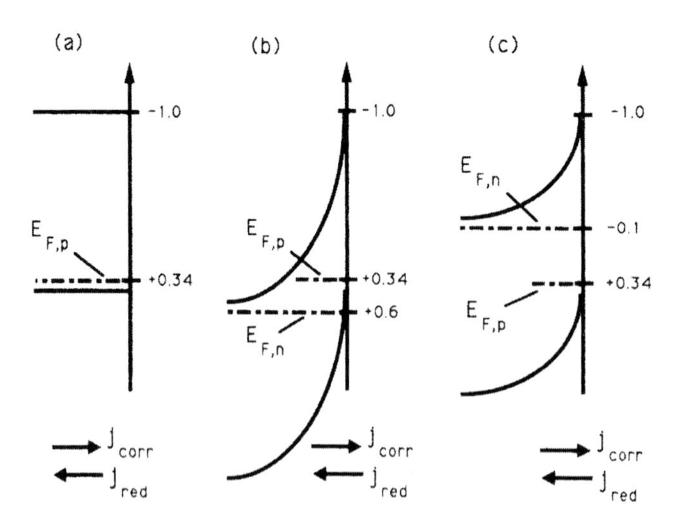

Fig. 7.41 Position of the quasi-Fermi levels in the presence of 0.7 mM Cu^{2+}, according to data given in Fig. 7.40 and 7.42: a) p-GaAs at the potential $U_{M,1} = +0.34$ V (saturated calomel electrode (SCE)); b) n-GaAs at $U_E = +0.6$ V; c) n-GaAs at $U_E = -0.1$ V. (After ref. [19])

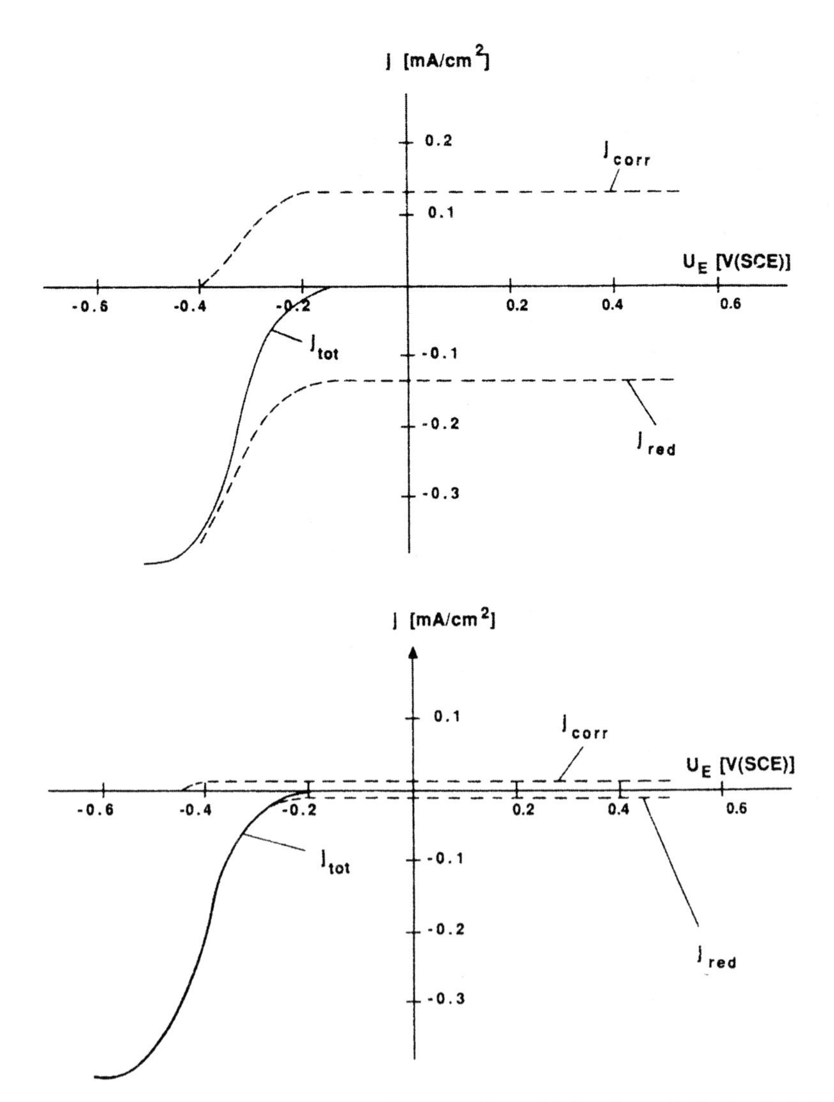

Fig. 7.42 Current–potential curve for a rotating n GaAs electrode in the dark in 6 M HCl with 0.76 mM Cu^{2+}. Dashed curves are the partial currents of the anodic decomposition (j_{corr}) and of Cu^{2+} reduction (j_{red}), as determined using a rotating ring disc electrode. (After ref. [54])

7.4.2 Application of the Concept to Photocurrents

Another important case is the comparison of anodic currents at n-type electrodes under illumination with the dark current at p-type, as illustrated in Fig. 7.44. The anodic dark current at the p-electrode is composed of corrosion (dashed curve) and the redox currents (dotted curve). The total current at the n-type electrode (solid curve) is now given by

$$j_{tot} = j_{ph} + j_0 + j_{rec} \tag{7.86}$$

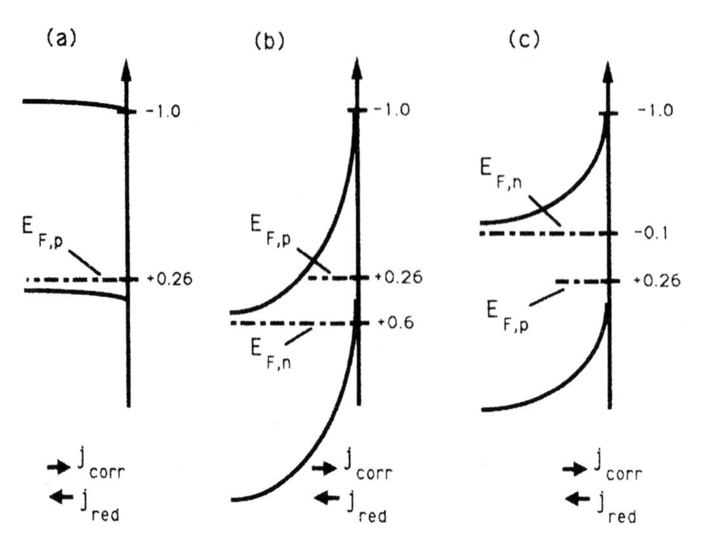

Fig. 7.43 Position of the quasi-Fermi levels in the presence of 0.7 mM Cu^{2+} and 50 mM Cu^{1+} according to data given in Fig. 7.40 and 7.42: a) p-GaAs at the potential $U_{M,1} = +0.26$ V (SCE); b) n-GaAs at the potential $U_E = +0.6$ V; c) $U_E = -0.1$ V. (After ref. [54])

in which j_{ph} is the photocurrent, j_0 the anodic dark current as defined by Eq. (7.65) and j_{rec} the recombination current. In general, the latter may be a complicated function of the quasi-Fermi level of electrons in the bulk, $E^b_{F,n}$, and the quasi-Fermi level of holes at the interface, $E^s_{F,p}$ [60, 61]. Considering, however, the most important processes, namely the diffusion mechanism derived by Shockley for p–ndiodes (see Section 2.3.3), and the Hall–Shockley–Read recombination (see Section 1.6), a simple calculation shows that the voltage applied across a solid state device is equal to the difference between the quasi-Fermi levels $E^b_{F,n}$ and $E^s_{F,p}$. Then the relation between the recombination current and the quasi-Fermi levels is given by

$$j_{rec} = -j_0 \exp\left(\frac{E^b_{F,n} - E^b_{F,p}}{nkT}\right) \tag{7.87}$$

in which n is the quality factor (see Chapter 2). $E^b_{F,n}$ is identical to the electrode potential of the n-type electrode $eU_E(n)$. Inserting this condition and Eq. (7.87) into (7.86), and solving Eq. (7.86) with respect to $U_E(n)$, one obtains

$$U_E(n) = U_E(p) - \frac{nkT}{e} \ln\left[\frac{j_{ph} - j_{tot}}{j_0} + 1\right] \tag{7.88}$$

in which we have assumed that $E^s_{F,p} = eU_E(p)$, where $U_E(p)$ is the electrode potential of the p electrode which exhibits the same j_{tot}. The current–potential curve for the illuminated n electrode has been calculated from the dark currents measured with the p-electrode. The results are given in Fig. 7.44. In the anodic range where the total cur-

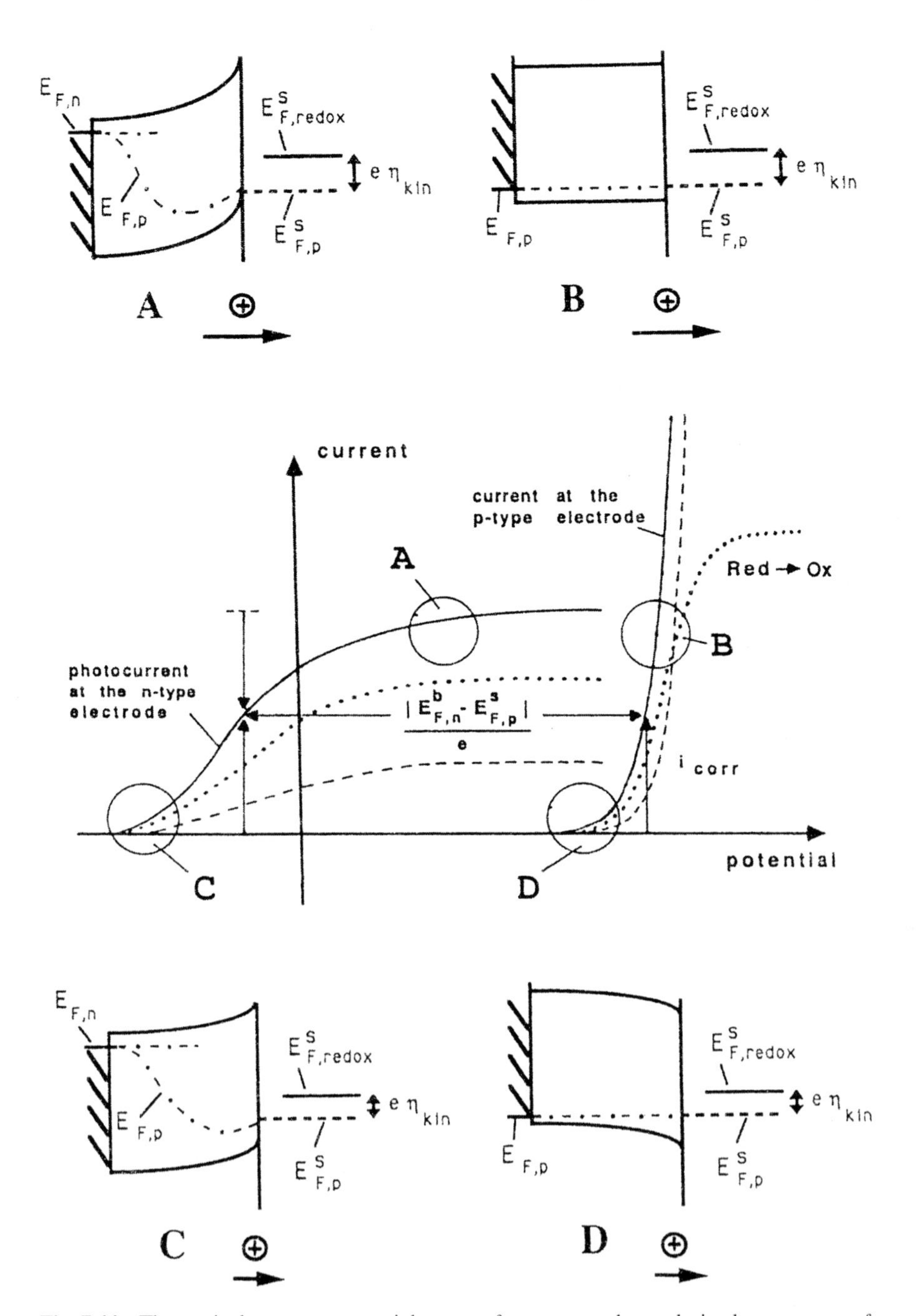

Fig. 7.44 Theoretical current–potential curve of an n-type electrode in the presence of an oxidized species of a redox system in the dark. It was calculated from an experimental current–potential curve measured with the corresponding p electrode (p-GaAs): solid lines, total current j_{tot}; dashed lines, partial current of anodic decomposition; dotted lines, partial current of reducing an Ox species (arbitrary units). Inserts A–D show energy schemes of the n- and p-type electrodes at potentials marked in the j–U_E curves. (After ref. [35])

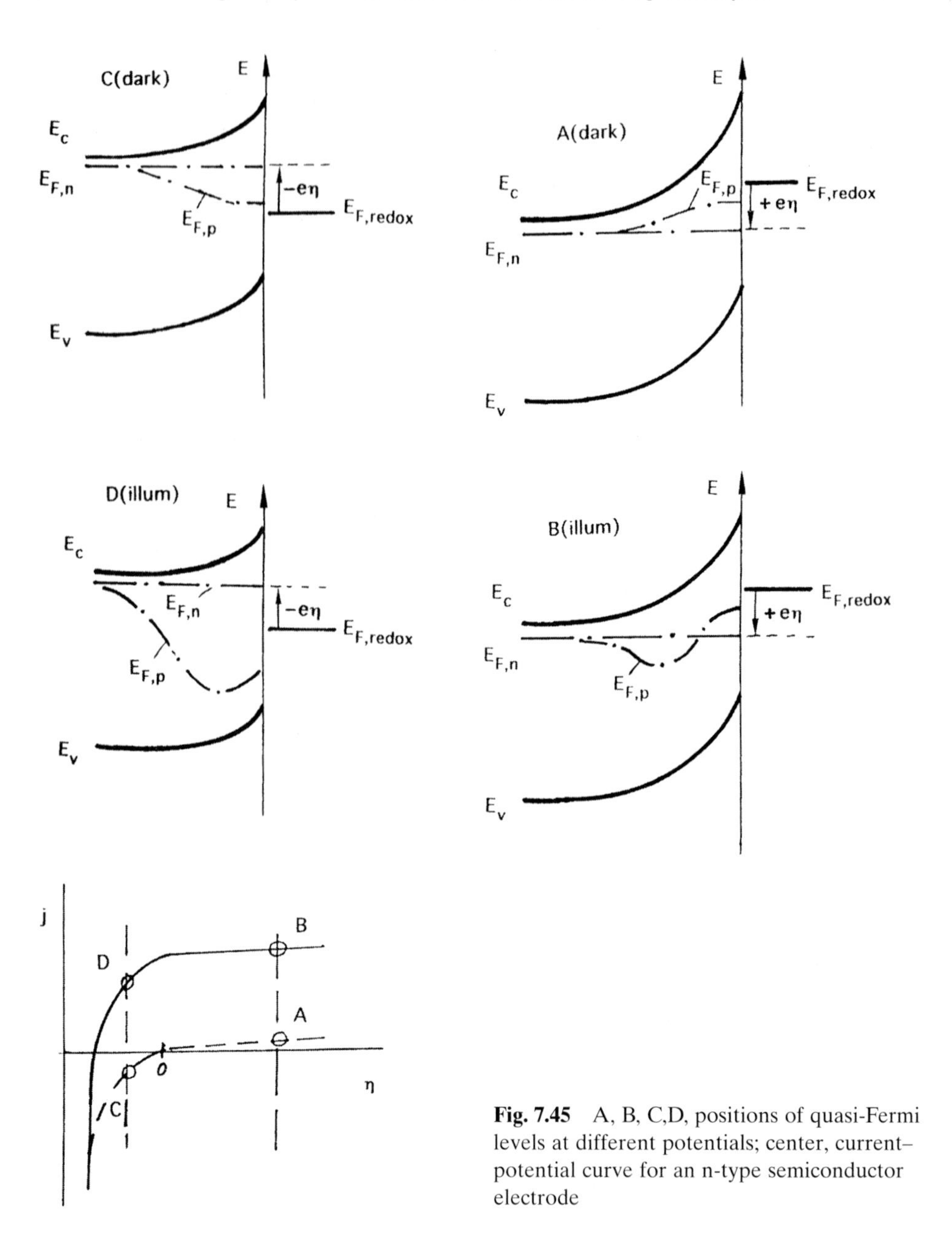

Fig. 7.45 A, B, C,D, positions of quasi-Fermi levels at different potentials; center, current–potential curve for an n-type semiconductor electrode

rent, which is mainly determined by the photocurrent, is constant, the quasi-Fermi level of holes, $E^{s}_{F,p}$, remains pinned (range A and B). In the range of the photocurrent onset (range C), the difference between the quasi-Fermi levels $E^{s}_{F,p}$ and $E^{b}_{F,n}$ remains nearly constant. Although $E^{s}_{F,p}$ is identical at the surface of n- and p-GaAs, the quasi-Fermi level $E_{F,p}$ may be different below the surface of n-GaAs due to the excitation profile (see A and C in Fig. 7.44).

Some authors have asserted that the quasi-Fermi level model requires a threshold with respect to light intensity. This problem was recently discussed for photoconversion systems, such as photoelectrolysis of H_2O, by Gregg and Nozik [62] and by Shreve and Lewis [63]. Since the discussion of the threshold problem has frequently lead to misinterpretations, we want to clarify the situation by considering a simple charge transfer between an n-type semiconductor and redox system as illustrated in Fig. 7.45. The system is at equilibrium ($j = 0$) if the overvoltage is zero ($\eta = 0$). Here the quasi-Fermi levels of electrons and holes are both equal to $E_{F,redox}$ (not shown). Assuming that the redox process occurs entirely via the valence band, then only the quasi-Fermi level of holes at the surface, $E^{s}_{F,p}$, is of interest. Polarizing the electrode anodically in the dark, a very small anodic current is found (see the j–η curve in the center of Fig. 7.45). As already mentioned in the previous section, $E^{s}_{F,p}$ is practically pinned to $E_{F,redox}$ or is slightly below it (Fig. 7.45A) whereas $E_{F,n}$ differs from $E_{F,redox}$ by $e\eta$. The position of $E_{F,n}$ is not of interest here. The anodic current is increased upon illumination and $E^{s}_{F,p}$ is shifted downward whereas $E_{F,n}$ remains unchanged (Fig. 7.45B). The shift is small at low intensities and large for high intensities. Since the dark position of $E^{s}_{F,p}$ already occurs below $E_{F,redox}$ there is no threshold for the anodic photocurrent with respect to light intensity. At cathodic bias, $E^{s}_{F,p}$ occurs above $E_{F,redox}$ because of the cathodic dark current (Fig. 7.45C). Here the difference between $E_{F,n}$ and $E^{s}_{F,p}$ is of interest insofar as it determines the recombination current as given by Eq. (7.88). During illumination an anodic current occurs as shown by the j–η curve. In this case, $E^{s}_{F,p}$ moves below $E_{F,redox}$ if the total current becomes positive (Fig. 7.45D). Accordingly, at low intensities j_{tot} remains negative, i.e. the hole injection still dominates. This is, however, not a real threshold. It must be remembered that the difference between $E^{s}_{F,p}$ and $E_{F,redox}$ is the driving force of the reaction and the relative position of the two Fermi levels determines the direction of the hole transfer.

7.4.3 Consequences for the Relation between Impedance and IMPS Spectra

The quasi-Fermi level concept makes it possible to quantitatively relate intensity-modulated photocurrent spectroscopy (IMPS) and impedance data as follows.

According to Eq. (4.15) the relative quantum yield, ϕ_r, obtained by IMPS measurements is given by

$$\phi_r = \frac{\Delta j}{\Delta j_{ph}} \tag{7.89}$$

(Here we express all currents in current densities.) The recombination current depends on the position of the quasi-Fermi levels in the semiconductor. Therefore, instead of Eq. (4.20), the current modulation can also be expressed in terms of $E_{F,n}$ and $E_{F,p}$, as given now by

$$\Delta j = \Delta j_{ph} - \frac{\mathrm{d}j_{rec}}{\mathrm{d}E_{F,n}}\Delta E_{F,n} - \frac{\mathrm{d}j_{rec}}{\mathrm{d}F_{F,p}}\Delta E_{F,p} \tag{7.90}$$

In Fig. 7.46 the modulation of photocurrent at an n-type electrode is illustrated. The electrode potential corresponds to the quasi-Fermi level of electrons (majority carriers), $E_{F,n}$. During the modulation of the light intensity $E_{F,n}$ remains constant when

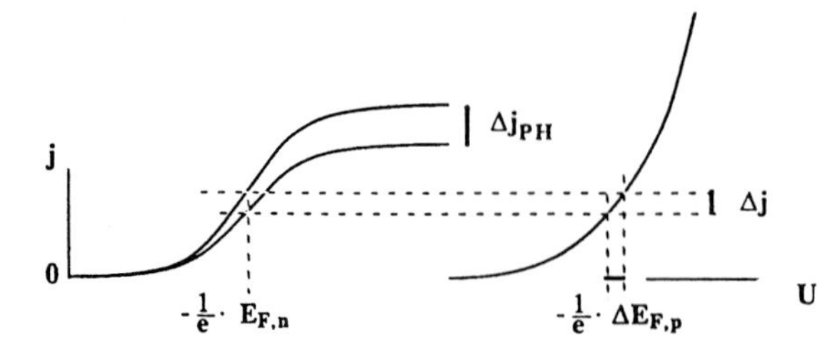

Fig. 7.46 Illustration of intensity-modulated photocurrent spectroscopy (IMPS) in terms of current–potential curves (see text)

potentiostatic conditions obtain. The photocurrent corresponds to a hole transfer via the valence band. The current–potential curve as measured with the corresponding p-type electrode in the dark (valence band process) is shown in the same figure. Here the same magnitude of current modulation can only be obtained by modulating the electrode potential, which means a modulation of the quasi-Fermi level of holes, $E_{F,p}$ (majority carriers in p-type) as also indicated in Fig. 7.46. In the latter case the potential modulation can be used for determining the impedance of a p-type sample. The variation of current upon modulating the light intensity leads to a modulation of j_{rec} at the n electrode and, therefore, of the quasi-Fermi level $E_{F,p}$ because j_{rec} is a function of the difference of $E_{F,n}$ and $E_{F,p}$ ($j = f(E_{F,n} - E_{F,p})$). Accordingly, we have

$$j_{rec} = f\left(E_{F,n} - E_{F,p}\right) \rightarrow \left(\frac{dj_{rec}}{-1/edE_{F,n}}\right) = -\left(\frac{dj_{rec}}{-1/edE_{F,p}}\right) = -\frac{1}{R_{rec}} \quad (7.91)$$

in which R_{rec} is the recombination resistance. The general condition required by the quasi-Fermi concept, is

$$\Delta E_{F,p}/\Delta j\big|_{\text{nelectrode}} = \Delta E_{F,p}/\Delta j\big|_{\text{pelectrode}} \quad (7.92)$$

Since in the case of IMPS at an n electrode $E_{F,n}$ is constant, Eq. (7.90) reduces to

$$\Delta j = \Delta j_{ph} - \frac{dj_{rec}}{dE_{F,p}}\Delta E_{F,p} \quad (7.93)$$

In addition, the Faraday impedance at the p electrode is given by

$$Z_{F,p} = \frac{-1/e\Delta E_{F,p}}{\Delta j} = \frac{\Delta U_E}{\Delta j} \quad (7.94)$$

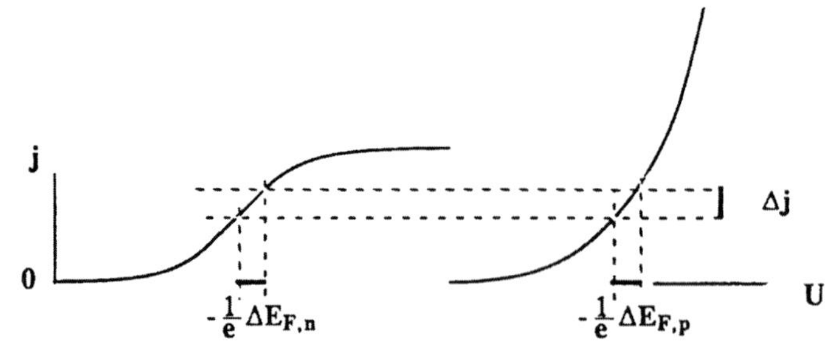

Fig. 7.47 Illustration of impedance measurements in terms of current-potential curves in the dark (p-type) and under illumination (n-type); see text

Substituting Eqs. (7.91), (7.93) and (7.94) in (7.89) one obtains

$$\eta_r^{-1} = 1 + \frac{Z_{F,p}}{R_{rec}} \tag{7.95}$$

One can also derive the Faraday impedance, $Z_{F,n}$, of the illuminated electrode. Here the electrode potential and consequently $E_{F,n}$ is modulated and $\Delta j_{ph} = 0$ (Fig. 7.47). Similarly to Eq. (7.94) we have then

$$Z_{F,n}(\text{illum}) = \frac{-1/e\Delta E_{F,n}}{\Delta j} \tag{7.96}$$

Inserting this equation into Eq. (7.90) and using Eqs. (7.91), (7.94) and (7.95) one obtains

$$Z_{F,n}(\text{ill}) = R_{rec}\phi_r^{-1} \tag{7.97}$$

$$Z_{F,n}(\text{ill}) = Z_{F,p} + R_{rec} \tag{7.98}$$

Corresponding measurements of spectra of ϕ_r, $Z_{F,n}$(ill) and of $Z_{F,p}$ have been performed with n- and p-type GaAs using the $Cu^{1+/2+}$ as a redox system (see above). One example, which proves the model rather well, is given in Fig. 7.48a and b.

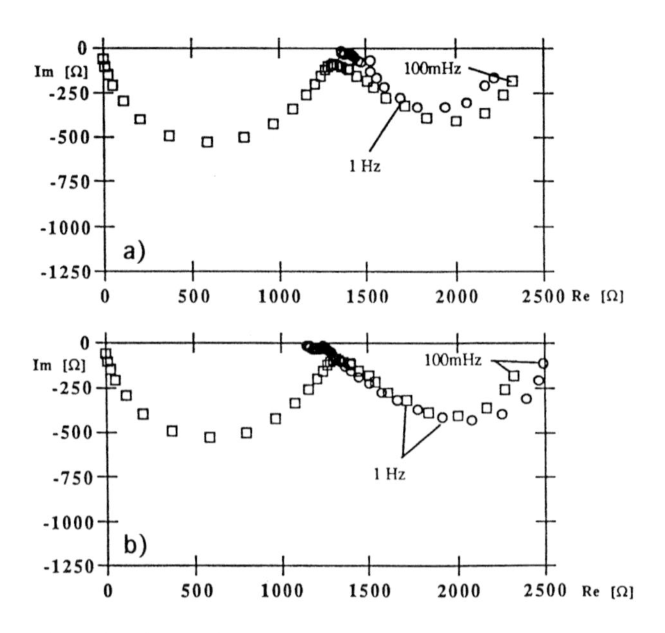

Fig. 7.48 a) Impedance Z (squares) and IMPS $1/\phi_r \times 1250\ \Omega$ (circles) spectra, at n-GaAs electrodes under anodic bias in the presence of 8 mM Cu^{1+} HCl during illumination; the $1/\phi_r$ data were multiplied by 1250 Ω to fit the two curves (see Eq. 7.97). b) Impedance spectra for n- and p-GaAs electrodes, Z_n(squares) and Z_p + 1250 Ω (circles) measured under the same conditions as above; to fit the two curves 1250 Ω were added to Z_p (see Eq. 7.98). (After ref. [110])

7.4.4 Quasi-Fermi Level Positions under High Level Injections

Tan et al. have studied charge transfer processes at fairly high-ohmic Si-electrodes (carrier density: 3×10^{13} cm^{-3}) in CH_3OH electrolytes [64]. They used thin electrodes, the thickness of which was smaller than the diffusion length. Accordingly, an excitation led to a smooth concentration of electron–hole pairs which was constant throughout the specimen, i.e. the quasi-Fermi levels, $E_{F,n}$ and $E_{F,p}$, were also constant. The thickness of the sample was also smaller than the theoretical thickness of the space charge layer, so that the electrical field was very small and the band bending could be neglected as shown in Fig. 7.49b–d. Accordingly, charge separation in this system had to rely on kinetic differences for the collection of photogenerated charge carriers.

The specimen is provided with n^+ and p^+ contacts at the rear of the electrode as illustrated in Fig. 7.49a and b. In the dark the two quasi-Fermi levels are equal. The barrier height at the n^+ contact is small so that the electron can easily be transferred across this contact in both directions. In the case of the p^+ contact, however, the barrier height is very large in the dark, i.e. it is a rectifying contact and holes can only be transferred from the n electrode to the p^+ contact. Since no holes are available in the n electrode, the dark current is negligibly small (Fig. 7.50b). The situation changes completely at the p^+ contact under illumination. Since the holes remain at equilibrium at this contact the quasi-Fermi level of the minority carriers is constant across the hole specimen, as shown in Fig. 7.49d. Hence, the holes can now travel across the contact in both directions. The most important point is that any potential measurement at the p^+ contact, here V_{p+}, gives quantitative information about the position of $E_{F,p}$. The corresponding current, j_{p+} as measured with dimethyl ferrocene ($Me_2Fc^{0/1+}$) is shown in Fig. 7.50b. All potentials are given vs. the redox potential, $E_{F,redox}(A/A^-)$. Under open circuit conditions, the potential at the p^+ contact is nearly zero; $E_{F,p}$ is close to $E_{F,redox}(A/A^-)$. At more positive potentials (i.e. when moving $E_{F,p}$ downward) the hole current strongly increases because holes can be moved from the p^+ contact through the illuminated electrode toward the Si–electrolyte interface. The current saturation at negative bias under illumination is only due to mass transfer limitation. In the case of the n^+ contact, a current–potential curve was measured which was typical for an n-type semiconductor (Fig. 7.50a). The photovoltage measured at the n^+ contact, corresponds to the difference of $E_{F,n}$ and $E_{F,redox}(A/A^-)$; in the case of Me_2Fc, $U_{ph} = 0.55$ V.

Using a redox system such as $Me_{10}Fc$ which has a more negative standard potential than Me_2Fc, one obtains current–potential curves as given in Fig. 7.50c and d. In this case the photovoltage is smaller because $E_{F,redox}$ is closer to $E_{F,n}$ ($U_{ph} = 0.37$ V). On the other hand, a photovoltage (–0.08 V) does now also occur at the p^+ contact because $E_{F,p}$ is located below $E_{F,redox}$ (see Fig. 7.49b). Using a redox couple such as cobaltocene, where its standard potential occurs close to the conduction band of Si, then U_{ph} becomes very small at the n^+ contact and large at the p^+ contact. According to these results, the illuminated material behaves like a p-type semiconductor electrode.

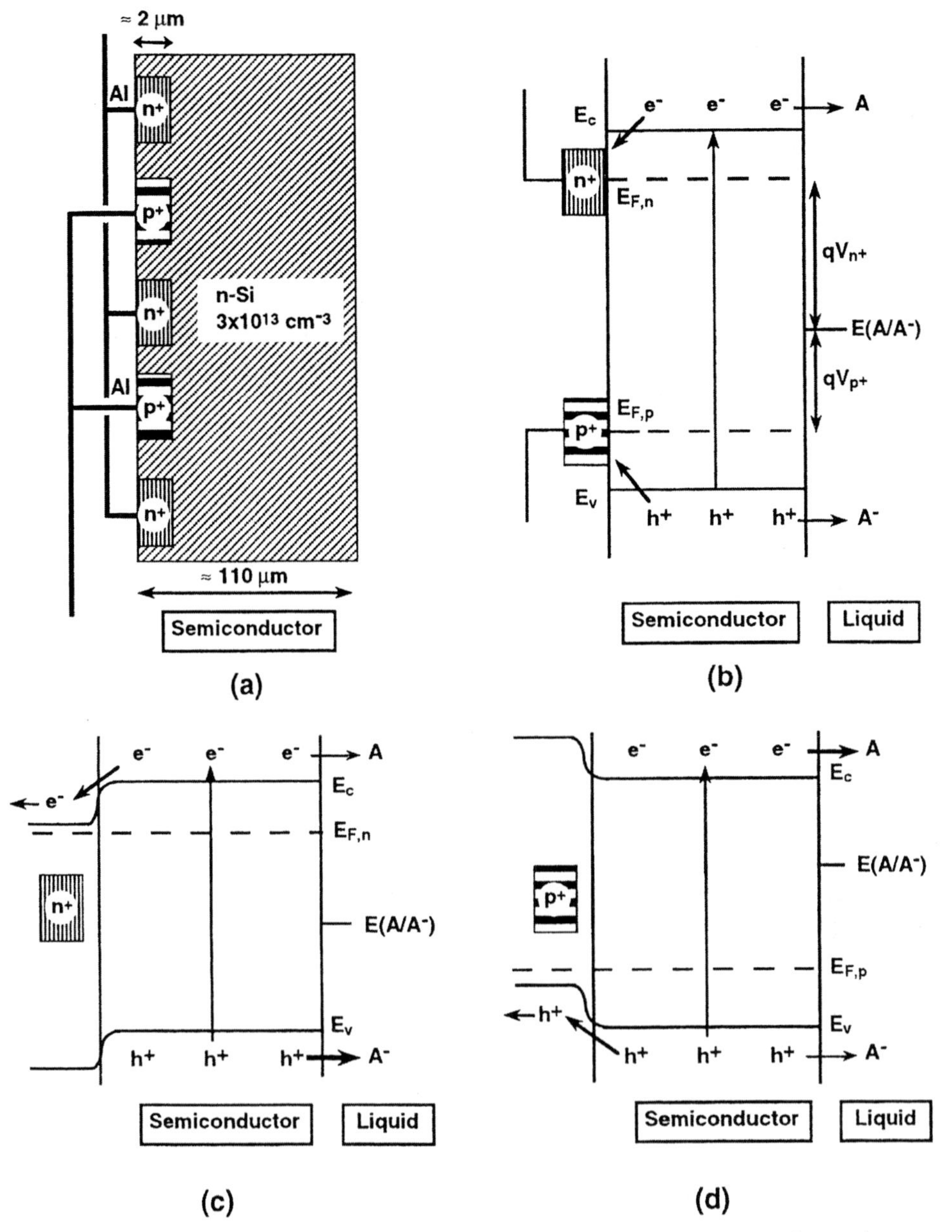

Fig. 7.49 a) Schematic representation of a high purity Si electrode showing the n^+ and p^+ at the rear of the sample. b) Band diagram of an Si–liquid junction containing a redox system (A/A^-) under high level injection (illumination), open circuit conditions; V_{n+} and V_{p+} are potentials measured at the n^+ and p^+ rear points against the solution potential ($E(A/A^-)/e$). c) Band diagram of the junction under potentiostatic control of the n^+ contact points. Electrons are collected from the back of the sample (n^+ region), and the holes are collected in the solution. Because of the lack of a significant electric field at the interface to effect charge separation, electrons can also react with the solution. d) Band diagram of the junction under potentiostatic control of the p^+ contact points. (After ref. [64])

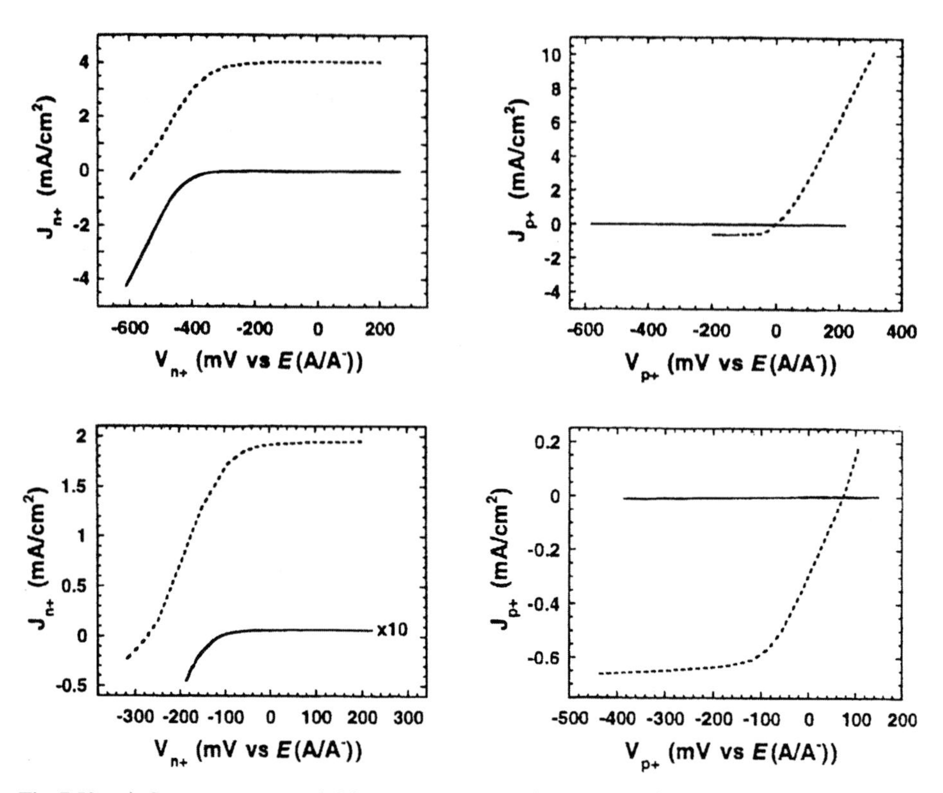

Fig. 7.50 a) Current vs. potential for patterned n-Si (see Fig. 7.49) electrodes in methanol containing methyl ferrocene ($Me_2Fc^{0/+}$), when the potential was applied to, and current collected at, the n^+ point contacts: solid line, in the dark; dashed line, under illumination. b) The same as a) when the potential was applied to, and the current collected at, the p^+ point contacts. c) The same as a) when using another redox system ($Me_{10}Fc^{0/+}$) which has a redox potential which is closer to the valence band of Si compared with the case of $Me_2Fc^{0/+}$. The potential was applied to and the current collected at n^+ point contacts. d) The same as c) when the potential was applied to, and the current collected at, p^+ point contacts. (After ref. [64])

7.5 Determination of the Reorganization Energy

As already mentioned in Sections 7.3.4 and 7.3.6, knowledge of the reorganization energy, λ, is important for the quantitative evaluation of charge transfer processes at semiconductor electrodes. According to Eq. (6.10) it can easily be calculated from $\Delta G^{\#}$ values. The latter have frequently been determined from measurements of the rate of electron exchange reactions between the oxidized and reduced species of a redox system in homogeneous solutions by isotopic tracer methods [66], or in some cases by using NMR techniques [67]. Some $\Delta G^{\#}$ values have been determined from measurements of the temperature dependence of the rate constant, others have been directly calculated from the rate constants. In the latter case, however, considerable errors can occur because the transmission factor κ in Eq. (6.11) is not known. Frequently, $\kappa = 1$ has simply been assumed. One example is the $Fe(H_2O)_6^{2+/3+}$ couple.

Here a value of $\Delta G^{\#} = 0.7$ eV is found, which leads to $\lambda = 2.8$ eV [66]. This rather large value is mainly due to an inner sphere reorganization (see Section 6.1.2). Much smaller values are obtained with pure outer sphere redox systems, for instance metallocenes. In the latter cases $\Delta G^{\#}$ values in the order of 0.2–0.26 eV have been reported [67], i.e. values which correspond to λ = 0.7–1 eV. There are other cases such $[Fe(CN)_6]^{3-/4-}$ where one would also expect an outer sphere reorganization but rather high values have been found ($\Delta G^{\#}$ = 0.55 eV; λ = 2.2 eV) [67]. In this context it should also be mentioned that modern theories on electron transfer at electrodes have shown that the λ values also depend on the distance of the electron acceptor or donor molecules from the electrode surface [65].

Concerning electrochemical charge transfer reactions at metal electrodes, the rate constants have been determined from exchange currents from which $\Delta G^{\#}$ values have again been calculated assuming $\kappa = 1$ [110], i.e. the same problem arises as with homogeneous solutions. As discussed in Section 6.1.2, reorganization energies are expected for heterogeneous processes which are half of those found for reactions in homogeneous solutions. In a few cases this condition seems to be fulfilled, in others not. From this point of view it was a big step forward when Miller and co-workers [53] succeeded in measuring kinetically controlled currents with Au electrodes, the surfaces of which were modified by a monolayer of ω-hydroxy thiol, over a large potential range (see Section 7.3.6). These authors evaluated the current–potential curves in the following way.

The current is given by the usual equation

$$j^- = ek^- c_{\mathrm{ox}} \tag{7.99}$$

in which the rate constant is given by

$$k^- = \kappa \nu T \int_0^{E_F} W_{\mathrm{ox}}(E)\mathrm{d}E \tag{7.100}$$

so that the derivative of the rate constant k^- with respect to E_F or to the overvoltage η is

$$\mathrm{d}k^-/\mathrm{d}E_F = 1/ec_{\mathrm{ox}}\mathrm{d}j^-/\mathrm{d}\eta = \kappa\nu T W_{\mathrm{ox}}(E) \tag{7.101}$$

in which κ is the transfer coefficient; ν is the frequency factor; T is the probability for electron tunneling through the thiol monolayer, and W_{ox} is the distribution function of density of states for the oxidized species of the redox system, as given by Eq. (6.35a).

According to Eq. (7.101) the distribution function can be obtained just from the derivative of the current–potential curve without the need for any information on $\kappa\nu$T. Various systems have been investigated and the results are presented in Fig. 7.51. The reorganization energy can be obtained from the maximum since here $\lambda = e\eta_{\mathrm{form}}$ (the overvoltage with respect to the standard potential). The λ values derived for various couples are given in Table 7.1. The potential range was limited to U_E = –0.8 V (vs. Ag/AgCl, saturated KCl) because of the stability problems of the tunneling layer. Accordingly, a complete distribution function could only be measured with redox couples of a rather positive standard potential and not excessively high λ value, e.g. $Mo(CN)_6^{4-/3-}$ (Fig. 7.38b) and $W(CN)_6^{4-/3-}$(Fig. 7.51C).

Investigations of reorganization energies at semiconductor electrodes have already been performed in the mid-1970s [27, 14]. In this case SnO_2 electrodes of relatively

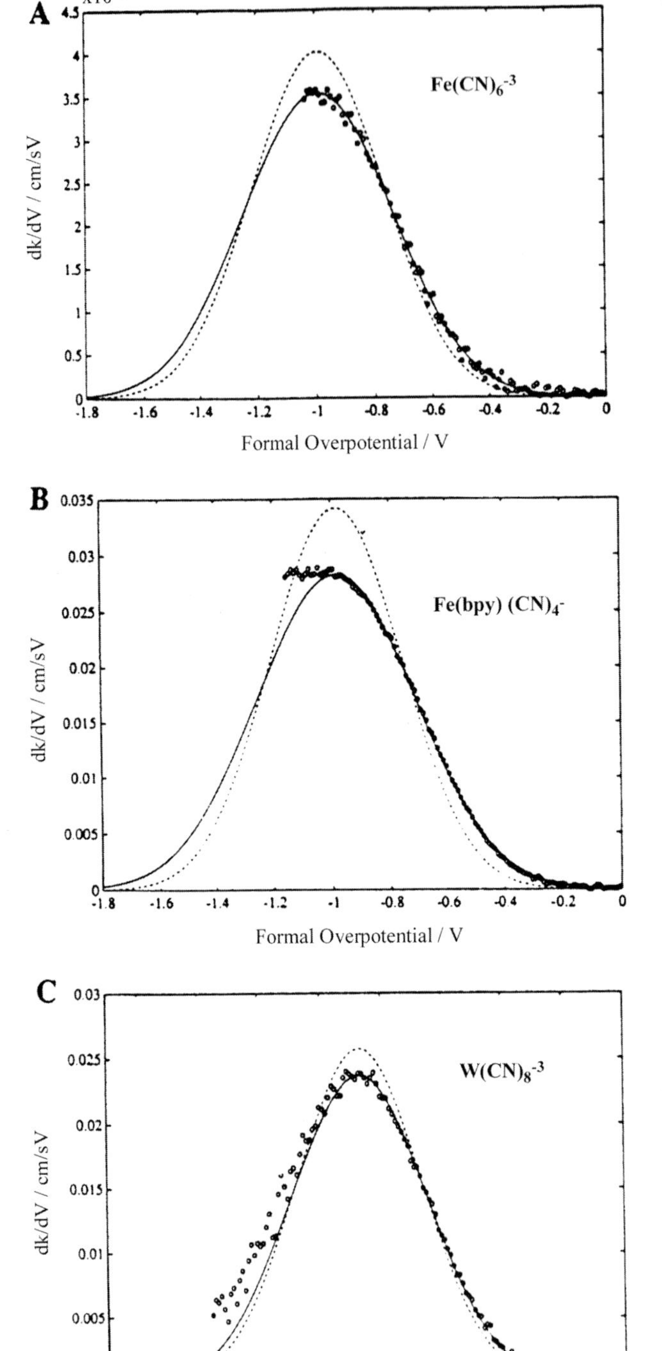

Fig. 7.51 Density of electronic states for various redox systems, as calculated from current–potential measurements at Au electrodes derivatized with a tetradecanethiol monolayer (compare with Figs. 7.37 and 7.38). (After ref. [53], for details concerning the experimental conditions, see this reference)

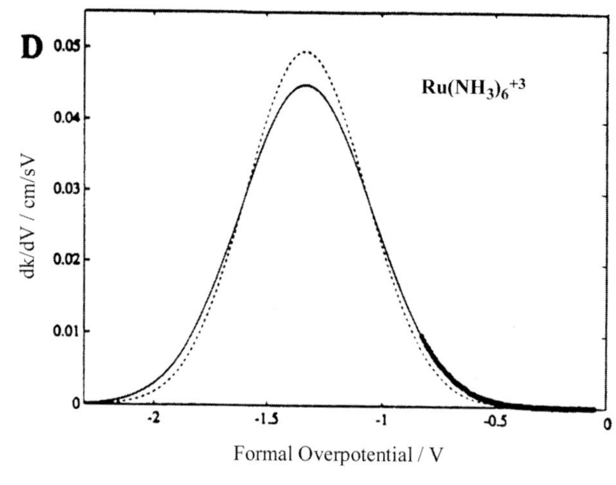

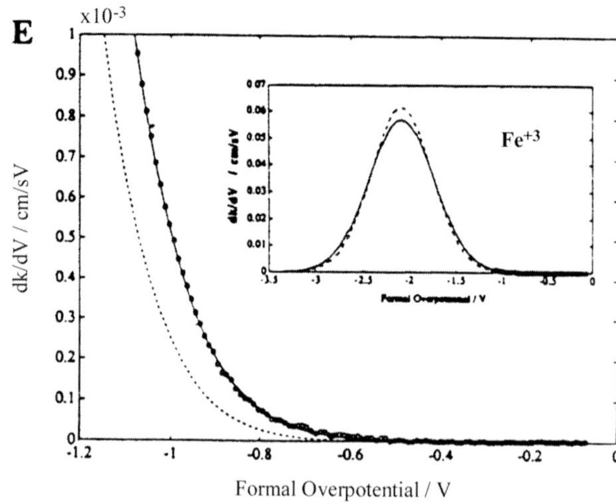

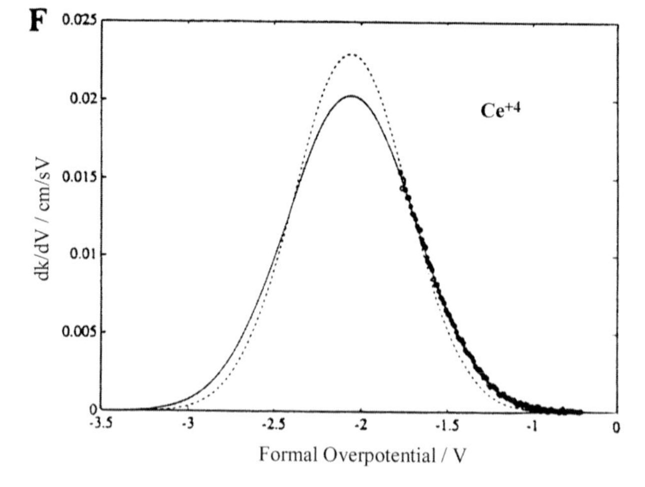

Fig. 7.51 (continued)

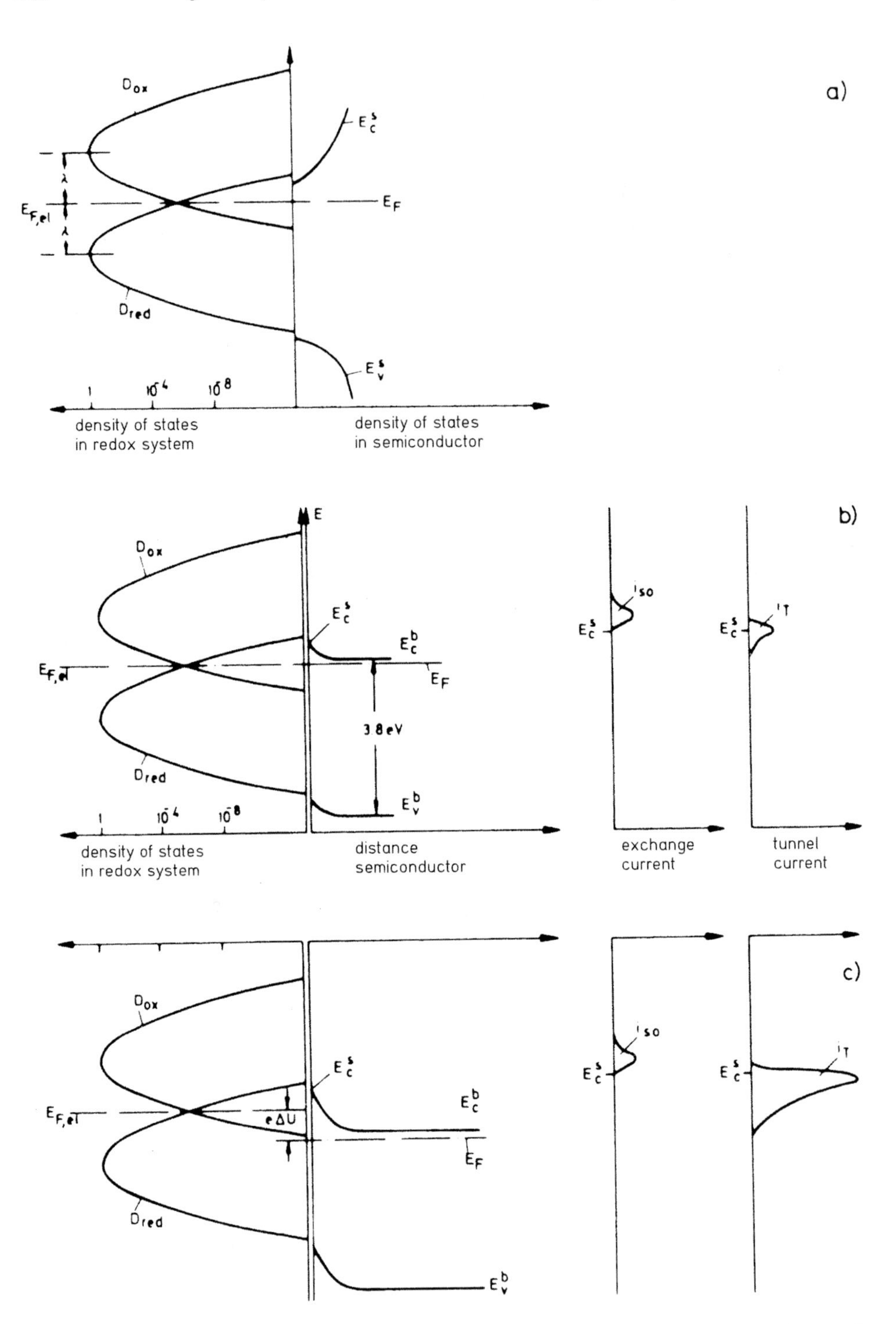

Fig. 7.52 Schematic energy diagram for the interface of a SnO_2–redox system: a) and b) at equilibrium; c) at anodic bias; i_{so}, exchange current; i_T tunneling current. (After ref. [14])

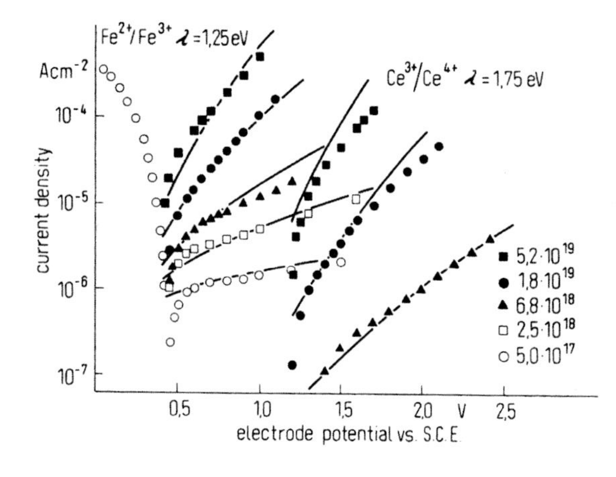

Fig. 7.53 Interfacial current vs. electrode potential for 0.05 M $Ce^{3+/4+}$ and 0.05 M $Fe^{2+/3+}$ at SnO_2 electrodes of different doping densities in 0.5 M. H_2SO_4; dots, experimental values; solid lines, theoretical curves. (After ref. [28])

high doping were used, so that electrons could tunnel through the space charge layer as illustrated in Fig. 7.52. Examples of the anodic oxidation of Fe^{2+} and Ce^{3+} are given in Fig. 7.53. According to this figure, the anodic current corresponding to the oxidation of Fe^{2+} is nearly independent of the electrode potential for a SnO_2 electrode of relatively low doping. Since the energy bands are fixed at the surface the electron transfer proceeds in the usual way via the conduction band, without any tunneling. Using electrodes of higher carrier densities, however, the current becomes potential-dependent because the space charge layer is sufficiently thin and tunneling can occur. In this case the anodic current is given by

$$j_c^+ = \frac{e}{(\pi kT\lambda)^{\frac{1}{2}}} \int_{E_c^b}^{E_s^b} k_c^+(E)T(E)N_c c_{red} dE \tag{7.102}$$

in which the denominator is a normalization factor; the rate constant is here energy-dependent as given by

$$k_c^+ = k_{c,max} W_{red}(E) \tag{7.103}$$

where the distribution function $W_{red}(E)$ is given by Eq. (6.36). The probability of electron tunneling through the space charge layer was obtained from

$$T(E) = T_0 \exp\left[-\frac{8\pi}{3h}(2m^*)^{1/2}\left(E_c^s - E\right)^{1/2}\kappa(E)\right] \tag{7.104}$$

where $(E_c^s - E)$ is the distance between the upper edge of the barrier E_c^s (lower edge of the conduction band) and the kinetic energy E of the tunneling electrons (see Fig. 7.52), and m^* is the reduced electron mass. Although the potential distribution across the space charge layer is not linear, the approximation by a triangular profile is adequate because the tunneling probability depends only slightly on the barrier shape. The length of the tunnel path is $z(E)$, and the relation between z and E was obtained by integrating the Poisson equation for the case of a depletion layer as given by

$$z(E) = \frac{1}{e}\left(\frac{2\varepsilon\varepsilon_0}{n_0}\right)^{1/2}\left[\left(E_c^s - E_c^b\right)^{1/2} - \left(E - E_c^b\right)^{1/2}\right] \tag{7.105}$$

Table 7.1 Reorganization energies

Redox system	U^0_{redox} (V) (SCE)	Solvent	λ (eV)		
			Homogeneous solutions [66]	Modified Au electrodes [53]	Heavily doped SnO_2 electrodes [28]
Fe^{2+}/Fe^{3+}	0.46	H_2O	2.8	2.1	1.2
Ce^{3+}/Ce^{4+}	1.20	"	3.3	2.1	1.75
$[Fe(CN)_6]^{4-/3-}$	0.26	"	2.2	0.99	0.4
$[Mo(CN)_8]^{4-/3-}$	0.61	"		0.9	
$[W(CN)_8]^{4-/3-}$	0.38	"		0.87	
$[Fe(CN)_4bipy]^{2-/1-}$	0.38	"		0.96	
$Ru(NH_3)^{2+/3+}$	–0.15	"		1.3	
Quinone/ hydroquinone	0.44	"			0.5
$Ferrocene^{0/1+}$	0.04	CH_3OH	1.0 [67]		0.5

Theoretical current–potential curves were determined by using Eqs. (7.102) to (7.105) as given by the solid curves in Fig. 7.53 which gave the best fit with experimental data. The theoretical curves, calculated for one redox system but with different dopings, were obtained with a single λ value.

The experimental values of the reorganization energy as obtained by various methods are listed in Table 7.1. Comparing these data, it is obvious that the values obtained with the modified Au electrode are considerably larger than those found with SnO_2 electrodes. The origin of this deviation is not clear at present. According to Section 6.1.2, the reorganization energy involved in an electron transfer process at an electrode, should be one-half of that in self-exchange reactions of the same redox system in homogeneous solutions. Although there is a tendency for λ to be considerably smaller for electrode processes, however there is no clear proof. Accordingly, data obtained for homogeneous solutions, should be treated with care when it is used for electrode reactions.

7.6 Two-step Redox Processes

Two-step redox processes are understood as reactions in which two electrons are transferred at the electrode, until a stable state of the redox system is reached. This occurs mainly in the oxidation and reduction of organic molecules, and also in the reduction of H_2O_2, $S_2O_8^{2-}$ and O_2. Frequently, it has been observed that the two reaction steps occur via different energy bands. In the first step, usually a very reactive radical is formed. This can be checked by current–potential measurements as shown for the cathodic reduction of H_2O_2 at n- and p-GaP in Fig. 7.54. According to this figure, the cathodic photocurrent at the p-GaP electrode is doubled in the saturation range upon addition of a sufficiently high concentration of H_2O_2[68]. Since in this range the photocurrent is only limited by the light intensity the *current doubling* means that one of the two electrons must be transferred without excitation. Accordingly, the first step requires that an electron is excited by light into the conduction band, leading to the

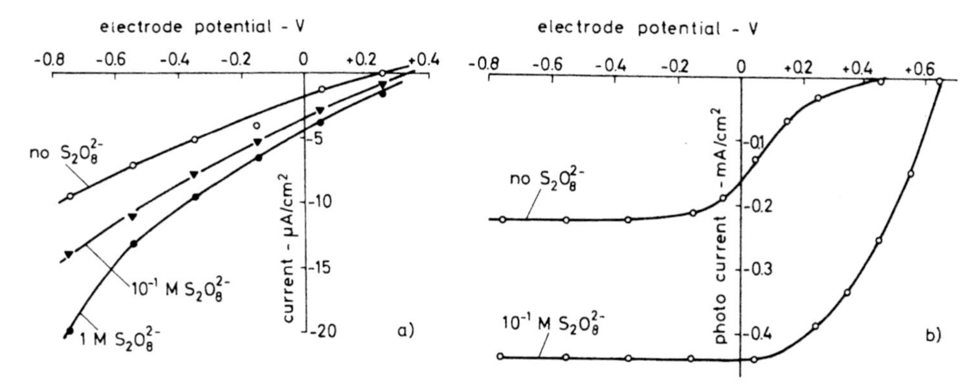

Fig. 7.54 Current vs. electrode potential (normal hydrogen electrode (NHE)) for a p-type GaP electrode in H_2SO_4 in the presence of $(NH_4)_2S_2O_8$ (redox system); a) in the dark; b) under illumination. (After ref. [68])

formation of an $OH^{\cdot}$ radical. In the second step an electron is transferred from the valence band to the $OH^{\cdot}$ radical, a process which takes place without any light excitation. The total reaction is summarized [68, 18] by

$$H_2O_2 + e^- \rightarrow OH^{\bullet} \rightarrow OH^- \tag{7.106a}$$

$$OH^{\bullet} \rightarrow OH^- + h^+ \tag{7.106b}$$

At n-type electrodes, the complete reaction already occurs in the dark because sufficient electrons are available in the conduction band. In the latter case the participation of the valence band has been proved by luminescence measurements. Since in the second reaction step electrons are transferred from the valence band to the $OH^{\cdot}$ radicals, hole are injected into the valence band of the n-type electrode which finally recombine with the electrons (majority carriers). In the case of n-GaP, this recombination is a light-emitting process, as has been found experimentally. The same result has been obtained with $S_2O_8{}^{2-}$ [68] and for quinones [69]. Since the reduction of H_2O_2 consists of two consecutive steps, it is reasonable to describe its redox properties by two standard potentials, given by

$$E_1 = eU_1 = eU_1^0 + kT \ \ln\left(\frac{c_{H_2O_2}}{c_{OH\bullet}c_{OH^-}}\right) \tag{7.107a}$$

$$E_2 = eU_2 = eU_2^0 + kT \ \ln\left(\frac{c_{OH\bullet}}{c_{OH^-}}\right) \tag{7.107b}$$

in which U_1^0 and U_2^0 are the standard potentials of each step, whereas the standard potential U_{redox}^0 of the whole system is an average value given by

$$E_{F,redox}^0 = eU_{redox}^0 = \frac{eU_1^0 + eU_2^0}{2} \tag{7.108}$$

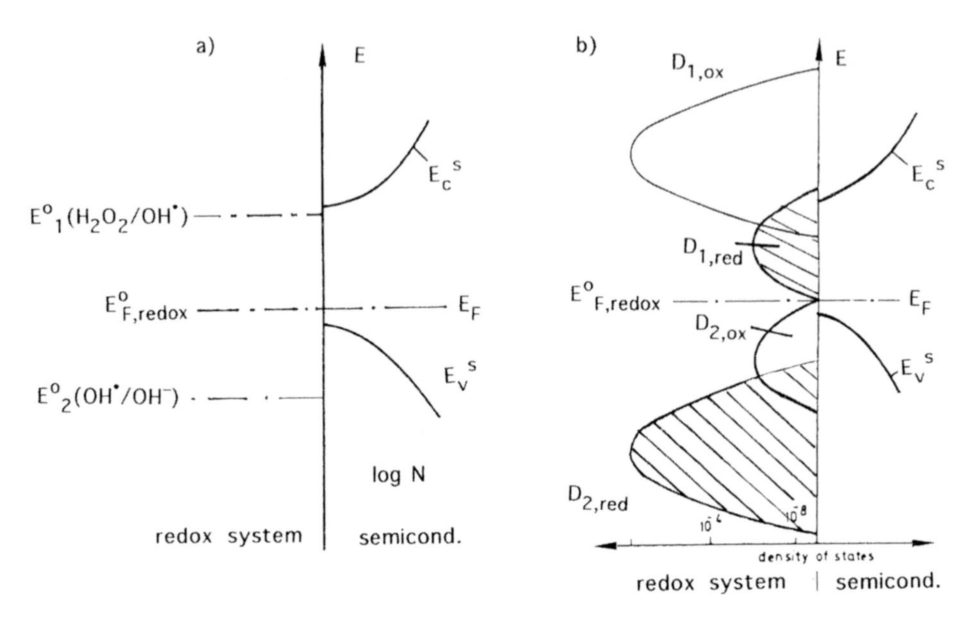

Fig. 7.55 Energy diagram for the interface of semiconductor–H_2O_2 (or $S_2O_8^{2-}$)

At equilibrium, the concentration of $OH^{\bullet}$ radicals is very low, i.e.

$$E = E_1 = E_2 = E_{F,redox} \tag{7.109}$$

so that according to Eq. (7.107)

$$E_1^0 - E_2^0 = e\left(U_1^0 - U_2^0\right) = kT\ \ln\left(\frac{C_{OH^{\bullet}}}{C_{OH^-}}\right) \tag{7.110}$$

The $OH^{\cdot}$ radical concentration is known to be extremely small at equilibrium, so that $E_2^0 \gg E_1^0$. From this follows that the energy levels of the half-system $H_2O_2/OH^{\bullet}$ occur at much higher energies than those of $OH^{\bullet}/OH^-$. Denoting the density of the occupied and empty states in the first system by $D_{1,red}$ and $D_{1,ox}$ (corresponding to the concentrations of $OH^{\cdot}$ and H_2O_2) and in the second system by $D_{2,red}$ and $D_{2,ox}$ ($OH^{\bullet}$ and OH^-), an energy scheme can be postulated as presented in Fig. 7.55. In Fig. 7.55a the standard
potentials, E_1^0, E_2^0 and $E_{F,redox}^0$ of the redox system and the band edges of GaP vs. density are given, whereas in Fig. 7.55b the corresponding distribution of energy states of the redox system are illustrated using the Gerischer model. The concentration of the intermediate states, $OH^{\cdot}$, should be very low so that the densities $D_{1,red}$ and $D_{2,ox}$ are very small. Since the actual redox potentials of the subsystems, E_1 and E_2, occur at energies where $D_{1,red} = D_{1,ox}$ and $D_{2,red} = D_{2,ox}$, respectively, E_1 and E_2 are very close to $E_{F,redox}$ which is in agreement with the Nernst law.

It should be emphasized that the description of such a redox system by two standard potentials is of general importance. The application of semiconductor electrodes, however, offers the possibility of proving the resulting two-step process. Since the reorganization energies are not known, values of the two formal potentials could only be estimated by comparing with other redox systems. Some standard potentials are given in Table 7.2. According to these values, the standard potentials differ considerably. Since the current-doubling effect was mainly observed at high pH values, E_1^0 and E_2^0 must be pH-dependent.

Table 7.2 Standard potentials of two-step redox systems (SCE)

Redox system	$E_{F,redox}^0$ (eV)	E_1^0 (eV)	E_2^0 (eV)
H_2O_2	1.77	≤0.6	≥2.9
$S_2O_8^{2-}$	2.0	≤0.6	≥3.4
Quinone	0.6	≤0.4	≥1.3

These results also lead to consequences concerning the electron transfer between these redox systems and metal electrodes. Since in a metal only the energy levels below the Fermi level are occupied, a rather small overlap between these levels and the occupied levels $D_{1,ox}$ in the redox system is expected at potentials corresponding to $E_{F,redox}^0$. Hence, the exchange current should be relatively small (compared with that found that with GaP). The rather large overpotential found for the cathodic reduction of H_2O_2 at metal electrodes supports this assumption [1, 14].

Current doubling has not only been observed for the reduction of H_2O_2 and $S_2O_8^{2-}$ but also for the cathodic reduction of O_2 at p-GaP in acid solutions [70]. The latter result is of interest insofar as the current doubling has been found to be intensity-dependent. The authors investigated the quantum efficiency, ϕ, over a large range of photon flux (four orders of magnitude) and found that ϕ decreased above a certain photon flux from 2 down to 1. The mechanism was quantitatively analyzed on the basis of the following kinetic scheme:

$$O_2 + H + e^- \rightarrow HO_2^\bullet \tag{7.111a}$$

$$HO_2^\bullet + H^+ \rightarrow H_2O_2 + h^+ \tag{7.111b}$$

$$HO_2^\bullet + H^+ + e^- \rightarrow H_2O_2 \tag{7.111c}$$

At low intensities holes are injected from the $HO_2^\bullet$ intermediate (Eq. 7.111b). At higher intensities the concentration of electrons in the conduction band becomes very large and the $HO_2^\bullet$ radical is reduced via a conduction band process (Eq. 7.111c).

A further interesting example from the fundamental point of view is the reaction of quinones, since simple quinones such as benzoquinone and duroquinones are reversible redox systems. Vetter has systematically analyzed the electrochemical reactions of quinones at Pt electrodes and postulated that two electrons are transferred in two consecutive steps [71]. Moreover, photochemical investigations with duroquinone have shown that an intermediate semiquinone exists [72]. Since the exchange currents at a Pt electrode were found to be relatively large [71], the difference in the two formal potentials is expected to be much smaller than that for H_2O_2 and $S_2O_8^{2-}$. According to photoelectrochemical investigations with various p-type semiconductor electrodes, such as Ge, Si and GaP, a current-doubling effect was found as described above [69]. The system could be analyzed somewhat quantitatively because the reorganization energy was also determined (see Table 7.1). The corresponding standard potentials, E_1^0 and E_2^0 are given in Table 7.2. In the case of quinones, however, the current-doubling depends strongly on pH, i.e. it occurs only at low pH values (Fig. 7.56). As already out-

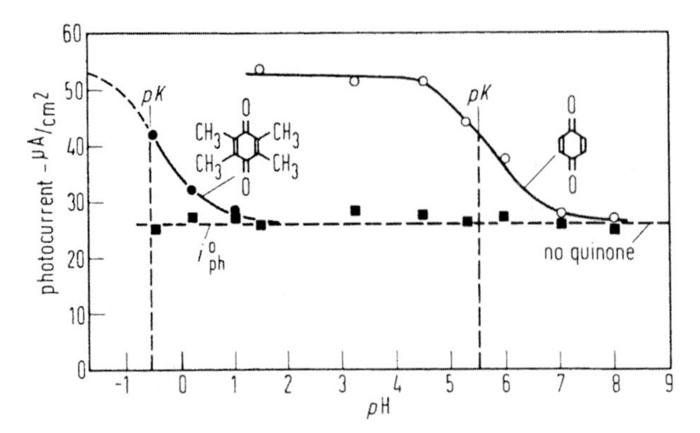

Fig. 7.56 The pH-dependence of the cathodic photocurrent at p-GaP for benzoquinone and duroquinone (current doubling). (After ref. [69])

lined above, only the protonation of the intermediate (semiquinone) is of interest here. There are two possibilities, a single ($QH^\bullet$) and a doubly protonated form ($QH_2^{\bullet +}$). Comparing the pK values from Fig. 7.56 with those derived from photochemical data, it was concluded that only the latter were reduced by a charge transfer via the valence band. The complete reaction scheme [69] is given by

$$Q + H^+ + e^- \rightarrow QH^\bullet \tag{7.112a}$$

$$QH^\bullet + H^+ \rightarrow QH_2^{\bullet +} \tag{7.112b}$$

$$QH_2^{\bullet +} \rightarrow QH_2 + h^+ \tag{7.112c}$$

Two-step reduction processes and the corresponding current-doubling effects have also been found with Br_2 [73], ClO^- [74] and BrO_3^- [73] at p-GaAs.

An analogous description can be given for the oxidation of various organic compounds. It is clear that an anodic photocurrent and its doubling can only be observed with n-type electrodes. For instance, Morrison and co-workers [75, 15] have studied the oxidation of formic acid at n-ZnO electrodes and found a current doubling upon addition of HCOOH. The basic photocurrent without any HCOOH was due here to the dissolution of ZnO. Morrison was actually the first to realise the importance of the current-doubling effect. He interpreted this result in terms of the reaction

$$HCOO^- + H^+ \rightarrow HCOO^\bullet \tag{7.113}$$

$$HCOO^\bullet \rightarrow CO_2H^+ + e^- \tag{7.114}$$

In this case, a hole created by light excitation is transferred in the first step where an electron is injected into the conduction band in the second step. Similar observations have been made for the oxidation of alcohols [19] at CdS [76], ZnO [77] and TiO_2 electrodes [78]. The latter case is illustrated in Fig. 7.57. This example is of special interest because the current doubling decreases if O_2 is flushed through the solution. This has been explained by a reaction of O_2 with the ethyl radical formed in the first electrode process, leading to a radical chain reaction (Fig. 7.57) [78]. This kind of reaction plays an important role in reactions at semiconductor particles (Chapter 9).

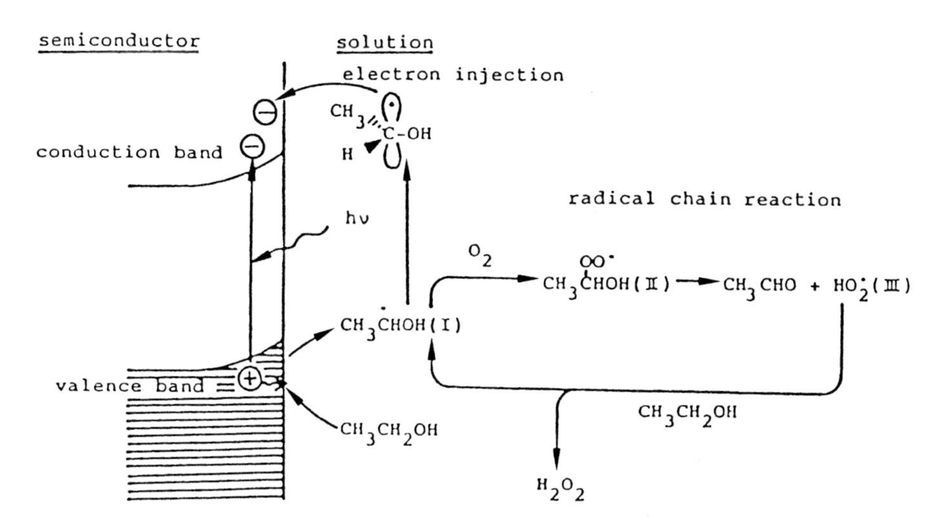

Fig. 7.57 Reaction mechanism of ethanol oxidation at illuminated TiO_2 in the presence of O_2. (After ref. [78])

7.7 Photoluminescence and Electroluminescence

Luminescence can be generated in a semiconductor electrode either, (a) by exciting an electron from the valence band to the conduction band by light absorption, or (b) via injection of minority carriers in an electrochemical process. In general, it has been observed with solid state devices that the luminescence originates from a radiative transition in the bulk. In the case of semiconductors with a direct bandgap, for example, GaAs, InP and CdS (see Appendix D), the luminescence corresponds mostly to a

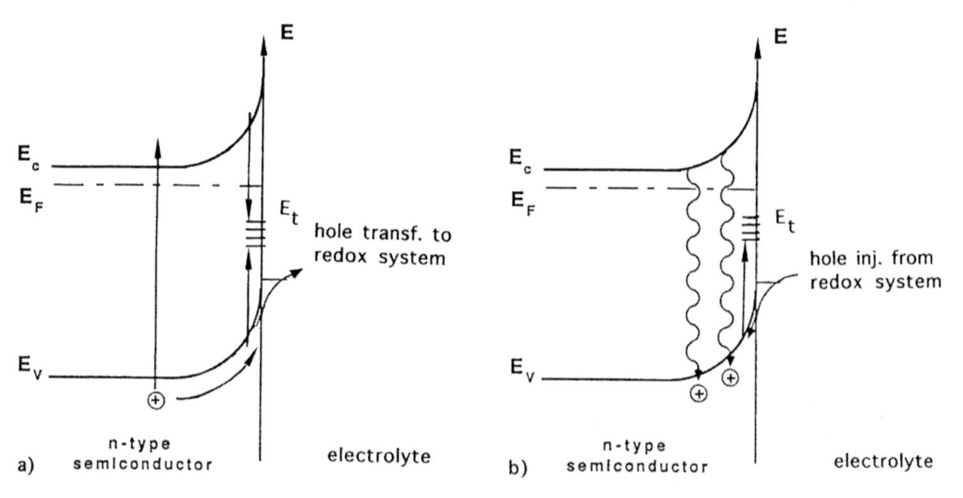

Fig. 7.58 a) Excitation of electron hole pairs and minority carrier transfer. b) Injection of minority carriers from the solution and recombination processes

band–band transition. In indirect bandgap materials (GaP, Si), essentially a recombination via special centers is involved. The latter process leads to an extrinsic emission spectrum and the quantum efficiencies are rather low. In the case of semiconductor–liquid junctions, there can be considerable competition by radiationless recombination via surface states as also indicated in Fig. 7.58. In addition, a charge transfer across the interface can affect the luminescence intensity. Since the types of physical information which can be obtained from photo- and electroluminescence, are somewhat different, both types will treated separately.

7.7.1 Kinetic Studies by Photoluminescence Measurement

As mentioned above, the luminescence created by light excitation, depends strongly on the interface parameters. This offers the possibility of determining kinetic parameters of the charge transfer by using stationary or pulse excitation methods. These effects become very pronounced if the semiconductor is illuminated through the solution by light which is highly absorbed so that the penetration length in the electrode is very small. This can be accomplished by using short-wavelengh monochromatic light for excitation. Concerning luminescence quenching, it is clear that this decreases to a very low level if the minority carriers are transferred across the interface to some acceptor in the solution. Accordingly, one might expect at first sight, that a high inten-

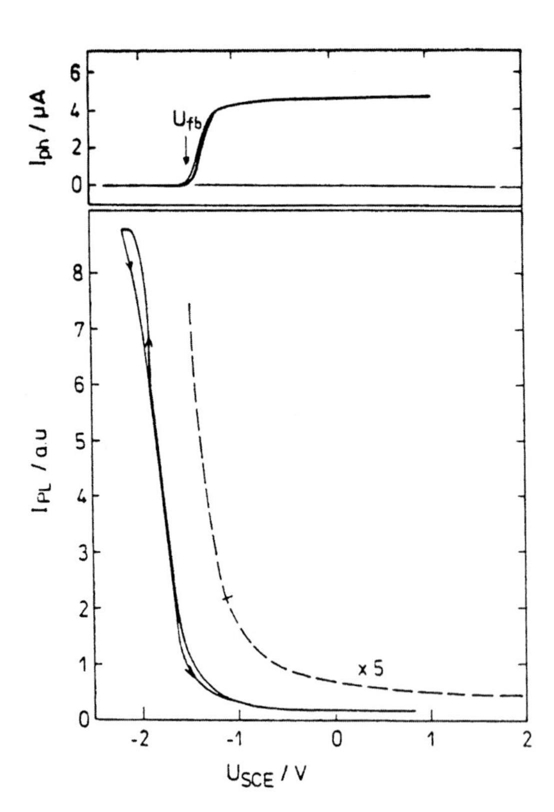

Fig. 7.59 Photoluminescence and photocurrent at CdS in sulfide (1 M)/polysulfide (0.3 M) electrolyte (0.1 M NaOH); excitation wavelength, 474 nm. (After ref. [81])

sity would already be reached at the flatband potential. However, several systems have been studied, but this prediction has been fulfilled with none of them. In all cases, the luminescence was still low, and a strong increase was only observed in the accumulation region, as illustrated in Fig. 7.59 for an n-CdS electrode in a sulfide/polysulfide electrolyte [79]. This is one of the most reliable results because any sulfur formed during the anodic polarization, is dissolved by the polysulfide [80]. Hysteresis effects occurred in many other systems which made a quantitative analysis rather obscure. In addition, in the case of the CdS/(S^{2-}/S_n^{2-}) system the flatband potential also remained pinned during illumination [80]. Chmiel et al. [79] analyzed their data as follows.

Theoretically, the luminescence intensity is proportional to the rate of electron–hole recombination, i.e. $I \propto n(z)p(z)$. In the simplest case it is assumed that the majority carrier density is not substantially changed upon illumination. Then the luminescence intensity is given by

$$I_{\mathrm{pl}} = \frac{\phi_{\mathrm{pl}}}{\tau} \int_0^{\infty} p(z)\mathrm{d}z \tag{7.115}$$

in which $\phi_{\mathrm{PL}}$is the quantum yield of radiative transition; τ the lifetime, and $p(z)$ is the hole profile which can be derived by integrating the continuity equation (Eq. 7.62) using the boundary conditions

$$ps = D \operatorname{grad} p\big|_{z=0} \tag{7.116}$$

One obtains then

$$p(z) = \frac{\alpha I_0 \tau}{1-\alpha^2 L_{\mathrm{p}}^2}\left[\exp(-\alpha z)\right] - \frac{\alpha L_{\mathrm{p}} + s}{1+s} \exp\left(\frac{z}{L_{\mathrm{p}}}\right) \tag{7.117}$$

where s is the sum of the surface recombination velocity (s_{r}) and the velocity (not the rate constant) of the hole transfer (s_{t}); both quantities are in units of cm s^{-1}. The integration leads to

$$I_{\mathrm{pl}} = \phi_{\mathrm{pl}} \frac{I_0}{1-\alpha^2 L_{\mathrm{p}}^2}\left[1 - \alpha L_{\mathrm{p}} \frac{\alpha L_{\mathrm{p}} + (s_{\mathrm{t}} - s_0)\, L_{\mathrm{p}}/D}{1 + (s_{\mathrm{t}} + s_0)\, L_{\mathrm{p}}/D}\right] \tag{7.118}$$

in which α is the absorption coefficient and L_{p} the diffusion length of holes. This equation is similar to that derived in ref. [81]. The velocity of the hole transfer, s_{t}, is proportional to the photocurrent; it is zero at the flatband potential and rises during anodic polarization. The surface recombination velocity, s_{r}, is also potential-dependent because the recombination depends not only on the minority but also on the majority carrier density (see Section 1.6), and the latter varies with polarization. Accordingly, s_r should become zero at larger anodic potentials whereas a maximum is expected around U_{fb}. Large values of s_{r} are expected for a high density of surface states. If s_{r} is large around U_{fb} and since s_{t} rises during anodic polarization, their sum, $s_{\mathrm{r}} + s_{\mathrm{t}}$, can remain constant when the electrode potential is varied from U_{fb} toward anodic potential. With respect to Eq. (7.118), s is large if $(s_{\mathrm{r}} + s_{\mathrm{t}}) \gg D/L_{\mathrm{p}}$. Eq. (7.118) then simplifies to

$$I_{\mathrm{pl,lim}} = \phi_{\mathrm{pl}} \frac{I_0}{I + \alpha L_{\mathrm{p}}} = I_{\mathrm{pl,fb}} \tag{7.119}$$

This is an equation which does not contain any further potential-dependent quantities. It has to be remembered, however, that the luminescence originates from a radiative transition in the bulk, and it has been assumed in this derivation that no recombination occurs within the space charge region, because all minority carriers reaching the space charge region are moved toward the surface where they either recombine or are transferred across the interface. On the other hand, the thickness of the space charge region is given by (see Eq. 5.31)

$$d_{sc} = \left(\frac{2\varepsilon\varepsilon_0 \Delta\phi_{sc}}{eN_D} \right)^{1/2} \tag{7.120}$$

It increases with the square root of the potential across the space charge region, ϕ_{sc}. Accordingly, the minimum luminescence should further decrease with increasing anodic potentials. This effect can be quantified by performing the integration in Eq. (7.115) under the condition that $s \gg D/L$ within the range of $d_{sc} \geq z \geq \infty$ (instead of $0 \geq z \geq \infty$). Instead of Eq. (7.119) we obtain then

$$I_{pl,lim} = \phi_{pl} \frac{I_0}{I + \alpha L_p} \exp(-\alpha d_{sc}) = I_{pl} \exp(-\alpha d_{sc}) \tag{7.121}$$

Accordingly, Eq. (7.119) is the luminescence at U_{fb} ($d_{sc} = 0$). Chmiel et al. analyzed their photoluminescence data on the basis of this model [79]. For instance in the case of the data presented in Fig. 7.59, a plot of $\{\ln(I_{PL,fb}/I_{PL})\}^2$ vs. ϕ_{sc} yielded a straight line as expected according to Eqs. (7.121) and (7.120). The same results have been found with various other systems which proves the dead layer model in the range of the depletion layer. There is a remarkably steep growth of the luminescence in the accumulation region beyond U_{fb} (Fig. 7.59). This result has been interpreted as a decrease of s_r caused by a decrease of the hole density at the surface [79]. The luminescence should finally saturate at higher cathodic polarization. Unfortunately, H_2 evolution occurred in that range which made any further measurements impossible.

More recently time-resolved techniques have been applied for studying photocarrier dynamics at the semiconductor–liquid interface. One of the main motivations is that such studies can lead to an estimation of the rate at which photo-induced charge carriers can be transferred from the semiconductor to a redox acceptor in the solution. This method is of great interest because rate constants for the transfer of photocarriers cannot be obtained from current–potential curves as in the case of majority carrier transfer (Section 7.3.5). The main aim is a detailed understanding of the carrier dynamics in the presence of surface states. The different recombination and transfer processes can be quantitatively analyzed by time-resolved photoluminescence emitted from the semiconductor following excitation by picosecond laser pulse. Two examples are shown in Fig. 7.60 [82, 83].

The figure shows how the luminescence decay observed with a p-GaAs electrode, varies upon addition of an electron acceptor such as cobaltocene ($CoCp_2{}^+$) and ferrocene ($FeCp_2^+$). The GaAs surface was passivated in order to keep the surface recombination low. The evaluation of such a decay curve is not easy. In principle the continuity equation (Eq. 7.62) must be solved again, this time under conditions of short pulse illumination; the resulting minority carrier density $p(z,t)$ is inserted into Eq. (7.115), and the intensity I(t) is finally obtained after integration of the latter equation. However,

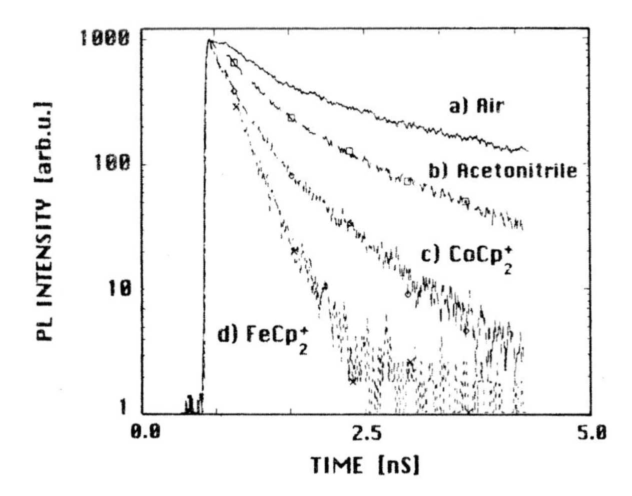

Fig. 7.60 Photoluminescence decay characteristics for p-GaAs passivated with Na_2S measured under the following conditions: a) in air; b) in acetonitrile; c) upon addition of 1 mM cobaltocenium as an electron acceptor; d) the same as c) but with 1 mM ferrocinium as an electron acceptor. (After ref. [82])

Eq. (7.62) is only valid for small changes in majority carrier densities, i.e. $\Delta p \ll p_0$. Since in most cases fairly intense laser pulses were used, a more complex equation must be used which finally leads to quadratic concentration terms in Eq. (7.115). The exact procedure for the evaluation cannot be given here and the reader is referred to

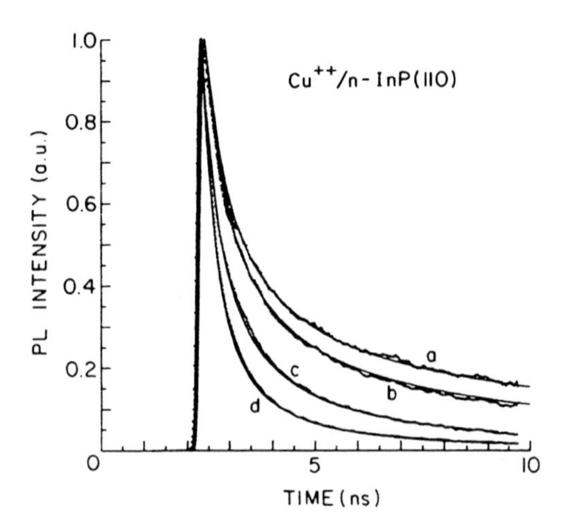

Fig. 7.61 Experimental (dotted curves) and calculated (solid curves) photoluminescence decay curves for n-InP: a = immersed in distilled H_2O, surface recombination velocity (SRV) <500 cm s^{-1}; b = immersed in 5×10^{-7} M $CuSO_4$ solution, SRV = 2×10^3 cm s^{-1}; c = immersed in 5×10^{-6} M $CuSO_4$ solution, SRV = 2×10^4 cm s^{-1}; d = immersed in 5×10^{-5} M $CuSO_4$ solution, SRV = 3.5×10^4 cm s^{-1}. (After ref. [83])

the literature [84]. The evaluation of the present example yielded s_t values from which a second-order rate constant k_{et} was calculated according to the relation

$$s_t = k_{et} c_A \tag{7.122}$$

in which c_A is the concentration of the electron acceptor. Calculation of k_{et} gave a value of 2×10^{-12} cm^4 s^{-1} which is a similarly high value to that determined from majority carrier (electron) processes at n-GaAs electrodes (see Section 7.3.4).

It is clear that the surface recombination velocity s_r can also be determined by photoluminescence decay measurements. One nice example (n–InP) is given in Fig. 7.61. InP is a semiconductor which exhibits in contact with H_2O a rather low surface recombination ($s_r < 500$ cm s^{-1}, curve a). After the electrode had been dipped into a solution of $CuSO_4$, Rosenwaks et al. found a considerably steeper decay (curves b–d) [83]. An excellent agreement between theoretical and experimental curves was obtained. Values of up to $s_r = 3.5 \times 10^4$ cm s^{-1} were reported for the surface recombination velocity. The same authors showed by capacity measurements that an additional capacity due to surface states occurs simultaneously, as already discussed in Section 5.2.4.

7.7.2 Electroluminescence Induced by Minority Carrier Injection

In solid state devices such as p–n junctions, luminescence is created by forward polarization in the dark. In such a minority carrier device, electrons move across the p–n interface into the p-type and holes into the n-type regions, where they recombine with the corresponding majority carriers. As already pointed out in Section 2.3, this kind of luminescence has not been found with semiconductor–metal junctions (Schottky junctions), as nobody has succeeded in producing a minority carrier device because of Fermi level pinning. Since the latter problem usually does not occur with semi-

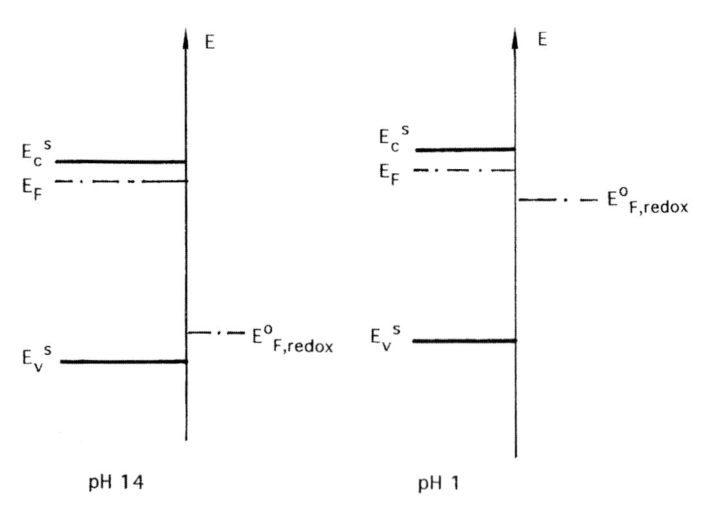

Fig. 7.62 Position of energy bands of GaP at the interface and of the standard redox potential of $[Fe(CN)_6]^{4-/3-}$ at pH 14 and 1

conductor–liquid systems, it is fairly easy to form a junction at which minority carriers are injected into the semiconductor.

Electroluminescence was first observed with n-GaP electrodes using hole donors such as $[Fe(CN)_6]^{3-}$ in alkaline or $S_2O_8^{2-}$ in acid solutions [112]. In these two cases the corresponding standard potentials occur at or even below the valence band edge (see Table in Appendix). In the case of $[Fe(CN)_6]^{3-}$ no luminescence was found in acid solutions although the current–potential curve indicates that the redox species is reduced. The differences between alkaline and acid solutions can be explained by the pH-dependence of the position of the energy bands at the surface, as shown in Fig. 7.62. Since E_v is far below $E^0_{F,redox}$ at pH 1 no charge transfer between the redox couple and the valence band is possible anymore, and the cathodic current is only due to an electron transfer via the conduction band.

Meanwhile electroluminescence has been observed with several other semiconductors such as GaAs and InP. Various authors have also studied the potential dependence of the emission in more detail. One example is the hole injection from Ce^{4+} ions into n-GaAs (Fig. 7.63) [85]. The emission sets in at the same potential where the interfacial current also occurs. The current becomes diffusion-limited with increasing cathodic potentials. The emission shows, however, a peak around –0.5 V. If the potential scan

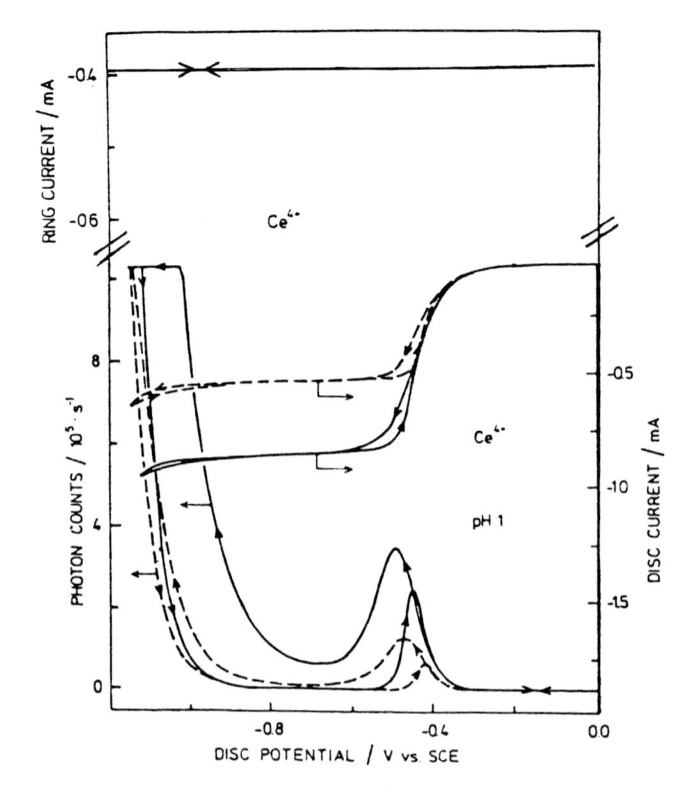

Fig. 7.63 Disc voltammograms and luminescence intensity vs. potential of an n-GaAs (ring disc electrode (RDE)) in the presence of 10^{-2} M Ce^{4+} at pH 1 at two different rotation velocities: 9 Hz (dashed) and 25 Hz (solid lines); scan rate: 10 mV s^{-1}. (After ref. [85])

was reversed at the peak then the emission curve could be retraced. As the potential was scanned to more negative potentials, however, the luminescence at first decreased, then increased again. In the reverse scan the signals were much lower. These hysteresis effects, which are also found with other systems, indicate that the potential-dependence of the luminescence is not only controlled by hole injection but also by some complex surface properties which are not yet understood.

The experimental results presented in Fig. 7.63, are also interesting from another point of view. They were obtained using a rotating ring-disc electrode assembly (RRDE technique, see Section 4.2.3) using an n-GaAs disc and a Pt ring. The ring potential is set to a value such that Ce^{3+} ions produced at the disc, are oxidized back to Ce^{4+}. The corresponding anodic ring current is then a measure of the reduction of Ce^{4+} at the GaAs disc. An anodic ring current occurs not only at rather negative potentials but also in the range where the disc current is zero. Accordingly, reduction of Ce^{4+} and also hole injection, occurs in the anodic range. At potentials above 0.4 V, however, the injected holes do not diffuse into the bulk (no disc current, no emission) but they are immediately used for the anodic dissolution reaction.

So far, we have described the emission due to hole injection into an n-type semiconductor electrode. The question arises concerning whether the counterpart, i.e. the injection of electrons into p-type electrodes, can also be realized. The only example reported in the literature, is the oxidation of $[Cr(CN)_6]^{4-}$ at p-InP electrodes [86]. This is a redox couple with a very negative standard potential ($U^0 = -1.4$ V (SCE)) with which electron injection into the conduction band of p-InP was possible (Fig. 7.64). Corresponding emission was observed. The same type of experiments with p-GaAs did not lead to any emission because the energy bands of GaAs are higher than those of InP.

These are very good examples from the fundamental point of view which could never be realized with pure solid state devices. In addition it should be emphasized that electroluminescence is a very useful in situ tool for the detection of minority carrier injection. In principle, the same type of information can be obtained by using the

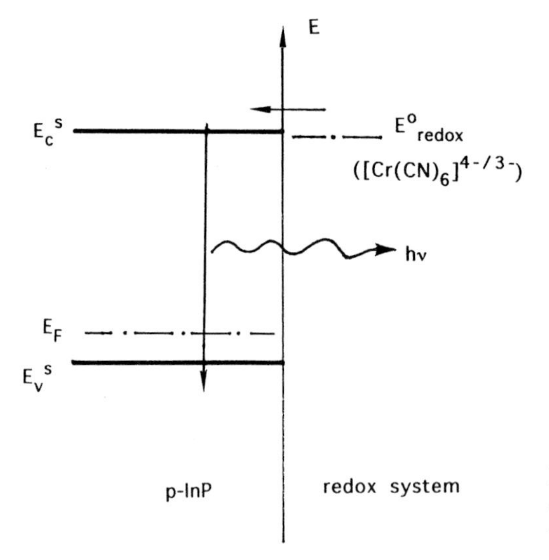

Fig. 7.64 Electron injection into the conduction band of p-InP and recombination. (After ref. [86])

thin slice method (see Section 4.3). However, the latter technique is much more difficult to realize in practise. One example which shows very nicely the possibilities of the electroluminescence method, is investigation of the anodic decomposition of p-InP. In this case, luminescence has been observed upon anodic polarization of this electrode in solutions free from any redox system. Since the emission yield (intensity divided by the interfacial current) remained constant, it was concluded that some of the decomposition steps occur via the conduction band [87]. Further details will be discussed in Chapter 8.

7.8 Hot Carrier Processes

In all light-induced reactions the photon energy required for excitation corresponds to the bandgap of the semiconductor. Photogenerated electrons and holes created by photons with energies higher than the bandgap create charge carriers with excess kinetic energy; these carriers are termed hot carriers. The hot carriers can dissipate their excess kinetic energy and cool to the lattice temperature through electron–phonon interactions; after complete relaxation, all the carriers are near their respective band edges. In general, because holes have greater effective masses than electrons, hot holes cool much faster than hot electrons, and the consideration of hot carrier effects can thus be restricted to electrons [20]. For bulk semiconductors, the relaxation takes about 2–10 ps, as determined for GaAs by photoluminescence decay measurements for moderate light intensities [88]. For quantized semiconductors (see Chapter 9), it was predicted [89, 90] and subsequently verified [88] that the electron cooling rate can be substantially reduced; characteristic values are 50–500 ps, which also depend on the light intensity [88].

Hot carrier charge transfer processes are important in solid state devices [52]. One fundamental question concerns whether these hot carriers can be transferred across the semiconductor–liquid junction, before they completely relax to the band edge [91]. This problem is not only of interest from a fundamental point of view but it has also some significance in the field of solar energy conversion [92]. Namely, if hot carriers can be transferred before they relax, then much higher efficiencies for the conversion of solar radiation into chemical or electrical free energy may be attained. Hot carrier processes can occur when the relaxation process is slow compared with the time required for the photogenerated charge carrier to traverse the space charge layer and be injected into the redox system. Fig. 7.65 illustrates two different possible hot carrier processes for an p-type semiconductor electrode [93]. In type I, electrons are fully relaxed to the bottom of the conduction band while still in the bulk, but are then driven across the depletion layer at an energy corresponding to the edge of the conduction band without further relaxation. In the type II process hot carriers are never fully thermalized and are injected into the solution with energies greater than the conduction band in the bulk. Taking InP as an example, the traverse time τ_{sc} required for an electron to pass through the space charge, can be estimated from the equation given by:

$$\tau_{sc} = d_{sc}/v_{sc} \tag{7.123}$$

in which v_{sc} is the maximum drift velocity across the space charge layer and d_{sc} the thickness of the space charge layer as given by Eq. (5.31). In the case of a doping den-

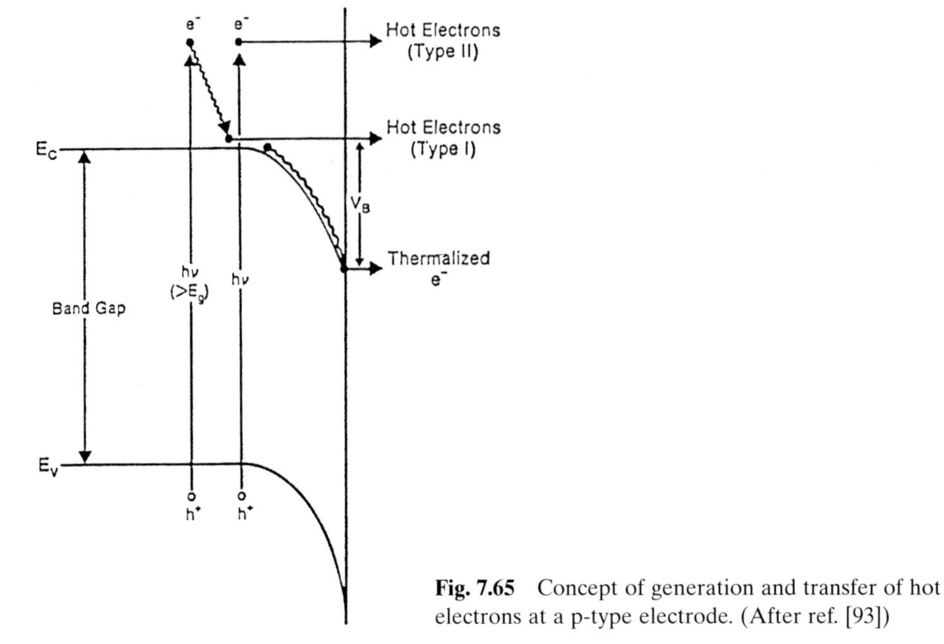

Fig. 7.65 Concept of generation and transfer of hot electrons at a p-type electrode. (After ref. [93])

sity of $2 \times 10^{18}\text{cm}^{-3}$ one obtains $d_{sc} \approx 200\text{Å}$. With $v_{sc} = 10^7\text{cm s}^{-1}$ the traverse time across the space charge layer is then, according to Eq. (7.123) $\tau_{sc} \leq 1$ ps, i.e. the hot electrons can pass across the space charge layer without any relaxation. This has been experimentally confirmed by Min and Miller [94].

The first experiments were reported by Nozik and co-workers, for p-GaP and p-InP liquid junctions [95, 96]. In particular, InP was a good candidate, because of its high electron mobility. The authors used p-nitrobenzonitrile ($U^0_{redox} = -0.86$ V (SCE)) as an electron acceptor, because the standard potential of this redox couple occurs 0.44 eV above the conduction band as determined by Mott–Schottky measurements. Photocurrent–potential curves in blank solutions were compared with those of solutions containing nitrobenzonitrile. The observation of increased cathodic photocurrent was reported as evidence for hot electron transfer.

The topic of hot carrier transfer remained controversal for many years. One reason was that it is difficult to find suitable redox systems with which such a process can convincingly be proved. In recent years Koval et al. have performed some very careful experiments [96]. These authors investigated corresponding cathodic reduction reactions at p-InP electrodes of different dopings ($p_0 = 2 \times 10^{18}$ and 2×10^{15} cm^{-3}). As outlined above, a hot electron transfer should not be possible at alow-doped electrode because τ_{sc} will be far too large. In one case, Koval et al. studied the electron transfer from p-InP to copper (trans-diene) complex during illumination [96]. This compound is reduced according to the following reactions:

$$\text{Cu(II)(trans-diene)}^{2+} + e^- \Leftrightarrow \text{Cu(I)(trans-diene)} \quad (7.124)$$

The reversible wave occurs at –0.97 V, i.e. slightly below the conduction band of InP. The Cu(I) complex is further reduced to Cu metal:

$$Cu(I)(trans\text{-}diene)^{1+} + e^- \Leftrightarrow Cu(0) + trans\text{-}diene \quad (7.125)$$

at –2.3 V at glassy carbon, and at –2.2 V at n-InP. A photocurrent was found at potentials negative of –0.5 V for low-doped p-InP. Interestingly, these authors found Cu(O) after illumination only with high-doped, but not with low-doped electrodes, the Cu metal being analyzed by an anodic stripping technique. This is an excellent proof for hot carrier ejection, indicating that an electron transfer is possible without thermalization, when the depletion layer is sufficiently thin.

Another type of hot electron transfer is illustrated in Fig. 7.66. When a Pt-electrode is polarized cathodically then electrons can be transferred to an Ox species of an organic redox system, such as thioanthrene (TH^+/TH) in acetonitrile (case B in Fig. 7.66). Using a metal electrode covered by a thin oxide film (here Ta_2O_5 on Ta), electrons can also be transferred to the organic acceptor molecule in the solution at a sufficiently high cathodic bias (case A). This process was found to be accompanied by electrogenerated chemiluminescence [97]. The complete reaction was interpreted as tunneling of electrons from the Ta substrate through the Ta_2O_5 film (thickness about 20 Å) into the conduction band of Ta_2O_5 ($E_g > 3.2$ eV) near the surface, from where the electrons are transferred into the empty upper level of the molecule, as indicated in case A of Fig. 7.66, leading to the formation of an excited molecule:

$$TH^+ + e^- \rightarrow TH^* \quad (7.126)$$

The luminescence is due to the recombination of the excited molecule TH* to its ground state. The fundamentals of these processes are given in Chapter 10.

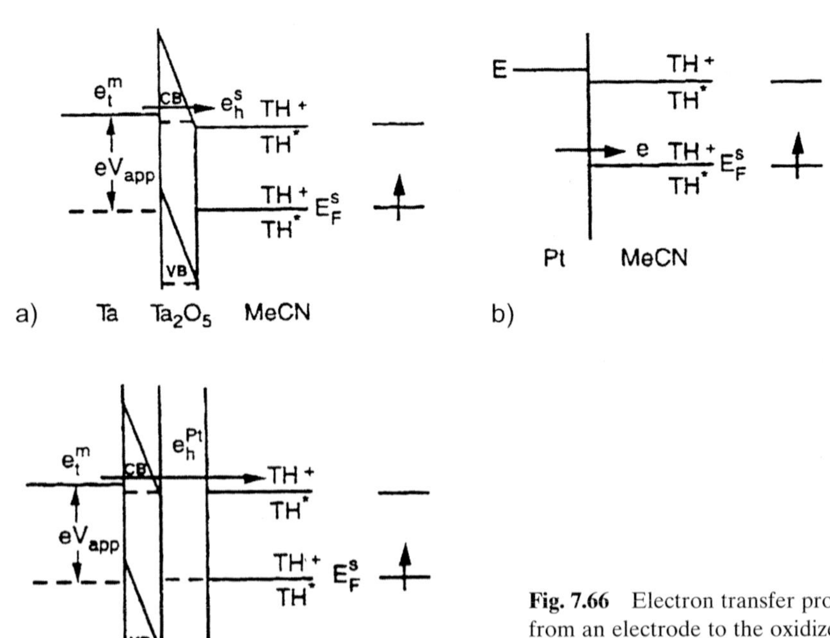

Fig. 7.66 Electron transfer processes from an electrode to the oxidized species TH^+ in solution at: a) Ta/Ta_2O_5; b) Pt, and c) $Ta/Ta_2O_5/Pt$ electrodes. (After ref. [98])

At Pt electrodes, the formation of an excited state is not possible, because the electron transfer occurs around the Fermi level which occurs close to the redox potential of the molecule in its ground state (case B). Interestingly, the electrogenerated luminescence intensity was enhanced by a factor of ~5 by deposition of a thin (<400 Å) Pt film on on the Ta/Ta_2O_5 electrode (case C). Obviously, the electrons injected from the conduction band of the Ta_2O_5 layer into the Pt film are transferred to the TH^+ molecules at a very high energy. Since these electrons kept their energy, this effect has been interpreted as hot electrons within the Pt film [98].

The question still arises, however, concerning which rate constant is expected in order to fulfil the condition that the hot electron is transferred before it relaxes to the bottom of the conduction band. In this context it should be mentioned that there may be quantization effects at the surface of highly doped III-V semiconductors because of the small space charge width (100 Å); the relaxation time may be longer at the surface compared with the bulk. Assuming this effect to be small, the relaxation time is in the order of 10 ps (see above). Then hot electron transfer can only occur if the time of the electron transfer from the semiconductor surface to an acceptor in the solution is comparable with this relaxation time. About 10 % of the excited carriers will be transferred as hot electrons if the transfer time is 100 ps. The rate at which electron transfer occurs is highly controversial at present, and is a critical issue with regard to the probability of hot electron transfer. As already outlined in Section 7.3.4, maximum k_{et} values extend over a relatively large range. According to calculations by B. B. Smith [99], theoretically k_{et} values as high as 10^{-14} to 10^{-13} $cm^4\ s^{-1}$ for non-adsorbed acceptors can be expected, and in the case of adsorbed redox systems such as $Co(Cp)_2{}^{1+/0}$ at GaAs electrodes, values such $k_c^- \approx 10^{-7} cm^3\ s^{-1}$, corresponding to $k_{et} = 10^{-14}\ cm^4\ s^{-1}$, have been determined experimentally [33]. In this case, a transfer velocity of $s_{et} = 5 \times 0^6\ cm\ s^{-1}$ was derived. It is still an open question as to what this velocity really means, and a conversion into a transfer time is not trivial. In a simple approach, s_{et} may be assumed to be the average velocity at which electrons are drawn from the electron pool in the semiconductor and placed on the acceptor. Assuming further that the electron pool is distributed over about 100 Å, the transfer time would be 1.6 ps [100]. This is very competitive with the relaxation time of hot electrons.

7.9 Catalysis of Electrode Reactions

Around 1975, investigations of photoelectrochemical reactions at semiconductor electrodes were begun in many research groups, with respect to their application in solar energy conversion systems (for details see Chapter 11). In this context, various scientists have also studied the problem of catalysing redox reactions, for instance, in order to reduce surface recombination and corrosion processes. Mostly noble metals, such as Pt, Pd, Ru and Rh, or metal oxides (RuO_2) have been deposited as possible catalysts on the semiconductor surface. This technique has been particularly applied in the case of suspensions or colloidal solutions of semiconductor particles [101]. However, it is rather difficult to prove a real catalytic property, because a deposition of a metal layer leads usually to the formation of a rectifying Schottky junction at the metal–semiconductor interface (compare with Chapter 2), as will be discussed below in more detail. Accordingly, the properties of a semiconductor electrode with a surface which is modified by a metal, and the nature of any possible catalytic action, can only be studied with extended electrodes. However, only a few research groups have really investigated the consequences of the surface modification on reaction routes and rates. In the following, only papers which gave some fundamental insight into these problems are mentioned.

One of the first results was reported by Heller [102], who studied the photoelectrochemical properties of single crystalline n-GaAs electrodes in alkaline solutions of the redox system Se^{2-}/Se_n^{2-}. He showed that the short circuit vs. photovoltage dependence (power plot, see also Chapter 11) became steeper and more rectangular after the electrode had been dipped in a solution of $RuCl_3$ before it was used in the cell. Heller interpreted this result as a reduction of surface recombination. It was assumed that the original surface states located around the midgap are shifted toward the energy bands due to a strong interaction between the deposited Ru and the GaAs surface. Near the edge of the energy bands surface states are no longer effective. According to this model, the hole transfer becomes more effective in comparison with surface recombination. The competition between recombination and hole transfer is illustrated in Fig. 7.67. This process was further investigated by Ming Tan et al. [103]. These authors essentially confirmed Heller's experimental result. They also made clear that there is not only a hole transfer from the valence band to the redox system but also a hole consumption at the surface due to the anodic decomposition (Fig. 7.67). A chemical analysis of the products formed during anodic polarization, showed that the rate for the oxidation of the redox system (k_{ox}) was increased with respect to that of the corrosion (k_{corr}) upon the surface modification by Ru. This result implies that Ru catalyses the hole transfer from GaAs to Se^{2-}.

Photocurrents due to the electrochemical reduction and oxidation of H_2O (H_2 and O_2 formation) usually occur at considerable overvoltages. Since this is an important problem for the solar production of a chemical fuel many researchers have tried to reduce the overvoltage by using a catalyst. In this case, it has to be realized again that the deposition of a metal monolayer on a semiconductor surface leads to the formation of a Schottky junction (see Section 2.2). Accordingly, the question arises whether there

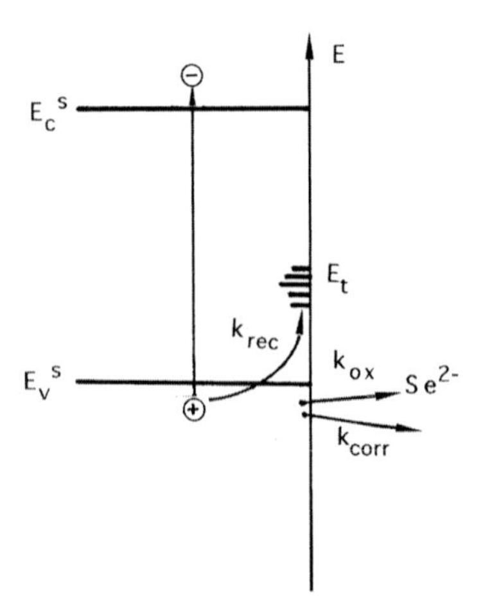

Fig. 7.67 Oxidation of a redox system and anodic corrosion by hole transfer at an n-type electrode during light excitation

is a true catalytic effect or whether a solid state photocell (Schottky junction) has been formed which is in contact with an electrolysis cell where hydrogen or oxygen is formed at the metal being deposited on the semiconductor surface. In this context several interesting results have been obtained. Heller et al. [104] have studied the photoelectrochemical reduction of H_2O at p-InP electrodes, the surface of which was modified by platinum group metal. They found that higher photocurrents at bare InP electrodes occur only at a considerable overvoltage which could be reduced by depositing Rh, Ru or Pt on the semiconductor. The actual photocurrent onset, however, was identical for bare and modified electrodes. Heller et al. interpreted these results as the formation of a semiconductor–Pt(H_2) Schottky junction. They proved this model by mea suring current–potential curves of dry InP/Pt Schottky junctions under illumination. The latter measurements indicated quite clearly a change of the barrier height and a corresponding shift of the metal work function when hydrogen was passed across the interface.

Another interesting approach was published by Tsubomura and Nakato [105]. They have investigated n-type Si electrodes covered by a Pt layer or by small Pt islands in aqueous solutions of various redox systems. The corresponding photocurrent–potential curves as measured with a Si electrode completely coverved with a thin Pt layer (curve a) or covered by Pt islands having a size of $\leq$10 nm (curve b) are shown in Fig. 7.68. The islands were produced by etching the platinized Si samples. The authors interpreted these results by using a model introduced by Nosaka et al. [106], as illustrated in Fig. 7.69. The deposition of Pt islands leads to a modulation of band positions along the surface. In the case of large islands (d > 10 nm), the barrier height at the semiconductor–metal interface is given by $e\phi_B$ whereas that of the semiconductor–liquid interface is given by $e\phi_B'$ (Fig. 7.69a). The barrier height $e\phi_B$ underneath the metal (Schottky junction) is entirely determined by the semiconductor–metal contact. It is independent of the redox system in the electrolyte. The barrier height of the free surface, $e\phi_B'$, however, depends on the redox couple. In aqueous solutions an oxide layer exists on free surface which limits any charge transfer to a very low level. If the width of the free surface is not too large, photoholes reaching the surface can enter the metal via quantum mechanical penetration through the potential well, as indicated by an arrow in Fig. 7.69a. Since the maximum photovoltage is only determined by $e\phi_B$ it is independent of the redox potential, as found experimentally. In the case of much

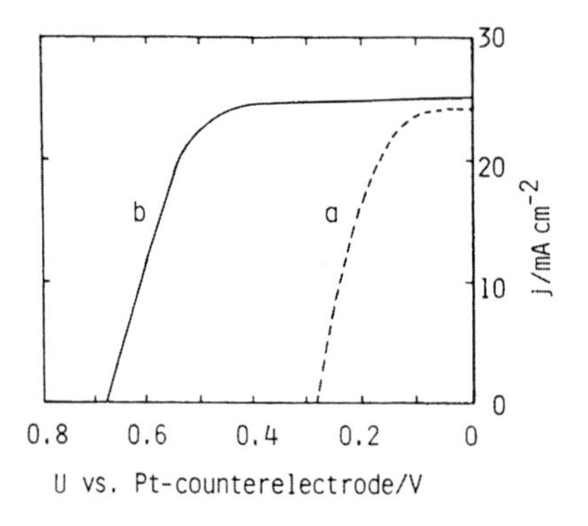

Fig. 7.68 Photocurrent–potential curves of Pt-coated n-Si electrodes in Br^-/Br_2 solutions under illumination. Curve a) Si covered by a Pt film, curve b) Si covered by Pt islands. (After ref. [105])

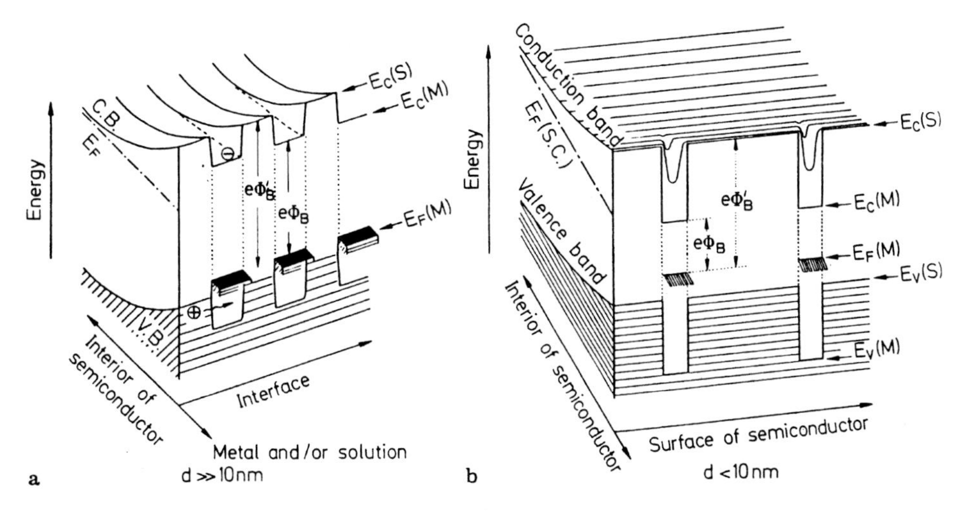

Fig. 7.69 Schematic energy diagram of the n-type Si electronic bands at the solid–liquid interface modulated by discontinuous metal coating. (After ref. [105])

smaller islands (<10 nm), the quantized levels of the electrons in the potential well in the conduction band become considerably higher than the bottom of the well (Fig. 7.69) so that the electrons have a higher kinetic energy. Since the potential well is very narrow, the band modulation diminishes rapidly toward the interior of the crystal. These two effects make the effective barrier height little different from that of the naked surface, i.e. $e\phi_B \rightarrow e\phi_B'$ (Fig. 7.69b) and $e\phi_B'$ now determines the photovoltage. This has been found experimentally as shown by the large increase in the photovoltage (curve b in Fig. 7.68). Since $e\phi_B'$ depends on the standard potential, U^0_{redox}, of the redox couple, the photovoltage should depend now on U^0_{redox} This has also been proved experimentally by Tsubomura et al. (not shown).

This model has been further tested by Meier and Meissner [107] using p-GaAs electrodes covered by gold layers or islands. The electrochemical behavior of these electrodes depended strongly on the deposition technique. For instance, if the gold was deposited from Au^{3+}-containing solutions under cathodic polarization, and if the resulting coverage was less than a monolayer, then there was no catalytic activity with respect to light-induced hydrogen evolution. According to impedance measurements, a high density of metal-induced surface states formed. These states are also effective recombination centers which explains the poor photoelectrochemical properties.

In another method, the same authors prepared gold colloids, where the particle size could be varied over a range from 6 to 100 nm. These Au colloids were deposited on the surface of a p-GaAs electrode in concentrations ranging from 10^8 to 10^{11} colloids per cm^{-2}. Multiple nano contacts (MNCs) were produced in a very defined way, i.e. a large number of contacts of equal diameter were formed. With respect to light-induced H_2 formation, experiments yielded a catalytic activity which increased with an increasing density of colloids as shown by curves a–e in Fig. 7.70. The catalytic activity was further increased by platinizing the Au colloids as shown by curve f in Fig. 7.70. The authors showed that the photocurrent occurred almost entirely across the MNCs even at low coverages (e.g. about 90 % at a coverage of 1.5 %). The evaluation of exchange

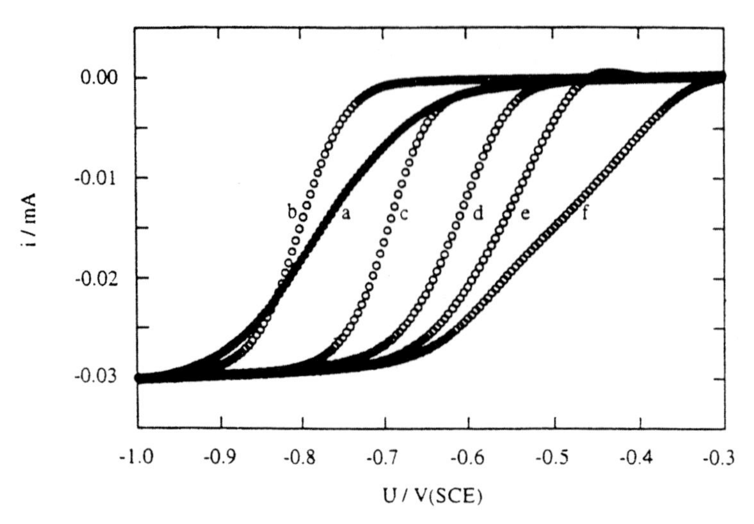

Fig. 7.70 Cathodic photocurrent–potential curves at p-GaAs in aqueous solutions (pH 3) under illumination: a) bare p-GaAs rotating disc electrode; b) gold-plated p-GaAs covered with 0.7 monolayers; c–e) p-GaAs covered with 6, 40 and 100 nm large Au particles, respectively; f) p-GaAs covered with platinized Au particles (40 nm); rotation velocity, 200 rpm; scan rate, 10 mV s^{-1}. (After ref. [107])

currents yielded results which support the Tsubomura model. For details the reader is referred to the relevant papers [108].

Other authors have also tried to carry out quantitative investigations on the nature of catalytic effects at modified surfaces of TiO_2. Unfortunately, the deposition of noble metals on single crystalline TiO_2 led mostly to the formation of ohmic contacts which made further investigations impossible [109].

8 Electrochemical Decomposition of Semiconductors

Under anodic polarization, many semiconductors, especially those having smaller bandgaps ($E_g < 3$ eV), undergo anodic decomposition. In such a reaction several charges per molecule are always involved. Since valence bonds are broken, most steps occur by hole consumption via the valence band. Accordingly, anodic dissolution occurs at p-type semiconductors in the dark and at n-type during illumination. In some cases, decomposition has also been found during cathodic polarization. In the following sections, the reaction routes and mechanisms are described for the most important semiconductors. The most detailed studies have been published for germanium and, especially, for silicon.

8.1 Anodic Dissolution Reactions

8.1.1 Germanium

The anodic dissolution of semiconductors, especially of germanium and silicon, was the subject of intensive investigation in the early stages of semiconductor electrochemistry. This great interest in corrosion arises because of technological problems with devices, such as surface stability and etching of device structures. Some essential investigations have been performed at the Bell Laboratories, namely by Turner (see, e.g., ref. [1]) and by Brattain and Garrett [2, 3]. These authors have already obtained essential data about the dissolution of germanium. According to their results, the overall reaction is given for alkaline solutions by

$$\mathrm{Ge} + \gamma \mathrm{h}^+ + 6\mathrm{OH}^{2-} \rightarrow \mathrm{GeO}^{2-} + (4-\gamma)\mathrm{e}^- + 3\mathrm{H_2O} \tag{8.1}$$

In total, four charges are required for the dissolution of one Ge atom so that $0 \leq \gamma \leq 4$. As determined with the thin slice method (see Chapter 4), γ is 2.4, i.e. this process does not entirely occur with holes via the valence band, but partially via the conduction band by injection of electrons [3]. It has later been found that γ increased with increasing anodic potentials, finally reaching a value close to $\gamma = 2$ [4].

A mechanism has been postulated by Beck and Gerischer [5, 6], as given in Fig. 8.1 in a slightly modified version. In the first step (a) a hole is trapped at the surface leading to the breaking of a surface Ge–Ge bond. This step is reversible, i.e. the trapped hole can be thermally regenerated and moves to another surface area. During the time interval in which the hole remains trapped, the corresponding surface group can swing away from the surface (step b), which is actually the rate-determining step. Then a second hole is trapped so that the unpaired electron disappears (step c). In a further step two OH groups are bonded (step d). Finally two other charges are required to separate the surface group from the Ge crystal (step e). Valence electrons involved in the Ge–Ge bonds, are usually located in the valence band (see Chapter 1). Since the single bond between the Ge atom of the surface group which is finally dissolved in step (e) and the corresponding Ge of the crystal is much weaker than the other Ge–Ge bonds, the corresponding valence electrons are in an energy state which is above the valence band as indicated in Fig. 8.2. There are now two ways of separating the surface group from the crystal (see step e): Either two further holes are transferred from the bulk of the crystal into these surface states, or the electrons in this state are thermally excited into the conduction band from where they recombine with holes in the bulk of a p-type electrode. In the latter case we would have $\gamma_e = 0$, and with respect to the overall reaction, $\gamma = 2$. Since values of $\gamma > 2$ have been obtained experimentally, both routes are possible. Although the energy position of this surface state (see Fig. 8.2) is not known, an electron excitation from the surface states is feasible because the bandgap of Ge is rather small ($E_g = 0.65$ eV). In the case of Si, which has a higher bandgap, such a process no longer seems to be possible (see below).

Concerning the current–potential curves given in Fig. 8.3, it is interesting to see that the current rises exponentially with increasing anodic potentials for a p-type Ge elec-

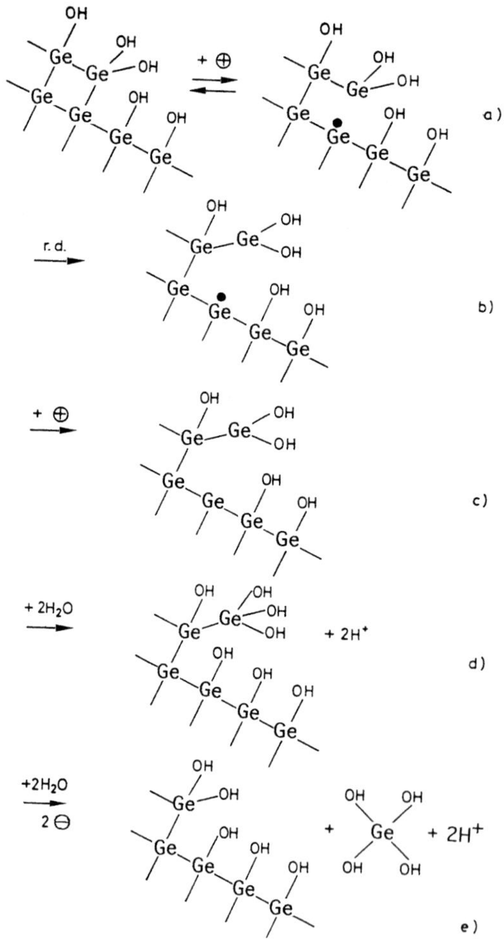

Fig. 8.1 Mechanism of the anodic dissolution of germanium. (Modified version after ref. [6])

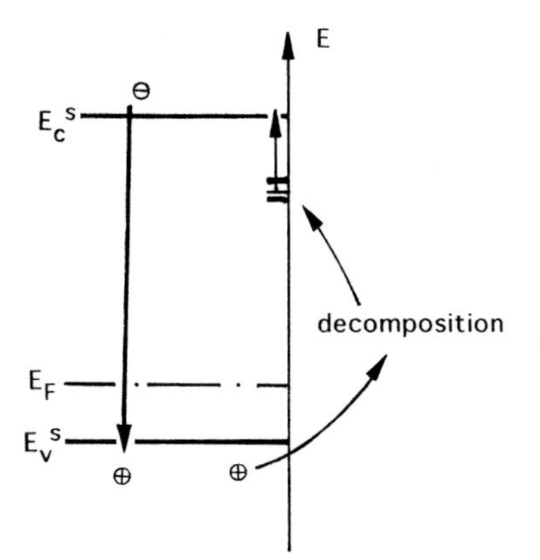

Fig. 8.2 Formation of intermediate states (surface states) during the anodic dissolution of germanium

trode. A quantitative evaluation of this current–potential curve yielded a slope of around 80 mV per one decade of current increase (not shown here). Since holes are required for the dissolution process one would expect theoretically that the rate-determining step is a one-hole process, i.e. the current should be proportional to the hole density at the surface ($j_{diss} \approx p_s$). Since $p_s = p_0 \exp e\Delta\phi_{sc}/kT$ one would expect that the j_{diss} vs. U_E curve would rise with 60 mV/decade. Obviously, some film formation at the Ge surface influences the potential distribution.

Concerning the current–potential curve as measured with an n-type electrode, one can observe that the saturation current–potential curve is not only determined by the light intensity but also by the doping of the n-type Ge electrode. As illustrated in Fig. 8.3, the anodic dark current measured with an n-type electrode, also depends on the doping. It increased with the resistivity of the n-type material which means that it

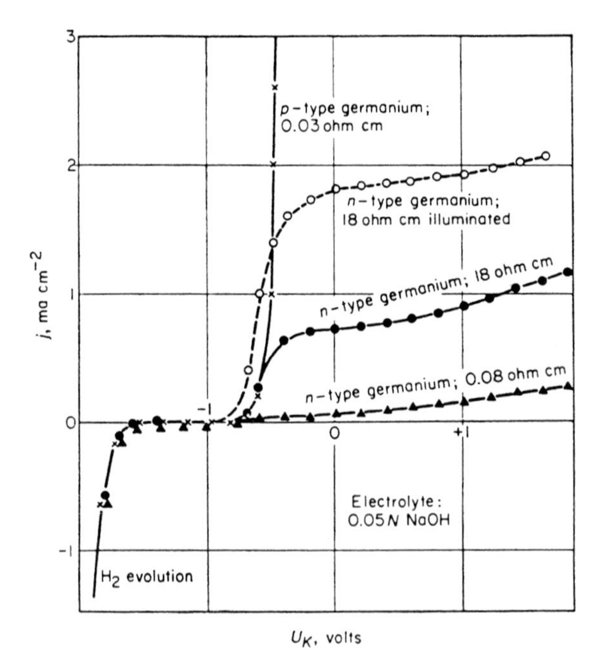

Fig. 8.3 Current–voltage curves at p- and n-germanium electrodes in alkaline solutions. (After ref. [5])

increased with decreasing electron density. It has been found that there is actually a linear relation between the anodic limiting current and the ratio of hole density and diffusion length (p_0/L_p). This is reasonable because the saturation current of a minority carrier device is just proportional to this quantity (see Eq. 2.32). An evaluation of the experimental data shows that the anodic saturation current agrees with that estimated from Eq. (2.32), at least within one order of magnitude.

Interestingly, the anodic dark current at n-Ge electrodes increases considerably upon addition of the oxidized species of a redox system, for instance Ce^{4+}, to the electrolyte, as shown in Fig. 8.4 [7]. The cathodic current is due to the reduction of Ce^{4+}. The latter process occurs also via the valence band (see Chapter 7), i.e. since electrons are transferred from the valence band to Ce^{4+}, holes are injected into the Ge electrode. Under cathodic polarization these holes drift into the bulk of the semiconductor where they recombine with the electrons (majority carriers) and the latter finally carry the cathodic current. In the case of anodic polarization, however, the injected holes remain at the interface and are consumed for the anodic decomposition of germanium, as illustrated in the insert of Fig. 8.4. Accordingly, the cathodic and anodic current should be compensated to zero. Since, however, the anodic current is increased upon addition of the redox system there is obviously a current multiplication involved, similarly to the case of „two-step“ redox processes (see Section 7.6). Thus, in step (e) (Fig. 8.1) electrons are injected into the conduction band. This experimental result is a very nice proof of the analytical result presented by Brattain and Garrett [3].

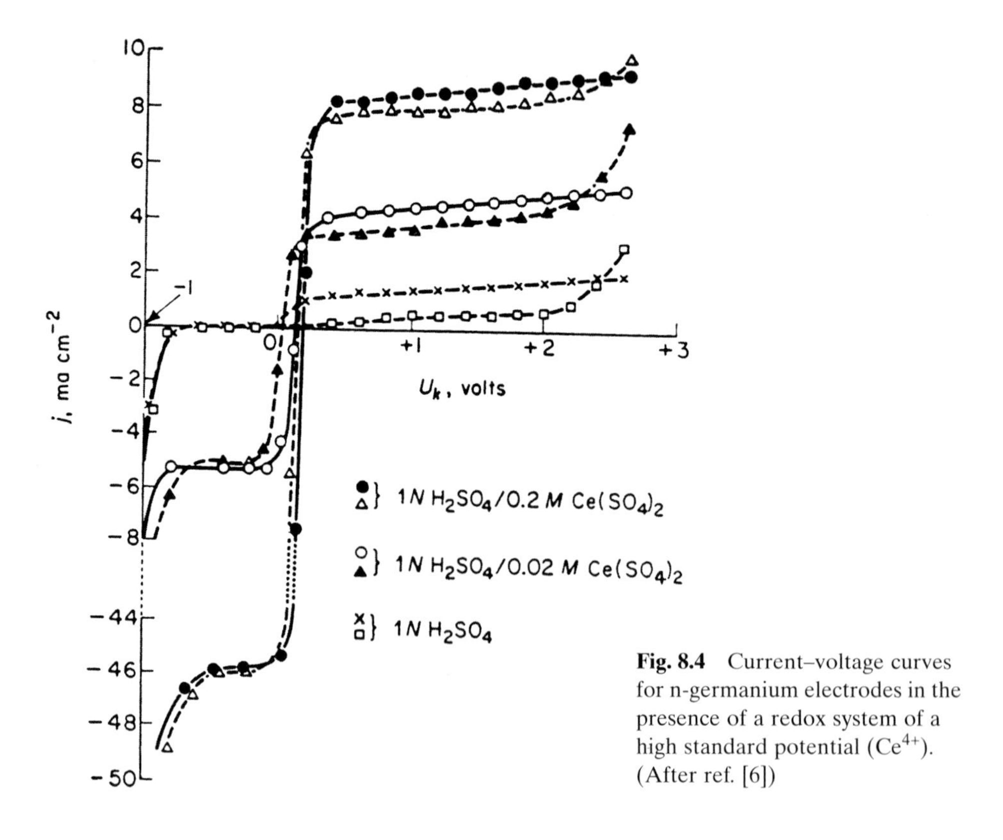

Fig. 8.4 Current–voltage curves for n-germanium electrodes in the presence of a redox system of a high standard potential (Ce^{4+}). (After ref. [6])

8.1.2 Silicon

The anodic dissolution of silicon is more complex because an oxide film is easily formed on the Si surface which is not soluble in water. This high stability of the oxide film makes Si particularly interesting for many applications. The oxide formation can be avoided, however, by working in hydrofluoric acid solutions. Since the results are different for concentrated and dilute solutions, these are treated separately.

a) Anodic Dissolution in Concentrated HF

The current–potential curve for n- and p-type electrodes look similar to those given in Fig. 7.10, i.e. the anodic current increases exponentially with potential for a p-type electrode and it saturates at a low value for an n-type electrode in the dark. A quantitative evaluation showed that the slope of the current–potential at a p-type electrode exhibits an ideal slope of 60 mV/decade as illustrated by a semilogarithmic plot of the current potential curve (Fig. 8.5) [8]. This is an ideal situation insofar as the current is proportional to the hole density at the surface, as already discussed in detail in Chapter 7. Using the thin slice method, it was shown that the oxidation of the Si electrode occurred entirely via the valence band and that there was no injection of electrons into the conduction band. In addition it was found by coulometric analysis that two and not four charges were required for the dissolution of one Si atom [8, 9]. Whereas about

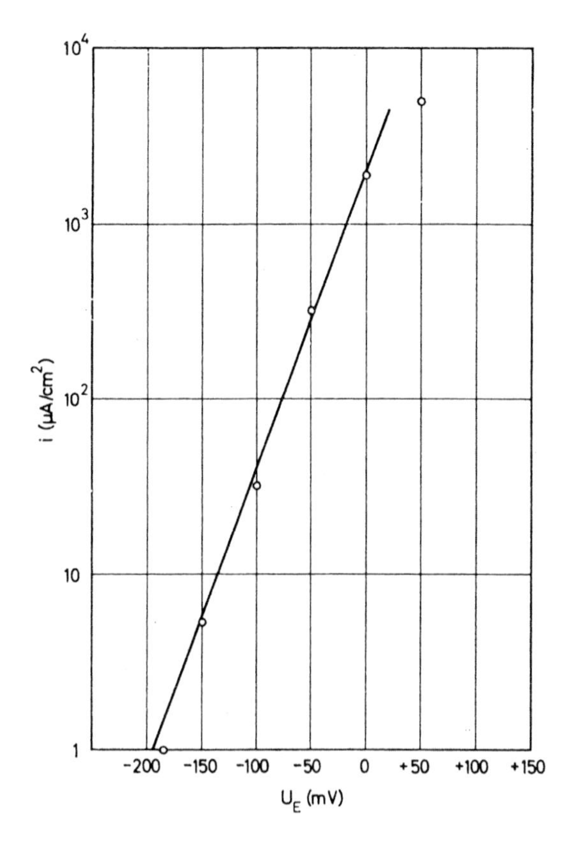

Fig. 8.5 Current–potential curve for p-silicon in concentrated HF solutions under anodic bias. (After ref. [8])

20 % of the dissolved Si was used for the formation of amorphous Si, the rest (80 %) stayed in solution. Simultaneously with the formation of the anodic amorphous film there was a strong evolution of a gas at the anodic silicon anode, which was identified as hydrogen. The amount of H_2 evolved at the electrode increased linearly with time.

As already shown by Uhlir [10] and by Turner [9], the dissolution mechanism for Si in concentrated HF is quite different from that of Ge in aqueous solutions. The fact that Si was dissolved in the divalent state is a little surprising because, in general, the stability of a divalent state of an element decreases in the IVth group in the direction Pb–Sn–Ge–Si–C. Therefore, a divalent silicon ion or silicon compound is expected to be unstable. Taking into account this instability of divalent Si, a dissolution mechanism of Si in concentrated HF solutions has been originally proposed as:

$$Si + 2HF + 2h^+ \rightarrow SiF_2 + 2H^+ \tag{8.2}$$

The divalent dissolution of Si was then described by a two-dimensional reaction scheme at a kink site on a silicon surface (not shown) assuming the surface to be covered by fluorine instead of hydroxyl groups [8]. It was further postulated that the unstable silicon difluoride changes into a stable tetravalent form by a disproportionation reaction [9]:

$$\begin{array}{llll} 2SiF_2 \rightarrow & Si^0_{amorphous} & + \; SiF_4 & \\ & \downarrow + 2H_2O & \quad \lfloor +2HF \longrightarrow & H_2SiF_6 \\ & SiO_2 + 2H_2 & & \\ & \lfloor +6HF \longrightarrow & H_2SiF_6 + 2H_2O & \end{array} \tag{8.3}$$

Since H_2 was formed even when the anodic polarization had been stopped, the oxidation of amorphous Si and the accompanying H_2 formation must be a slow chemical step.

b) Dissolution at Lower HF Concentration ($<$0.1 M)

The anodic behavior of p-type Si electrodes is quite different for lower HF concentrations. The current increases, but not really exponentially, with rising anodic polarization, it passes a maximum and increases again slowly at higher anodic potentials [8] (Fig. 8.6). The current increases with the rotation speed ω of the electrode. Since the current does not follow a $\omega^{1/2}$ dependence (Levich relation [11]) the relationship cannot be determined entirely by diffusion. At electrode potentials below the peak, silicon is dissolved again in the divalent state, as already reported above in the case of high HF concentrations. Here also H_2 formation was observed. At electrode potentials beyond the current peak, as shown in Fig. 8.6, the dissolution was found to occur via the tetravalent state of Si and the H_2 evolution disappeared at p-type electrodes [8]. These results were confirmed 25 years later [12]. Experiments performed using the thin slice arrangement (see Chapter 4) have shown that the anodic reactions occur only via the valence band at all electrode potentials [8].

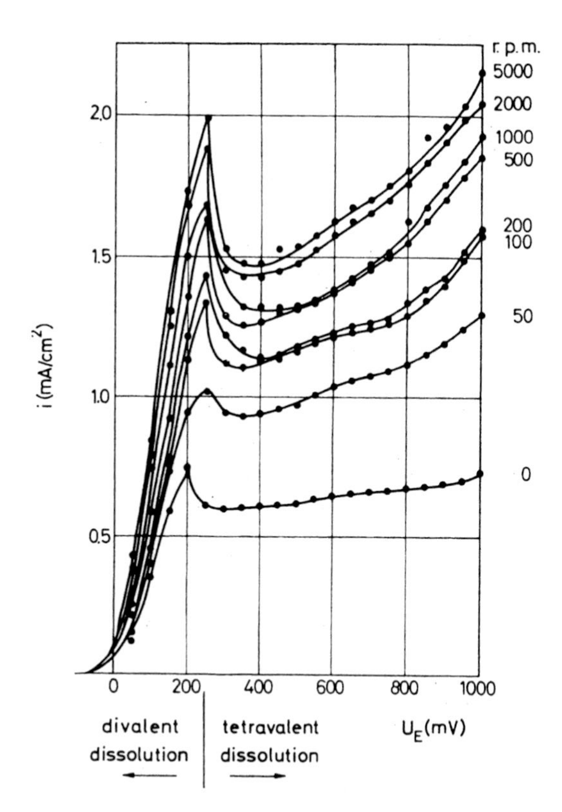

Fig. 8.6 Current–potential curves for a rotating p-type silicon electrode in 0.1 M hydrofluoric acid solutions at different rotation velocities. (After ref. [8])

In the early work published about 30 years ago (see above), the mechanism of the divalent dissolution at lower HF concentration was originally described by an electrochemical reaction in which SiF_2 was formed as an intermediate which disproportionated into amorphous Si and SiF_4 again [8]. Early infrared measurements performed by Beckmann in 1965 [13], had already shown that amorphous films of polymerized silicon hydride were formed. This result was previously not really understood. Since the Si–F bond is much more stable than the Si–OH bond one would expect that etching of a oxide-covered surface in HF leads to a surface consisting of Si–F_2 groups. It was thus very surprising to find, by ex situ IR absorption studies, that the surface of microcrystalline Si consisted of hydrogen and not of F nor of O containing ligands after the removal of the oxide [14]. This observation stimulated a large number of investigations of single crystalline Si surfaces by various surface spectroscopic techniques [15–17] which cannot be discussed here in detail. In situ IR studies of the Si–HF interface revealed that the Si surface consists predominantly of a Si hydride even under anodic polarization [18].

These observations have led to a more adequate model of electrochemical divalent Si dissolution in the low potential range, as illustrated in Fig. 8.7 [19]. This model starts with a kink site of a (111) surface in which the Si atom has two bonds to the lattice and the two remaining bonds are saturated by H ligands. If a hole arrives at the surface it weakens one of the Si–H bonds of the kink site atom and this H atom reacts with a F ion in the solution, i.e. a process by which the surface H atom leaves its electron at the

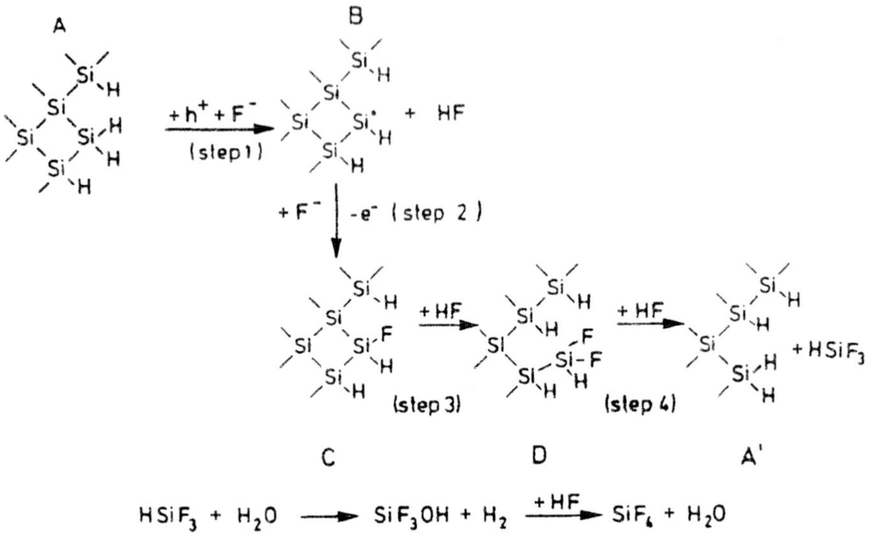

Fig. 8.7 Two-dimensional description of the reaction steps at a kink site on a silicon surface for the oxidation initiated by a hole. (After ref. [19])

Si atom (step 1 in Fig. 8.7). In a second charge transfer step (hole or electron) a new stable Si–F bond is formed. When one of the Si–H bonds is substituted by a Si–F bond the two Si back bonds to the lattice are a little polarized due to the very polar Si–F bond. This increases their chemical reactivity, and one of the two back bonds is split by reaction with HF (step 3). The remaining back bond is now much weaker and will be split by another HF molecule (step 4). The sequence of steps 1–4 in Fig. 8.7 corresponds to the overall reaction

$$Si + \gamma h^+ + 3HF \rightarrow HSiF_3 + (2 - \gamma)e^- + 2H^+ \tag{8.4}$$

The $HSiF_3$ molecule is finally oxidized by water to the tetravalent oxidation state, i.e. we have

$$HSiF_3 + H_2O \xrightarrow{+HF} SiF_3OH + H_2 \rightarrow SiF_4 + H_2O \tag{8.5}$$

SiF_4 reacts further to H_2SiF_6 as described in reaction (8.3).

As already described above, measurements with the thin slice method had shown that the anodic dissolution at p-Si electrodes in the dark occurs entirely via the valence band [8], as has been recently confirmed by Cattarin et al. [20]. About 20 years later Matsumura and Morrison investigated the current–potential behavior of n-type Si electrodes in acidic fluoride media at different light intensities [21]. According to their results, the quantum yield ϕ of the anodic photocurrent varied from $\phi = 4$ at low light intensities to $\phi = 2$ at high light intensities (Fig. 8.8), i.e. a current doubling or multiplication took place (see also Section 7.6). At the same time H_2 formation occurred, as also shown in Fig. 8.8 [18]. This result was highly surprising for two reasons. First, a quantum yield of $\phi > 1$ indicates that only a fraction of the multi-electron reaction occurs with holes via the valence band. Secondly, a value of $\phi = 4$ means that only one

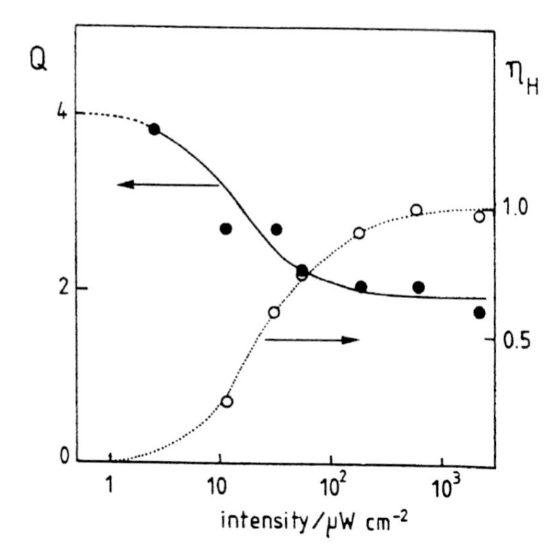

Fig. 8.8 Intensity dependence of the quantum yield Q for n-Si (111) and the corresponding hydrogen production efficiency η_H measured in a ring disc arrangement. Electrolyte, 1 M NH_4F, pH 4.7. (After ref. [18])

hole is required for the first step and three electrons are injected into the conduction band of the n-Si electrode in three subsequent steps. These three latter steps require, of course, no light excitation. In addition, a quantum yield of $\phi = 4$ indicates that four charges are required for the complete dissolution process. L.M. Peter et al. have further analyzed this process by intensity-modulated photocurrent spectroscopy (IMPS; see Section 4.5) [18, 22]. In the range of low light intensities, they found experimentally three semicircles (Fig. 8.9, upper curve), where the two high frequency semicircles strongly overlap. This experimental result was sufficiently well simulated (lower curve in Fig. 8.9) on the basis of a reaction scheme given by Eq. (8.6)

$$\mathrm{Si(0)} \underset{+h^+}{\longrightarrow} \mathrm{Si(I)} \overset{k_a}{\underset{-e^-}{\longrightarrow}} \mathrm{Si(II)} \overset{k_b}{\underset{-e^-}{\longrightarrow}} \mathrm{Si(III)} \overset{k_c}{\underset{-e^-}{\longrightarrow}} \mathrm{Si(IV)} \tag{8.6}$$

Here it has been assumed that the first step (Si(0) → Si(I)) occurs via the valence band where the three other subsequent steps proceed either under hole consumption or electron injection. The k_is are first-order rate constants having a dimension of s^{-1}. The best fit was obtained with $k_a = 2 \cdot 10^4\ s^{-1}$, $k_b = 500\ s^{-1}$ and $k_c = 0.5\ s^{-1}$. These values of the first-order rate constants are very small. They were interpreted in terms of relatively large activation energies in the order of about 0.5 eV.

On the other hand, the decrease of ϕ from 4 down to 2 is much more difficult to interpret. As already mentioned above, gravimetric experiments have shown that the decrease is due to a change in the dissolution valency of Si from IV to II. This analytical result is supported by the fact that H_2 is formed as soon as the dissolution valency changes (Fig. 8.9), because H_2 formation under anodic bias is only possible if Si is dissolved in the divalent state (see Eq. 8.3). Very recently, a model has been presented in which it is assumed that Si(I) (which is just a mobile surface radical) catalyses the divalent dissolution [23].

These results obtained with n-Si, seem to be in contradiction to those obtained with the p-type electrode. However, a quantum yield of $\phi = 4$ was found with n electrodes only at very low light intensities (<10 μW) which corresponds to a current of <5 μA cm^{-2}. An analysis of the dissolution valency at p-electrodes, however, was not possible

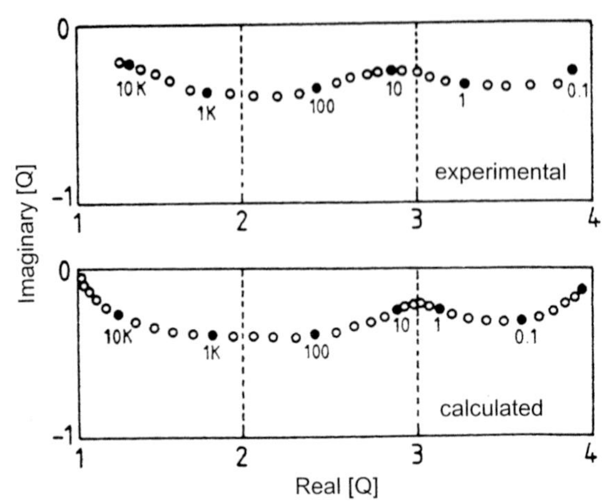

Fig. 8.9 Comparison of experimental and calculated intensity-modulated photocurrent spectroscopy (IMPS) plots for the photodissolution of n-Si (111) in 6.5 M NH_4F at low light intensities (photo quadrupling regime) The data used for the theoretical plot are: $k_a = 2 \cdot 10^4\ s^{-1}$; $k_b = 500\ s^{-1}$; $k_c = 0.5\ s^{-1}$. A gaussian distribution of activation energies for electron injection with a standard deviation of 1.5 kT and a pre-exponential factor of $10^{12}\ s^{-1}$ were used to simulate the flattening of the IMPS. (After ref. [18])

at such low currents. Accordingly, it is reasonable to assume that p-Si is also dissolved in the tetravalent state at such low currents. On the other hand, a more severe problem is the result that a current doubling ($\phi = 2$) was found with n-electrodes whereas measurements at p-electrodes, using the thin slice method (see Section 4.3), indicated that the divalent dissolution is a pure valence band process [8]. The differing behavior of n- and p-Si can only be explained by assuming that in the case of p-Si also an electron is injected into the conduction band which recombines efficiently with holes via surface states so that they could not be detected on the rear p–n junction when the thin slice method was used. This model was supported by Blackwood et al. [18] who found strong surface recombination in the corresponding potential range.

In HF solutions, p-Si begins to be electropolished when the critical current density is reached (current peak in Fig. 8.6). The critical current is interpreted as the point at which HF begins to be consumed by the anode process as fast as it reaches the surface. At potentials anodic with respect to the current peak (electropolishing region), a thin SiO_2 film is present on the surface which remains there until sufficient HF reaches the surface by mass transport to form the water-soluble fluoride complex [8, 9]. Si dissolved here in the tetravalent state, as given by the reaction:

$$\begin{aligned} Si + 2H_2O + 4h^+ \rightarrow\ & SiO_2 + 4H^+ \\ & \downarrow +6HF \\ & H_2SiF_6 + 2H_2O \end{aligned} \tag{8.7}$$

The negative resistance portion of the j–U_E curve around the critical current, in Fig. 8.6, is characteristic of the electropolishing and passivating process. It can be explained as follows. Below the critical current there is considerable gassing (H_2 formation) which has a stirring effect at the surface. When the critical current is reached and electropolishing sets in, virtually all gassing stops. This eliminates the stirring effect caused by gas evolution and the rate of mass transfer of HF to the surface decreases. In the electropolishing region the current density flow is controlled by the resistance of the oxide film and the rate of arrival of HF at the surface. Thus if the rate of mass transfer of HF decreases the current is also reduced.

8.1.3 Anodic Formation of Amorphous (Porous) Silicon

As already mentioned in Section 8.1.2, in situ IR measurements have proved that the silicon surface is terminated by hydrogen when an Si electrode (p-type in the dark, n-type at illumination) in HF solutions is polarized anodically in the range between the open circuit potential and the anodic peak. In addition an amorphous layer consisting mainly of Si hydrides is formed in the potential range of divalent dissolution of Si. A closer inspection of the surface has shown that this process is accompanied by the formation of deep pores which are filled with small Si crystals of a size smaller than 13 Å [24]. The morphology of the pores depends strongly on the doping type and density, the anodizing current and the composition of the etching solution.

Porous Si has been shown to have striking physical properties. The optical absorption of the semiconductor is shifted into the visible region [25] while strong photoluminescence was also observed in the same range [26]. In addition, electroluminescence was generated in the same wavelength range by injection of holes from a redox system such as $S_2O_2^{2-}/SO_4^{2-}$ [27]. Such results have been attributed by a number of groups to quantization effects resulting from the very small diameter of the Si particles in the pores [25, 26]. On the other hand, Brandt et al. suggested that the visible luminescence is due to siloxenes which may be formed during porous Si preparation [28]. However, a growing consensus has emerged for explaining the luminescence by quantum confined structures because the luminescence from porous Si passivated with oxygen rather than hydrogen has ruled out SiH_x species [29], and the absence of Si–O bonding in x-ray absorption data from porous Si has ruled out siloxenes [30]. Further details on quantization effects are given in Chapter 9.

The exciting prospect of incorporating optical functions into Si integrated circuitry has led to the great interest in porous Si. Besides investigations of the physical properties, the mechanism of pore propagation has also been studied (see e.g. [31]). On the other hand, there is little information about pore initiation on the flat surface of a Si single crystal. A detailed discussion of these effects is beyond the scope of this chapter.

8.1.4 Compound Semiconductors

The anodic dissolution of compound semiconductors has only been studied with a few materials, e.g. SiC, various III-V compounds, CdS and $MoSe_2$. The essential reactions are given below.

SiC is a very hard and stable material. It exists in many different modifications. Only a few experimental results are available and many of them have not been sufficiently reproducible. One reason is that reliable single crystals (mostly as thin layers on Si) have become available only rather recently. At first sight one may expect anodic behavior similar to that of Si. However, in contrast to Si, the anodic current at a p-type SiC electrode in H_2SO_4 was found to remain stable and did not essentially decrease with polarization time, and a white SiO_2 film was formed after prolonged polarization [32]. The head space analysis of gaseous products revealed CO and CO_2 in a ratio close to 1:1. The formations of these products were explained by an eight-hole process [33] and a six-hole reaction [32] which proceed at the same rate, both being given by

$$SiC + 8h^+ + 4H_2O \rightarrow SiO_2 + CO_2 + 8H^+ \tag{8.8}$$

$$SiC + 4h^+ + 2H_2O \rightarrow SiO + CO + 4H^+ \tag{8.9}$$

Since the oxide does not limit the interfacial current it was concluded that it is very porous. The oxide is very soluble in HF so that SiC can be etched which is important for the production of devices [34]. The anodic decomposition of SiC was also studied in concentrated HF solutions. In this case, a yellow-brown compound was formed which was insoluble in HF. This film consists of amorphous SiC [32]. This behavior is similar to the formation of porous Si on Si (see Section 8.1.3).

Concerning III-V semiconductors, mainly GaAs, GaP and InP have been studied. In all cases it has been found that six charges are required for the anodic decomposition of one semiconductor molecule. As an example taking GaAs, with which most investigations have been performed, the overall reaction is given by [35]

$$GaAs + \gamma h^+ + 10OH^- \rightarrow GaO_2^- + AsO_3^{3-} + 5H_2O + (6 - \gamma)e^- \tag{8.10a}$$

in alkaline solutions and

$$GaAs + \gamma h^+ + 3H_2O \rightarrow Ga^{3+} + H_3AsO_3 + 3H^+ + (6 - \gamma)e^- \tag{8.10b}$$

in acid solutions.

In this case it was found that $\gamma \approx 6$ [35, 36], i.e. the reaction occurs essentially via the valence band. In addition, an ideal slope of 60 mV/decade was found with p-GaAs at anodic polarization [37]. From this result it must be concluded that only one hole is involved in the rate-determining step. Investigations with the electrodes with, in one experiment the Ga (111) face and in a second experiment the As $(\bar{1}\bar{1}\bar{1})$ face contacting the electrolyte, yielded no difference in dissolution rate [35]. This result is an indication that the dissolution occurs mainly at steps as illustrated for GaP in Fig. 8.10. The influence of steps is discussed in detail by Morrison in his book [39]. On the other hand, the flatband potential and the onset potential of the photocurrent do depend on the crystal faces. The difference in flatband potential is in the order of about 0.1 V [39]. Accordingly, the Helmholtz layer is somewhat different for the two surfaces.

Similar results have been obtained with GaP [40, 41]. In this case it also has been made clear that the anodic dark currents at p-GaP are limited to a rather low value

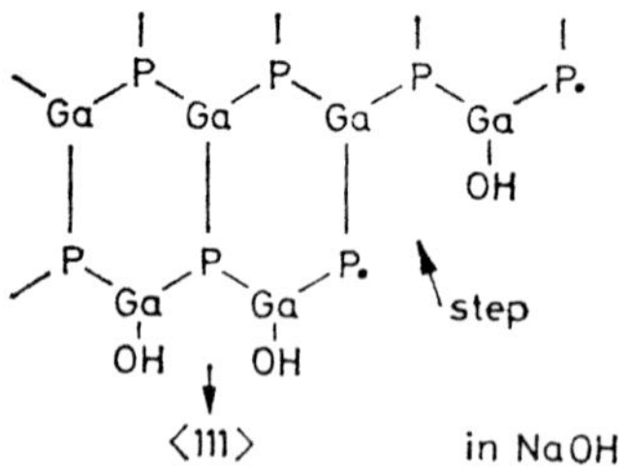

Fig. 8.10 Surface structure of a GaP lattice in NaOH. (After ref. [40])

because of the formation of Ga_2O_3 which is not soluble in solutions of intermediate range (see e.g. Pourbaix [42]).

In principle, the same observations have been made with InP. In this case, however, it is more difficult to find the proper conditions for electrochemical measurements. This is caused by the fact that InP is rather easily reduced under cathodic polarization, leading to the formation of metallic indium at the surface. Subsequent anodic polarization leads then to the formation of In_2O_3 which can easily influence the corresponding current–potential curves [43]. The anodic corrosion of InP was investigated with respect to the participation of electrons in the dissolution mechanism, including IMPS measurements [44]. These measurements have yielded $\gamma = 3$ at low intensities. Accordingly, three electrons are injected into the conduction band besides the consumption of three holes, whereas at higher intensities five holes are used for the dissolution process and one electron is injected. This result is similar to that obtained with Si (see Section 8.1.2). The interpretations, however, were different. In the case of n-InP, the decrease of photocurrent was interpreted as a competition between hole and electron transfer, whereas for n-Si the decrease was explained by a change of valency. The advantage of applying IMPS is that kinetic data such as rate constants for several steps can be obtained. A phase shift occurs if an electron is injected into the conduction band. If the subsequent steps are hole transfer processes, then they are in phase with the light modulation and the semicircle would be reduced to a single point on the real axis in the complex plane. The analysis of the relevant results is rather complicated [44] and cannot be discussed here in further detail.

Another class of semiconductors comprises II-VI compounds, mainly sulfides and oxides, which also undergo anodic decomposition. A typical example is CdS (E_g = 2.5 eV; only available as n-type material) which has been the subject of many investigations. As shown by Meissner et al., anodic dissolution can occur under illumination in two ways depending on the oxygen concentration in the electrolyte [45, 46]. In the presence of oxygen:

$$CdS + 4h^+ + 2H_2O + O_2 \rightarrow Cd^{2+} + SO_4^{2-} + H^+ \tag{8.11a}$$

and in O_2-free solutions:

$$CdS + 2h^+ \rightarrow Cd^{2+} + S \tag{8.11b}$$

Since usually not much O_2 is dissolved in H_2O, the second reaction dominates. The sulfur formed in the second reaction remains at the surface because it is insoluble in H_2O. This has severe consequences for many other investigations, especially on Mott–Schottky measurements as already mentioned in Section 5.3.4. In the case of electrodes with sulfur on the surface, the Mott–Schottky curves are not straight and become frequency-dependent, a typical sign of a contaminated surface. Even etching CdS in HCl did not lead to a clean surface as shown by photoelectron spectroscopy (XPS) [46]. Cleaning of the surface was possible by means of cathodic polarization in O_2-saturated solutions. It has been assumed that, in the anodic decomposition and the cleaning in O_2-saturated solutions, an essential role is played by the formation of an $S^{\bullet -}$ at the surface, according to the following reactions

$$CdS + h^+ \rightarrow Cd^{2+} + S^{\bullet -} \tag{8.12a}$$

$$S_{surf} + e^- \rightarrow SO_2^{\bullet -} \tag{8.12b}$$

In the presence of O_2 one obtains

$$S^{\bullet -} + O_2 \rightarrow SO_2^{\bullet -} \tag{8.13}$$

The latter product is then further oxidized to SO_4^- with the consumption of three further holes. The formation of sulfur on the surface can also be avoided by working in a sulfide-containing electrolyte in which sulfur is soluble under the formation of polysulfide (S_n^{2-}). According to IMPS measurements, there is a small probability of electron injection in the subsequent dissolution steps [47].

The transition metal chalcogenides, e.g. $MoSe_2$, WSe_2 or the corresponding sulfides, represent a particular class of materials which form layer compounds as shown in Fig. 5.11. In each layer the metal is shielded by the sulfur or selenium atom. Therefore the individual layers are kept together just by Van der Waals forces. If the basal planar surfaces are contacting the solution then the metal is also shielded from the liquid. The metal can interact with H_2O only at steps (Fig. 5.16), i.e. a dissolution occurs at such sites. In addition, Tributsch originally expected a high stability of these materials against corrosion, because the electronic states of the valence band are formed by non-bonding d-electron states of the metal which do not participate in the bonding [48–50]. A more detailed consideration has proved, however, that the valence band consists of a mixture of d and p electrons. Accordingly, anodic dissolution rather than oxygen formation was experimentally observed [48]. It has been suggested by Gerischer that H_2O is oxidized to some radical intermediate which in turn reacts with selenium atoms of the layer [51]. This would explain that not only selenium or sulfur but mainly SO_3^{2-} (SeO_3^{2-}) or SO_4^{2-} (SeO_3^{2-}) were formed. Nevertheless, a relatively high overvoltage for the onset of photocurrent accompanied by a shift of energy bands was observed (see impedance measurements in Chapter 5) [52] [53]. Investigations of these electrodes where the planar surface contacts the electrolyte, have also shown that the flatband potential is independent of pH for a surface with a low density of steps [54]. All of these observations indicate that the planar surface is fairly stable and that many redox processes can compete with the anodic dissolution.

The reaction products of various semiconductors are listed in Table 8.1.

Table 8.1 Decomposition products of various semiconductor electrodes during anodic and cathodic polarization

Electrode material	E_g (eV)	Anodic decomposition products	Cathodic decomposition products
Ge (n, p)	0.8	Ge^{4+}	H_2
Si (n, p)	1.1	Si^{2+}; Si^{4+} (in HF)	H_2
GaAs (n, p)	1.4	Ga^{3+}, AsO_3^{3-}	AsH_3, H_2
CdSe (n, p)	1.75	Cd^{2+}, Se^0	Cd, (H_2Se)
GaP (n, p)	2.35	Ga^{3+}, P^0, P^{3+}	H_2, PH_3
CdS (n)	2.4	Cd^{2+}, S^0	Cd, H_2S
WO_3 (n)	2.5	O_2	H_xWO_3, H_2
ZnO (n)	3.2	Zn^{2+}, O_2	Zn, H_2O

8.2 Cathodic Decomposition

Under cathodic polarization hydrogen is usually formed in aqueous solutions. Since the valence band of all the semiconductors studied so far, occurs considerably below the H^+/H_2 standard potential, this reaction is a conduction band process. Besides this reaction, a number of compound semiconductors are reduced or decomposed under cathodic polarization. A selection of semiconductors is also given in Table 8.1. These reactions have never been investigated quantitatively. In general, it was found that with semiconductors which decompose during cathodic polarization, a metal layer was formed on the surface. As already mentioned in the previous section, such a metal layer can strongly influence the electrochemical behavior of a semiconductor electrode because it may not necessarily be dissolved again during a subsequent anodic cycle. For instance, a metal oxide can be formed which, depending on the pH of the solution, remains stable during an anodic potential sweep, as observed with InP.

8.3 Dissolution under Open Circuit Conditions

Anodic decomposition is also possible in the presence of a suitable redox system in the solution without external voltage. Taking a redox couple of a very positive standard potential, such as Ce^{4+}/Ce^{3+}, then holes are injected from Ce^{4+} ions into the valence band of the semiconductor. These holes are available for the anodic decomposition. The corresponding j–U_E curves (with and without the redox system) for a p-type electrode are shown in Fig. 8.11. The cathodic reduction of the redox system sets in at U_{redox}, i.e. when the quasi-Fermi level of holes passes E_{redox} (dotted line). In the cathodic range the current is diffusion-limited. The resulting j–U_E curve (dashed curve) passes the potential axis at a value at which cathodic and anodic currents are equal (compare also with Section 7.4.1). The rate of the dissolution current is controlled by the redox system. In the case of a very stable semiconductor, the decomposition current is much smaller and the two partial currents would be equal at more positive potentials.

The same reactions occur at the corresponding n-type electrode (Fig. 8.12). In the anodic range, the total current remains very small upon addition of the redox system. However, it is determined by two partial currents, namely the cathodic reduction current and the anodic decomposition current, as can be proved analytically. A description in terms of quasi-Fermi levels has already been given in Section 7.4.1.

According to Figs. 8.11 and 8.12, a decomposition under open circuit conditions occurs with p-type as well as with n-type semiconductors. Besides this electrochemical decomposition some semiconductors can also be dissolved chemically as, for instance, GaAs in the presense of Br_2 [55] or H_2O_2 [56]. Both are strong oxidizing agents which exhibit, current-doubling behavior, however, at GaP and GaAs electrodes during cathodic polarization (compare with Section 7.6). Accordingly, the electrochemical reduction of Br_2 and H_2O_2 occurs at the p-type electrodes only under illumination. Pure chemical decomposition has not been found at n-GaAs electrodes but only at p-GaAs in the dark. Interestingly, this chemical etching occurred only in that potential range where

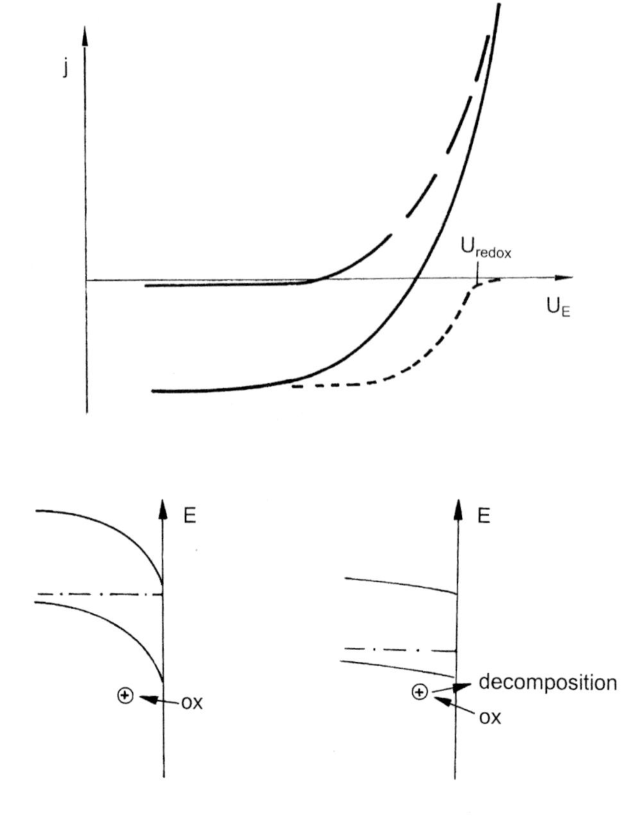

Fig. 8.11 Top: theoretical current–potential curves at a p-type semiconductor electrode in the presence (solid curve) and absence (long-dashed curve) of a redox system with a very positive standard potential; short-dashed curve, cathodic partial current for a redox system which is reduced by an electron transfer via the valence band of a semiconductor. Bottom: energy diagrams for cathodic (left) and anodic (right) polarization

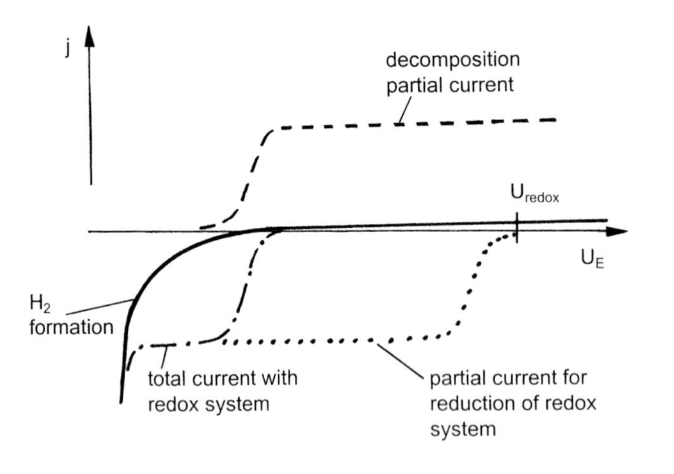

Fig. 8.12 Theoretical current–potential curves at an n-type semiconductor electrode in the presence of a redox system of a high standard potential (similarly as in Fig. 8.11)

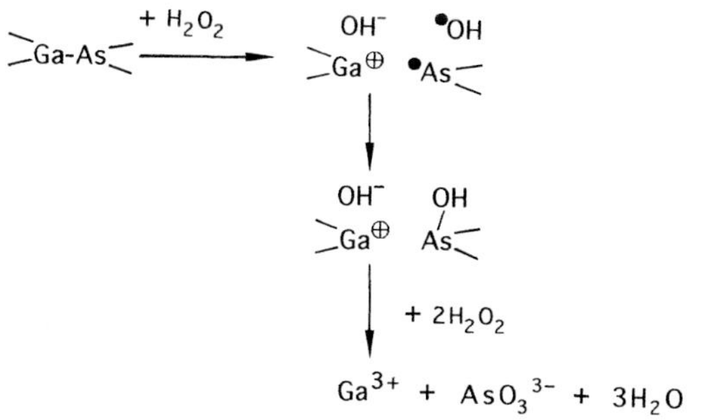

$$\geq Ga\text{-}As\leq \xrightarrow{+H_2O_2} \geq Ga^{\oplus}\,OH^- \quad {}^{\bullet}As\leq\,{}^{\bullet}OH$$

$$\downarrow$$

$$\geq Ga^{\oplus}\,OH^- \quad As(OH)\leq$$

$$\downarrow + 2H_2O_2$$

$$Ga^{3+} + AsO_3^{3-} + 3H_2O$$

Fig. 8.13 Model for the chemical dissolution of GaAs in the presence of H_2O_2. (After ref. [36])

a cathodic photocurrent was measured upon illumination. The chemical etch rate was reduced to a low value as soon as the p electrode was illuminated [57]. This result shows that there is a rather complex interdependence between electrochemical and chemical processes. A simplified version of the chemical dissolution is shown in Fig. 8.13. In the first step one bond is broken and a radical is formed. In several further reaction steps the other corresponding bonds are broken. The radical (surface state) is essential insofar as it can trap an electron from the conduction band which means a repairing of the original bond. This is possible at n-GaAs and at p-GaAs under illumination. Minks et al. [57] have derived a complete reaction scheme which still contains some reaction routes.

8.4 Energetics and Thermodynamics of Corrosion

The question arises whether it is useful to define a corrosion potential which, as in the case of standard potentials for redox systems, can be described within the general energy level model in order to compare it with the energy of electrons and holes in the semiconductor. Theoretically this is possible by calculating the free energy of the corrosion reaction. Various authors have formulated corresponding expressions [39, 58–60]. Following Gerischer's derivations, two simple reactions at a semiconductor MX are considered:

$$\mathrm{MX} + z\mathrm{e}^- + \mathrm{solv} \rightarrow \mathrm{M} + \mathrm{X}^{z-}_{\mathrm{solv}} \tag{8.14}$$

for a cathodic reaction and

$$\mathrm{MX} + z\mathrm{h}^+ + \mathrm{solv} \rightarrow \mathrm{M}^{z+}_{\mathrm{solv}} + \mathrm{X} \tag{8.15}$$

for an anodic reaction

in which MX is the compound semiconductor, M the metal and the term „solv" represents the complexing of the elements. The free energy of these reactions can be obtained from the appropriate handbooks or from ref. [61]. In electrochemistry, data are usually referred to the H^+/H_2 standard potential. In order to get the same reference point here we have to write the corresponding reaction for hydrogen as given by

$$\tfrac{1}{2}z\mathrm{H}_2 + \mathrm{solv} \rightarrow z\mathrm{H}^+_{\mathrm{solv}} + z\mathrm{e}^- \tag{8.16}$$

The difference between Eqs. (8.16) and (8.14) or (8.15) then yields the corresponding free energy values, ${}_{n}\Delta G_{sH}$ and ${}_{p}\Delta G_{sH}$, respectively. The decomposition potentials are then given by

$$ {}_{p}E_{decomp} = {}_{p}\Delta G_{sH}/z \tag{8.17a}$$

for the oxidation and

$$ {}_{n}E_{decomp} = -{}_{n}\Delta G_{sH}/z \tag{8.17b}$$

for the reduction of the semiconductor.

Possible positions of the electron-induced corrosion potential ${}_{n}E_{decomp}$ and of the hole-induced corrosion value ${}_{p}E_{decomp}$ are illustrated in Fig. 8.14. If both energies occur within the bandgap (case b in Fig. 8.14) then reduction as well as oxidation is thermodynamically possible. In the case of oxidation, the electrode has to be polarized to the extent that the quasi-Fermi level of holes occurs between E_v and ${}_{p}E_{decomp}$; and for reduction the quasi-Fermi level of electrons must be between E_c and ${}_{n}E_{decomp}$. If an E_{decomp} value occurs outside the bandedges then the semiconductor should be fairly stable. A decomposition then only happens in a potential range where the externally applied voltage already occurs across the Helmholtz layer. A selection of decomposition potentials is given in Fig. 8.15. The reactions on which these calculations are based are given in the upper part of Fig. 8.15.

The practical use of these calculations is limited, however, because the kinetics of a reaction can play an important role. This becomes quite obvious for layer compounds such as MoS_2. The kinetics may be controlled by adsorption, surface chemistry, surface structure and crystal orientation. According to Fig. 8.15, ${}_{p}E_{decomp}$ is close to the conduction band, i.e. MoS_2 is rather easily oxidized. In the case of a flat basal surface, it has been observed with several transition metal chalcogenides that the photocurrent onset at n-electrodes occurs with high overvoltages accompanied by a shift of U_{fb}.(see Section 5.3). Since this is caused by an accumulation of holes at the surface the hole transfer is kinetically inhibited.

The various dissolution processes discussed above, play an important role in semiconductor etching for device technology (see Chapter 11).

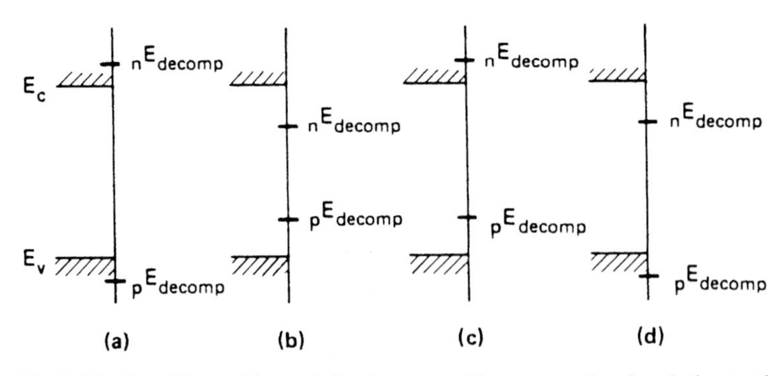

Fig. 8.14 Possible positions of the decomposition energy levels relative to the bandedges: a) for a relatively stable semiconductor; b) implies instability if either electrons or holes reach the surface; c) and d) imply instability with regard to holes and to electrons, respectively. (After ref. [38])

$SnO_2 + 4HCl \cdot aq \rightarrow SnCl_4 \cdot aq + O_2 + 2H_2$
$SnO_2 + 2H_2 \rightarrow Sn + 2H_2O$

$WO_3 + H_2O \rightarrow WO_2 + (1/2)O_2 + H_2$
$WO_3 + 3H_2 \rightarrow W + 3H_2O$

$ZnO + 2HCl \cdot aq \rightarrow ZnCl_2 \cdot aq + (1/2)O_2 + H_2$
$ZnO + H_2 \rightarrow Zn + H_2O$

$TiO_2 + 4HCl \cdot aq \rightarrow TiCl_4 \cdot aq + O_2 + 2H_2$
$TiO_2 + 2H_2 \rightarrow Ti + 2H_2O$

$Cu_2O + H_2O \rightarrow 2CuO + H_2$
$Cu_2O + H_2 \rightarrow 2Cu + H_2O$

$CdS + 2HCl \cdot aq \rightarrow CdCl_2 \cdot aq + 2S$
$CdS + H_2 \rightarrow Cd + H_2S \cdot aq$

$MoS_2 + 2H_2O \rightarrow MoO_2 + 2S + 2H_2 ({}_pE_{dec})$
$MoS_2 + 2H_2 \rightarrow Mo + 2H_2S \cdot aq$

$GaP + 6H_2O \rightarrow Ga(OH)_3 + H_3PO_3 \cdot aq + 3H_2$
$GaP + (3/2)H_2 \rightarrow Ga + PH_3$

$GaAs + 5H_2O \rightarrow Ga(OH)_3 + H_3AsO_3 \cdot aq + eH_2$
$GaAs + (3/2)H_2 \rightarrow Ga + AsH_3$

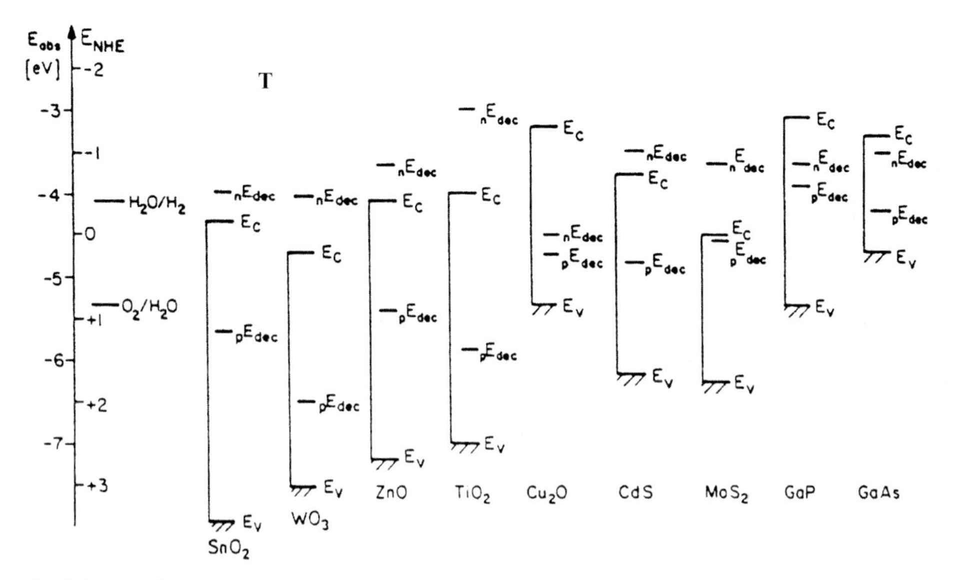

Fig. 8.15 Positions of bandedges and decomposition Fermi levels for various semiconductors. (After ref. [51])

8.5 Competition between Redox Reaction and Anodic Dissolution

The competition between redox reaction and anodic dissolution became very important in the development of stable regenerative solar cells on the basis of semiconductor–liquid junctions. As shown in the previous section, it is determined by the thermodynamic and kinetic properties of the processes involved. Information on the competitions between these reactions cannot be obtained entirely from current–potential curves, because in many cases they do not look very different upon addition of a redox system, especially if the current is controlled by the light intensity. Therefore, a rotating ring disc electrode (RRDE) assembly consisting of a semiconductor disc and a Pt ring is usually applied, i.e. a technique which makes it possible to determine separately the current corresponding to the oxidation of a redox system [62, 63].

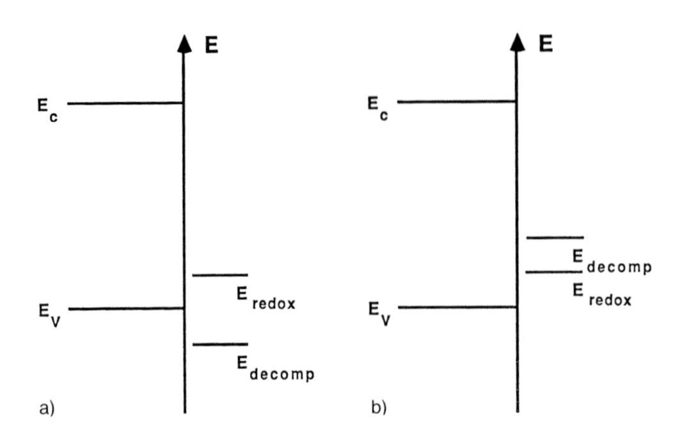

Fig. 8.16 Decomposition and redox potentials with respect to the position of energy bands

In this case, the oxidized species of a redox couple produced at the semiconductor disc, can be collected at the Pt ring, provided that the ring is polarized negatively with respect to the standard potential of the redox system. (For details of the RRDE technique see Section 4.2.3). Usually, the competition between the redox process and the dissolution is quantitatively described by the stability factor, as defined by [64]

$$s = \frac{j_{ox}}{j_{tot}} = \frac{j_{ox}}{j_{ox} + j_{corr}} \tag{8.18}$$

in which j_{ox} and j_{corr} correspond to the oxidation and corrosion current, respectively.

From the thermodynamic point of view, a redox process would preferably proceed if the redox potential E_{redox} is located above E_{decomp} as illustrated in Fig. 8.16a and, conversely, the decomposition reaction should dominate if E_{redox} occurs below E_{decomp} (Fig. 8.16b). Many experimental investigations have shown, however, that such a picture is far too simple because the kinetics of both processes play a dominant role. Accordingly, it is very difficult to predict whether corrosion or the redox process will dominate under given circumstances.

Most experiments were performed with n-type semiconductors because it was easier to detect small ring currents upon addition of a redox system to the solution by using a modulation technique (see Section 4.2.3). A typical result, as obtained with chopped light, is shown in Fig. 8.17. A selection of data evaluated from RRDE measurements is given in Fig. 8.18 [60]. These are examples where E_{decomp} is higher than E_{redox}, (comparable to case b in Fig. 8.16). Nevertheless, the ratio j_{ox}/j_{tot} can reach a value which is equivalent to $s = 1$. These results also show that the pH of the solution plays an important role. Other authors have shown that the stabilization factor s decreases with increasing light intensities, as shown first for n-GaAs electrodes by Frese et al. [65]. They interpreted their result by assuming the following reaction scheme:

$$S + h^+ \rightarrow S^{\bullet +} \tag{8.19}$$

$$S^{\bullet +} + (m-1)h^+ + mX^- \rightarrow SX_m \tag{8.20}$$

$$Red + h^+ \rightarrow Ox \tag{8.21}$$

$$S^{\bullet +} + Red \rightarrow S + Ox \tag{8.22}$$

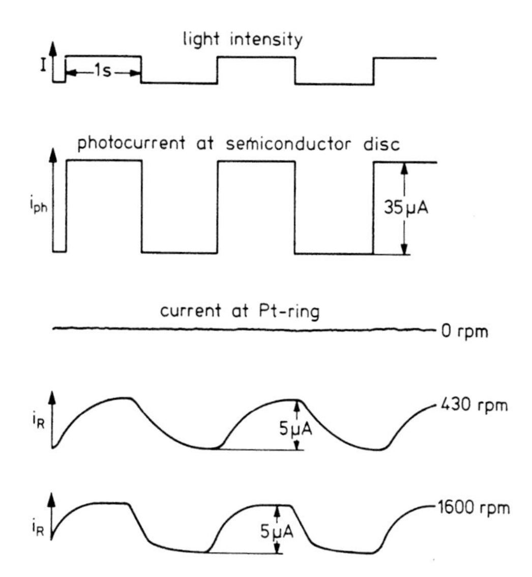

Fig. 8.17 Photo- and ring current at rotating electrodes. (After ref. [60])

in which S represents a semiconductor surface molecule, $S^{\bullet+}$ a surface radical, X^- a species in the solution (e.g. OH^-), and m the number of holes required for the complete dissolution of one surface molecule. According to the reactions presented in Eqs. (8.21) and (8.22), it is assumed that the oxidation of the redox system not only occurs via hole transfer from the valence band to the Red form of the redox system but also by an electron injection from the Red form to the surface radical $S^{\bullet+}$. As previously discussed, such a radical was already assumed in various decomposition reactions as an intermediate state (see e.g. Figs. 8.10 and 8.13). Actually, the surface bond broken in the first step (Eq. 8.19), is repaired via the last reaction (Eq. 8.22) which is an essential step for the stabilization of a semiconductor electrode. The intensity dependence of s has been particularly studied for GaP and GaAs [65–67]. Gomes and co-workers, especially, have investigated the intensity dependence in detail and have derived a reaction model [68]. The models presented by Frese et al. [65], Gerischer [69] and Gomes [68],

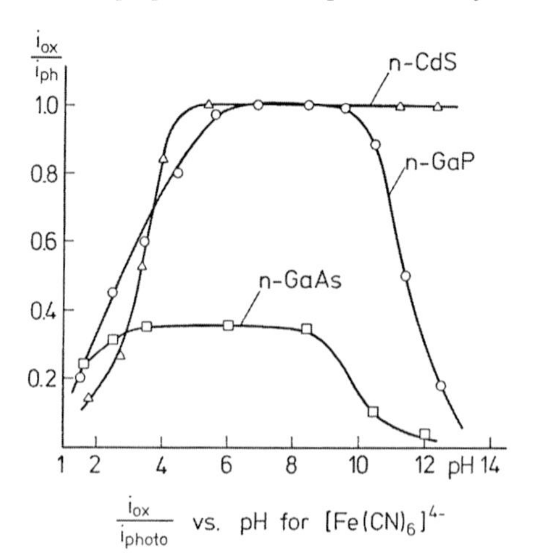

Fig. 8.18 i_{ox}/i_{ph} vs. pH for $[Fe(CN)_6]^{4-}$ as measured for different n-type semiconductors. (After ref. [60])

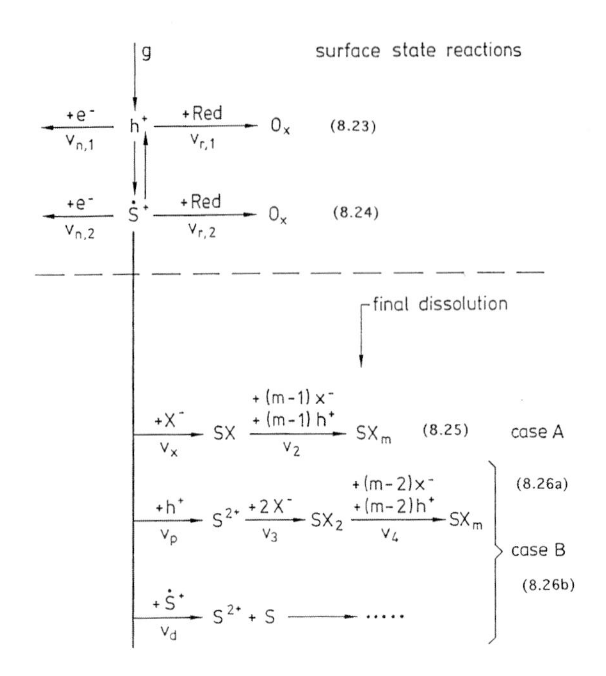

Fig. 8.19 General scheme of a dissolution reaction. (After ref. [70])

can be summarized in one scheme, as given in Fig. 8.19, in which the various v_n represent reaction rates.

The lower part of the reaction scheme describes three possible dissolution pathways (Eqs. 8.25 and 8.26). Only the first step in each sequence is essential. It is important to note that in one pathway (case A) the first step is a pure chemical reaction, whereas in the others (case B) a hole transfer is involved. According to the kinetics of this reaction scheme, an intensity dependence of the stability factor s or the ratio j_{ox}/j_{tot} occurs only for case B [70]. Most experiments were performed with n-semiconductors under anodic bias and illumination. A quantitative relation between s and light intensity was derived, and experimental data published by various authors seem to fit this relation quantitatively [71]. The same conclusion was arrived at by Gomes et al., who used a slightly different model [68]. The corresponding kinetic equations are not presented here because evaluation of the experimental data is not unambiguous. The reason for this is that the formation of surface layers may also be responsible for the decrease of s when the light intensity is increasing, which was not considered in the evaluations. Such an effect is not visible in the anodic photocurrent because the latter is only determined by the light intensity. This can be much better controlled by studying the redox reaction and the decomposition at a p-type electrode. At this electrode, both reactions are majority carrier processes, i.e. the corresponding dark currents are potential-dependent. If a surface film is formed upon addition of a redox system the slope of the current–potential curve should be lowered. This actually happened, for instance, with p-GaAs electrodes after addition of Fe^{2+} ions [72]. For further details the reader is referred to the relevant literature [64, 70]

A particular class of electrode materials is represented by the transition metal chalcogenides, such as n-WSe_2, n-$MoSe_2$ and others, which form layer crystals. As already

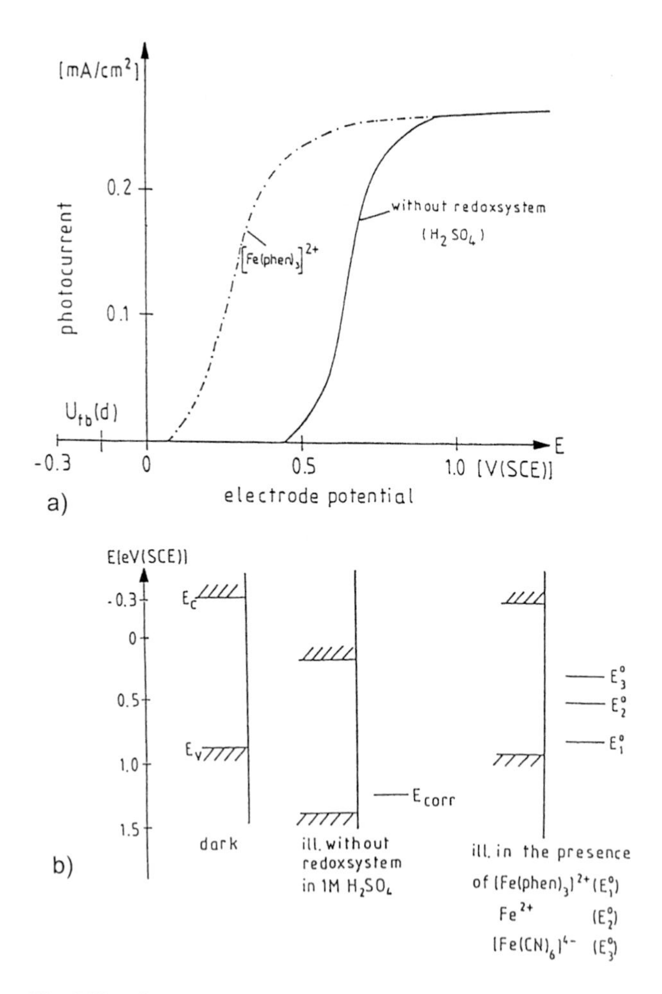

Fig. 8.20 a) Anodic photocurrent vs. electrode potential for an n-WSe_2 electrode in the absence and in the presence of a redox system. b) Position of energy bands at the surface of WSe_2 in the dark and under illumination. (After ref. [64])

mentioned in Section 8.1.4, the basal planar surfaces of these electrodes (perpendicular to the *c*-axis) are relatively stable. In consequence, holes created by light excitation, are not transferred and accumulate at the surface. This leads to a large downward shift of the energy bands, as found by Mott–Schottky measurements [52] and as illustrated in Fig. 8.20b (left and middle). The photocurrent is due to dissolution and it starts to rise around $U_{fb}(h\nu)$ (Fig. 8.20a). Upon addition of a suitable hole acceptor, such as $[Fe(phen)_3]^{2+}$, the position of the energy bands at the surface remains nearly constant when the light is switched on (right side of Fig. 8.20b), and the photocurrent onset occurs near U_{fb}(dark) (Fig. 8.20a). The anodic current is only due to the oxidation of $[Fe(phen)_3]^{2+}$, i.e. also at a large anodic bias where anodic decomposition is found without any redox system, as analyzed by using the RRDE technique [53]. In this case, the stability factor is $s = 1$ because a shift of bands is avoided.

9 Photoreactions at Semiconductor Particles

During the last 15 years, many investigations have been performed with semiconductor particles or nanocrystals, either dissolved as colloids or used as suspensions in aqueous solutions. Recently films of nanocrystalline layers have also been produced which were used as electrodes in photoelectrochemical systems. Essential results have already been summarized in various reviews [1–9]. All kinds of systems, containing small or large particles have been used in various investigations. In this chapter, the essential properties of semiconductor particles will be described. Small particles, i.e. nanocrystals, are of special interest because of quantum size effects.

9.1 Quantum Size Effects

The most striking observation with small semiconductor nanocrystals is the spectacular change in their absorption spectra when the size of the particles comes into the nanometer range. This becomes quite obvious with semiconductors with a low bandgap. For instance, Cd_3As_2 has a bulk bandgap of 0.13 eV and the absorption edge appears at $\lambda = 1250$ nm. With corresponding nanoparticles with a diameter of about 300 Å, however, the absorption edge was found at around $\lambda = 300$ nm. As is known from elementary quantum mechanics, when the sizes of semiconductor particles are comparable to or smaller than the de Broglie wavelength, the density of electronic states decreases, i.e. the allowed energy states become discrete rather than continuous. The critical dimension for quantization effects to appear depends on the effective mass, m^*, of the electronic charge carrier. For instance, the critical dimension is about 300 Å for $m^* \sim 0.05$; it decreases more or less linearly with increasing m^*.

Charge carriers in semiconductors can be confined in one spatial dimension (1D), two spatial dimensions (2D), or three spatial dimensions (3D). These regimes are termed quantum films, quantum wires, and quantum dots as illustrated in Fig. 9.1. Quantum films are commonly referred to as single quantum wells, multiple quantum wells or superlattices, depending on the specific number, thickness, and configuration of the thin films. These structures are produced by molecular beam epitaxy (MBE) and metalorganic chemical vapor deposition (MOCVD) [2]. The three-dimensional quantum dots are usually produced through the synthesis of small colloidal particles.

Some fundamental differences exist for the three types of quantization. In particular, the densities of electronic states (DOS) as a function of energy are quite different, as illustrated in Fig. 9.2. For quantum films the DOS is a step function, for quantum dots there is a series of discrete levels and in the case of quantum wires, the DOS distribution is intermediate between that of films and dots. According to the distribution of the density of electronic states, nanocrystals lie in between the atomic and molecular limits of a discrete density of states and the extended crystalline limit of continuous bands. With respect to electrochemical reactions or simply charge transfer reactions,

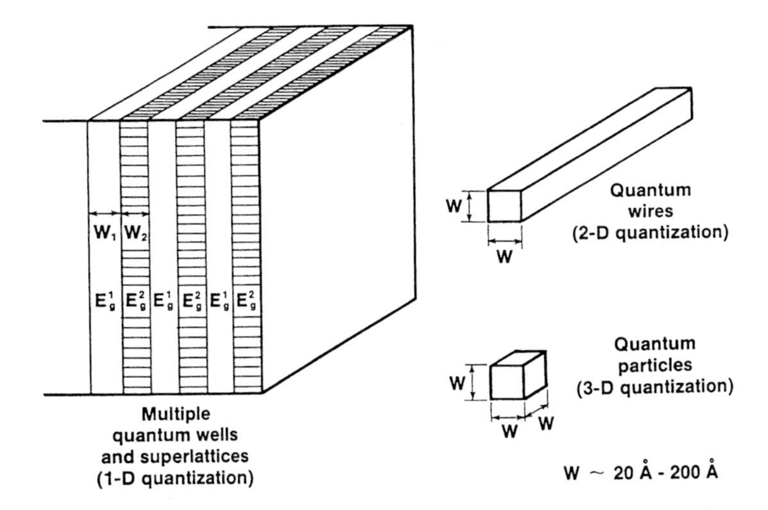

Fig. 9.1 Quantization configuration types in semiconductors depending upon the dimensionality of carrier confinement. (After ref. [2])

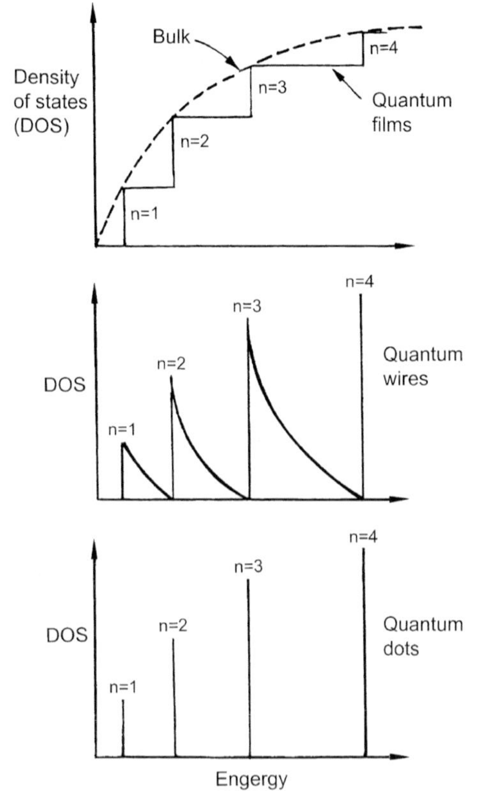

Fig. 9.2 Density of states (DOS) functions for quantum films, quantum wires and quantum dots. (After ref. [2])

quantum films and dots are of especial interest. The quantization effects for these types will be derived in more detail in the next two sections.

9.1.1 Quantum Dots

In extended semiconductors, the energy states of electrons form energy bands. Light excitation leads to the formation of electron–hole pairs. As already discussed in Section 1.2, electrons and holes undergo a Coulomb interaction and can form excitons (Wannier excitons). Their energy states are located just below the conduction band (Fig. 1.9). In bulk materials, the bonding energy of these excitons is small (usually <0.03 eV) and their radius is large (see below). At room temperature, the excitons easily dissociate into free carriers so that electrons and holes move approximately independently within the crystal. Brus and others [11–14] have reasoned that photogenerated electron–hole pairs can be physically confined, i.e. the radius of the exciton becomes smaller than the particle diameter. The radius of an exciton in the bulk can be calculated from the relation $R_{exc} = (h^2\varepsilon/4\pi^2e^2)(1/m_e^* + 1/m_h^*)^{-1}$. (The exciton radius corresponds to the Bohr radius of an electron in a H atom modified by introducing a dielectric constant and a reduced mass.) Typical values for some bulk semiconductors are: 43 Å (Si), 28 Å (CdS), 125 Å (GaAs), 100 Å (InP). In the case of very small particles, strong confinement leads to a raising of the electronic energy levels, and discrete energy states instead of bands are formed. These energy states are also usually termed excitons although the electrons and holes move freely within the particle.

The size effect is generally described by the quantum mechanics of a "particle in a box". The Schrödinger equation used in the simplest approach is given by

$$\frac{\partial^2\psi}{\partial r^2} = -\frac{8\pi m}{h}(E - \infty)\psi = 0 \tag{9.1}$$

where the wavefunction is zero outside the box, and inside of the box

$$\Psi = C_1 \exp(i\xi) + C_2 \exp(-i\xi) \tag{9.2}$$

with

$$\xi = (2mE)^{1/2}\,\frac{2\pi r}{h} \tag{9.2a}$$

Solving Eq. (9.1) yields the familiar result

$$E = \frac{h^2}{8mR^2}n \qquad n = 1, 2, 3\ldots \tag{9.3}$$

when using spherical coordinates (m = electron mass; R = particle radius).

This equation satisfies the boundary condition of continuity at $r = R$. The change of energy levels when particles are made small, is schematically shown in Fig. 9.3. A full quantum mechanical treatment of the problem was reported by Brus [13]. Taking into account that the exciton consists of an electron–hole pair, it is advantageous to formulate the Schrödinger equation as follows:

$$\left[-\frac{h^2}{8\pi^2 m_e}\nabla_e^2 \frac{h^2}{8\pi^2 m_h^2}\nabla_h^2 + V_0\right]\Phi = E\Phi \tag{9.4}$$

Using the wave function $\Phi = \psi_i(r_e)\psi_i(r_h)$ one can solve Eq. (9.4). Taking the vacuum level as a reference value at infinity ($V_0 = \infty$) one obtains for the energy of the lowest excited state (equivalent to the lower edge of the conduction band)

$$E(R) = E_g + \frac{h^2}{8m_0R^2}\left[\frac{1}{m_e^*} + \frac{1}{m_h^*}\right] - \frac{1.8e^2}{\epsilon R} \tag{9.5}$$

in which m_0 is the electron mass in vacuum, and m_e^* and m_h^* the reduced effective mass, respectively. This derivation is the so-called effective mass approximation in which the confined exciton is treated as one particle with the reduced mass $m^* = (1/m_e^* + 1/m_h^*)^{-1}$. The use of an effective mass concept is very common in solid state physics, as discussed in Chapter 1. Brus has also considered the Coulomb interaction between electron and hole as given by the last term in Eq. (9.5) (for details of the derivation see ref. [13]). As mentioned before, it is assumed in Eq. (9.5) that the semiconductor particle consists of a spherical box with an infinite potential drop at its wall. In the real world, however, this potential drop must be finite as pointed out by Weller and Henglein [15]. These authors assumed an energy difference between the lower edge of the conduction band of the particle and the vacuum level of $V_0 = 3.8$ eV in aqueous systems. This is actually an upper limit because water has acceptor levels for excess electrons below the vacuum level. The introduction of $V_0 = 3.8$ eV reduces slightly the apparent particle size required to see the same effect as with $V_0 = \infty$.

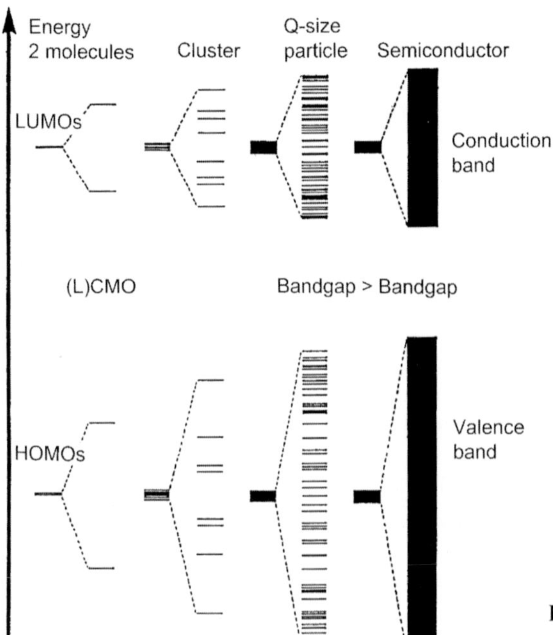

Fig. 9.3 Molecular orbital model for different particle sizes. (After ref. [84])

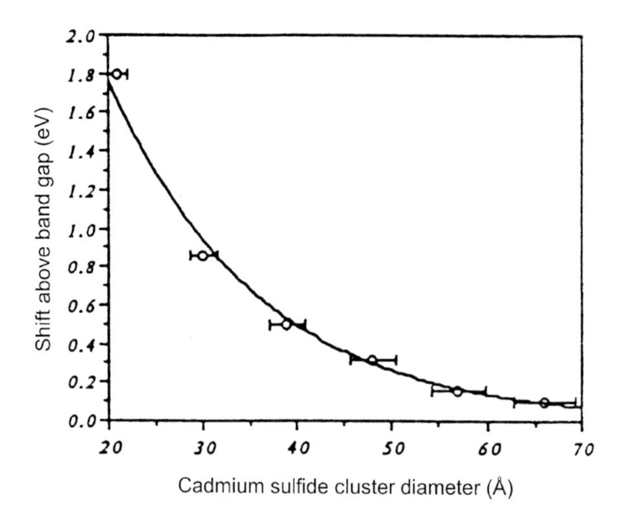

Fig. 9.4 The solution of Eq. (9.4) plotted as the shift above E_g = 2.8 eV for CdS particles (solid line). Data points are experimental results. Particle size was determined from transmission electron microscopy studies. CdS was prepared as an aqueous colloid stabilized with sodium hexametaphosphate. (After ref. [2])

When experimental data are compared with the theory then values of the effective masses, m_e^* and m_h^*, as determined for extended semiconductors, are usually inserted into Eq. (9.5). Taking CdS as an example, the shift of the bandgap, $\Delta E = (E(R) - E_g)$, is given for several particle sizes in Fig. 9.4. The theoretical curve (solid line) was calculated using E_g = 2.42 eV, m_e^* = 0.21, m_h^* = 0.8, ε = 5.4 [2]. Similar results have been obtained by Weller et al. [15]. According to Eq. (9.5), large shifts of bandgaps are only expected for semiconductors having an effective mass considerably smaller than unity. This is not always the case; some materials have m^* values where $m^* > 1$, such as TiO_2 with m_e^* = 30. In the latter case no quantization was observed. This is important to realize because many photocatalytic reactions have been investigated only with regard to TiO_2.

There is still much controversy in the literature on the applicability of the effective mass model to small clusters [2]. Some authors claim that it is applicable for clusters containing as few as 100 atoms whereas others have shown that large errors occur between theoretical and experimental values of the bandgap shift. The effective mass approximation overestimates the bandgaps, and the error can be quite large ($\gg$1 eV) at very small particle sizes (<15–20 Å). Another approach to the problem is still to use the effective mass approximation but to adjust the effective mass in the region of small-sized particles. The effective mass model cannot take account of sharp changes in potentials and thus it cannot address a microscopic atomic structure. Other models which use pseudopotentials [17] or tight binding calculations [18], provide better agreement between predicted and experimental bandgaps for CdS and CdSe quantum dots at very small cluster sizes ($R <$ 15–20 Å) (see e.g. ref. [21]). Zunger discussed the problem of applying the effective mass model [16]. He emphasized that the effective mass model is very successful in describing spectroscopy and transport phenomena in two-dimensional quantum well structures. Its success in describing quantum dot properties, while impressive in some cases, is frequently clouded because there are too many adjustable parameters. Zunger and his group developed a model using the same conceptual methods with which bulk solids have successfully been treated in the past [19].

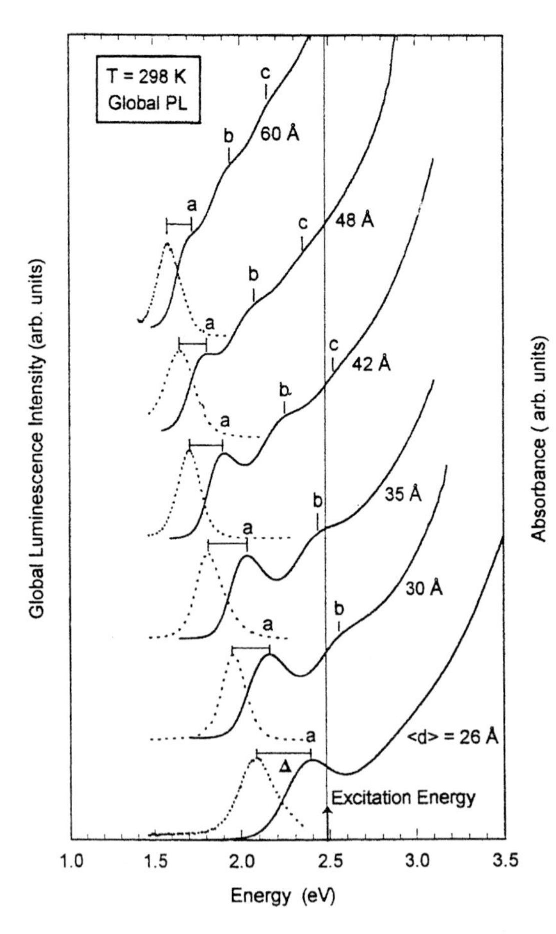

Fig. 9.5 Absorption (solid line) and global photoluminescence (dotted line) spectra at 298 K for an aqueous colloidal solution of InP with different mean diameters. Photoexcitation at 2.48 eV. (After ref. [23])

They included a real atomistic surface of the nanostructure in the description instead of an infinite potential barrier. They created an artificial 3D periodic lattice so that ordinary band-theory methodology could be applied (see e.g. refs. [16, 19, 20]). This model describes, for instance, the dependence of bandgap vs. particle diameter very well for CdSe, InP and Si [19–20] as proved by appropriate experimental results. These models cannot be discussed here in more detail; further information is given in refs. [2] and [16].

It should be emphasized here that experimental data about bandedge shifts or the complete spectra can only be obtained with colloidal solutions which have a sufficiently narrow size distribution. Monodispersive samples were obtained by selecting a proper synthesis method (see e.g. refs. [22, 23]) or by size fractionations such as gel-electrophoresis and selective precipitation [87, 88]. One example is the absorption of InP colloids with different diameters [23] (Fig. 9.5). In many cases the particle sizes were determined by transmission electron microscopy (TEM). It should also be mentioned here that in most cases cluster growth occurs, i.e. the bigger particles grow at the expense of the smaller ones, which can easily be detected in the absorption spectrum. In order to avoid agglomeration, most semiconductor clusters were prepared in the

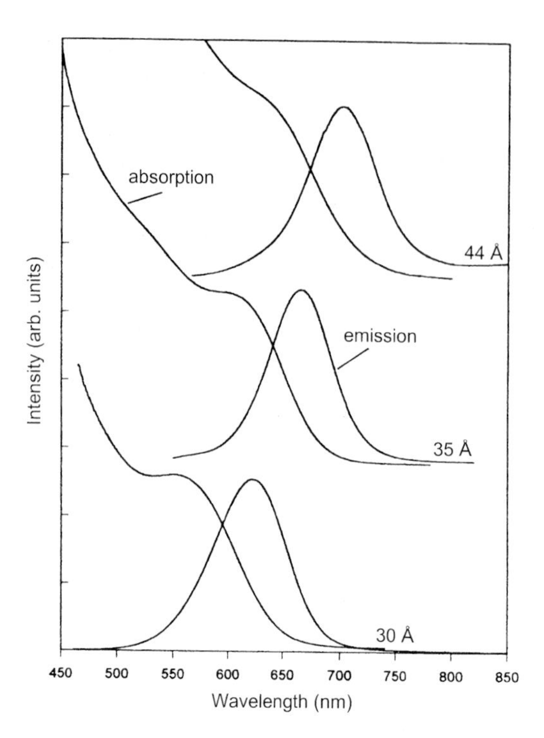

Fig. 9.6 Absorption and emission spectra of HF-treated InP quantum dots at 300 K with different particle diameters. (After ref. [24])

presence of a polyanionic stabilizer such as polyphosphate or colloidal SiO_2 [1, 2, 7]. In the first case, the ionic end is locked to the semiconductor surface. If reactions between particles and organic substances are being studied, then long-chain hydrocarbon stabilizers are avoided. In this case one uses semiconductor particles grown on SiO_2 colloids, a process which seems to work without many problems.

Particles of different sizes also have different fluorescence spectra. In general the emission is shifted toward higher energies with decreasing particle size as shown for bandedge emission at InP dots in Fig. 9.6 [24]. The mechanism of excitonic emission seems to be complex [7] and cannot be treated here. For instance, the single emission peak in Fig. 9.6 was only observed after etching the dots in a methanolic solution containing 5 % HF. It was assumed that fluoride ions filled phosphorus vacancies on the InP surface. The etching also led to an increase of the fluorescence quantum yield up to 30 %. In addition, an electron or hole can be trapped in surface states which leads to radiationless recombination. Fortunately, in some materials there are ways of blocking these surface sites. Taking CdS quantum dots as an example, it has been found that a strong fluorescence of a high quantum yield ($\phi = 0.2$) and a narrow excitonic type of spectrum occurred when the pH of the solution was increased to pH = 11 and Cd^{2+} was added. This effect was interpreted as the bonding of the Cd^{2+} at $S^{\bullet -}$ surface sites, a process by which the surface state disappeared [25]. Other examples are given in Section 9.2.4.

Another interesting material is silicon because of its indirect bandgap. The question is whether the indirect gap develops with size in the same way as the direct gap. According to quantitative studies performed by Brus et al., the bandgap and luminescence energy correspondingly increased to about 2 eV as the size of the Si particles decreased to 10–20 Å in diameter [26]. The absorption spectrum also remained completely structureless in the range around 3.5 eV where the absorption of bulk Si shows a maximum (direct gap). Since the square root of absorption was linear with increasing photon energy it was concluded that the absorption remained indirect-gap-like (see also Chapter 1). This conclusion was supported by the result that phonon structures in the absorption and luminescence spectra were found which are typical for indirect gap transitions. It is important to realize that the spectroscopic and dynamic data of porous Si are similar to the data of Si nanocrystals, as already briefly mentioned in Section 8.1.3. Both show indirect-gap-type excitation and size-selective luminescence spectra in the range around 700 nm [27, 28].

The quantization effect in Si is smaller than in CdS or CdSe because Si has a larger electron effective mass (see Appendix D). The highest possible gap is about 2.5 eV for a 10-Å Si particle. There has been a goal has been of making porous Si samples which luminesce in the green and blue region (2.5–3.5 eV). However, according to the data given above, this is not possible.

9.1.2 Single Crystalline Quantum Films and Superlattices

Quantum films are usually produced in a periodic sequence of sandwich type of two semiconductors, one with a small, the other with a large bandgap, as illustrated in Fig. 9.7a. They can be produced with much higher precision than quantum dots of a narrow

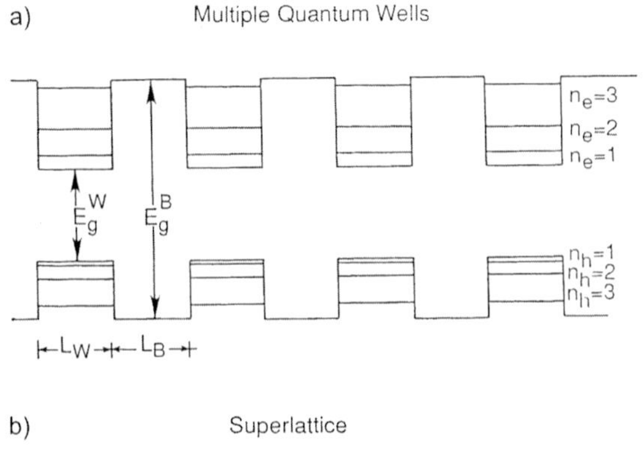

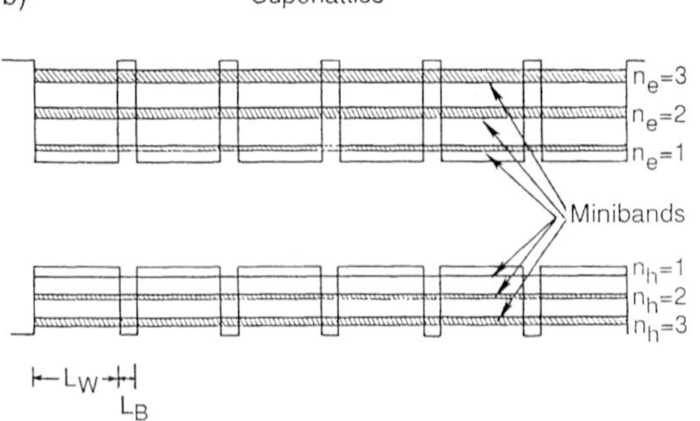

Fig. 9.7 Difference between multiple quantum well (MQW) structures (barriers >40 Å) and superlattices (barriers <40 Å); miniband formation occurs in the superlattice structure. (After ref. 22])

size distribution. This is possible by using molecular beam epitaxy (MBE) for deposition of the layers [2]. The resulting heterojunctions and their corresponding bandedge discontinuities produce a potential well of thickness L_w and a barrier of thickness L_b. Quantization effects have been found for film thicknesses ranging around 15–300 Å. Discrete energy levels are formed for electrons and holes in their respective quantum wells, in accordance with the solution of the Schrödinger equation. If a system is produced with thick barrier layers, the electronic wave functions in the wells do not interact; i.e. the wells are electronically decoupled. Such structures are referred to as multiple quantum wells (MQW) [2, 6]. If the barrier layers are made sufficiently thin so that electron tunneling can occur, then the wave functions of each well interact and the quantum states of each well couple to produce delocalized but quantized states across the whole structure. This coupling leads to minibands as illustrated in Fig. 9.7b. This configuration is termed a superlattice [7].

Quantum films represent a beautiful physical example of the particle in the box problem. The energy levels of the conduction band electrons in the electron well (MQW) can be easily calculated using the envelope function or the effective mass approximation [6]. The electron wave function Ψ_n is then [29–31]

$$\psi_n = \sum_{w,b} e^{ikr} U_k^{w,b}(r)\phi_n(z) \tag{9.6}$$

where w and b refer to well and barrier, respectively; k is a transverse electron wave vector (see Chapter 1); z is the growth direction, and $\phi_n(z)$ is the envelope wave function. In this case, an energy profile, $U_k^{w,b}(r)$, in the well or in the barrier is assumed as given by the Bloch function (compare with Section 1.2). The envelope wave function is determined from the Schrödinger equation

$$\left(\frac{\hbar^2}{2m^*}\frac{\delta^2}{\delta z^2} + V_c(z)\right)\phi_n(z) = E_n\phi_n(z) \tag{9.7}$$

where $V_c(z)$ is the potential barrier function and E_n is the quantized energy levels in the well. Assuming an infinitely deep well, the solution of Eq. (9.7) is very simple since the wave function must be zero at the well–barrier interfaces. This leads to the well-known solutions for the particles in a box:

$$\Psi_n = A\sin(n\pi z/L_w) \tag{9.8}$$

and for the energy levels

$$E_n = \frac{\hbar^2}{2m^*}\left(\frac{n\pi}{L_w}\right)^2 \qquad n = 1, 2, 3\ldots \tag{9.9}$$

The calculation of hole levels is much more complicated because the band structure of most semiconductors shows hole bands of fourfold degeneracy at $k = 0$. This leads to light and heavy holes with different effective masses (see Section 1.2). Consequently, a double set of energy levels is formed in a quantum well.

A further theoretical analysis of the quantum film shows that the density of states (DOS) is given by [2]

$$N(E)_{\mathrm{qf}} = \frac{nm^*}{\pi\hbar^2} \tag{9.10}$$

(Compare with the DOS of an extended semiconductor as given by Eq. (1.24)). A plot of $N(E)_{\mathrm{qf}}$ vs. energy consists here of a series of steps (Fig. 9.2). Since the absorption is proportional to the density of states in the conduction and valence band, the absorption spectrum of a quantum film should consist of a series of steps. The position of these steps corresponds to transitions between quantum states in the valence and the conduction band following the selection rule $\Delta n = 0$. Here the exciton binding energy is also considerably increased (compare with quantum dots in Section 9.1.1) and becomes stable at room temperature because the width of the well is smaller than the exciton diameter. Thus the absorption spectra of quantum wells can be expected to show exciton peaks which occur at energies below the step. Such spectra have indeed been found by several researchers [32–34]. Corresponding spectra were analysed using the photoreflectance method or by measuring the excitation spectrum of the luminescence as shown in Fig. 9.8. In the latter case, the luminescence was detected at $E_{\mathrm{ph}} = 1.53$ eV whereas the excitation energy was varied over the range 1.52–1.72 eV. The peaks correspond to transitions from discrete levels in the valence band to empty levels in the conduction band. (Compare with the quantum well structure in Fig. 9.7).

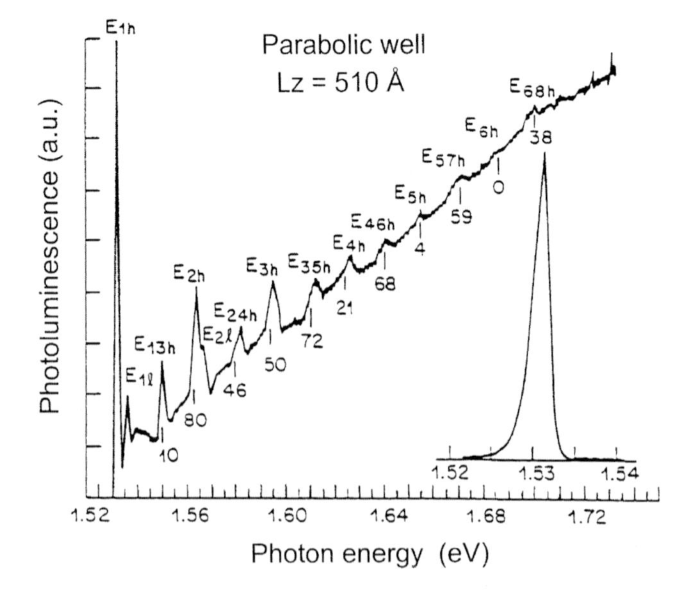

Fig. 9.8 Photoluminescence excitation spectrum of a quantum well structure with parabolic potential barriers. (After ref. [85])

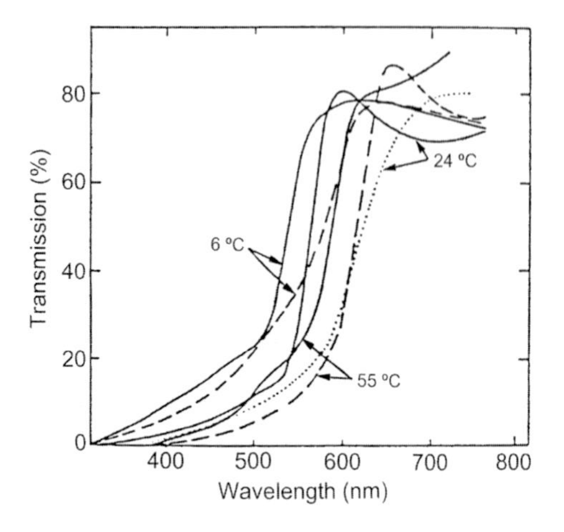

Fig. 9.9 Optical transmission spectra of CdSe films deposited at different temperatures in the presence and absence of illumination: solid curves, 6, 24 and 55 °C, dark; dashed curves, 6 and 55 °C, illuminated; dotted curve, 24 °C, illuminated. (After ref. [35])

9.1.3 Size Quantized Nanocrystalline Films

There are also various techniques for producing semiconductor films consisting of nanocrystalline particles. These films may exhibit size quantization characteristic of the individual particle, depending on the effective mass of the semiconductor as described above. For instance, CdS and CdSe films have been prepared by chemical deposition or by electrodeposition [35, 36]. Interestingly, rather thick layers (100–300 nm) containing particles with a diameter of 5–10 nm, were obtained. According to TEM measurements, the particle size decreased when the substrate temperature was lowered during deposition [35]. A corresponding shift of the absorption due to size quantization was still visible as shown in Fig. 9.9. Also, thermal treatment of layers led to stronger aggregation and crystallite growth.

9.2 Charge Transfer Processes at Semiconductor Particles

9.2.1 Reactions in Suspensions and Colloidal Solutions

In principle, the same electron reactions should occur at particles and bulk electrodes. One essential advantage of using particles is the large surface. The photogenerated charge carriers in a particle can easily reach the surface before they recombine, so that fairly high quantum yields can be expected. However, one difficulty arises insofar as two reactions, an oxidation and a reduction must always occur simultaneously, as indicated in Fig. 9.10. Otherwise, the particles would be charged up, which would lead to a complete stop of the total reaction. Accordingly, the slowest process determines the rate of the total reaction. A particle actually behaves like a microelectrode, always kept under open circuit potential, at which the anodic and cathodic currents are equal. At extended electrodes, the partial currents are mostly rather small under open circuit conditions, as the majority carrier density at the surface is small because of the deple-

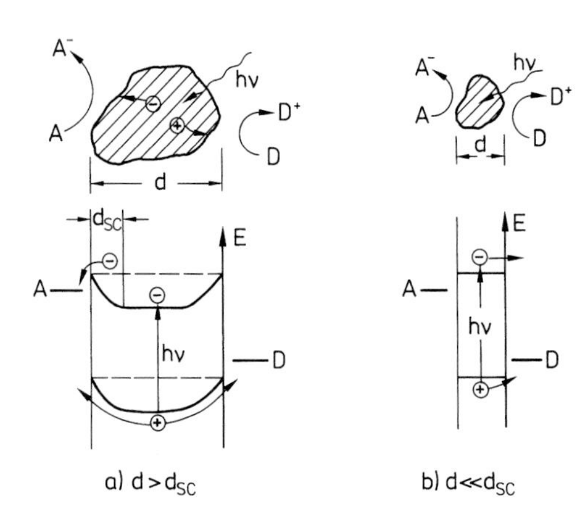

Fig. 9.10 Electron and hole transfer at large (a) and small (b) semiconductor particles to an electron acceptor A and donor D

tion layer below the semiconductor surface, as illustrated in Fig. 9.10a. The thickness of such a space charge layer depends on the doping and on the potential across the space charge layer (see Eq. 5.31). Taking a typical value such as $n_0 = 10^{17}$ cm^{-3} and $\phi_{sc} = 1$ V, the thickness of the space charge layer is about 10^{-5} cm. No space charge layer exists, of course, in much smaller particles of diameter $d \ll d_{sc}$ (Fig. 9.10b). At first sight, one may expect that the electron transfer at the interface of bigger particles would be limited to a very low rate because of the large upward band bending. This is not true because upon light excitation, a few holes may be transferred to an electron acceptor in the solution, which leads to a negative charging of the particle which reduces the positive space charge. The latter effect causes a flattening of energy bands (see dashed line in Fig. 9.10a), which is equivalent to a negative shift of the rest potential of an extended electrode upon light excitation This leads to a higher electron density at the surface and to a correspondingly greater transfer rate [5]. Most investigations have been performed with much smaller particles ($d \ll d_{sc}$) which can be treated almost like molecules. However, the positions of energy bands are still determined by the interaction of the semiconductor with the solvent and are still in most cases dependent on pH as described in Chapter 5. Whether the stabilizer frequently used for small particles influences the band position has not been investigated.

When electron–hole pairs are produced by light excitation in a small particle ($d \ll d_{sc}$), electrons and holes can easily be transferred to an electron and hole acceptor, respectively, provided that the energetic requirements are fulfilled. The quantum efficiency of the reaction depends on the transfer rate at the interface, on the recombination rate within the particle and on the transit time. The latter can be obtained by solving Fick's diffusion law. The average transit time within a particle of a radius R has then been obtained as [37]

$$\tau_{tr} = R^2/\pi^2 D \tag{9.11}$$

Taking typical values of $D \approx 0.1$ cm^2 s^{-1} and $R = 10$ nm, the average transit time is only about 1 ps. This time is much shorter than the recombination time so that electrons and holes can easily reach the surface.

As discussed in Chapter 7, fundamental aspects of charge transfer reactions at extended electrodes have mainly been studied by using simple one-step redox systems which are reversible. There is no point in studying such reactions at colloidal particles, because a redox system being oxidized by a hole would be immediately reduced by an electron transferred from the conduction band. Therefore, only irreversible reactions of organic compounds have been investigated. As mentioned before, stabilizers such as SiO_2 or polymers have frequently been used in order to avoid conglomeration of particles in the solution. The usage of polymers may lead to problems in studies of charge transfer processes because they can also be oxidized by hole transfer.

Charge transfer reactions have mainly been studied with regard to colloidal particles of TiO_2, and to some extent with regard to metal sulfides such as CdS. Illumination of CdS colloids in aqueous solutions saturated with oxygen, have led to a relatively fast anodic corrosion according to the reaction [38]

$$CdS + 2O_2 + h\nu \rightarrow Cd^{2+} + SO_4^{2-} \tag{9.12a}$$

The role of O_2 is here twofold: first it acts as an electron acceptor, secondly it is involved in the anodic corrosion reaction leading to the formation of SO_4^{2-}. This is in agreement with the results obtained with CdS electrodes as discussed in Section 8.1.4. Interestingly, the corrosion rate was considerably reduced after treatment of the particles in a solution of Cd^{2+} ions, which led to a blocking of S^{1-} or HS radical sites at the surface, as already described in Section 9.1.1 [39]. Obviously, the formation of S^{1-} radicals by holes is hindered.

In the absence of O_2, CdS was not photodissolved although it is possible from the point of view of energetics that H_2 was formed instead of reducing O_2 in the cathodic reaction. This could be either due to a low rate or to the formation of sulfur on the CdS surface which also affects further hole transfer. In addition it has been shown that the dissolution rate even in the presence of O_2 is strongly increased upon addition of methyl viologen as a further electron acceptor [40]. On the other hand, CdS undergoes cathodic dissolution in the absence of O_2 during illumination when sulfite, for example, is used as a hole acceptor. The overall reaction is then [41]:

$$CdS + SO_3^{2-} + H_2O \rightarrow Cd^0 + SO_4^{2-} + SH^- + H^+ \tag{9.12b}$$

These examples show clearly how reaction routes vary upon changes in the composition of the solution and that it is rather difficult to get information on the rate-limiting step.

CdS particles or suspensions were the subject of many investigations for a period of time (1980–1985) because Grätzel et al. concluded from their experiments that photoelectrolysis of H_2O occurred at CdS particles [42]. The relevant experiments were performed with particles loaded with two kinds of catalysts, namely with Pt for H_2 formation and RuO_2 for the formation of O_2. It was shown, however, that O_2 could not be formed at CdS and that corrosion is the only anodic reaction [43].

TiO_2 is a very suitable material because it is more stable than CdS. The quantum size effect is small because of the large effective mass. The associated colloids have mainly been used in studies of the oxidation and reduction of organic compounds, and TiO_2 colloidal solutions or suspensions are applied in the photocatalytic mineralization of organic waste material, as described in Section 11.2. In order to get more insight into the primary reaction steps, the formation of intermediate states at the surface of the TiO_2 particles was studied using photoflash techniques (see Section 4.6). The experi-

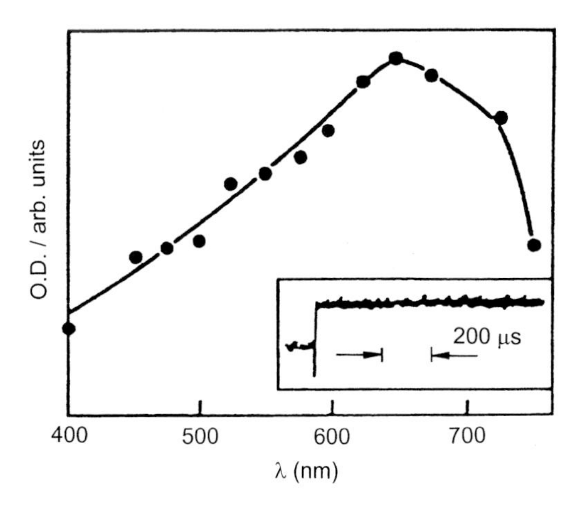

Fig. 9.11 Transient absorption spectrum of trapped electrons in 6.3×10^{-3} M TiO_2 colloids in solutions containing 5×10^{-3} M polyvinyl alcohol (hole acceptor) at pH 10 after a laser flash. Insert: time profile of the absorption. (After ref. [44])

ments were performed with low laser intensities to keep the density of photons absorbed by the colloidal solution smaller than the density of particles. The average particle size used in these experiments, was usually 2–5 nm in diameter. In the reactions involved, the organic molecules were oxidized by transfer of photoexcited holes and O_2 was reduced by transfer of photoexcited electrons.

Two rather featureless transient spectra were found upon a nanosecond laser flash, namely one peaking around 650 nm if polyvinyl alcohol was used as a hole scavenger (Fig. 9.11), and another one peaking around 430 nm when platinized particles were used in the presence of O_2 (Fig. 9.12) [44]. Since the long wavelength absorption around 650 nm was reduced by adding an electron acceptor such as O_2 to the solution or by depositing Pt islands on TiO_2, this absorption was interpreted as the excitation of electrons trapped in surface states. Similarly, the short wavelength peak in Fig. 9.12 was related to the absorption of trapped holes [44]. Concerning the nature of electron and hole trapping centers, several problems arise. Assuming that the absorptions of trapped electrons around 650 nm (1.7 eV) and of holes at 430 nm (2.7 eV) correspond

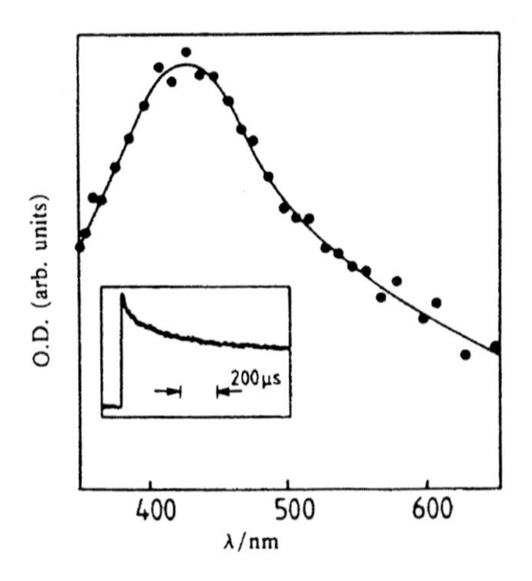

Fig. 9.12 Transient absorption spectrum of trapped holes in TiO_2 particles loaded with Pt islands (3.6 M TiO_2 and 1.6 M Pt) at pH 2.5 after a laser flash. Insert: time profile of the absorption. (After ref. [44])

to transitions between the classical surface states and the conduction and valence band, respectively, these surface states should be located near the middle of the bandgap (E_g = 3.1 eV). This assignment, however, cannot be correct for the following reason. Experimentally it has been found that the reduction of O_2 occurs via transfer of trapped electrons (see below). This process would not be possible thermodynamically if the electrons originate from an energy state which is 1.7 eV below the conduction band of TiO_2 because the surface state would be located below the standard potential of the couple O_2/H_2O. Accordingly, it has been assumed that the absorption is due to an excitation of a trapped electron within a surface molecule such as a hydrated Ti(III) molecule ($t_{2g} \rightarrow e_g$ transition) [45]. In the case of the trapped holes, the corresponding optical transition was correlated to an excitation within a peroxide formed at the surface [46].

According to investigations with an ultrafast laser excitation, the analysis of the transient spectra in the picosecond range has shown that electrons and holes produced by light excitation were trapped very rapidly [47, 48]. These trapped electrons and holes recombine if no corresponding acceptors are present in the solution. In the case of pure TiO_2 particles, the transient absorption due to excitation of trapped electrons was found to decay exponentially in the microsecond range. Since this decay time decreased with increasing O_2 concentrations it was concluded that the trapped electrons were transferred to O_2 [45]. The latter process is rather slow; a second-order rate constant of about 7×10^7 l mol^{-1} s^{-1} was obtained. The quantitative analysis was difficult because of the large overlap of the two transient spectra.

The kinetics of hole transfer is somewhat different. It could be studied very well with Pt/TiO_2 because here the electrons are rapidly transferred to the Pt islands so that a pure hole spectrum (Fig. 9.12) remains after about 1 μs. The corresponding decay curves in the presence of dichloroacetate (DCA) as a hole acceptor are given in Fig. 9.13 for different DCA concentrations [45]. Interestingly, the absorption signal decreased when the DCA concentration was increased. The decay time, however, remained unchanged which could be seen after the signal had been normalized to equal heights. The decay occurred over a period of hundreds of microseconds; it could not be fitted by a simple exponential rate law. Since the decay time was not affected by the hole acceptor (DCA), holes trapped in states at the surface could not be transferred to DCA. Accordingly, only free holes primarily produced by light excitation, must be transferred to the DCA molecules before they are trapped [45]. Hence, the

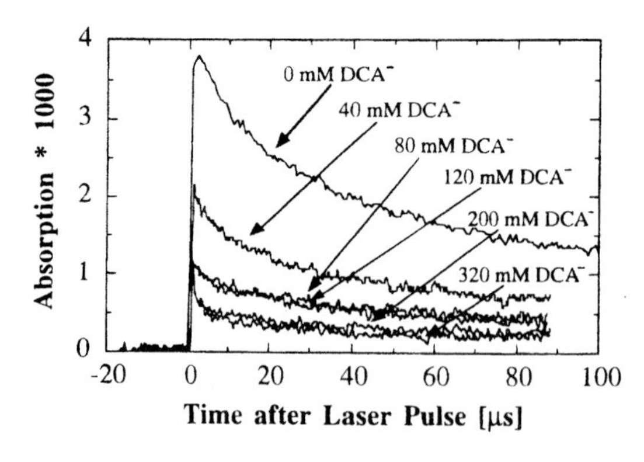

Fig. 9.13 Transient absorption vs. time observed upon laser excitation (λ_{ex} = 355 nm) in TiO_2 colloids loaded with Pt at 500 nm, in the presence of various concentrations of dichroroacetate (DCA) (hole acceptor): pH 2; aqueous air saturated solution, 1.0 M colloidal TiO_2/Pt (1 %) particles; absorbed photon concentration per pulse, 1.6×10^{-5} M. (After ref. [45])

hole transfer to DCA must be very fast. Further kinetic investigations have shown that the high rate was due to DCA molecules adsorbed on the TiO_2 particles. In summary we have then

$$TiO_2 + h\nu \rightarrow e^- + h^+ \tag{9.13}$$

$$e^- \rightarrow e^-_{tr} \tag{9.14a}$$
$$h^+ \rightarrow h^+_{tr} \tag{9.14b}$$

The recombination of the trapped species is given by

$$e^-_{tr} + h^+_{tr} \rightarrow TiO_2 \tag{9.15}$$

The cathodic and anodic reactions are given by

$$e^-_{tr} + O_2 \rightarrow O_2^- \tag{9.16a}$$
$$h^+ + DCA \rightarrow DCA^{\bullet +} \tag{9.16b}$$

According to this scheme, reaction (9.16b) competes with (9.14b).

As illustrated for various colloidal solutions, two acceptors are always required, one for electrons and the other for holes, when charge transfer processes are studied. Some are fairly effective, others not. This is not only determined by the appropriate constants, but adsorption seems to play an essential role (see also Section 9.2.4). In another experiment two Pt electrodes were installed in the colloidal solution, with one electrode (collector electrode) acting as an electron acceptor as illustrated in Fig. 9.14. Corresponding experiments were performed by Chen et al. using FeS_2 colloids [49]. Here the hole produced by light excitation was transferred to a hole acceptor (R) whereas the electron remained on the particle until the particle reached the Pt electrode via diffusion (case A in Fig. 9.14). The currents resulting from the electron transfer from the colloids to the electrode, depended strongly on the hole acceptor added to the solution. Accordingly, the kinetics of the hole transfer were rate-determining in this case. For low rates, most electron–hole pairs recombine. Relatively large photocurrents were obtained with S^{2-} as a hole acceptor in the solution. In the case of other hole acceptors, such as tartrates, the current was very low and could only be increased by also adding an electron acceptor like methyl viologen (MV^{2+}) to the solution. The redox couple $MV^{2+/1+}$ then acted as a mediator (case B in Fig. 9.14) [49].

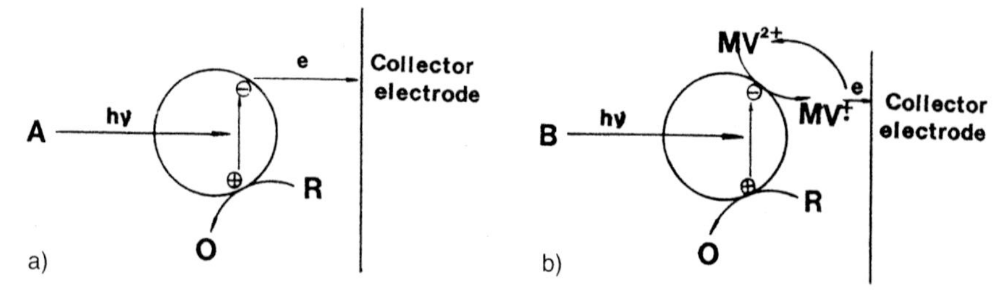

Fig. 9.14 Scheme for an electron transfer from an excited particle to a metal collector electrode in the absence (a) and in the presence (b) of a redox couple. (After ref. [49])

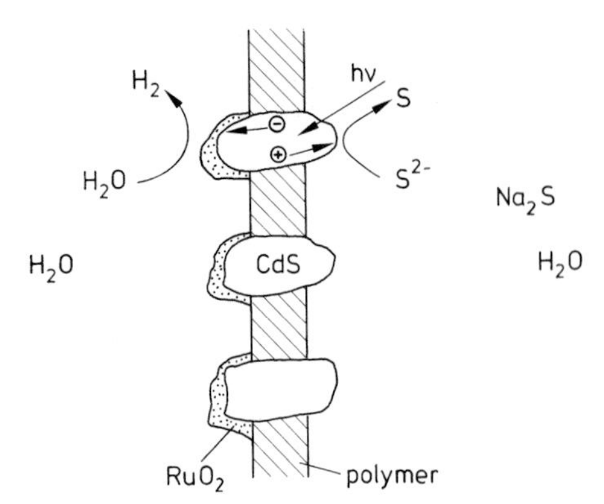

Fig. 9.15 Reactions at semiconductor monograin membranes. (After ref. [43])

As already mentioned, photoreactions have frequently been studied at semiconductor particles loaded with a catalyst such as a metal (Pt) or metal oxide (e.g. RuO_2). Such catalysts enhance the reaction rate, and sometimes other products were formed [50]. Although it is generally assumed that noble metals, Pt for instance, catalyse a reduction process such as H_2 formation, it is impossible to obtain an experimental proof of whether the reduction or oxidation occurs via the catalyst. Several years ago, a semiconductor monograin membrane technique was developed by which the particles were fixed as illustrated in Fig. 9.15 [51]. This technique made it possible to load one side of the membrane with a catalyst or both sides with different catalysts. Using a two-compartment cell in which both are separated by the membrane, it is possible to determine whether the product is formed on the free or on the catalyst-loaded side. One example is the charge transfer at a CdS membrane where one side is loaded with RuO_2. S^{2-} ions were only added to one compartment, either to the left or to the right side of the membrane (Fig. 9.15). The electrons produced by light excitation, were used for the formation of H_2 whereas the corresponding holes oxidized S^{2-}. According to the analysis of the products during and after illumination, H_2 was only found in the compartment with the catalyst side of the membrane. The highest yield was obtained with the bare CdS surface contacting the S^{2-} solution [51]. It is interesting to note that RuO_2 acts here as an electron acceptor although it is a typical catalyst for O_2 formation. However, it has to be realized that the catalyst is in contact with the semiconductor which influences the barrier height. Obviously, the barrier height is rather small at the CdS–RuO_2 interface so that electrons are easily transferred to the catalyst and from there on to protons (Fig. 9.16).

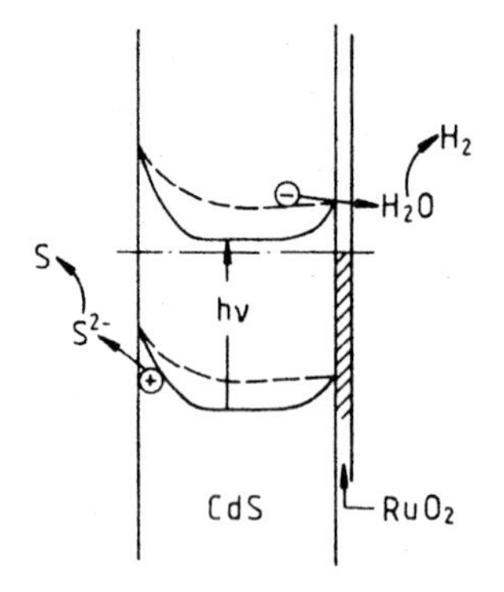

Fig. 9.16 Energy band model for particles embedded in a monograin membrane. (Compare with Fig. 9.15). (After ref. [43])

The monograin technique has only been applied so far to CdS and SiC. Since the production of the monograin membranes is not easy and rather big particles ($\approx$ 10 μm) are required, this technique has not been used much yet.

9.2.2 Photoelectron Emission

When solutions of CdS colloids containing no additional electron and hole acceptor in the solution, are exposed to a high intensity laser flash, a rather large absorption of an intermediate is observed around 700 nm, similarto that described for the laser excitation of TiO_2 in the previous section. The absorption spectrum of the intermediate is given in Fig. 9.17 [52]. It is not due to trapped electrons and holes but it is identical with to the well-known spectrum of hydrated electrons as proved by radiolysis experiments [52]. The half-life of the hydrated electrons is a few microseconds. In the presence of typical hydrated electron scavengers, such as oxygen, acetone or cadmium ions, the decay of the intermediate became much faster.

This is a rather surprising result because the standard potential of e^-_{aq} is $E^0 = -2.9$ eV (vs. the normal hydrogen electrode (NHE)) whereas the conduction band was found between $E_c = -0.9$ and -1.5 eV depending on the purity of the CdS surface (see Section 8.1.4). Accordingly, the excitation of a single electron in a particle should never reach the energy required for an electron transfer into the solution. It is interesting to note that this process can compete with an electron transfer to a proton which is even more favorable from the energy point of view because the conduction band occurs above $E^0(H^+/H_2)$. This estimate was a first indication for a two-photon excitation mechanism.

Photoelectron emission was only found in aqueous solutions. No e^-_{aq} formation was observed in acetonitrile or alcohol solutions. It also has been reported that the electron emission occurred only with CdS colloids stabilized by polyphosphates or colloidal SiO_2. The negative charge of the stabilizer prevents the emitted electron from rapid return to the particle by electrostatic repulsion [3].

A quantum yield of about 0.07 electrons emitted into the solution per absorbed photon was found. Interestingly, this value exceeds, by several orders of magnitude, the

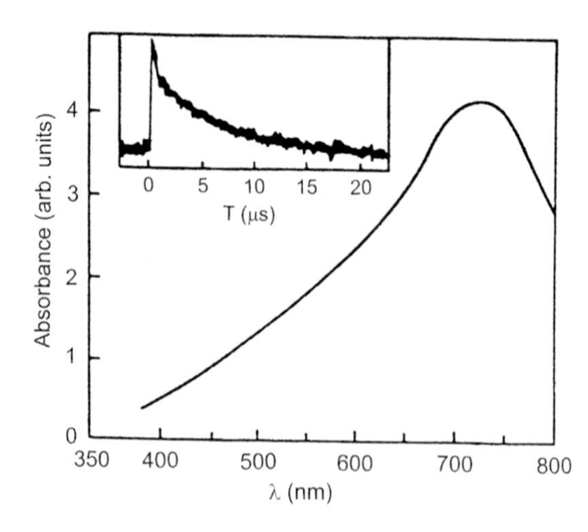

Fig. 9.17 Transient spectrum of hydrated electrons produced by a strong laser pulse in ZnS/CdS (3:1) co-colloids in the presence of Na_2S as a hole acceptor. Concentrations: 4×10^{-4} M ZnS/CdS and 2×10^{-3} M Na_2S in H_2O. Insert: decay of absorption signal after the laser pulse. (After ref. [54])

yields encountered in photo-emission experiments with compact semiconductor electrodes [53]. This result indicates that the particle size may be important and indeed it was found that the e^-_{aq} absorption occurs only with small particles (nm range) and the absorption coefficient increased with decreasing particle size [54]. Henglein et al. related the latter result to the quantization effect insofar as the rate of the competing thermalization of the electron within the particle was also reduced with decreasing particle size, because the density of energy states becomes low in very small particles.

The absorbance (log I_0/I) and therefore the concentration of e^-_{aq} increased with increasing density of absorbed photons as measured immediately after the laser flash (Fig. 9.18). Since the extinction coefficient is well known, the concentration could easily be determined from absorption measurements. In comparison, the absorption of holes trapped at the particle surface is also shown in Fig. 9.18. The latter has a spectrum peaking around 600 nm and has a long lifetime (>1 ms), and was obtained after the e^-_{aq} absorption had been quenched by adding acetone to the solution. Details are given in ref. [54]. The absorption of the trapped holes saturates at higher intensities. The holes were probably used in an anodic corrosion reaction.

Concerning the nature of the excited state, it must be first realized that at high doses each colloidal particle absorbs many hundred photons during one flash. From their measurements, Henglein et al. concluded that the following mechanism occurs [3, 54]:

$$\begin{array}{ccccccc} (\mathrm{CdS})_n & \xrightarrow{h\nu_1} & (\mathrm{CdS})_n\,(\mathrm{e}^- + \mathrm{h}^+) & \xrightarrow{h\nu_2} & (\mathrm{CdS})_n(\mathrm{e}^- + \mathrm{h}^+)_2 & \xrightarrow{k_e} & (\mathrm{CdS})_n(\mathrm{h}^+) + \mathrm{e}^-_{aq} \\ & & \downarrow k_1 & & \downarrow k_2 & & \\ & & (\mathrm{CdS})_n & & (\mathrm{CdS})_n(\mathrm{e}^- + \mathrm{h}^+) & & \end{array} \tag{9.17}$$

where $(CdS)_n$ represents one particle. The absorption of the first photon leads to the formation of one electron–hole pair in the particle. It is rather unlikely that two photons are absorbed simultaneously. The excited particle may relax to the ground state or absorbs a second photon to receive a second excited state. The doubly excited particle may lose its second excitation or emit an electron. The specific rate of the second recombination, k_2, is much greater than that of the first recombination, k_1, and of emis-

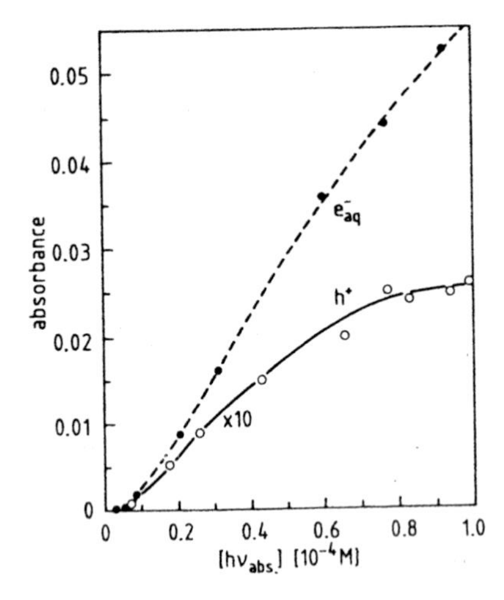

Fig. 9.18 Laser-induced photoelectron emission from CdS particles: absorption of emitted electrons and remaining holes as a function of absorbed photon concentrations (After ref. [54])

sion, k_e. The interaction of the two excited states leading to emission can be regarded as an Auger recombination, i.e. the annihilation energy of a recombining charge carrier pair was transferred to the electron of another pair in the same particle. The Auger effect is frequently observed in compact semiconductors and is used for the analysis of semiconductor surfaces (see e.g. ref. [55]). As mentioned above, the annihilation process is very effective in small particles because the density of energy states is rather small which reduces the thermalization rate. It should be mentioned that the emission of solvated electrons has also been found with ZnS [3].

9.2.3 Comparison between Reactions at Semiconductor Particles and at Compact Electrodes

As already mentioned before, mainly irreversible reactions with organic compounds have been investigated at semiconductor particles. When organic molecules, for example alcohols, are oxidized by hole transfer, O_2 usually acts as an electron acceptor or in the case of platinized particles, protons or H_2O are reduced. A whole sequence of reaction steps can occur, which are frequently difficult to analyze because cross-reactions may also be possible at particles and a new product could be formed. Concerning the primary electron and hole transfer, certainly there should be no difference between particles and compact electrodes. Since sites at which reduction and oxidation occur are adjacent at a particle, the final product may be different. An interesting example is the photo-Kolbe reaction, studied for TiO_2 electrodes and for Pt-loaded particles. Ethane at extended electrodes and methane at Pt/TiO_2 particles have been found as reaction products upon photo-oxidation of acetic acid [56, 57]. The mechanism was explained by Kraeutler et al. as follows.

At the TiO_2 electrode and the separated Pt electrode, both of which are short-circuited (Fig. 9.19a), the reactions are [56]

$$2CH_3COOH + 2h^+ \rightarrow 2CH_3^{\bullet} + 2CO_2 + 2H^+ \quad (9.18a)$$

$$2CH_3^{\bullet} \rightarrow C_2H_6 \quad (9.18b)$$

at the illuminated TiO_2 electrode and

$$2H^+ + 2e^- \rightarrow H_2 \quad (9.19)$$

at the Pt counter electrode.

In the case of particles, TiO_2 and Pt are also short-circuited because both are in direct contact. The primary steps are the same as above, i.e. [57]

$$CH_3COOH + h^+ \rightarrow CH_3^{\bullet} + CO_2 + H^+ \quad (9.20a)$$

at a TiO_2 surface site of the particle (Fig. 9.18b) and

$$H^+ + e^- \rightarrow H_{ad} \quad (9.20b)$$

at a Pt site. At the adjacent TiO_2 and Pt sites, CH_4 is formed according to the reaction

$$CH_3^{\bullet} + H_{ad} \rightarrow CH_4 \quad (9.21)$$

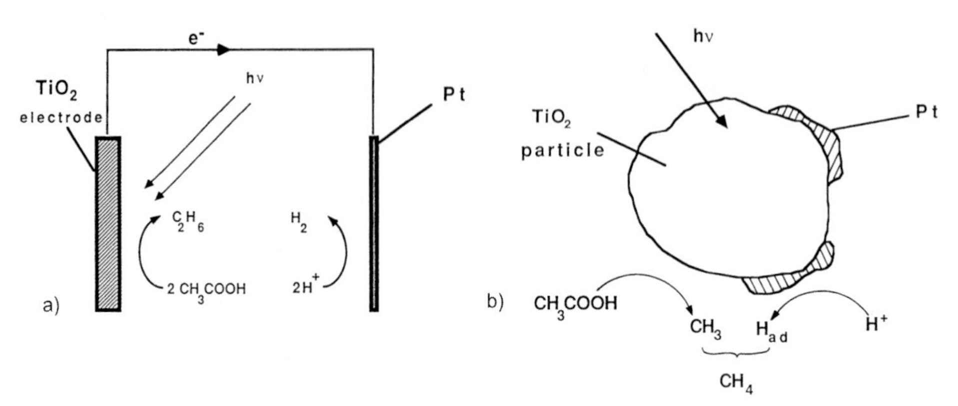

Fig. 9.19 Photo-oxidation of acetic acid (photo-Kolbe reaction) at TiO_2 electrodes (a) and TiO_2 particles (b). (After refs. [56, 57])

Another example will be discussed in Section 9.2.6.

9.2.4 The Role of Surface Chemistry

There are many indications in the literature that surface chemistry plays an important role in photoelectrochemical reactions at extended electrodes and at particles. One example has already been given in Section 9.1.1, where it has been shown for CdS colloids that surface states could be blocked by adding Cd^{2+} to the solution. There are, however, only a few quantitative investigations on this problem [5, 59], probably due to the lack of sufficiently sensitive methods. In the case of metal oxide particles, the adsorption of H_2O plays an important role. Due to the amphoteric behavior of most metal hydroxides, two surface equilibria have to be considered [59]:

$$-\text{M–OH} + \text{H}^+ \Leftrightarrow -\text{M–OH}_2^+ \qquad (pK_1) \tag{9.22}$$

$$-\text{M–OH} \Leftrightarrow -\text{M–OH}^- + \text{H}^+ \qquad (pK_2) \tag{9.23}$$

The zero point-of-charge (pH_{zpc}) of the metal oxide at the surface is defined as the pH where the concentrations of protonated and deprotonated surface groups are equal, i.e.

$$\text{pH}_{\text{zpc}} = \tfrac{1}{2}(\text{pK}_1 + \text{pK}_2) \tag{9.24}$$

According to the equilibria given in Eqs. (9.22) and (9.23), the surface is predominantly positively charged below pH_{zpc} and negatively charged above this value. The influence of the surface charge on the rate of a photoreaction is demonstrated in Fig. 9.20, according to which the oxidation of an anion such as trichloroacetate at TiO_2 particles is only observed at low pH and of a cation (chloroethylammonium) at high pH values [60]. This interpretation is entirely based on an electrostatic model, according to which the reaction rate is reduced to a very low value if the surface charge and that of the ions involved in the reaction have equal signs. In such a case, the photoexcited charge carriers either recombine, or the holes are used for the oxidation of H_2O, lead-

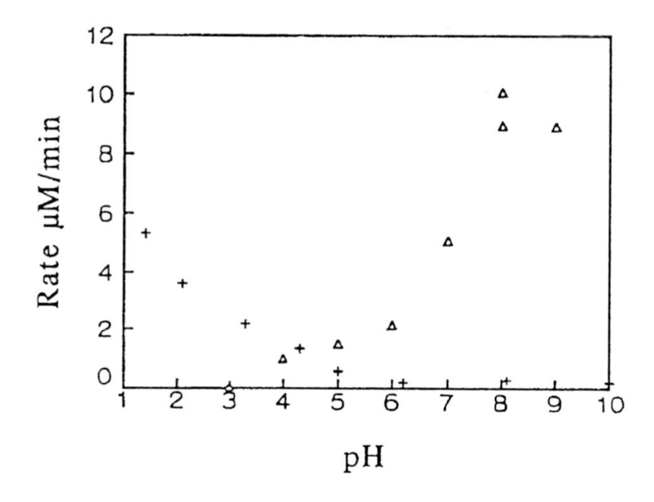

Fig. 9.20 pH-dependence of the degradation of trichloroacetate (+) and chloroethylammonium (Δ) during illumination of TiO_2 suspensions. (After ref. [86])

ing to the formation of O_2. Since O_2 is reduced again by the photoexcited electrons, no overall reaction occurs, i.e. we have a recombination via the solution. In principle, the same process should occur at compact electrodes in dilute solutions. At an anodically polarized n-type electrode, however, the holes are forced across the interface of the electrode because of the large band bending whereas the electrons occur at the rear contact.

There are also reports in the literature that surface chemistry, i.e. surface composition, can play an important role in the emission properties particles [61], and also in the reaction routes as found for the oxidation of ethanol at ZnS colloids [62]. This topic cannot be treated here.

9.2.5 Enhanced Redox Chemistry in Quantized Colloids

If the bandgap of small semiconductor particles is increased by decreasing the particle size, electrons in the lowest level of the conduction band and holes in the highest level of the valence band reach higher negative and higher positive potentials, respectively, as schematically illustrated in Fig. 9.21. In other words, electrons have a higher reduction power in smaller particles. In consequence, an acceptor molecule may be reducible only at a very small and not at a large particle if the acceptor level E_A is located between $E_{c,1}$ and $E_{c,2}$ (Fig. 9.21). On the other hand, the reverse process, i.e. an electron transfer from a donor into the lowest empty state ($E_{c,1}$), may be possible with large but not with small particles if the electronic energy of the donor occurs above $E_{c,1}$ but below $E_{c,2}$.

In order to quantify this effect it is necessary to have a technique for controlling this property. Before discussing various methods, the question arises: what happens if an electron is transferred to or taken away from a small semiconductor particle? This problem has been studied in two ways. In the first, electrons were injected into the conduction band of a semiconductor particle from hydrated electrons, the latter being generated in H_2O by pulse radiolysis [63]. In the second, the colloidal solution is investigated in an electrochemical cell, using two inert Pt electrodes. At the negatively polar-

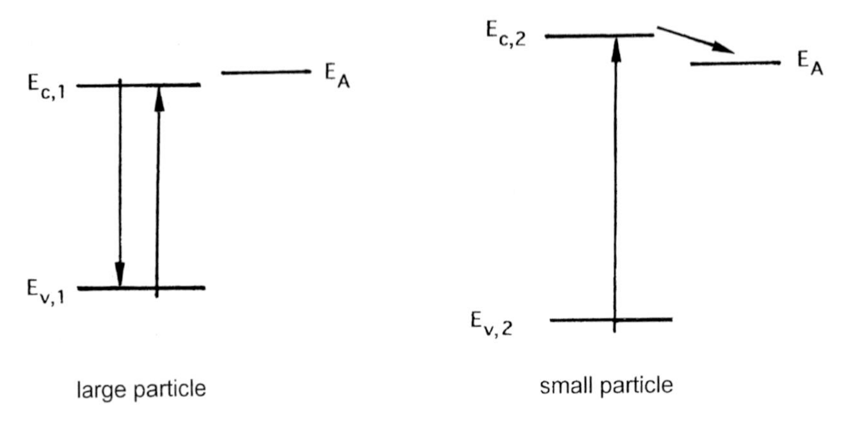

Fig. 9.21 Electron transfer from small and large particles to an electron acceptor after photoexcitation

ized electrode, electrons are transferred from the Pt electrode to the particles [64]. Such an electron injection into a particle has led to an absorption change insofar as the low energy range (high wavelength) was decreased, which is ascribed to the first excitonic transition within the particle as shown for ZnO colloids in Fig. 9.22.

There are two different models for explaining this effect, which has also been found with various other colloidal materials. The first model is the so-called Burstein shift [65]. This is a band-filling model which has been derived in solid state physics. For instance, if an n-type bulk semiconductor is heavily doped then the lowest levels of the conduction band are filled with electrons, so that light absorption can only lead to an excitation of electrons from the valence band into higher levels of the conduction band. Accordingly, the absorption is blue-shifted. In the case of small particles, this effect is more severe; not so much because of doping but the density of states at the lower edge of the conduction band is much lower than for bulk material. With very small particles, even the transfer of only one electron may lead to a complete occupation of the lowest level and to a subsequent shift of the absorption. The second model is based on the semiconductor Stark effect [8, 63]. The Stark effect is the shift and splitting of atomic and molecular energy states in the presence of a strong electric field, as first detected by absorption measurements in the gas phase. In the corresponding absorption spectra, considerable line broadening is observed. In the case of semiconductor particles, it is assumed that an electric field is produced by injection of an additional charge which is treated as a localized point charge at the surface of a particle. Due to the resulting inhomogeneous charge distribution within the particle, a strong electric field is established. The latter leads to a polarization of the exciton resulting in a broadening of the excitonic band. The corresponding excitonic band has been seen to decrease with the applied field in the region of the absorption maximum while at longer and shorter wavelengths an absorption increase could be observed, i.e. typical spectral features when line broadening occurs. This was found in the spectra given in Fig. 9.22. Therefore, Hoyer et al. attributed the spectral changes to the influence of electrons trapped in a surface state [64].

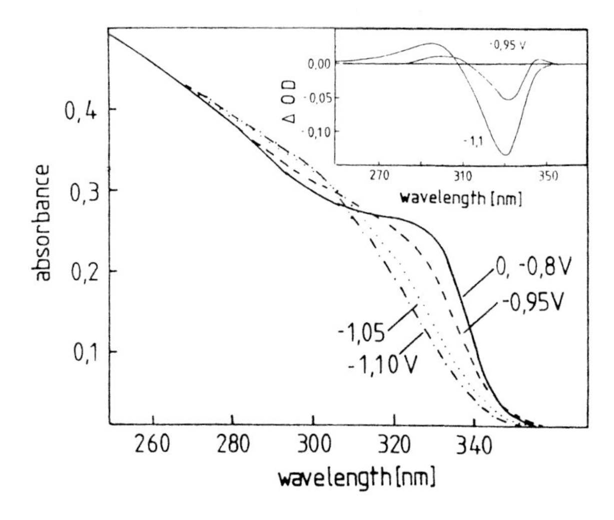

Fig. 9.22 Absorption spectrum of ZnO colloids (diameter 29.3 Å) in ethanol at different electrode potentials (reference electrode: Ag/AgCl). The light was transmitted through an optically transparent electrode (indium tin oxide (ITO) layer on glass). Insert: difference spectra between –0.6 and –0.95 V and –0.6 and –1.1 V. (After ref. [64])

As mentioned above, the reduction and oxidation power of electrons and holes depend on the position of the conduction and valence band of the particle, respectively. The variation of the conduction band with decreasing particle size was studied by measuring the electron transfer from a metal electrode to the colloidal particle (here ZnO) in an electrochemical cell [64]. The reduction of the particle upon negative polarization of the metal electrode, was detected by following the long wavelength excitonic absorption as described above. The onset of the absorption change, which corresponds to the onset potential (critical potential) of electron transfer from the metal electrode to ZnO particles, versus particle diameter is given in Fig. 9.23. In addition, the solid line in Fig. 9.23 represents the shift of the conduction band edge with the variation of the particle diameter as calculated via Eq. (9.5). The latter curve was adjusted to the experimental values at large diameters. It was concluded from the fairly good agreement between experimental values and the theoretical curve that the change of the critical potential was caused by a corresponding shift of the conduction

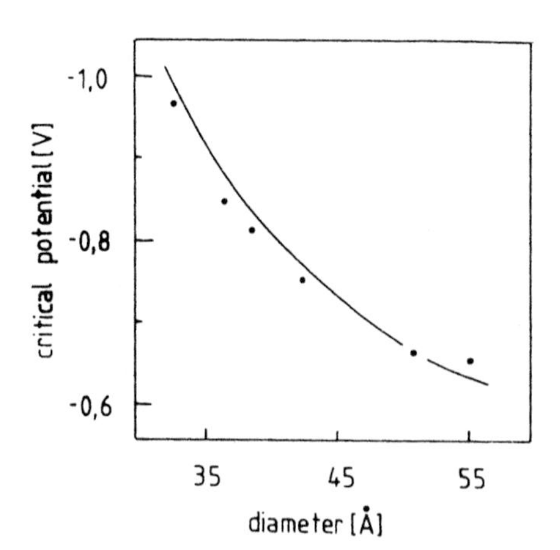

Fig. 9.23 Critical potential (onset of absorption change) vs. diameter of ZnO colloids. Dots, experimental data (see Fig. 9.22); solid line, from quantum mechanical calculations. (After ref. [64])

band. The same authors found that about one electron per particle was transferred [64]. An anodic polarization led to a re-oxidation of the particles. Absolute values of the conduction band E_c can be estimated from these electron injection measurements if E_c for the corresponding bulk material is known.

Band positions of quantized particles can also be determined via redox reactions. Considering, for instance, the electron transfer from the reduced species, $R^{(n-1)+}$, of a redox system into a particle according to the reaction

$$R^{(n-1)+} + \text{colloid} \rightarrow R^{n+}\text{colloid}(e^-) \tag{9.25}$$

then this process is only possible if the standard potential is close to or above the conduction band of the particle. This method was applied to 30-Å HgSe and PbSe particles using a variety of redox systems as listed in Fig. 9.24 [66]. According to absorption spectra, the bandgap of these HgSe colloids was shifted by about 2.8 eV with respect to that of the bulk bands (0.35 eV). Since an electron injection was possible from the reduced species of methyl viologen (MV^{1+}) but not from aldehyde (CH_2O), the lowest level of the conduction band of the HgSe particle must occur between the standard potential of these two redox couples as shown in Fig. 9.24. The uncertainty in the electronic energy position of the semiconductors is also indicated in Fig. 9.24. The position of the conduction band of 50-Å PbSe particles was determined using the same type of procedure. Here, the band positions of small and large particles were compared. For instance, MV^+ could not inject electrons into 50-Å particles but could do so into large particles (>1000 Å). This result proves again that the conduction band edge of the small particles occurs at higher energies than for large particles. In all the cases reported here, any change of the radical concentration (i.e. MV^{1+} concentration) was detected by absorption measurements [67]. The MV^{2+}/MV^{1+} redox couple is very convenient in these studies because the strong absorption of MV^{1+} and its change can be followed easily.

These are important examples because certain photoreduction processes may only be achieved with small particles of a given material. This has been demonstrated for H_2

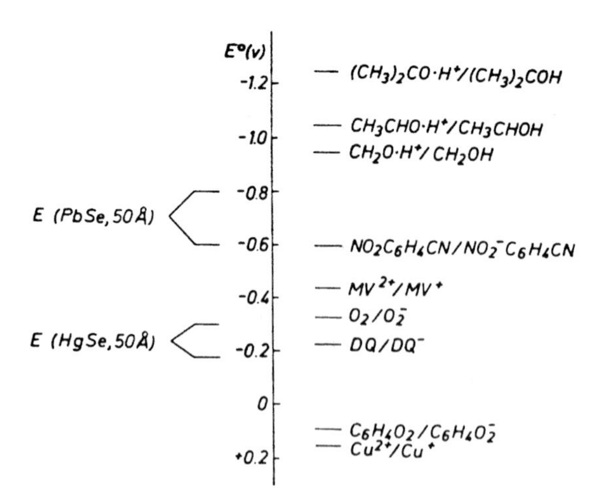

Fig. 9.24 Energy position of the lowest empty electronic state for PbSe and HgSe colloids (particle size, 50 Å) and redox couples in aqueous solution. The uncertainty in the electronic energy positions of the semiconductors is indicated. (After ref. [66])

evolution at 50-Å PbSe and HgSe colloids in the presence of a hole scavenger such as EDTA or S^{2-}, which has not been observed with large particles [66]. It should be emphasized, however, that this method is not very accurate compared with Mott–Schottky measurements at compact semiconductor electrodes (see Section 5.2), because the reorganization energy is not known and several kinetic factors may influence the results.

9.2.6 Reaction Routes at Small and Big Particles

Another interesting aspect of particle size effect is related to the density of photons absorbed by semiconductor particles, in comparison with the density of particles in a solution. Considering two solutions containing colloids of different sizes, in one case for instance 3-nm and in the other 4-μm particles, then many more particles are present in the solution of 3-nm colloid than in that of 4-μm colloid, provided that the total concentration of the semiconductor material is identical in both solutions. The two solutions differ only insofar as the same material is distributed over a small density of large particles (4 μm) or over a high density of small particles (3 nm). As can be easily calculated, a time interval of 5.4 ms exists between the absorption events of two photons in one individual 3-nm particle for a photon flux of 4×10^{17} cm^{-2} s^{-1}, assuming that all photons are absorbed in the colloidal solution [62]. In the case of the 4-μm particles, the time interval between two absorption events is only about 20 ps for the same photon flux, i.e. it is shorter by a factor of 10^8, compared with the time interval estimated for the 3 nm particle. This difference can be important for reactions where two or more electrons are involved, typically in many oxidation and reduction reactions with organic molecules.

This problem has been analyzed by studying the oxidation of ethanol at ZnS [62]. This semiconductor was selected because the oxidation of alcohol to acetoaldehyde occurs entirely by hole transfer in two subsequent steps upon illumination. In the first step, a radical is formed by hole transfer at the surface of an individual particle after absorption of one photon (Eq. 9.26). Since it takes in the average several milliseconds before another hole is generated by photon absorption in the same individual 3-nm particle, the radical diffuses into the solution before another hole is created in the same particle. In the solution, the radicals formed at different particles can disproportionate (Fig. 9.25a) and subsequently dimerize. The whole sequence of possible reactions is given by

$$CH_3CH_2OH + h^+ \rightarrow CH_3CHOH^\bullet + H^+ \tag{9.26}$$

$$\left.\begin{array}{l}(CH_3CHOH^\bullet)\\ \\ (CH_3CHOH^\bullet)\end{array}\right] \rightarrow \left[\begin{array}{ll}\xrightarrow{\text{disproportionation}} CH_3CH_2OH + CH_3CH{=}O & \text{(a)}\\ \\ \xrightarrow{\text{dimerization}} H_3C{-}CH(OH){-}CH(OH){-}CH_3 & \text{(b)}\end{array}\right. \tag{9.27}$$

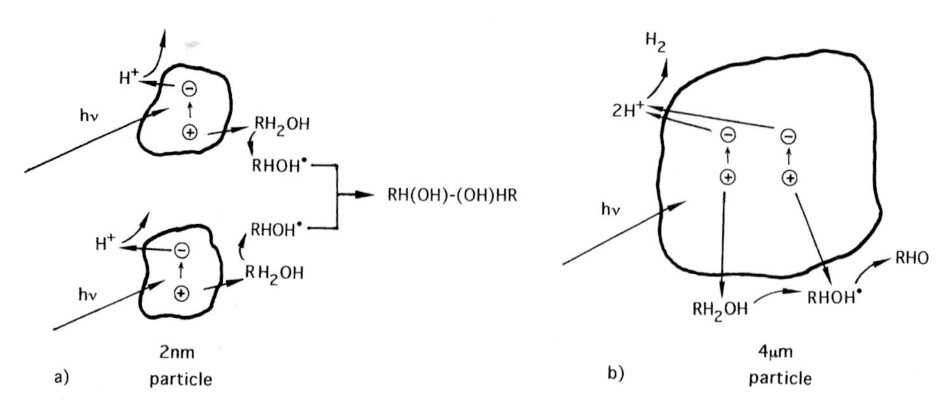

Fig. 9.25 Reaction routes at 2-nm and 4-μm ZnS particles (oxidation of alcohol)

The disproportionation leads to aldehyde and ethanol (Eq. 9.27a) whereas dimerization leads to butanediol (Eq. 9.27b and Fig. 9.25a). These products have indeed been found (concentration ratio of acetaldehyde to butanediol, 2.5) during illumination of a solution of 3-nm colloid [62]. Besides butanediol and acetaldehyde H_2 was also detected, the latter being formed in the corresponding cathodic reaction. When the same experiment was performed with much larger particles (4 μm), no butanediol was found [62]. Since the time interval between the absorption events of two photons in one 4-μm particle is only 20 s, the radical can be oxidized to acetaldehyde by a further hole transfer at the same particle as illustrated in Fig. 9.25b [5]. Instead of Eq. (9.27) we have then

$$CH_3CHOH^\bullet + h^+ \rightarrow CH_3CH{=}O + H^+ \tag{9.28}$$

This example illustrates quite nicely that reaction routes may depend on the particle size. Most other semiconductors, for example TiO_2, would not be suitable for the investigation discussed above. The reason is that only the first oxidation step occurs via the valence band (transfer of a hole produced by light excitation) leading to the formation of an ethyl radical. This radical is then usually further oxidized by the injection of an electron into the conduction. Accordingly, only one photon is required for the two-step oxidation of alcohol; the second step occurs immediately after first step at the same particle and aldehyde is the only product ("current-doubling" effect, see Section 7.6). In the case of ZnS the oxidation can occur entirely via the valence band because the conduction band is located at rather high energies (Fig. 5.20) which makes an electron injection into the conduction band impossible.

9.2.7 Sandwich Formation between Different Particles and between Particle and Electrode

In Sections 9.1.1 and 9.2.4 some surface modifications of colloidal particles have already been discussed. Another kind of modification can be obtained by forming a sandwich between two particles of different materials such as TiO_2 and CdS. Corresponding structures were formed spontaneously, when the separately prepared solutions of the colloids were mixed under certain conditions [3, 68, 69]. Such sandwich

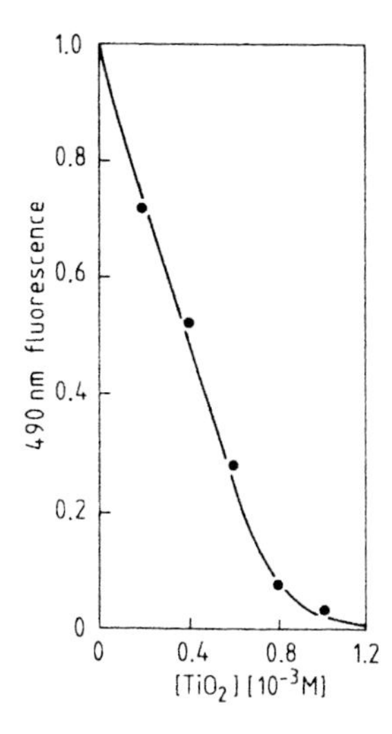

Fig. 9.26 Fluorescence quantum yield of CdS colloids as dependent on the concentration of added TiO_2 colloids. (After ref. [68])

structures have also been labeled "photochemical diodes" [6, 71]. In the latter case, however, two different materials are usually bonded together through ohmic contacts which are deposited on each semiconductor before the bonding (details are given in Section 11.2).

The formation of sandwiches between TiO_2 and CdS particles was recognized by measuring the quenching of the fluorescence of CdS. In Fig. 9.26 the fluorescence intensity as dependent upon the concentration of TiO_2 particles is shown as measured upon excitation of the CdS colloid [68, 69]. According to Spanhel et al., the quenching cannot be explained by diffusion of TiO_2 particles to the CdS particles (fluorescence lifetime, $\approx$ 10 ns) because the diffusion process would be far too slow, i.e. by at least five orders of magnitude, to explain the strong fluorescence quenching in CdS [68]. The effective fluorescence quenching can only be understood by an electron transfer from the excited CdS particle to the TiO_2 within a sandwich structure as illustrated in Fig. 9.27. Similar observations have also been made with Cd_3P_2– and ZnO particles [7, 70].

As a result of the very efficient primary charge separation within the CdS/TiO_2 sandwich, certain electron transfer reactions induced by light excitation within CdS, are expected to occur at a much higher yield when combined within a CdS/TiO_2 sand-

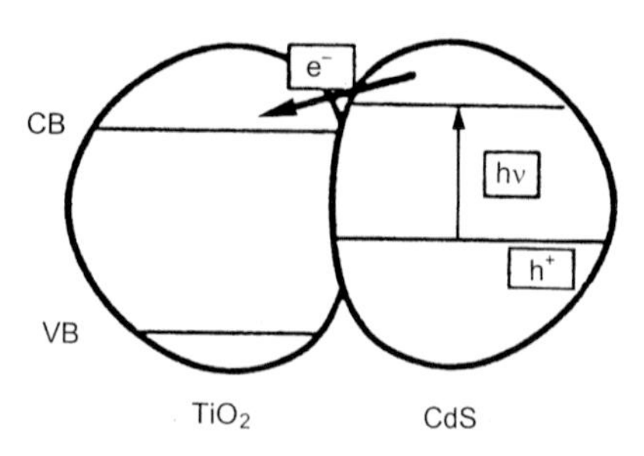

Fig. 9.27 Energy levels of a TiO_2/CdS sandwich colloid. (After ref. [7])

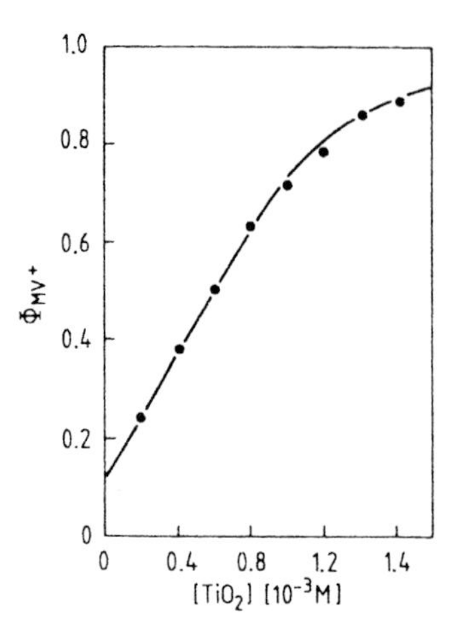

Fig. 9.28 Quantum yield of methylviologen (MV^{2+}) reduction in a CdS colloidal solution as dependent on added TiO_2 particles. (After ref. [3])

wich than in the case of a single CdS particle. This has been shown for the reduction of methyl viologen (MV^{2+}) in solutions containing different amounts of TiO_2 particles using methanol as a hole scavenger [68]. The quantum yield for the formation of the MV^+ radical, which can be readily detected by its strong blue color, was about 10 % in the absence of TiO_2 and increased to almost 100 % in the presence of TiO_2 (Fig. 9.28). This high efficiency is quite understandable on the basis of the energy diagram in Fig. 9.27, according to which electrons are forced to move toward the TiO_2 leading finally to a electron transfer to MV^{2+}, whereas the holes remain at the CdS side of the sandwich and are consumed in the oxidation of the alcohol. Similar results have been obtained with ZnO/Cd_3P_2 sandwiches [70]. According to these results, the TiO_2/CdS or ZnO/Cd_3P_2 sandwiches act as a heterojunction (see Chapter 2) of almost molecular dimensions. The electron transfer within the sandwich, however, is kinetically controlled here, and not by the field across a space charge layer.

The same kind of sandwich can also be formed by depositing colloidal particles onto a compact semiconductor electrode. This has been demonstrated, for instance, with PbS particles on a TiO_2 electrode [7, 72]. In this case TiO_2 electrodes with a very rough surface were used. (The surface was about 400 times larger than the geometric surface). The PbS particles were directly formed at the TiO_2 surface by dipping the electrode first into a lead salt solution and then washing it in a Na_2S solution. Particle sizes ranging between 30 and about 100 Å were obtained, depending on the deposition parameters. These configurations have the great advantage that the electrons injected from the excited PbS particles into the conduction band of TiO_2 can be detected as an interfacial current. In this case a typical electrochemical cell was used having a redox system in the electrolyte. The excitation spectra of the photocurrent normalized to the incident photons (photocurrent yield) are given in Fig. 9.29 for different sizes of PbS particles [7]. The excitation spectra correspond roughly to the absorption spectra of PbS colloids. Fairly high yields of up 70 % were obtained with small particles of 30–40 Å in diameter.

According to Fig. 9.29, the onset of photocurrent occurs at longer wavelengths when the particle size is increased. The bandgap of bulk PbS is only 0.45 eV. According to absorption measurements, large quantization effects have been found for small par-

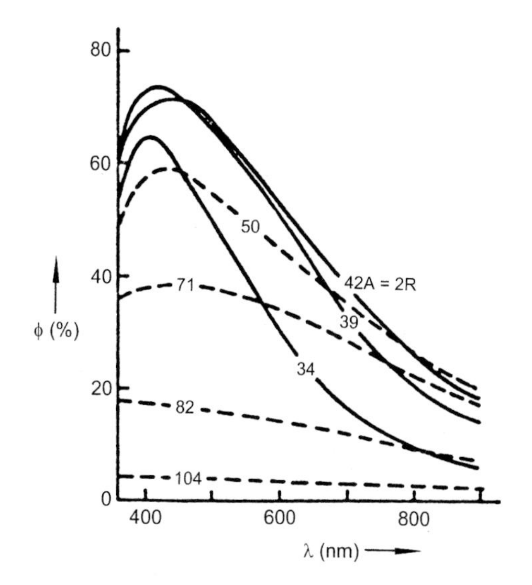

Fig. 9.29 Photocurrent yield vs. wavelength at porous TiO_2 electrodes, with PbS particles of different sizes adsorbed on the electrode surface, in 0.1 M Na_2S solutions. Photocurrent yield defined as number of transferred electrons per incident photon. (After ref. [7])

ticles. Interestingly, the yield of photocurrent decreases with particle sizes above 40 Å and becomes very low (<5 %) for a particle diameter of about 100 Å (Fig. 9.29). The latter result was interpreted as a downward shift of the lowest unoccupied state or of the conduction band edge of the PbS particles to below the conduction band edge of the TiO_2 electrode when the particle size increased and the bandgap decreased, as illustrated in Fig. 9.30 [7]. In order to keep a constant photocurrent these measurements were performed in an electrolyte containing S^{2-}/S_n^{2-} as a redox system. Electrons were transferred from S^{2-} ions into the valence band of PbS, which prevented a charging of the PbS particles or their corrosion when electrons were transferred from the particle to the TiO_2 electrode during illumination (Fig. 9.29).

The light-induced charge transfer between a semiconductor electrode and a semiconductor particle adsorbed on the electrode surface, is also termed sensitization because the same type of effect occurs with organic molecules adsorbed onto a semiconductor electrode. These processes are described in Chapter 10.

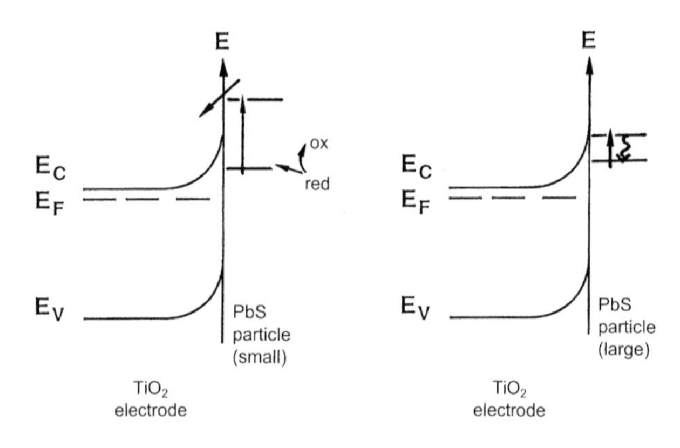

Fig. 9.30 Position of electronic energy levels for differently sized PbS particles on a TiO_2 electrode

9.3 Charge Transfer Processes at Quantum Well Electrodes (MQW, SQW)

Photo-induced electron transfer reactions from quantum well electrodes into a redox system in solution represent an intriguing research area of photoelectrochemistry. Several aspects of quantized semiconductor electrodes are of interest, including the question of hot carrier transfer from quantum well electrodes into solution. The most interesting question here is whether an electron transfer from higher quantized levels to the oxidized species of the redox system can occur, as illustrated in Fig. 9.31. In order to accomplish such a hot electron transfer, the rate of electron transfer must be competitive with the rate of electron relaxation. It has been shown that quantization can slow down the carrier cooling dynamics and make hot carrier transfer competitive with carrier cooling.

The carrier thermalization process can be divided into two phases. The first, which occurs within a few hundred femtoseconds, results from electron–electron and intervalley scattering events that equilibrate the electrons among themselves to form a hot carrier plasma with a Boltzmann-like distribution. This process, properly termed "thermalization", permits the assignment to the hot carrier plasma of a temperature that is higher than the lattice temperature. The second phase of carrier relaxation involves the cooling of the hot carrier plasma to lattice temperature through electron–phonon interactions. The first phase does not result in energy loss, but rather involves a redistribution of electron energy and momentum. The second phase leads to the conversion of the excess electron kinetic energy into heat via phonon excitation.

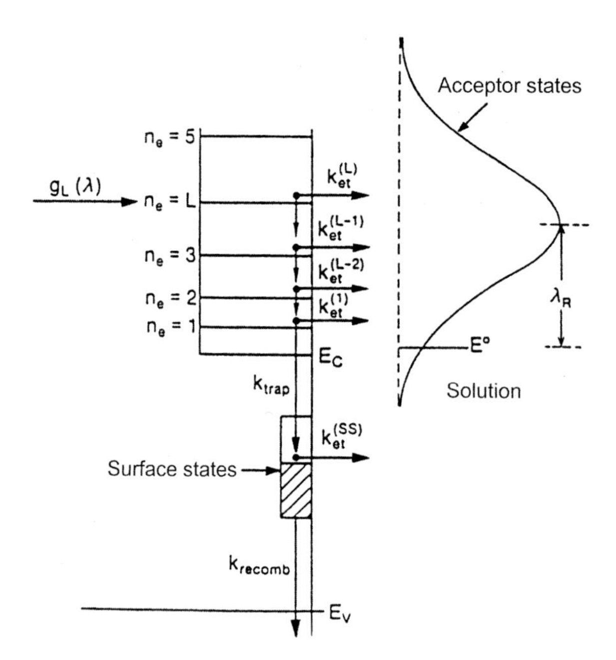

Fig. 9.31 Kinetic model for electron transfer from a quantum well (QW) electrode into a redox electrolyte. E^0 = standard potential of the redox couple; λ the reorganization energy. The acceptor states correspond to the oxidized species of the redox system. (After ref. [2])

As mentioned above, in quantized semiconductors it was predicted [73, 74], and experimentally verified by analyzing time resolved photoluminescence spectra [2, 75–78], that the electron cooling rate can be substantially reduced in quantum wells. For instance, in the case of multiple quantum wells (MQWs), consisting of 250 Å/250 Å $GaAs/Al_{0.38}Ga_{0.62}As$, characteristic times of 50–350 ps for relaxation from higher levels to the bottom of the conduction band within the GaAs well have been found, compared with 5–40 ps for bulk GaAs. The long times were determined for a relaxation from the highest energy levels. On the other hand, the hot carrier cooling rates have been found to depend on the density of the photogenerated carriers: the higher the carrier density or the absorbed photons, the slower the cooling rate [2, 76–78]. This effect is also found for bulk GaAs, but is much higher for quantized GaAs. However, a substantial difference between bulk and quantized GaAs was only found for light-induced carrier densities of $n \geq 10^{19}$ cm^{-3}. Accordingly, an efficient transfer of hot electrons from a quantized film to a redox system can only be expected for suffiently high light intensities. The latter effect is not easy to understand. It is attributed to a "hot phonon bottleneck" caused by a non-equilibrium phonon population that is created at high photo-excited carrier densities [80]. This non-equilibrium phonon population slows down the hot electron cooling rates because the hot phonons are re-absorbed by the electrons, thus reheating the electron population.

Time-resolved measurements of electron transfer times for quantum well photoelectrodes which can be compared with hot electron relaxation times, have not yet been reported. Only some excitation spectra, i.e. photocurrent vs. photon energy for MQWs and single quantum wells (SQWs), have been published so far [2]. In both cases, the photocurrent spectra show distinct structures corresponding to transitions between the hole and electron wells as shown for SQW electrodes in Fig. 9.32. The

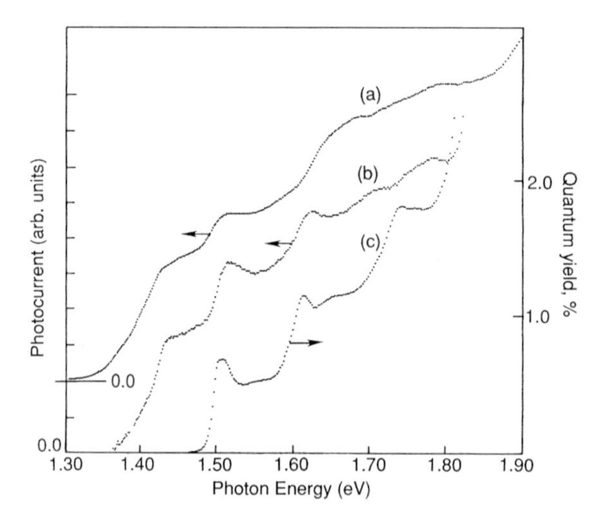

Fig. 9.32 Photocurrent action spectrum for GaAs single quantum well (SQW) electrodes at room temperature as a function of inner barrier thickness, L_b^i: a) L_b^i = 170 Å; b) L_b^i = 1.5 μm; c) L_b^i = 2.5 μm. For all three samples, the outer barrier thickness is 270 Å and the nominal well width is 130 Å. For (a) and (b), the peak at about 1.43 eV is due to the GaAs buffer layer. The zero baseline for curve (a) is offset for clarity. (After ref. [79])

MQW electrodes are not very suitable because the externally applied potential occurs across many quantum wells which leads to a misalignment of the energy levels in the electric field, as discussed in more detail in ref. [2]. This problem could be avoided by using a SQW photoelectrode. Such an electrode was prepared by first depositing an inner barrier layer ($Al_{0.3}Ga_{0.7}As$) of different thicknesses (see Fig. 9.33 legend) on a p-type GaAs substrate, then the single GaAs well and finally an outer barrier layer. In the case of a rather thin inner barrier layer (curve (a) in Fig. 9.32) the structure of the transitions is almost obscured by a large background photocurrent contribution from the bulk GaAs (substrate). The unwanted photocurrent could be reduced by using thicker inner barrier layers (curves (b) and (c) in Fig. 9.32) [79]. In this case the energy conditions at the surface were fairly well defined and could be determined by Mott–Schottky measurements which yielded straight lines.

The electrons excited into the different levels within in the single well, could be transferred to an acceptor molecule in the electrolyte either by thermionic emission across the outer barrier layer (j_{therm}) or via tunneling through it (j_{tun}) (Fig. 9.33). The photocurrent spectrum does not give any information about whether a hot electron was transferred. The observed structure in these spectra could in principle be caused simply by quantized absorption followed by a complete hot carrier relaxation and electron transfer from the lowest quantum level.

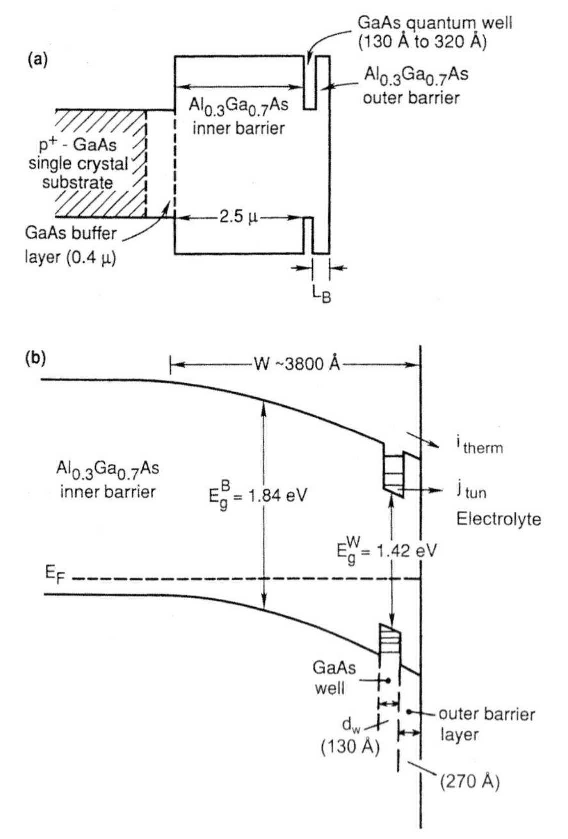

Fig. 9.33 a) Single quantum well (SQW) electrode structure (not to scale); b) energy level diagram for SQW under reverse bias. (After ref. [80])

Since in this case electrons could only be excited in a single well the photocurrent was small. On the other hand, the quantum yield, i.e. the number of transferred electrons per absorbed photons, reached values of up to $\phi = 0.63$ [80]. This might appear surprisingly high for a relatively thick outer barrier layer. However, calculations and measurements of the temperature dependence of the photocurrent showed that at room temperature the mechanism of electron transfer out of the well was thermionic emission over the barrier [80]. The rate of thermionic emission at lattice temperatures in the range of 200–300K was sufficient to keep up with the measured rate of interfacial electron transfer. Studies with very thin outer barriers (20 Å) have shown that the mechanism of charge transfer was field-assisted tunneling, and the photocurrent was then independent of temperature.

Recently, the rate of electron transfer at GaAs SQWs has been investigated quantitatively by measuring the fluorescence lifetime and its decrease upon addition of an electron acceptor [81]. The authors used an electrode structure similar to that shown in Fig. 9.33, with a well width of 50 Å but without a top barrier layer. The experiments were performed in acetonitrile using ferrocenium ions $(FeCp_2)^+$ as electron acceptors. The experimental values of the transfer rate were expressed in terms of capture cross-sections. Very high values of about $\sigma_{et} = 2 \times 10^{-15}$ cm^2/molecule were reported. This result indicates a high degree of coupling between $(FeCp_2)^+$ and GaAs which is larger than the minimum required for the adiabatic condition. Similar observations have been made in measurements of current–potential curves at standard n-type GaAs electrodes (see Section 7.3). These results were rather surprising because redox systems such as $(FeCp_2)^+$ and other metallocenes were considered to be classical examples of outer sphere redox systems (Chapter 1 in ref. [2]). Corresponding investigations at GaAs electrodes have shown that the redox system is adsorbed because of the strong interaction [82]. Details of the analysis are discussed in Section 7.3.

9.4 Photoelectrochemical Reactions at Nanocrystalline Semiconductor Layers

As already described in Section 9.1.3, techniques exist for preparing fairly thick layers of small semiconductor particles which exhibit quantum size effects. Electrochemical experiments were performed in which layers were deposited on a conducting substrate. Since primarily photoeffects at nanocrystalline semiconductors of interest heavily doped In_2O_3/SnO_2 layers are used as a conducting substrate which is sufficiently transparent. The heavily doped SnO_2, also called ITO (indium tin oxide), is a degenerated semiconductor, so that space charge effects at the SnO_2 surface do not influence the electron transfer between the nanocrystalline layer and the underlying SnO_2. The generated SnO_2 behaves like a metal.

The photoeletrochemical behavior has been extensively studied by Hodes [35, 36]. Surprisingly, the corresponding current–potential curve measured with such an electrode (CdSe) in contact with an electrolyte containing, for example, S^{2-}/S_n^{2-} as a redox system, showed a typical diode characteristic (Fig. 9.34). On the other hand, when a gold layer was deposited on the CdSe nanocrystalline film instead of making a contact

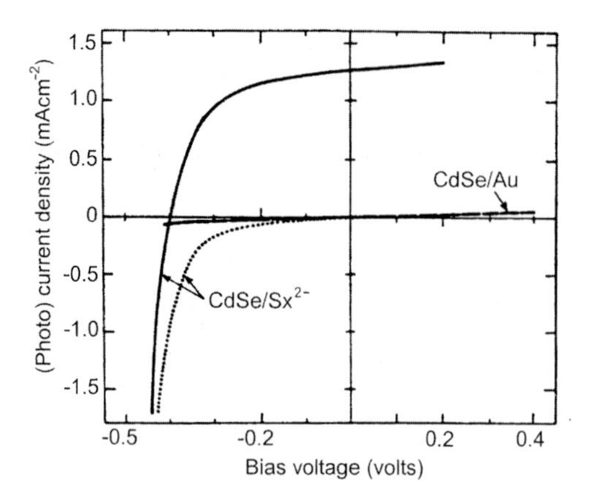

Fig. 9.34 Current–voltage curve for a nanocrystalline CdSe layer on Ti in an aqueous solution of 4 M K_2S + 2 M S. Solid line, under illumination; dotted line, in the dark; Layer thickness ≈ 0.7 μm. (After ref. [36])

with the liquid, a very small current and no rectification was observed (see also Fig. 9.34). This electrochemical behavior of the nanocrystalline film was interpreted on the basis of a model as illustrated in Fig. 9.35. It is assumed that the liquid contacts many particles because of the porous structure of the film. A photon absorption within one particle leads either to a recombination within the corresponding particle or the electron–hole pairs are separated at the surface of an individual particle. In contrast to a bulk semiconductor where the field across the space charge layer is responsible for the separation of electrons and holes, here no space charge layer exists and kinetic properties must be responsible for the charge separation. It is clear from Fig. 9.35 that a hole is transferred to the reduced species of a redox system in the liquid leaving the corresponding particle negatively charged. The electron must then travel rather a long way to the rear electrode, depending on the distance of the individual particle from the rear electrode. Accordingly, there is a competition between electron–hole recombination and a charge transfer at each individual particle. This model is supported by excitation spectra of the photocurrent; one was taken by illuminating the system through the electrolyte (at the front), the other through the conducting rear electrode. The photocurrent, normalized to the incident density of photons, became constant at shorter wavelengths (small penetration depths) if the cell was illuminated through the

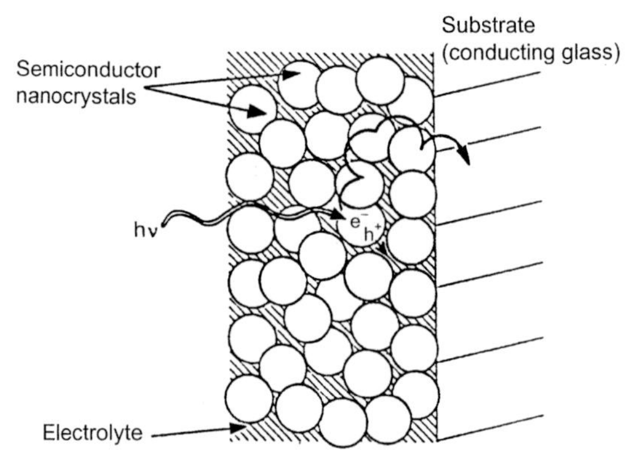

Fig. 9.35 Schematic model of a porous nanocrystalline CdSe film–electrolyte system under illumination. (After ref. [35])

rear electrode. Illumination through the front, however, led to a decrease of the photocurrent when the wavelength was varied toward shorter wavelengths. In the latter case, light was mainly absorbed by the front layer. Since recombination is the dominating process here, this layer acted only as an absorbing filter and the quantum yield for charge transfer was low [35].

The photocurrent–potential dependence in Fig. 9.34 has a shape which is typically found for valence band processes with bulk n-type semiconductor electrodes. As mentioned above, these photoelectrochemical properties of nanocrystalline films can only be discussed in terms of a kinetic separation of charges at the interface rather than in terms of space charge effects and doping, i.e. terms such as n- and p-type are not relevant. The question arises, however, as to why is there a preferential hole transfer to the electrolyte. According to investigations by Hodes [35], this depends on the nature of the semiconductor nanocrystal surface and on the electrolyte. For instance, the sign of the photoresponse is changed to the opposite direction after CdS or CdSe have been treated with mild etching (5–10 % HCl). Then, electrons are preferably transferred to the electrolyte. On the other hand, the large increase of the cathodic dark current in Fig. 9.34 may also be due to a reduction of the redox system at the substrate because the electrolyte also certainly contacts the ITO electrode.

In the late 1990s nanocrystalline TiO_2 layers have been used for many applications because of the large bandgap and the large surface area [83]. Electron transfer processes between excited molecules and a porous TiO_2 electrode were particularly studied, as will be described in Chapter 10. It also should be mentioned that porous Si electrodes belong to this class of material, as has already been discussed in Chapter 8.

10 Electron Transfer Processes between Excited Molecules and Semiconductor Electrodes

Photoreactions between a solid and a dye adsorbed at the surface of the solid, are also frequently termed sensitization. The latter process usually implies an extension of photosensitivity toward longer wavelengths where the solid itself is not sensitive to light excitation. This phenomenon was at first discovered by Vogel [1] as early as 1873, using organic dyes deposited on photosensitive silver halide particles. The sensitization of a silver halide grain via electron injection from a highly absorbing organic dye molecule is the basic process in commercially important photography [2, 3]. Sensitization of large bandgap semiconductors was viewed as a model system of the photographic process. Photoelectrochemical techniques are attractive for studying the fundamentals of this process because the photocurrent is a directly measurable parameter at semiconductor electrodes and can be studied as a function of dye coverage, excitation wavelength and potential.

Several sensitization effects observed with semiconductors are based on phenomena found with excited dye molecules in homogeneous solutions [4, 5]. From these it is well known that an excited molecule generated by light absorption exhibits a higher reduction power in the case of electron acceptors and also a higher oxidation power in the case of electron donors. This change of the electron donor and acceptor properties upon light excitation is the driving force for reactions in homogeneous solutions and also, as will be shown in this chapter, in heterogeneous processes.

10.1 Energy Levels of Excited Molecules

In photochemistry the energy levels of organic molecules are usually presented in a molecular orbital scheme as illustrated in Fig. 10.1. From the highest occupied singlet state, denoted as S_0, electrons can be excited into the lowest unoccupied singlet state, S_1, or even into a higher singlet state, such as S_2. The excited molecule can return to the ground state S_1 by emitting light (fluorescence) or by a non-radiative recombination process. The rate constants are usually in the order of nanoseconds. Besides recombination, the molecule can also be transferred from the S_1 state into a triplet state which means a spin conversion; the spin of the excited electron is then parallel to the spin of the remaining electron in the ground state. From the lowest triplet state, the molecule returns back, again by intersystem crossing, directly to the ground state S_0 via emission or by a non-radiative process (see Fig. 10.1). Such an intersystem crossing is a forbidden transition. Accordingly the corresponding rate constant k_{isc} should be very small. However, there are certain spin orbital coupling effects, especially in the presence of a heavy metal, which lead to a considerable increase of the intersystem crossing rate [6]. Since the intersystem crossing rate is usually small the lifetime of the triplet is rather long and the decay of the corresponding $T_1 \rightarrow S_0$ emission (phosphorescence) very

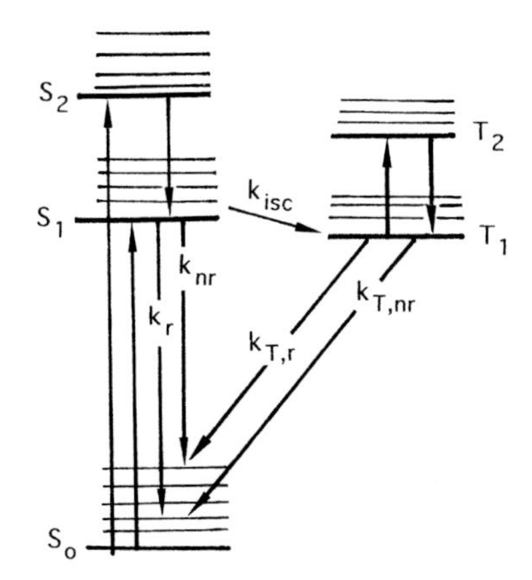

Fig. 10.1 Singlet (S_0, S_1, S_2) and triplet states (T_1,T_2) of a molecule; k_r, $k_{T,r}$, rates for radiative transitions; k_{nr}, $k_{T,nr}$, rates for non-radiative transitions; k_{isc}, rate for intersystem crossing

slow. The triplet state is efficiently quenched by oxygen. In the absence of O_2, the lifetime of a triplet state ranges usually from 1 μs up to 0.1 s. The lifetime of an excited singlet state can be determined by exciting the molecule by a short laser light pulse and measuring the decay time of the corresponding fluorescence. The lifetime of a triplet state can be measured in the same way, as long as the $T_1 \rightarrow S_0$ transition is a radiative process. Mostly the non-radiative recombination dominates due to side reactions. Therefore, the triplet lifetime is usually determined by using a double-beam technique, i.e. the molecules are excited by a short laser pulse leading to some intersystem crossing, and the triplets and their decay are determined by detecting the $T_1 \rightarrow T_2$ absorption in a suitable wavelength range by using a stationary light beam (see Section 4.6).

In the case of organic molecules, for instance dyes, several oxidation states generally exist. Each electron transfer step can be correlated qualitatively with the energy of molecular orbitals as illustrated in Fig. 10.2 (triplets are neglected here). A reduction of a molecule M can only occur by electron transfer from an electron donor to an unoccupied level of M. On the other hand, an oxidation of M is only possible by an electron transfer from the lower lying occupied state to a suitable acceptor molecule. These two processes must be described by two different redox potentials, which cannot be

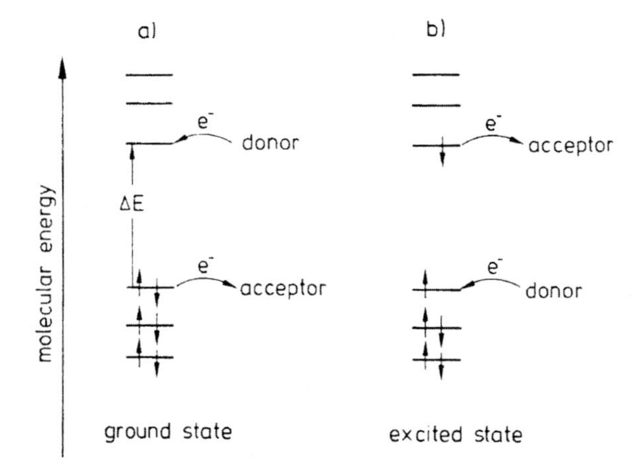

Fig. 10.2 Electron transfer between an organic molecule and an acceptor or donor

derived from the molecular orbital scheme. One can only roughly estimate the difference of the two redox potentials which should be in the order of the difference of the lowest unoccupied and the highest occupied states in the molecule (ΔE in Fig. 10.2).

It is well known that an excited molecule is more easily reduced or oxidized because the excitation energy ΔE^* is stored in the molecule. Possible reactions are

$$M^* \rightarrow M^+ + e^- \tag{10.1}$$

and

$$M^* + e^- \rightarrow M^- \tag{10.2}$$

This is immediately clear from the molecular orbital scheme in Fig. 10.2b. Here, an electron transfer is possible from a higher level to an acceptor, or from a donor to a level of M* which is only half-filled.

In Sections 3.2.4 and 6.2.1 the energy states of a redox couple in the dark were derived in terms of a Fermi level, $E_{F,redox}$, (redox potential) and of occupied and empty states (E_{red} and E_{ox}). According to the molecular energy scheme, one would expect that the standard redox potentials, $E^0_{F,redox}(M/M^+)$ and $E^0_{F,redox}(M/M^-)$, corresponding in the first case to the oxidation and in the second to the reduction of the dye molecule M in the dark, should differ by about ΔE (see Fig. 10.2a), i.e.

$$E^0_{F,redox}(M/M^+) - E^0_{F,redox}(M/M^-) \approx \Delta E \tag{10.3}$$

Using the normal electrochemical scale, $E^0_{F,redox}(M/M^+)$ occurs at more positive potentials than $E^0_{F,redox}(M/M^-)$. Eq. (10.3) has been verified by measurements of the reduction and oxidation potential of dyes. A plot of the difference of these potentials (which corresponds roughly to the difference of the standard potentials) for a large number of dyes, vs. the absorption maxima of the corresponding molecules, yielded a straight line with a slope of nearly unity [7, 21]. This result suggests that the reorganization energy must be nearly equal for all the dyes tested here.

The redox potentials of an excited molecule can now be derived as follows: The stored energy ΔE^* is given by the 0–0 transition between the lowest vibrational levels in the ground and excited state, i.e. $\Delta E^* = \Delta E_{0\text{–}0}$. The excited state of a molecule may be either a singlet or a triplet state. It is possible to estimate the redox potentials of excited molecules by adding or subtracting $\Delta E_{0\text{–}0}$ from the redox potential for the molecule in the ground state. One obtains

$$^*E_{F,redox}(M^*/M^+) = E_{F,redox}(M/M^+) - E_{0-0}(M/M^*) \tag{10.4}$$

or

$$^*E_{F,redox}(M^*/M^-) = E_{F,redox}(M/M^-) + \Delta E_{0-0}(M/M^*) \tag{10.5}$$

in which $^*E_{F,redox}(M^*/M^+)$ and $^*E_{F,redox}(M^*/M^-)$ represent, respectively, the redox potentials of the two couples M^*/M^+ and M^*/M^- in the excited state. It is clear from Fig. 10.2 that the first one, $^*E_{F,redox}(M^*/M^+)$, is expected to occur at much more negative potentials or energies than the second one, with respect to the usual electrochemical reference electrodes such as the normal hydrogen electrode (NHE). This is illustrated in the electron energy scheme given in Fig. 10.3. According to the foregoing dis-

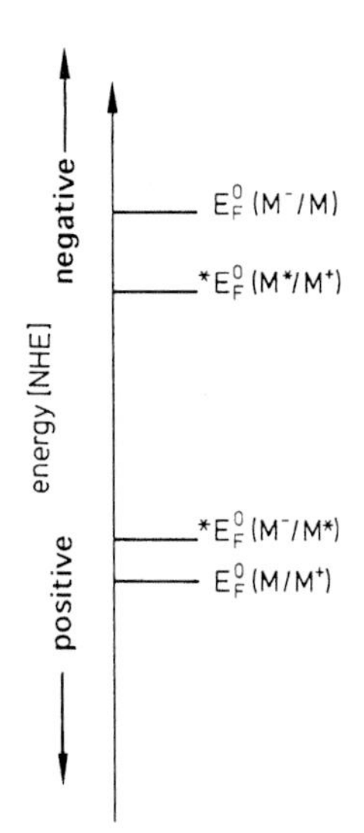

Fig. 10.3 Fermi levels of a redox system in its ground and excited state. NHE, normal hydrogen electrode

cussion, $*E_{F,redox}(M^*/M^+)$ must be rather close to $E_{F,redox}(M/M^-)$ and $*E_{F,redox}(M^*/M^-)$ to $E_{F,redox}(M/M^+)$. Differences do occur, of course, because the reorganization energies may be different. This has been taken into account in Fig. 10.3.

Many of the dark redox potentials were determined by polarographic methods. One can further easily derive, from Eqs. (10.4) and (10.5), the relative positions of the redox potentials of a molecule in its excited state, i.e. $*E_{F,redox}(M^*/M^+)$ should occur at higher (or more negative) values compared with $E_{F,redox}(M/M^+)$ and $*E_{F,redox}(M^*/M^-)$ at lower (or more positive) energies with respect to $E_{F,redox}(M/M^-)$, as shown in Fig. 10.3.

The situation can be very well illustrated by taking as an example M = Ru(II)(bipy)$_3^{2+}$, the photochemical and photoelectrochemical properties of which have been extensively studied (see refs. [9–11]). This compound is a transition metal complex which has been widely used for this kind of research because it is stable, the ruthenium can exist in various oxidation states (mainly in the +3 and +2 state), and most standard potentials in the ground state are known, i.e.: $E^0_{F,redox}(Ru^{2+}/Ru^{3+})$ = 1.26 eV; $E^0_{F,redox}(Ru^{1+}/Ru^{2+})$ = –1.28 eV vs. the NHE, and $\Delta E_{0\text{–}0}$ = 2.12 eV [11]. The calculated $*E_F$ values are given in Fig. 10.4. They were experimentally confirmed by investigation of electron transfer reactions of excited *Ru(bipy)$_3^{2+}$ with other redox systems which acted as electron donors or acceptors in solution [11–13].

As previously discussed, in the Gerischer model (see Section 6.2) a redox system is also characterized by occupied and empty states which are distributed over a certain energy range above and below the corresponding standard potential. Introducing the corresponding states into Fig. 10.4, one obtains Fig. 10.5. The two possible reactions, Eq. (10.1) for oxidation in the dark and under illumination, and Eq. (10.2) for the reduction in the dark and under illumination, must be treated separately, because the Fermi levels of the redox system considered here are different. In Fig. 10.5a, for instance, the redox couple M/M$^+$ is described. In the dark, the Fermi levels in the solid, E_F, and in the redox system, $E_{F,redox}(M/M^+)$ must be equal at equilibrium. Since in the experiments generally only the reduced species (M in Fig. 10.5a) is added to the elec-

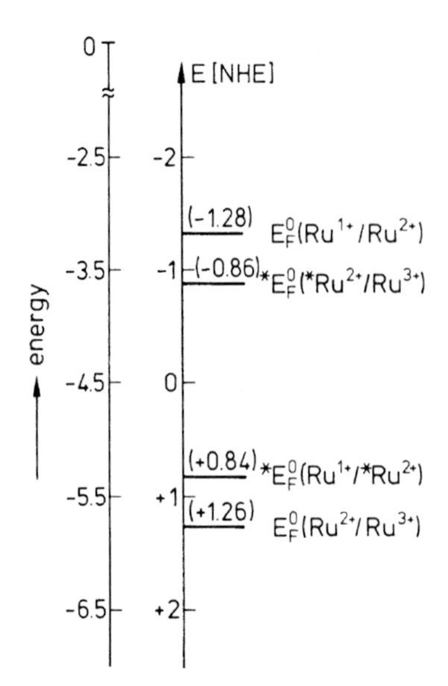

Fig. 10.4 Fermi levels of $Ru(bipy)_3^{2+}$; i.e. of the Ru(III)/Ru(II) and Ru(II)/Ru(I) couples of the complex in their ground and excited states

trolyte, the density of empty states (D_{ox}) is much smaller than the density of occupied states (D_{red}), as indicated by the different heights of the gaussian distribution curves in Fig. 10.5. The empty and occupied states of the excited molecule (M*) are also illustrated in a similar way. The excitation of the molecule M leads to the creation of occupied states of the excited molecule. From this state, an electron transfer can occur to a corresponding acceptor, e.g. to the conduction band of a semiconductor. In Fig. 10.5b,

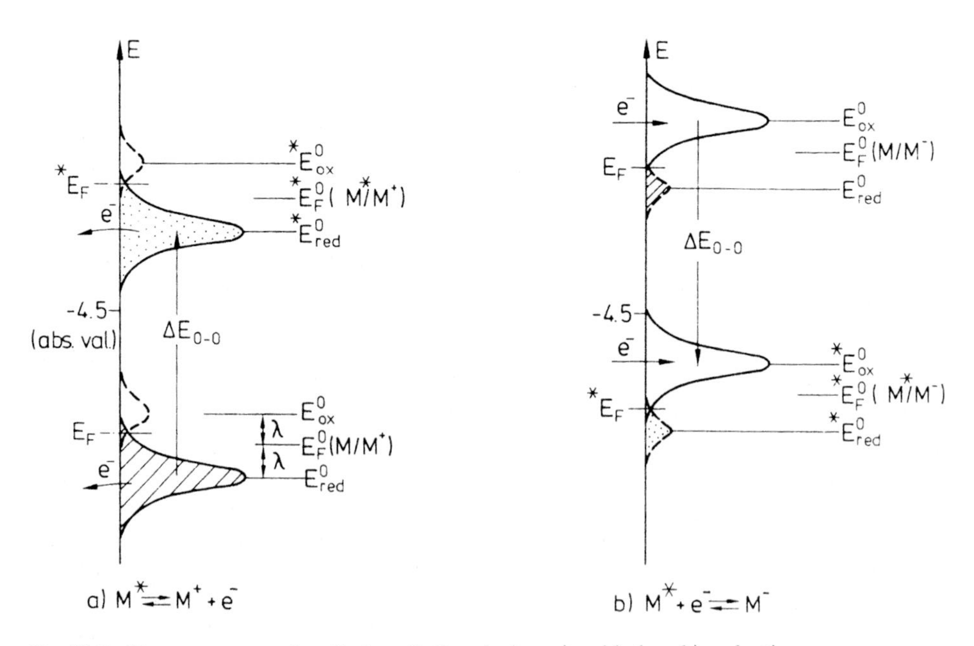

Fig. 10.5 Energy states and optical excitation during: a) oxidation, b) reduction

the energy states are illustrated for a molecule which is reduced by electron transfer from a donor to the molecule. In this case, the empty and occupied states of the unexcited redox couple M/M^- occur at higher energies near the states of the M^*/M^+ couple (compare a) and b) in Fig. 10.5), whereas the corresponding states of the excited couple M^*/M^- are located at much lower energies. Here, the concentration of empty states dominates, and excitation leads to the formation of empty states of the excited couple which can easily accept an electron from the semiconductor.

There are many organic molecules such as $Ru(II)(bipy)_3^{2+}$, which can be oxidized as well as reduced. If, in this case, only molecules in their original oxidation state are present in the solution, then both redox potentials, $E_{F,redox}(M/M^+)$ and $E_{F,redox}(M/M^-)$, should determine the equilibrium. Since the concentrations of M^+ and M^- are usually very small, the two redox potentials merge to one value as given by

$$\begin{aligned} E_{F,redox} &= E_{F,redox}(M/M^+) = E_{F,redox}(M/M^-) \\ &= 1/2\left\{E^0_{F,redox}(M/M^+) + E^0_{F,redox}(M/M^-)\right\} \end{aligned} \tag{10.6}$$

$E_{F,redox}$ occurs just in the middle of the two standard potentials. At equilibium $E_{F,redox}$ must be equal to E_F of the semiconductor. Since the concentrations of M^+ and M^- are very low under these conditions, slight changes of the M^+ or M^- concentration leads to dramatic variations in $E_{F,redox}$ and the equilibrium conditions.

10.2 Reactions at Semiconductor Electrodes

10.2.1 Spectra of Sensitized Photocurrents

During the first period of investigation experiments were performed with electrolytes containing a relatively low dye concentration and thin cells, in order to get the light at least close to the semiconductor surface and to avoid much absorption in the bulk of the solution. Typical excitation spectra of photocurrents (normalized to equal densities of incident photons), as obtained with $Ru(bipy)_3^{2+}$ at various semiconductor electrodes, are shown in Fig. 10.6 [14–16]. The excitation spectra roughly resemble the absorption spectra of this transition metal complex. Deviations are discussed in more detail in subsequent sections. In the case of TiO_2 and SiC, the photocurrent spectra are a little distorted because the absorption edge of the semiconductors occurs within the range of the $Ru(bipy)_3^{2+}$ absorption [16]. We have selected this dye as a first example because of its well-defined redox properties and defined potentials. This makes it possible to prove immediately the theoretical model derived in Section 10.1. Since the the energy positions of the semiconductors are known from measurements of the flatband potential (Section 5.3.4) a rather detailed energy scheme of the interface can be plotted using $\Delta E_{0-0} = 2.1$ eV (Fig. 10.7). According to this scheme, it is quite obvious that in the case of SnO_2 the redox couple $Ru(II/III)(bipy)_3^{2+/3+}$($M/M^+$ in Figs. 10.3 and 10.4) has to be considered. In the dark, the corresponding redox potential occurs at rather positive potentials, i.e. at low energies in the energy scheme (Fig. 10.7) and the redox potential of the excited system at relatively negative potentials, i.e. at high ener-

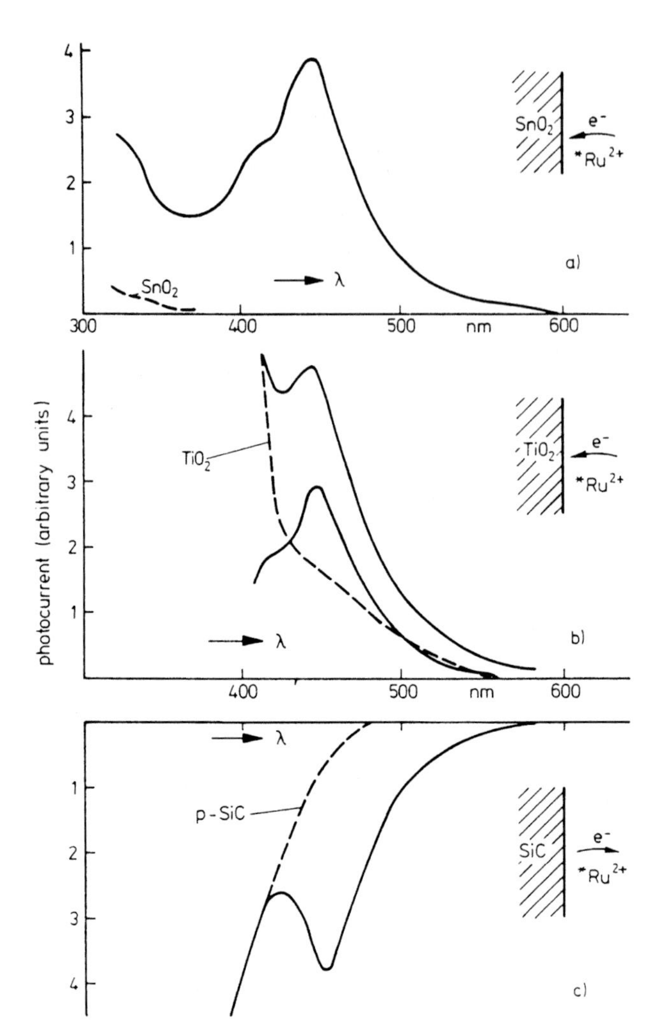

Fig. 10.6 Excitation spectra of photocurrents at various semiconductors for $Ru(bipy)_3^{2+}$ in 1 M H_2SO_4. (After ref. [9])

gies. The energy levels of the excited $^*Ru(II)(bipy)_3^{2+}$ overlap very well with the conduction band, leading to an electron transfer from these levels into the conduction band of SnO_2 (Fig. 10.7a) which corresponds to an anodic photocurrent, as observed experimentally [14]. The dye molecule is oxidized ($Ru(II) \rightarrow Ru(III)$). The same result is obtained with n-TiO_2 [15]. In the case of SiC, the energy bands are much lower so that an electron injection from the excited dye into the conduction band is no longer possible. On the other hand, a cathodic sensitization current has been observed with p-SiC (see lower part of Fig. 10.6), i.e. the Ru(II) complex is reduced. Accordingly, we must consider here the redox potential of $Ru(I/II)(bipy)_3^{1+/2+}$. In the dark, the redox potential occurs at rather negative potentials (high energies in Fig. 10.7d), and with the excited system it occurs at relatively positive potentials (low energies in Fig. 10.7d). The energy levels of the excited $Ru(II)(bipy)_3^{2+}$ occur at the energy of the valence band of SiC which makes an electron transfer possible from the valence band of SiC to the excited $^*Ru(II)(bipy)_3^{2+}$ (hole injection), which explains the cathodic photocurrent

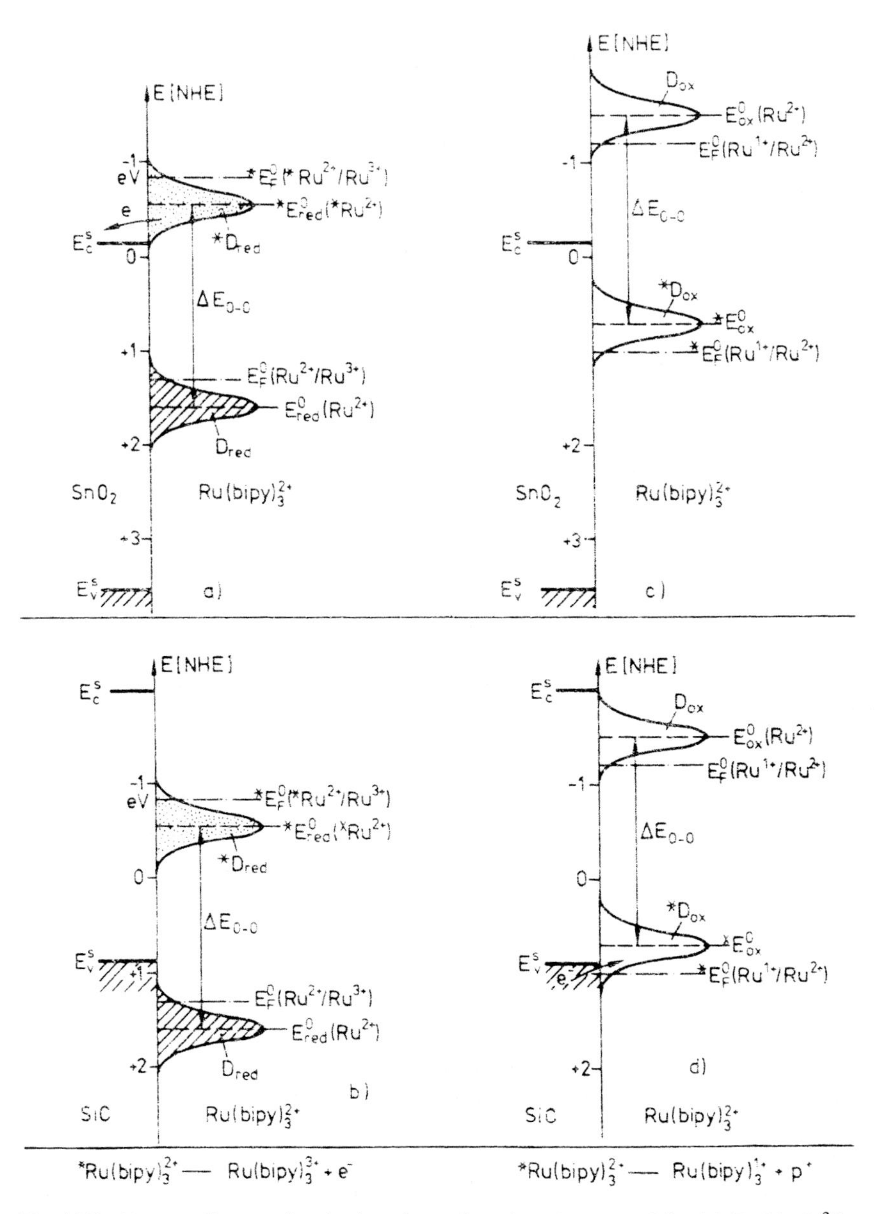

Fig. 10.7 Energy diagram for the interface of semiconductor and liquid $(Rubipy)_3^{2+}$): a) and c) n-SnO_2; b) and d) p-SiC. (After ref. [9])

at p-SiC [9]. A reduction via the valence band process is not possible with SnO_2 and TiO_2 electrodes (Fig. 10.7c).

In fact the very first investigations in this field were performed using ZnO, by Gerischer and Tributsch in 1968 [17, 18]. Besides ZnO, SnO_2, TiO_2 and SiC sensitization at a few other semiconductor electrodes, such as CdS [19], GaP [20] and SnS_2 [21] have been investigated. All the results obtained so far, can be explained on the basis of the energy diagrams as given in Figs. 10.4 and 10.7.

In one case, namely GaP, the excitation spectrum of the sensitization current was measured by using n- and p-type electrodes using rhodamine-B as a dye [20]. According to Fig. 10.8, cathodic photocurrents were observed upon excitation of the dye. This result is reasonable in terms of a corresponding energy scheme because the energy bands of GaP occur at rather high energies (see Fig. 5.21), similarly as with SiC. The cathodic current corresponds to an electron transfer from the valence band into the empty states of rhodamine-B. Such a valence band process is possible with both, n- and p-type electrodes. Further details concerning the potential dependence of the sensitization current are given in Section 10.2.3.

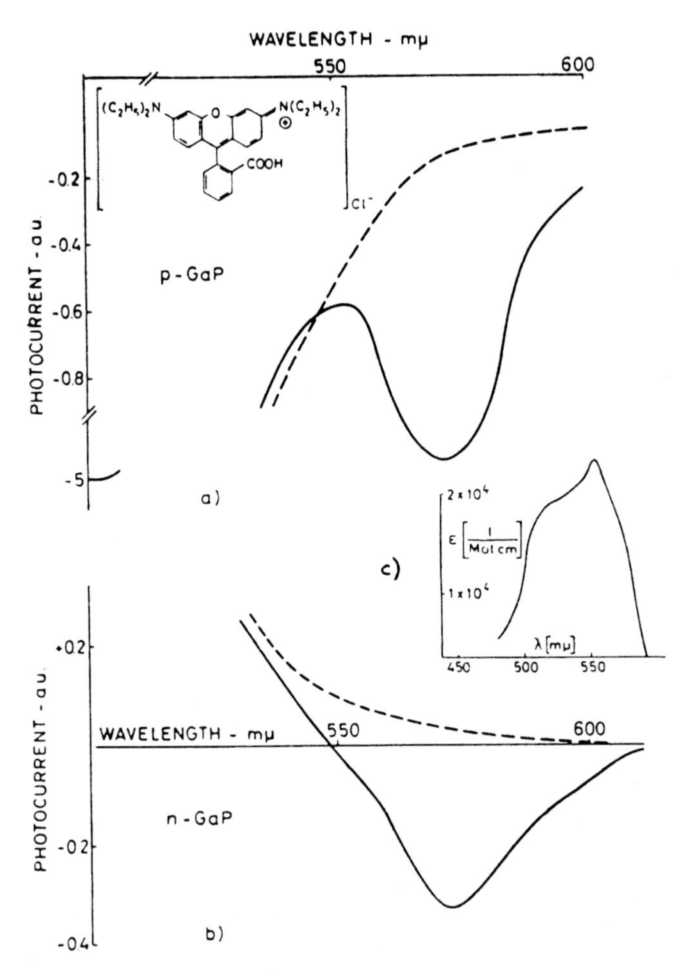

Fig. 10.8 Excitation spectra of the sensitized photocurrent for rhodamine-B (10^{-5} M) in 1 N KCl solution (dashed line without dye): a) p-GaP; b) n-GaP; c) absorption spectrum. (After ref. [20])

10.2.2 Dye Molecules Adsorbed on the Electrode and in Solution

As demonstrated in the previous section, photocurrents were observed upon excitation of dyes in a system where the dye was dissolved in the electrolyte. Since an excited molecule has a limited lifetime, the question arises regarding whether only adsorbed molecules or also those in the solution are involved in the charge transfer process. This depends on the nature of the excited state, i.e. whether the excited molecule remains in its singlet state or is converted into the triplet state before the electron transfer occurs. As already mentioned in Section 10.1., the singlet lifetime is mostly in the order of 10^{-9} to 10^{-8} s, whereas the triplet lifetime ranges from about 10^{-6} s up to some 10^{-1} s.

Since in most photoelectrochemical experiments the O_2 concentration was relatively high, the triplet lifetime was certainly never greater than 10^{-6} s. The diffusion length L of an excited molecule can be calculated by using Eq. (2.26) as given by

$$L = (D\tau)^{1/2} \tag{10.7}$$

in which D is here the diffusion constant and τ the lifetime of the excited dye molecule. Using $D \approx 10^{-5}$ cm^2 s^{-1}, one obtains for a molecule in its excited singlet state $L = 1$ nm, assuming a lifetime of $\tau = 10^{-9}$ s, and in its triplet state $L = 30$ nm ($\tau = 10^{-6}$ s). According to these values of L, molecules excited in the bulk of the solution, have practically no chance of reaching the electrode before being deactivated to their ground state. Hence, only dye molecules adsorbed at the electrode surface can be involved in the electron transfer process. When a dye solution is used, then some of the dye molecules must be adsorbed at the electrode.

This conclusion has been confirmed by several experimental results. First of all, the excitation spectra of the photocurrent are frequently shifted compared with the absorption spectra measured with dye solutions, as shown for oxazine adsorbed on an n-type SnS_2 electrode ($E_g = 2.22$ eV) in Fig. 10.9 [22]. This red shift corresponds to an energy difference of about 0.1 eV, which indicates a strong interaction between SnS_2 and oxazine (compare also with Section 10.2.4). The photoelectrochemical experiments were performed with very dilute solutions and the anodic photocurrent was found to saturate at a dye concentration of about 2×10^{-6} M [21]. Probably, just one complete monolayer was formed at this concentration.

In other cases, there is experimental evidence in the excitation spectra for aggregate formation such as dimers (double peak) which are not visible in the absorption spectra of the solutions with low dye concentrations. This has been confirmed by absorption measurements of adsorbed dye layers (crystal violet on ZnO electrodes [23]. Even polymers can be formed by adsorbed dye molecules, as found, for example, with pseudoisocyanine on ZnO [24]. One example is given in Fig. 10.10. This is a well-known phenomenon for cyanine dyes [25] where polymer bands have been found at high dye concentrations.

In most sensitization experiments it has been observed that the sensitization current decreased with long exposure to light. This result indicates that the reaction product, i.e. the oxidized or reduced dye, remained adsorbed at the electrode surface and was only slowly exchanged for other dye molecules from the solution. This result is also of interest from another point of view. At an early stage of investigation in this field, the

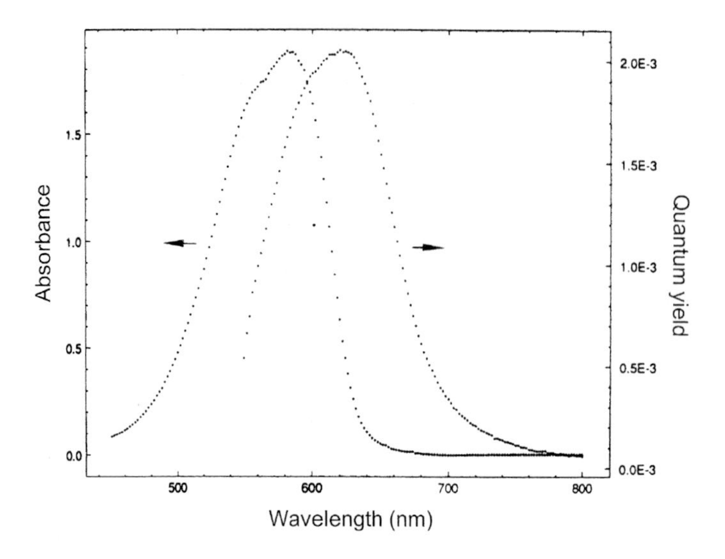

Fig. 10.9 The photocurrent action spectrum of 2 mM methanol solution of oxazine with 0.1 M LiCl at a SnS_2 electrode, and absorption spectrum of the dye. (After ref. [22])

question arose concerning whether the primary step is an electron transfer or possibly an energy transfer after excitation of the adsorbed dye. The latter process is an exchange of excitation energy between molecule and semiconductor. Since in all the experiments the bandgap of the semiconductor was larger than the excitation energy in the adsorbed molecule, an energy transfer would only have been possible if there were surface states and if the energy difference between surface states and the valence or the conduction band matched the excitation energy in the dye. In this case, either an electron in the conduction band and a trapped hole, or a hole and a trapped electron in the surface state, would be formed by an energy transfer from the adsorbed dye. The electron or hole produced by energy transfer, could then be used for further reduction or oxidation reactions. As far as the adsorbed dye is concerned, an energy transfer would lead to a complete deactivation of the dye, i.e. the dye would return to its origi-

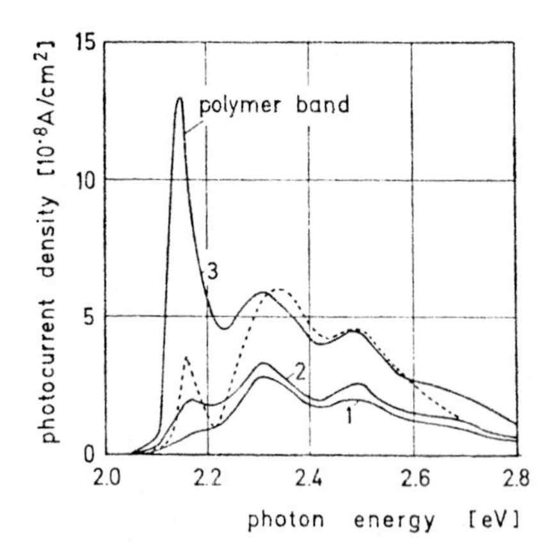

Fig. 10.10 Photocurrent action spectra for an n-ZnO electrode in aqueous solution (pH 5) with different amounts of adsorbed cyanine dye: curves 1–3, increasing concentration of adsorbed dye; dotted curve, absorption spectrum. (After ref. [28])

nal state. The experimental result that the sensitization current decreases with exposure time, is usually taken as a proof for an electron transfer rather than an energy transfer and there have been no other reports in the literature which would indicate energy transfer as a primary process. On the other hand, there is evidence that energy transfer leads to quenching of the excitation energy. For instance, it was observed that the sensitization current measured with p-SiC upon excitation of adsorbed eosin decreased immediately to a lower value with the addition of a second dye, such as methylene blue to the electrolyte, before the appearance of a photocurrent peak at wavelengths of the methyl blue absorption [26]. There was no photocurrent peak due to the excitation of methylene blue during the first period of illumination. A photocurrent peak related to methylene blue absorption rose only very slowly, due to a replacement of eosin by methylene blue.

It is also an interesting question whether the electron transfer between an excited dye molecule and a semiconductor electrode occurs via the singlet or the triplet of the molecule. This depends on thermodynamic and kinetic factors, as follows:

1. Since the optical energy is smaller for a triplet ($S_0 \rightarrow T_1$) compared with an excited singlet ($S_0 \rightarrow S_1$), the reduction potential of a triplet molecule is less negative and the oxidation potential less positive than those of an excited singlet molecule. An electron transfer can only occur if the corresponding energy states still overlap with the conduction or valence band of the semiconductor.
2. It depends on the yield of intersystem crossing ($S_1 \rightarrow T_1$).
3. The intersystem crossing rate (rate constant k_{isc}) has to be compared with that of electron transfer between the excited molecule in its S_1 state and the semiconductor.

It is reported that the oxidation of several excited xanthene dyes at n-ZnO electrodes occured via the triplet state. This was concluded from the result that the quantum efficiency of the sensitization current decreased upon addition of a typical triplet quencher [27]. Another interesting example is $Ru(bipy)_3^{2+}$, which shows in solution an intersystem crossing yield of unity [11]. Nearly all electron transfer reactions between the Ru complex and an electron acceptor in solution occur via the triplet state. In the case of adsorbed $Ru(bipy)_3^{2+}$, electron transfer from the excited singlet and from the triplet to the semiconductor electrode is possible depending on the coupling of the dye to the semiconductor (see Section 10.2.6). It should be mentioned here that the energy levels of the excited Ru complex given in Fig. 10.4, were calculated for a triplet state.

10.2.3 Potential Dependence of Sensitization Currents

Fig. 10.11 shows a typical photocurrent–potential dependence as measured with rhodamine-B at an n-type SnO_2 electrode, for three different light intensities. The photocurrent sets in around the flatband potential and is independent of potential for large anodic polarization. Around the flatband potential where no electric field occurs across the space charge layer, the injected electron must return to the oxidized molecule. This process will be discussed in more detail in Section 10.2.6. At more anodic potentials, the electron injected from the excited dye, is drawn towards the bulk of the semiconductor by the electric field across the space charge layer (see insert, Fig. 10.11).

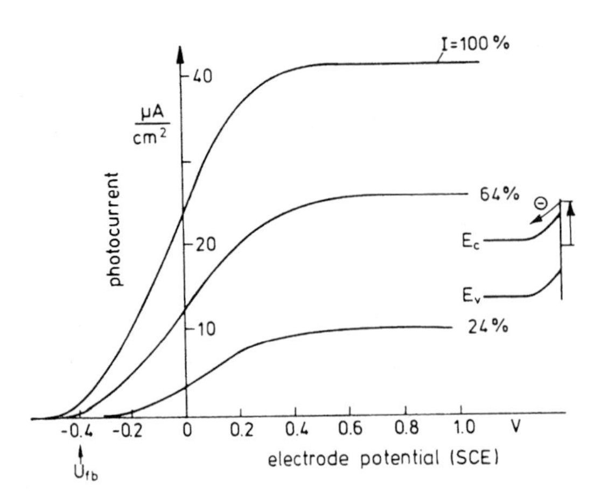

Fig. 10.11 Sensitized photocurrents vs. electrode potential for 10^{-3} M solution (pH 9) of rhodamine-B at an SnO_2 electrode (excitation wavelength: 570 nm) at different light intensities. SCE, saturated calomel electrode. (After ref. [9])

In the saturation range, all injected electrons are forced to move into the bulk and measured as a current. The current was found to be proportional to the light intensity. It should be emphasized that we have here a photocurrent caused by the injection of majority carriers, whereas an excitation within the semiconductor itself always leads to the transfer of minority carriers across a corresponding interface. The same results have been obtained with other semiconductor electrodes such as ZnO [28], TiO_2 and SnS_2 [21, 22].

Similar results were also obtained with p-type semiconductors, such as p-GaP (Fig. 10.12a) [20]. In this case, a cathodic sensitization process was observed which corresponds to a reduction of the excited dye via the valence band (see also excitation spec-

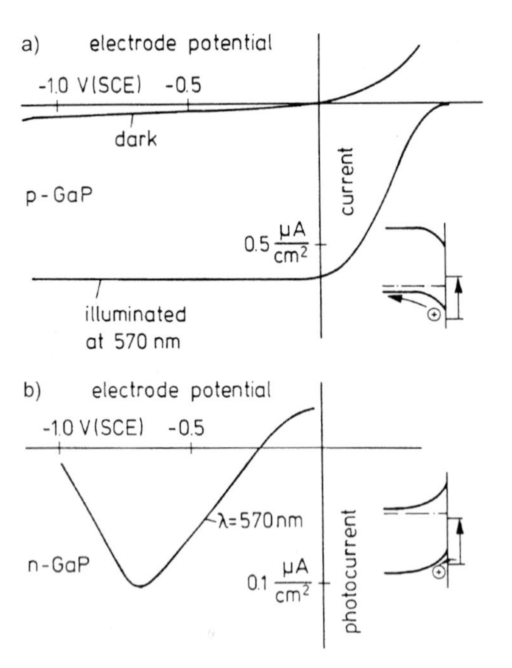

Fig. 10.12 Sensitized photocurrents vs. electrode potential for rhodamine-B at GaP electrodes in 1 M KCl solution (excitation wavelength: 570 nm). (After ref. [20])

trum in Fig. 10.8). Since the U_{fb} of p-GaP occurs at a very positive electrode potential, the cathodic photocurrent onset also occurs in a fairly positive potential range and remains constant far into the cathodic region. Here, the energy bands are bent downwards so that the holes injected into the valence band are pushed towards the bulk by the electric field across the space charge layer (insert, Fig. 10.12a). Here again, majority carriers are injected into the p-type semiconductor.

As already shown by the spectrum in Fig. 10.8, a cathodic photocurrent was also observed with n-GaP, its potential dependence being shown in Fig. 10.12b. Since the U_{fb} of n-GaP was found at a rather negative electrode potential, the energy bands are bent upwards at potentials positive of U_{fb}. Accordingly, the hole injected from the excited dye, cannot move into the bulk of the n-electrode (see insert of Fig. 10.12b) and are transferred back to the reduced dye molecule. The cathodic photocurrent shows a maximum, i.e. it decreases again at high cathodic polarization. This decrease is due to a reduction of the dye in the dark. The sensitization of the n-GaP electrode is, of course, a minority carrier process. The injected holes recombine with the electrons in the bulk and the current is carried by electrons. Many investigations have shown, however, that more reliable results could be obtained with majority carrier systems.

10.2.4 Sensitization Processes at Semiconductor Surfaces Modified by Dye Monolayers

As already discussed in the previous sections, sometimes problems arise because it is difficult to distinguish between electron transfer reactions between the semiconconductor electrode and adsorbed dye molecules or dye molecules in the solution. Therefore, various investigators have studied sensitization effects with monomolecular dye layers deposited on a semiconductor electrode and using electrolytes which are free from any more dye molecules. Two different methods have been applied for depositing a monolayer. The first method is the Langmuir–Blodgett technique [29], in which surfactant derivitized dye molecules are at first spread on water and then transferred to the electrode surface by dip-coating. In order to obtain a defined monolayer, the dye is mixed with arachidate molecules, so that finally a diluted monolayer is produced on the semiconductor surface. In the second method, the dye is transferred to the surface by a surface modification technique [30–32], in which a dye is covalently bonded to the semiconductor crystal.

With regard to the Langmuir–Blodgett technique, the sensitization of monomolecular layers at the SnO_2 electrode has been studied. Cyanine dyes with long hydrocarbon chains have been applied, mainly. Typical excitation spectra of the sensitized photocurrents, as obtained with a derivatized oxicarbocyanine, are shown in Fig. 10.13 [33]. These spectra were obtained by using an electrolyte containing no further dye. The photocurrent spectra show the typical dimer spectrum at higher dye concentrations as proved by absorption measurements with the same layer. In addition, the photocurrents decreased with increasing dye concentration. Obviously, the excitation energy was quenched; this was not investigated further. It should be mentioned finally that a stable photocurrent could only be measured in the presence of a suitable reducing agent such as thiourea, which reduced the oxidized dye to its original state (see also Section 10.2.5).

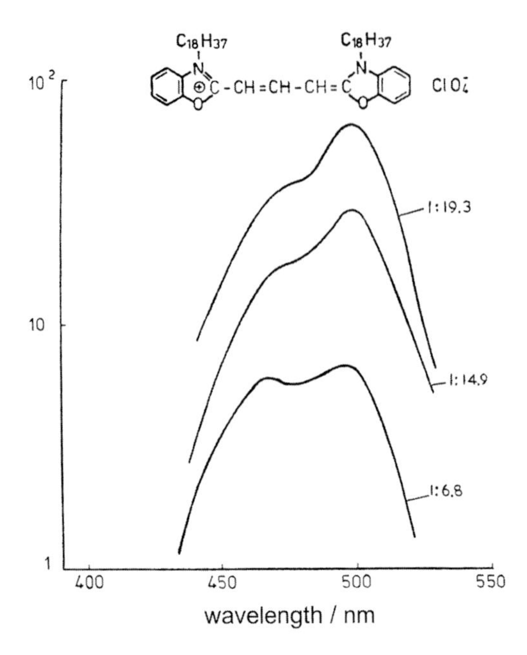

Fig. 10.13 Photocurrent action spectra for a monolayer of 3,3'-distearyloxa-carbcyanine on SnO_2, in 0.1 M KCl and 10 M allylthiourea as an electron donor; Parameters give the concentration ratios of the dye arachidic acid. (After ref. [33])

A very interesting example was published by Arden and Fromherz [34] who studied the performance of a multilayer carbocyanine dye structure. They used two types of cyanine dyes, A and D, incorporated into the multilayer structure as shown schematically in the upper part of Fig. 10.14. Dye A absorbs at around 420 nm and dye D around 370 nm. The photocurrent spectra presented in Fig. 10.14 exhibited the following features. If only one dye was used (curves I and II), the typical photocurrent was observed. In the case of curve I, the photocurrent was relatively low because the dye was separated from the electrode by an inert layer of arachidate molecules which inhibited the electron transfer process. If both dyes were used, in a structure given by III, the corresponding photocurrent exhibited a spectrum which was determined by both dyes. It should be noted that the spectrum of dye D was then much more pronounced compared with curve I. This effect was interpreted as energy transfer from D to A. The complete reaction scheme is then given by

$$D + h\nu \rightarrow D^* \tag{10.8a}$$

$$D^* + A \rightarrow D + A^* \tag{10.8b}$$

$$A^* \rightarrow A^+ + e^- \tag{10.8c}$$

$$A^+ + R \rightarrow A + R^+ \tag{10.8d}$$

in which R is the reducing agent in the solution.

The Langmuir–Blodgett technique has also been used for the deposition of chlorophyl monomolecular layers on SnO_2 electrodes. Such a system is of biological interest for studying photoeffects in an ordered structure, because the latter is a crucial factor in regulating energy migration between chlorophyls as well as promoting electron transfer processes in photosynthetic organisms. Honda and his co-workers have studied sensitization processes at SnO_2 in detail [35, 36]. Anodic photocurrents were

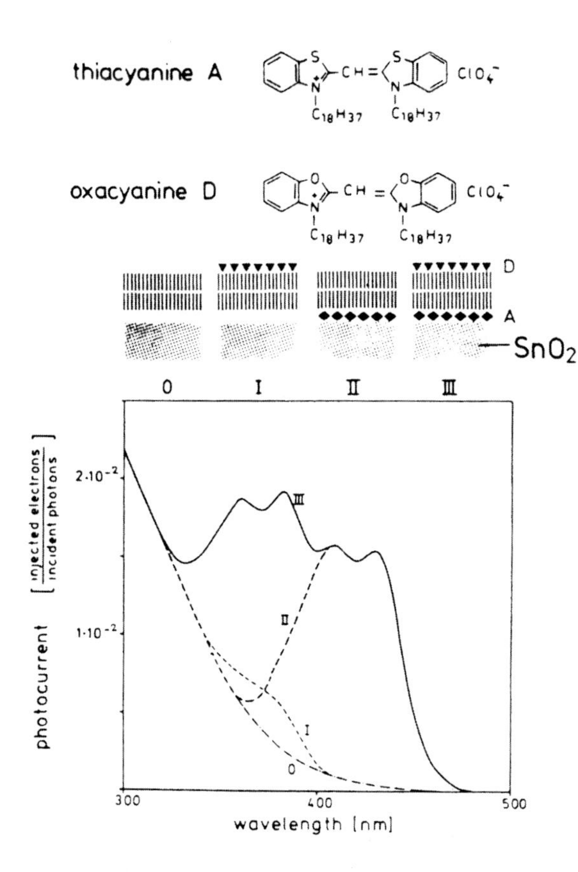

Fig. 10.14 Photocurrent action spectra for monomolecular layers of cyanine dyes at a SnO_2/In_2O_3 electrode at pH 10 with 0.5 M allylthiourea. Molar ratio of dye to arachidic acid, 1:5. The parameters for different curves are explained by the structure of dye layers shown at the top of the figure. (After ref. [34])

observed with chlorophyl-a and -b, indicating electron injection into the conduction band of SnO_2. Since the conduction band edge of SnO_2 occurs in neutral solutions at –0.35 eV vs. SCE, and the excited singlet and the triplet donor levels of chlorophyl-a are located around –1.3 eV and –0.75 eV vs. SCE, respectively, the conduction band is lower than the triplet and the excited singlet donor levels. Accordingly, it is not possible to distinguish between a singlet and a triplet mechanism [36]. Most of the sensitization experiments with chlorophyl monolayers on SnO_2 electrodes were performed with solutions again containing a reducing agent in order to keep the current constant. It is interesting to note, however, that a small sensitization current was also found without addition of a reducing agent. This result was tentatively interpreted as an electron transfer from H_2 to the oxidized chlorophyl [35]. A similar observation was made with monolayers of a Ru complex deposited on SnO_2 [37].

In contrast to cyanine dyes and Ru complexes, fairly concentrated chlorophyl monolayers could be deposited on SnO_2. In the latter case, the quantum yield of the photocurrent was determined as a function of the molar ration of chlorophyl and stearic acid (concerning the definition of quantum yield see Section 10.2.5). An optimum was obtained for a 1:1 ratio, whereas the quantum yield dropped by a factor of nearly 3 for a pure chlorophyl layer [35]. This effect was interpreted as concentration quenching which cannot be further discussed here. The same authors have also investigated chlorophyl multilayers, all of which were deposited using the Langmuir–

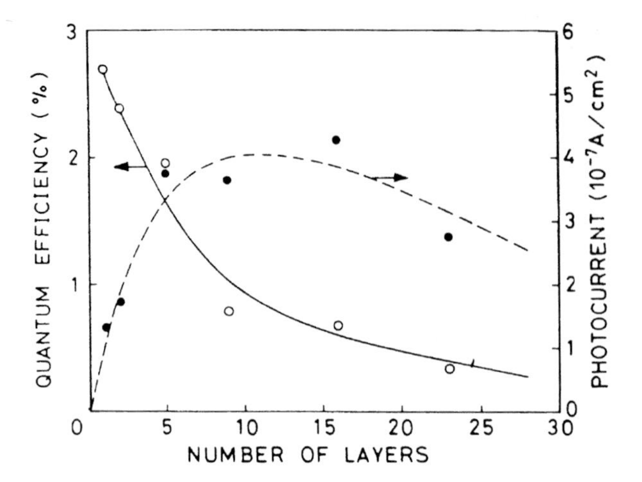

Fig. 10.15 Dependence of quantum efficiency (solid curve) and photocurrent (dashed curve) at 675 nm on the number of chlorophyl monolayers on SnO_2 at 0.1 V (SCE). Electrolyte, pH 6.9 + 0.05 hydroquinone. (After ref. [38])

Blodgett technique [38]. Since a single chlorophyl monolayer is about 14 Å thick, precise control of the thickness was possible. As shown in Fig. 10.15, the magnitude of the anodic current increased with the number of monolayers until it reached a maximum around 10 layers. The quantum yield, however, was very low and decreased with the number of layers. The latter phenomenon reflects enhanced quenching of the excitation energy and, perhaps, an increase in the electrical resistance of the multilayer with an increasing number of layers. The low quantum efficiency is probably related to back-reactions at the highly doped SnO_2 electrode, as will be described in Section 10.2.6. The photoelectrochemical properties of multilayers have also been studied with much better defined semiconductor electrodes. Here highly ordered phthalocyanine monolayers have been deposited on single crystalline SnS_2 electrodes in a vacuum by using an organic molecular beam epitaxy (MBE) [39]. These investigations have also shown that the quantum efficiency declined with more than one monolayer.

As already mentioned, a dye monolayer can also be deposited on a semiconductor by means of a surface modification technique. In particular, semiconductor oxides can

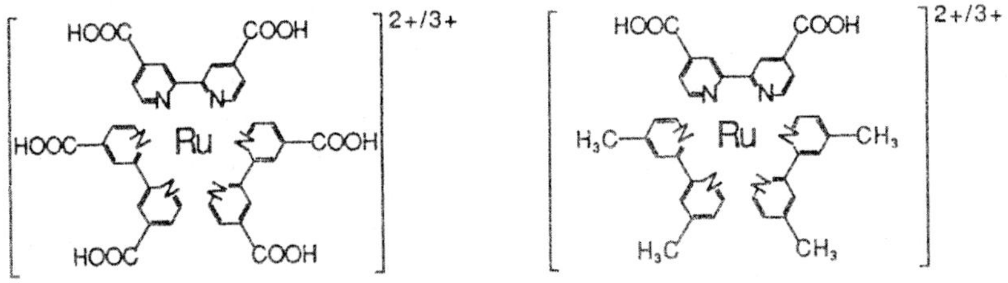

a) Ru(4,4'dicarboxy-2,2'bipyridine)$_3^{2+/3+}$ b) Ru(4,4'dimethyl-2,2'bipyridine)$_3^{2+/3+}$

Fig. 10.16 Molecular structure of $Ru(bipy)_3^{2+}$ complexes with several carboxy groups, suitable for anchoring the dye on oxide semiconductors

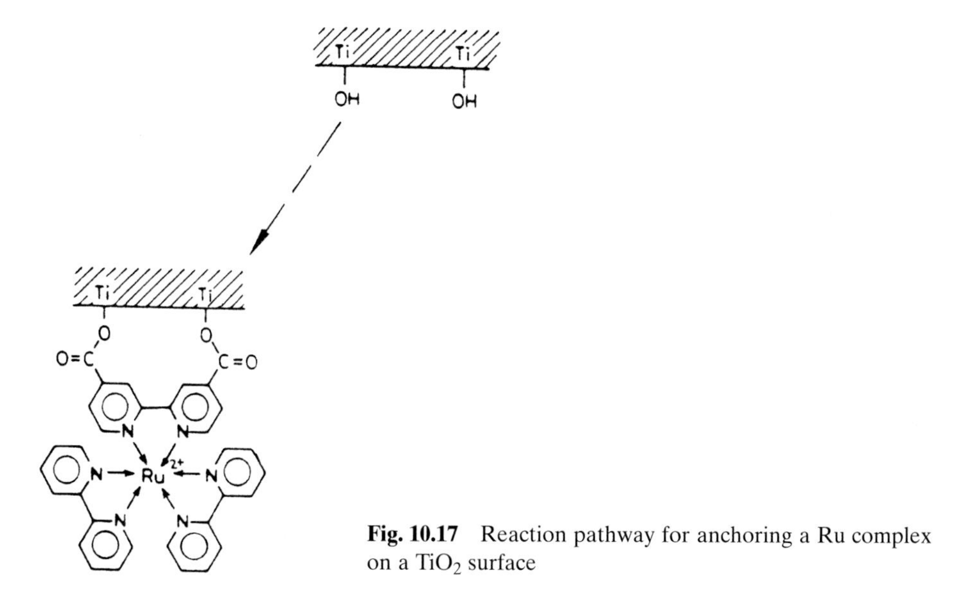

Fig. 10.17 Reaction pathway for anchoring a Ru complex on a TiO_2 surface

be fairly easily modified, as was first shown by S.Anderson et al. for Ru–bipyridyl complexes [30–32]. Suitable compounds include complexes where the 2,2'-bipyridine groups have carbonyl groups either only at one or at all bipyridine rings, as shown in Fig. 10.16. Such a molecule can form an ester bond between the oxide surface and the carbonyl group, as illustrated for one bipyridine group in the reaction scheme given in Fig. 10.17. This technique has been widely applied for various derivatives of the Ru complex, not only on TiO_2 [32, 40] but also on SnO_2 [41]. The excitation spectrum as found with Ru(4,4'-dicarboxy-bipy)$_3^{2+}$ on TiO_2 single crystals is considerably red-shifted with respect to the absorption spectrum of the Ru complex measured in H_2O (Fig. 10.18) [42]. Accordingly, there is a rather strong interaction between TiO_2 and the dye. In addition, there is only a weak minimum at the low wavelength side of the photocurrent maximum and there is a long tail at the red end of the excitation spectrum. These observations can be taken as an indication for two overlapping spectra, i.e. one is the typical dye excitation spectrum and the other a structureless absorption edge. Such behavior may be explained by two transitions: one within the dye molecule, the other a direct transition from the dye ground state to the conduction band, as indicated by the arrows in Fig. 10.19. Such an interpretation has also been suggested by other authors, for instance by Goodenough et al. for small coverages of a Ru-bipy-derivative anchored on TiO_2 [30]. In the latter case, there was even no maximum at all found in the excitation spectrum [30, 31]. The possibility of a direct electron transition from a dye (oxasine) directly to the conduction band of a semiconductor (SnS_2) has also been suggested by Parkinson [21].

Sensitization by different Ru complexes has been studied, mainly by using nanocrystalline TiO_2 layers (see also 10.2.7 and Section 11.1.3). It is not clear, however, whether the application of a Ru complex with only one bipyridine derivatized by two carboxyl groups (Fig. 10.16b) is advantageous, or whether the use of a completely derivatized system (Fig. 10.16a) would be more successful.

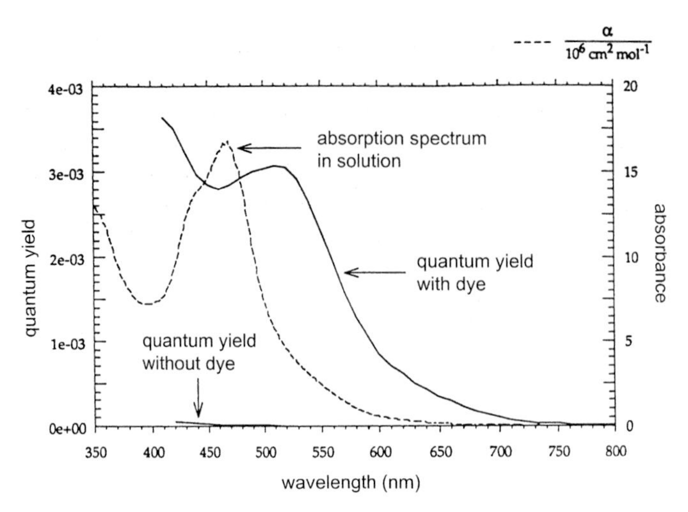

Fig. 10.18 Quantum yield of the sensitization current vs. wavelength for a monolayer of Ru(4,4'-dicarboxy-bipy)$_3^{2+}$ anchored on the surface of a single crystalline TiO_2 electrode in aqueous solution (pH 2.4), using iodide as an electron donor. (After ref. [42])

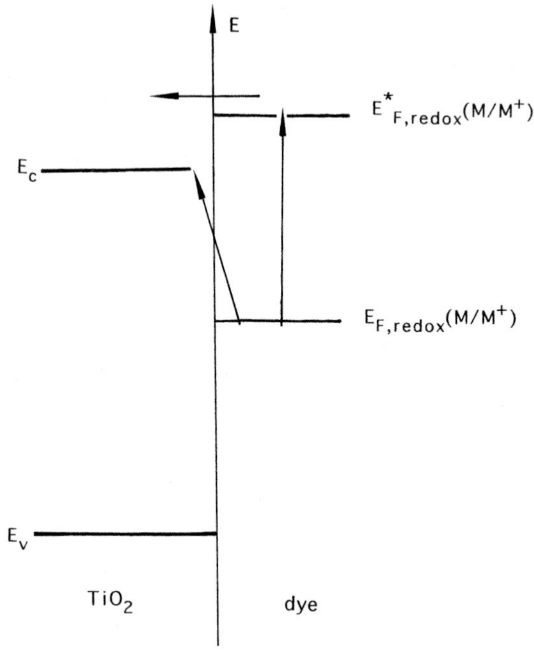

Fig. 10.19 Transitions and electron injection from a Ru complex, anchored on the surface of a single crystalline TiO_2 electrode, into the conduction band (for details see text)

10.2.5 Quantum Efficiencies, Regeneration and Supersensitization

As already mentioned in Section 10.2.2, an electron transfer occurs only from an excited molecule adsorbed at the electrode surface. Since the exchange of oxidized molecules by new dye molecules in solution is rather slow, the sensitization decreases with time (Fig. 10.20). The relaxation time depends on the light intensity and on the rate at which the oxidized molecules are exchanged for dye molecules from the solution. Such a decrease, however, can be avoided by adding a suitable reducing agent (electron donor D) to the electrolyte which is capable of reducing the oxidized species of the adsorbed dye, i.e. regenerating it to its original state. The reactions involved are given by

$$S_{ad} + h\nu \rightarrow S^*_{ad} \tag{10.9}$$

$$S^*_{ad} \xrightarrow{k_F} S_{ad} \tag{10.10}$$

$$S^*_{ad} \xrightarrow{k_{tr}} S^+_{ad} + e^- \tag{10.11}$$

$$S^+_{ad} + D \xrightarrow{k_g} S^+_{ad} + D^+ \tag{10.12}$$

At a fairly early stage of sensitization experiments at the semiconductor–electrolyte interface, it was found that the stationary anodic sensitization current at n-type semiconductor electrodes often undergoes a remarkable increase upon addition of a reducing agent [43]. This phenomenon was frequently termed supersensitization by analogy with supersensitization in photographic systems [44]. This effect was not only observed with ZnO [43] but also with CdS [45] and SnO_2. Reducing agents were applied which were known from photochemical studies to be capable of reducing excited dyes. Originally, supersensitization was interpreted as the reaction

$$S^*_{ad} + D \rightarrow S^-_{ad} + D^+ \tag{10.13}$$

$$S^-_{ad} \rightarrow S_{ad} + e^- \tag{10.14}$$

Accordingly, the excited sensitizer was reduced by an electron donor before the electron transfer occurred. This process competes with the recombination given by Eq. (10.10). Such a supersensitization was never really proved. It is assumed that the

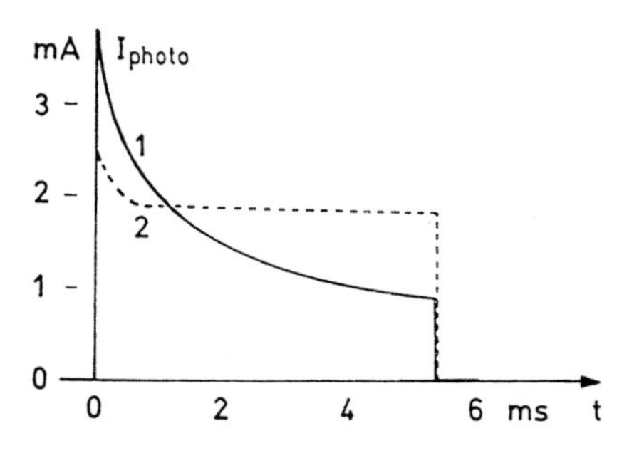

Fig. 10.20 Photocurrent during light pulse (514 nm) at a ZnO electrode in a 10^{-4} M aqueous solution of rhodamine-B, in the absence (1) and in the presence (2) of hydroquinone. (After ref. [28])

increase of the stationary photocurrent upon addition of a reducing agent is entirely due to the regeneration of the adsorbed dye (Fig. 10.20).

The quantum efficiency ϕ defined as the ratio of the number of electrons transferred across the interface, and of the number of photons absorbed by the adsorbed dye layer, is not easy to determine. This is on account of the problems of measuring the light absorption by one monolayer or separating it from the absorption by the dye solution. Therefore, many scientists prefer to give values of a quantum yield, ϕ_{in}, defined as the number of injected electrons per incident photons which is easy to measure. These two quantum yields are related by the equation

$$\phi_{in} = \phi(1 - 10^{-\epsilon'(\lambda)\gamma}) \tag{10.15}$$

where the term in brackets gives the ratio of absorbed photons. In this equation, ε' is given in units of $cm^2\ mol^{-1}$ (i.e., $\varepsilon' = 10^3\varepsilon$) and Γ as the number of mols per cm^2, i.e. $\Gamma = \chi(6.6 \times 10^{23} a_{dye})^{-1}\ mol\ cm^{-2}$, in which a_{dye} is the area covered by one dye molecule, and χ is the ratio of effective to geometric surface area. Assuming for instance that $a_{dye} = 1\ nm^2$ and $\chi = 1$, one has $\Gamma = 1.5 \times 10^{-10}\ mol\ cm^{-2}$, and with $\varepsilon' = 4 \times 10^7\ cm^2\ mol^{-1}$ one obtains $\phi_{in} = 0.014\phi$. Accordingly, if the quantum yield would be unity, ϕ_{in} is only about 1.4 % for a monolayer.

Quantitative investigations by Parkinson [21] and Spitler [47] have shown that such high ϕ_{in} values and consequently quantum yields of nearly unity were only obtained with two-dimensional dichalcogenide electrodes, such as $MoSe_2$, WS_2, WSe_2 and SnS_2, with their perfect van der Waals surfaces in contact with a dye solution (e.g. oxazine). As already mentioned in Section 10.2.2, here a dye monolayer was formed on the electrode surface at rather low dye concentrations in solution. With oxide semiconductors in contact with dye solutions, much lower values were obtained which usually did not exceed quantum efficiencies of a few percent ($\phi \leq 5\,\%$) [48]. According to the photocurrent–potential curves (Fig. 10.11), one would expect at first sight that in the range of saturated photocurrents all injected electrons would be driven across the space charge layer and detected as current in the external circuit. Since the luminescence is quenched also, one has to assume that a considerable fraction of electrons transferred from an excited dye molecule into the conduction band of an oxide such as TiO_2 and ZnO, are efficiently trapped in surface states from where the electrons are

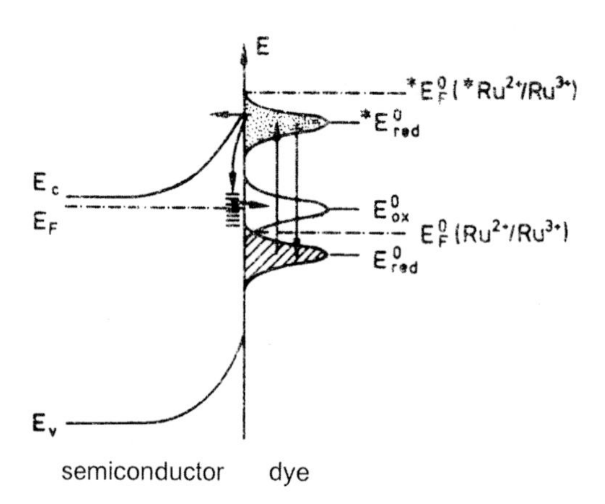

Fig. 10.21 Quenching of photocurrent by electron transfer via surface states. (After ref. [16])

transferred back to the oxidized dye, as illustrated in Fig. 10.21 for SnO_2 as a semiconductor and an Ru complex as a dye [8, 16]. Since the dichalcogenide electrodes exhibit nearly perfect van der Waals surfaces, and therefore a very low density of surface states, much higher quantum yields can be attained [22]. It is interesting to note that with single crystalline TiO_2 electrodes also, quantum yields in the order of 30–50 % were obtained if the electrode surface was modified by a Ru complex via an ester bond (see Fig. 10.17) [42]. Accordingly, the density of surface states was reduced by making additional bonds between the oxide surface and the bipyridine groups of the Ru complex.

10.2.6 Kinetics of Electron Transfer between Dye and Semiconductor Electrode

According to the results presented in the previous sections, it is quite clear that an electron or a hole can be efficiently transferred from an excited molecule adsorbed at the electrode surface into the conduction or valence band, respectively. The dynamics of electron transfer have been measured quantitatively with oxazine adsorbed at atomically smooth SnS_2 electrodes, by studying the fluorescence decay of the dye [48]. The adsorbed oxazine was excited by dye laser pulses using a system with a response time of about 40 ps. The fluorescence of oxazine on SnS_2 showed strong quenching when compared with oxazine on a tape as a reference, as shown in Fig. 10.22. The decay curves were normalized at their maximum, i.e. the fluorescence signal found on SnS_2 was multiplied by a factor of 125. When excited at 640 nm, the sensitized SnS_2 showed a very weak fluorescence with a spectrum similar to the oxazine on tape. This indicates that this fluorescence originated from molecules which were not strongly coupled to the surface, since the red shift in the emission spectrum associated with surface adsorption was not present (compare also with the sensitization spectrum in Fig. 10.9). However, when the oxazine was excited near the absorption maximum for the adsorbed dye, no fluorescence was observed. The decay of the oxazine on tape samples was a single-exponential decay with a lifetime of 2.6 ns. The decay for the sensitized semiconductor was more complex and was fitted by two exponential curves: the shorter was an instrument-limited decay of 40 ps, the longer lifetime was similar to that found on tape.

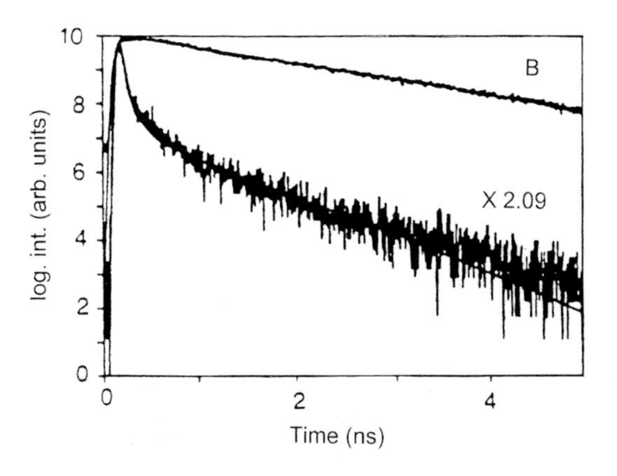

Fig. 10.22 Fluorescence decay of oxazine adsorbed on SnS_2. This measurement was performed in a dry atmosphere after the dye had been adsorbed. The upper curve (smooth line) was obtained with oxazine adsorbed on tape. For comparison, the signals obtained with SnS_2 were multiplied by 125. For details see text. (After ref. [48])

Since the latter virtually disappeared at low coverages, this component was ascribed to the weakly adsorbed dye molecules.

The shorter decay (instrument-limited) was attributed to the fluorescence from adsorbred molecules strongly coupled to the surface, indicating that the electron transfer rate is much faster than 40 ps. In order to get a more accurate value for the injection rate, the integrated fluorescence intensities of the fast component of the sensitized semiconductor decay curve were compared with the intensity of the oxazine on the tape reference sample [48]. This analysis yielded a quenching factor of about 10^5. From this an electron injection rate of 3×10^{13} s^{-1}, corresponding to an electron transfer time of around 40 fs, was obtained.

The same authors also investigated the bleaching and recovery of the ground state of the dye after a 1-ps laser excitation [48]. Since these experiments were performed under open-circuit conditions, all the electrons transferred from the excited dye into the conduction band of SnS_2, had to return to the oxidized dye. The corresponding ground state recovery was found to occur within about 10 ps. Other investigations on transfer rates will be discussed in Section 10.2.7.

Frequently, semiconductor electrodes of relatively high doping have been used in sensitization experiments. In this case, electrons injected from the excited dye into the conduction band of a semiconductor electrode can tunnel back through the space charge layer to the oxidized dye molecule, as illustrated for a SnO_2 electrode in Fig. 10.23a and b. The thickness of a space charge layer, d_{sc}, can be calculated according to Eq. (5.31) and one obtains, for instance, for a donor density of 10^{20} cm^{-3}, $d_{sc} = 30$ Å at $U_E = 1$ V; i.e. a barrier through which electrons can easily tunnel. The latter process

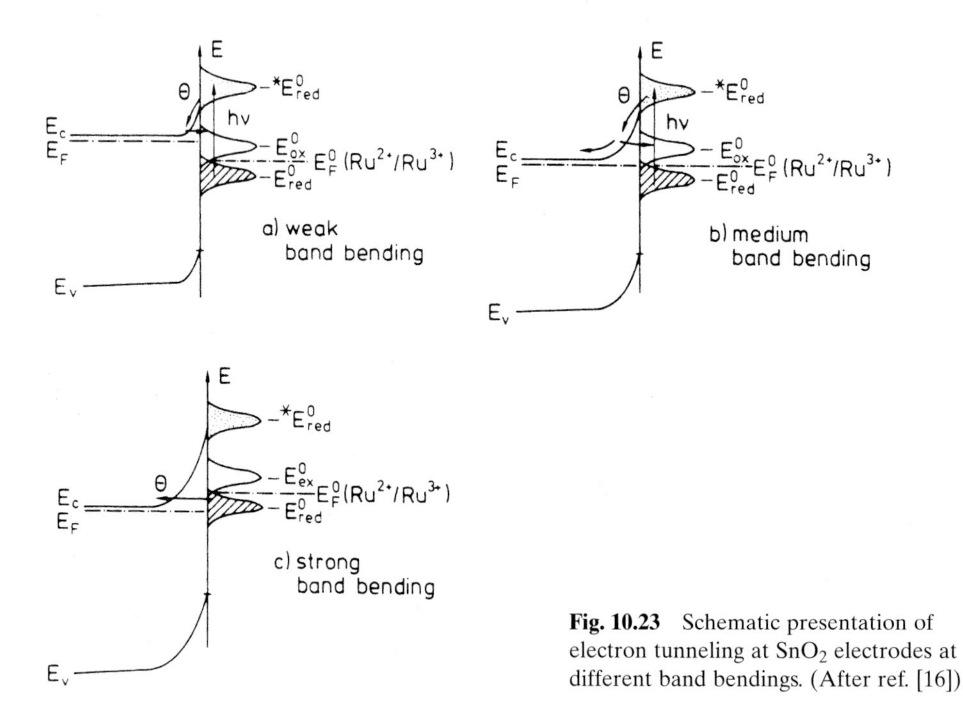

Fig. 10.23 Schematic presentation of electron tunneling at SnO_2 electrodes at different band bendings. (After ref. [16])

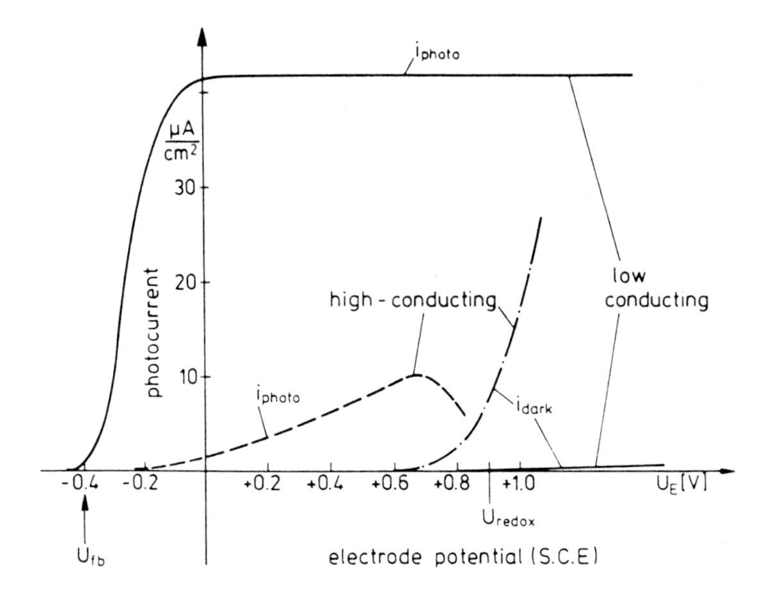

Fig. 10.24 Current–potential dependence for differently doped SnO_2 electrodes in aqueous solutions of $Ru(bipy)_3^{2+}$ at pH 9; excitation wavelength, 488 nm. (After ref. [16])

leads to a quenching of the photocurrent, as shown for highly doped SnO_2 in comparison with weakly doped SnO_2 in Fig. 10.24 [16]. The small photocurrent even passes a maximum at a potential at which the dark current sets in. This dark current at heavily doped SnO_2 occurs just at the redox potential of $Ru(bipy)_3^{2+}$ which was used as a dye. It corresponds to the oxidation of the dye, which is achieved by tunneling of electrons in the opposite direction, i.e. from the dye into the conduction band as illustrated in Fig. 10.23c. This tunneling is, of course, only possible at potentials at which the conduction band occurs below the redox potential of the dye. The sensitization current decreases in this range because fewer dye molecules are available at the surface.

Similar effects have also been found with heavily doped ZnO electrodes [49]. In this case, the kinetics have been studied in detail by using light pulses of a duration of some milliseconds for dye excitation. The corresponding transient behavior of the photocurrents for moderately and heavily doped ZnO electrodes is shown in Fig. 10.25. Since a relatively strong light source (argon laser, 514 nm) and an electrolyte without any reducing agent were used, a slight decay of the photocurrent within 5 ms was found for moderate electrodes because a considerable amount of the adsorbed dye was oxidized and could not be replaced sufficiently quickly by dye molecules from the solution. It should be noted that the photocurrent decreased immediately to zero when the light was turned off. In the case of a heavily doped electrode, the photocurrent decreased considerably more quickly under the same excitation conditions. Here, a cathodic current was found as soon as the light was switched off, which finally decayed to zero within about 1 ms (lower part of Fig. 10.25). This cathodic dark current was related to tunneling of electrons from the conduction band to the oxidized molecule (Fig. 10.23b). Following the treatment of Pettinger et al., the kinetics can be described as follows [49].

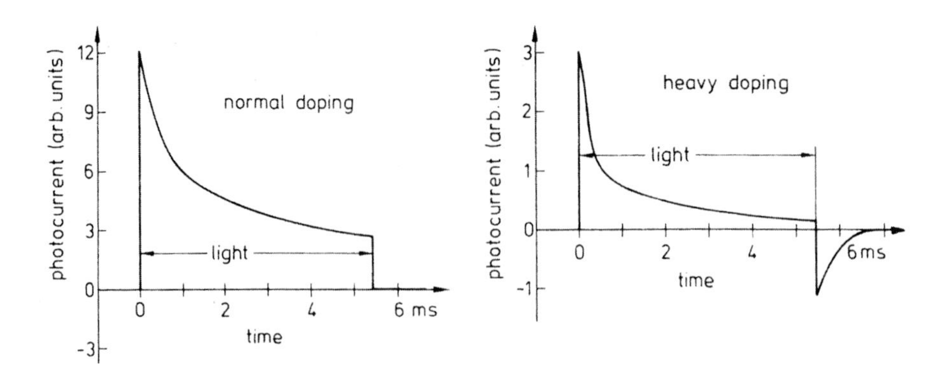

Fig. 10.25 Relaxation of sensitization current at ZnO electrodes in aqueous solutions of rhodamine during laser pulse, at pH 1; $U_E = 0$ V (SCE). (After ref. [49])

The decay rate of dye molecules adsorbed at the surface is given by

$$\frac{dc_{a,sd}}{dt} = -k_{tr}\alpha I c_{s,ad} + k_T c_{ox,ad} + k_{ex} c^+_{s,ad} \tag{10.16}$$

in which k_{tr} is the rate constant for the electron injection into the conduction band, k_T the rate constant for the tunneling process and k_{ex} determines the exchange of oxidized dye molecules adsorbed at the surface with dye molecules in the electrolyte. Since the dye concentration in the solution is constant we have $k_{ex} = k^0_{ex} c_s$. If $c_{s,ad}$, c_s and $c^+_{s,ad}$ represent the concentration of the adsorbed dye molecules, of the dye molecules in solution and of the adsorbed oxidized dye molecules, respectively, and α = absorption coefficient, I_0 = light intensity, the total current is then given by:

$$\begin{aligned} j(t) &= j^+_{ph} + j_T \\ &= e k_{tr} \alpha I_0 c^0_s \frac{k_2}{k_1} (\exp -k_1 t) + \frac{k_{ex}}{k_1} \end{aligned} \tag{10.17}$$

in which

$$(k_{tr} \alpha I_0 + k_T + k_{ex}) = k_1 \tag{10.18}$$

$$(k_{tr} \alpha I_0 + k_T) = k_2 \tag{10.19}$$

and

$$c^0_s = c_{s,ad} c^+_{s,ad} \tag{10.20}$$

It should be mentioned that direct recombination from the excited molecule to the ground state, and also the ground state recovery due to recombination and transfer of electrons via surface states (compare with Fig. 10.21), have been neglected in this derivation. According to Eq. (10.17), the exponential decay is determined by k_1. Since in heavily doped electrodes, k_1 includes the tunneling rate constant k_T, the decay should be faster than for moderate dopings as found experimentally (Fig. 10.25). After the light is switched off, the total current is given by ($j^{+}_{\mathrm{ph}} = 0$)

$$j(t) = ek_{\mathrm{T}} c_{\mathrm{ox,ad}}(t = t_{\mathrm{e}}) \exp\left[(k_{\mathrm{T}} + k_{\mathrm{ex}})(t - t_{\mathrm{e}})\right] \tag{10.21}$$

for $t > t_{\mathrm{e}}$.

Here t_{e} is the time at which the light was turned off. It is clear from the above equation that the relaxation of the cathodic dark current is determined by k_{T} and k_{ex}.

10.2.7 Sensitization Processes at Nanocrystalline Semiconductor Electrodes

During the 1990s, many research groups became interested in sensitization processes at nanocrystalline semiconductor electrodes because of the possible applications in photoelectrochemical solar cells (see Section 11.1.1). The first experiments were performed by O'Reagen and Grätzel who prepared nanocrystalline TiO_2 layers (anatase modification) by depositing TiO_2 particles from colloidal solutions on a conducting glass [50]. Electronic contact between the particles is made by brief annealing at 450 °C. The internal surface area of the film is controlled by the particle size and the pores. Huge surface areas can be obtained with this sponge-like structure; assuming a close packing of 15-nm sized spheres to a 10 μm thick film, then a 2000-fold increase in surface area can be expected. Meanwhile various techniques have been applied for making nanocrystalline layers which cannot be described here (see also Section 9.4). The dye is usually adsorbed by immersing the nanocrystalline electrode in a corresponding dye solution. In most experiments the semiconductor surface has been modified by anchoring a Ru complex to the particles (Fig. 10.26).

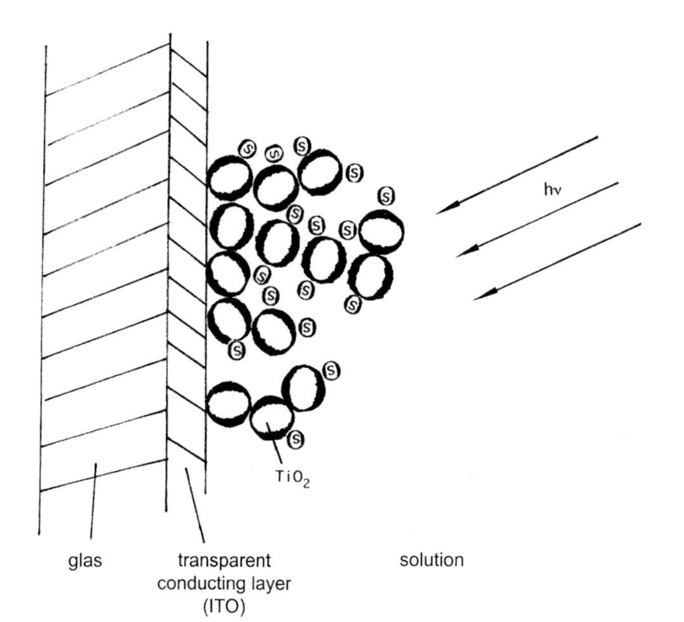

Fig. 10.26 Schematic presentation of the structure of a nanocrystalline TiO_2 layer on a transparent electrode; the surface of the TiO_2 particles is modified by a sensitizer (s)

Many interesting experiments became possible because the absorption is high even if all the dye molecules are anchored to the particle surface. For instance, optical densities of nearly unity are obtained at the wavelength of the absorption maximum for a film thickness of about 1 μm. Various research groups have applied laser flash techniques in order to obtain information on the intermediate states involved in the electron transfer processes and on the rate of electron injection. However, there is considerable controversy and confusion in the literature concerning the timescale and the nature of the primary electron transfer from a Ru complex into the conduction band. Therefore, only few essential experimental results are presented here. For instance, Hannappel et al. reported on laser flash experiments performed with TiO_2 nanocrystalline layers loaded with a Ru complex (here: Ru(II)cis-di-(isothiocyanato)bis(2,2'-bipyridyl-4,4'-dicarboxylate), abbreviated as N3) [51]. After anchoring the dye to the TiO_2 surface and carefully rinsing the system in ethanol, these authors performed laser flash experiments with the electrode in ultrahigh vacuum (UHV). The dye was excited by a short laser pulse (75–150 fs halfwidth of the laser pump pulse) in a wavelength range where it absorbs light. The bleaching of the S_0–S_1 transition and the excitation of free electron absorption within the TiO_2 was followed by a light probe signal as illustrated in Fig. 10.27. The result of the bleaching as measured as a transmission increase after various delay times, is shown in Fig. 10.28b (open symbols). For comparison, the absorption spectra of the dye N3 and its oxidized form, both dissolved in ethanol, are given in Fig. 10.28a. Additional transient absorption of the excited state of the N3 dye could not be detected. From this result the authors concluded that the lifetime of the excited state of the anchored dye is shorter than 25 fs; at least this is the shortest time constant which the authors could extract from a deconvolution fit.

This interpretation is supported by the observation that a transient absorption was detected in the near-infrared (1100 nm) [51]. The analysis of these curves yielded an

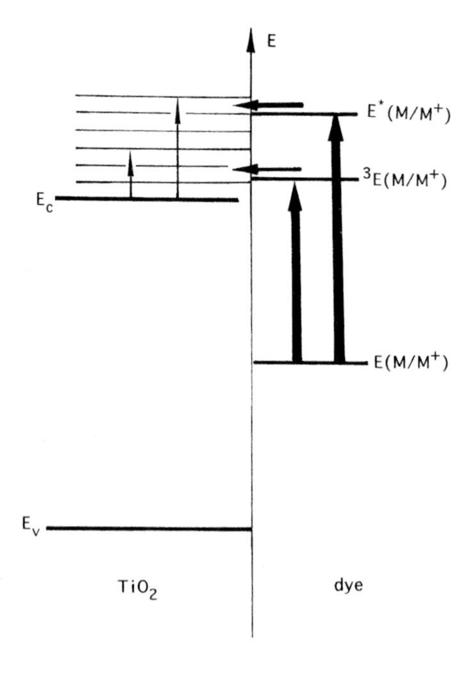

Fig. 10.27 Excitation and electron injection from a dye into TiO_2 particles (nanocrystalline layer); large arrows, excitation of the dye by a pump laser; thin arrows correspond to an excitation within the conduction band by an additional light beam for probing the injected electrons

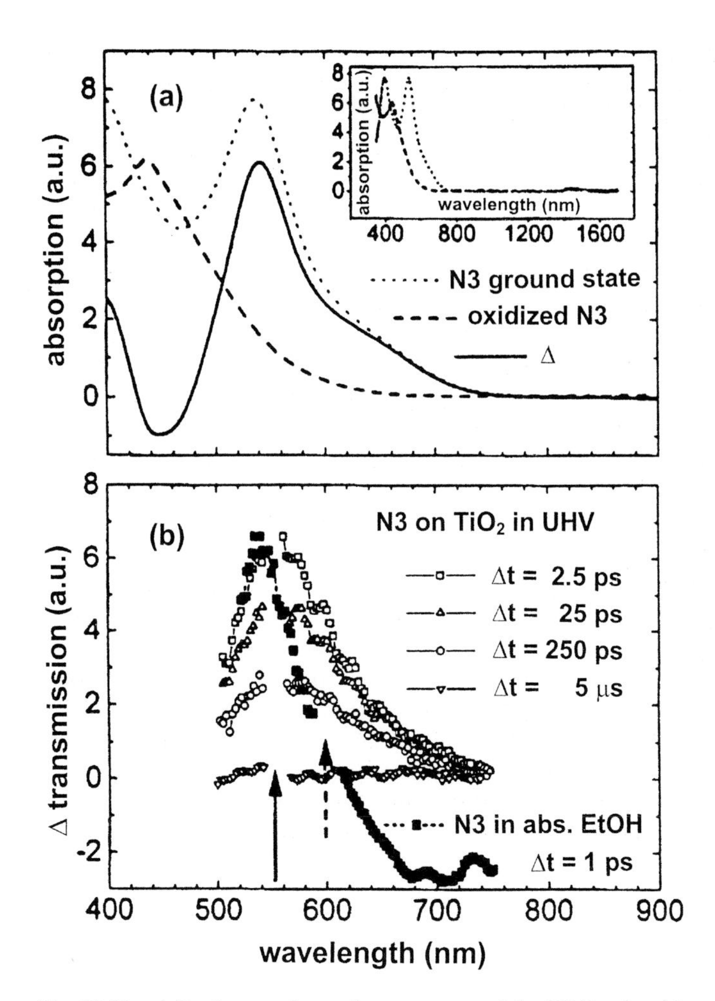

Fig. 10.28 a) Stationary absorption spectrum of the N3 Ru dye (dotted curve), of the oxidized N3 dye (dashed curve), both in absolute ethanol, and the difference of these curves (solid curve). b) Transient transmission signal due to N3 dye molecules anchored to the colloidal TiO_2 electrode in ultrahigh vacuum. The parameter is the time delay between pump pulse and probe pulses. (The results were nearly the same for electrodes immersed in the solution.) The absorption spectrum of the excited triplet state of the N3 dye, measured in ethanol, is shown for comparison (solid squares). (After ref. [51])

apparent rise time which was also <25 fs. This transient absorption was assigned to intraband absorption of electrons injected from the excited dye into the conduction band of TiO_2 (free carrier absorption, see Chapter 1) which is an important result.

As already mentioned above, these investigations were performed in UHV which had the advantage that side-reactions were excluded. Some other investigations have been carried out with the same dye (N3) anchored to TiO_2 in ethylene carbonate/propylene carbonate solutions [52]. Tachibana et al. also predicted a fast electron injection (<150 fs) [52]. They did not measure the intraband absorption, but detected an intermediate which they related to the absorption of the excited dye. There are still some discrepancies between the results obtained by the two groups. However, the results and interpretation of Hannapel et al. have since been supported by investigations carried out by Ellingson et al. [53]. These authors performed similar investigations to those published by Hannapel et al., but with electrodes which were in contact with an electrolyte, i.e. by probing the time-resolved IR absorption of the injected electrons. They also observed IR transitions in the range of 4–7 μm, due to the absorption of injected electrons within TiO_2 (Fig. 10.29). In addition, these authors also used ZrO_2 insulator electrodes on which the same dye was deposited. As expected, in the latter case no free electron absorption was found (see also Fig. 10.29). This is an additional proof for the free carrier absorption model.

The ultrashort time constants of <25 fs for electron transfer, found by Hannappel et al., indicate a different reaction mechanism for the electron transfer process [53]. The electron transfer occurs more quickly than the vibrational relaxation within the dye molecule, i.e. the electron is transferred from any excitation level directly into a corresponding energy level in the conduction band. Because of this strong coupling, this electron transfer cannot be described by the simple Marcus–Gerischer model which is only valid for comparably weak interactions.

Other research groups have reported on transient formation and luminescence decays of nanocrystalline TiO_2 and SnO_2 electrodes, modified by similar Ru complexes, in the pico- and nanosecond range [54–56]. The transient corresponds to a triplet–triplet absorption and the luminescence to a triplet→singlet transition of the dye. In these cases, it has been clearly shown that electron transfer into the semiconductor leads to corresponding bleaching of the triplet absorption and to quenching of the luminescence, or to a faster decay of the emission. Accordingly, the dye molecule

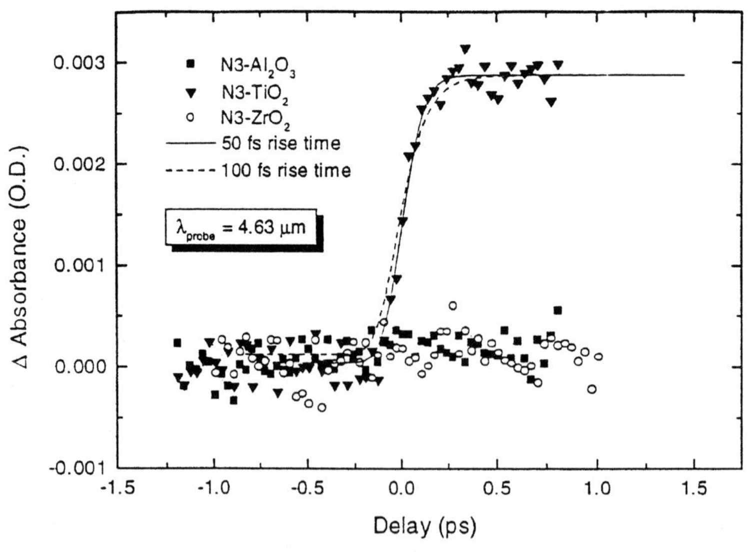

Fig. 10.29 Time-resolved infrared absorption data for a probe wavelength of 4.63 μm, following photoexcitation (400 nm) of the N3 Ru dye anchored on different oxide surfaces. (Duration of pump laser pulse, 100 fs). (After ref. [53])

relaxed to its triplet state before any electron transfer took place. These rather slow processes (nanosecond range) may be assigned to an electron transfer from an excited molecule (triplet state) at larger distances from the electrode. Interestingly, these slow effects depend on electrode potential. For instance, Kamat et al. have shown for SnO_2 electrodes that the triplet absorption was only quenched at anodic bias [56]. Qualitatively, quenching and photocurrent show a similar potential dependence.

Further studies on the mechanism of electron transfer from an excited dye to a nanocrystalline semiconductor are necessary for a full understanding of the intermediates and products involved in these processes.

10.3 Comparison with Reactions at Metal Electrodes

Sensitization effects have also been studied with metal electrodes. No photocurrents were observed with dye solutions without any reducing agent [15]. This result can be easily understood in terms of an energy diagram as given in Fig. 10.30. In this figure, the empty and occupied energy states in the metal are shown and the redox potentials of an excited dye, $^*E_F(M^*/M^+)$ for oxidation, and $^*E_F(M^*/M^-)$ for reduction are given. It is quite clear from this scheme that two processes can occur at a metal electrode: first, an electron transfer from the excited dye to empty states of the metal (oxidation) and, secondly, an electron transfer from occupied levels in the metal to the excited dye (reduction), as indicated by arrows. Since the corresponding currents just cancel, no photocurrent would be expected. One also could describe the behavior by energy transfer from the excited dye to the metal which leads to quenching of the excitation energy, as demonstrated with dye monolayers by Kuhn [57].

In some cases, however, an anodic sensitization current was observed if a reducing agent was added to the dye solution contacting the metal electrode. This has been found, for instance with rhodamine-B as a dye and allylthiourea as a reducing agent [58]. The reaction is given by

$$S + h\nu \rightarrow S^* \tag{10.22}$$

$$S^* \rightarrow {}^3S \tag{10.23}$$

$$^3S + D \rightarrow S^{\bullet -} + D^+ \tag{10.24}$$

in which ^3S is the triplet and D the electron donor. These processes occur in the bulk of the solution. Since the lifetime of the radical $S^{\bullet -}$ can be relatively large, it can diffuse to the electrode where it is re-oxidized according to

$$S^{\bullet -} \rightarrow S + e^- \tag{10.25}$$

The corresponding current value is mainly determined by the back-reaction between $S^{\bullet -}$ and D^+. This kind of photocurrent is based on photochemical effects in solution and is also known as the photogalvanic effect, already reported in 1961 [59]. Very large photocurrents have been obtained by using the leuco dye (SH_2) as an electron donor.

Then we have instead of Eq. (10.24)

$$^{3}S + SH_2 \rightarrow 2S^{\bullet -} + 2H^+ \qquad (10.26)$$

Here the radical is the only product of the reaction. Since the radicals may exhibit a large lifetime in O_2-free solution, many of them reach the electrode which leads to a large current [58].

10.4 Production of Excited Molecules by Electron Transfer

As studied by Bard and co-workers, the chemiluminescence of $Ru(bipy)_3{}^{2+}$ can be observed by electron transfer at metal electrodes when alternatively polarizing the Pt electrode between potentials corresponding to the redox potentials of Ru^{2+}/Ru^{3+} and Ru^{1+}/Ru^{2+} [60]. The emission was interpreted as an electron transfer according to the reaction:

$$Ru^{1+} + Ru^{3+} \rightarrow {}^{3}Ru^{2+} + Ru^{2+} \qquad (10.27)$$

followed by

$$^{3}Ru^{2+} \rightarrow Ru^{2+} + h\nu \qquad (10.28)$$

In this case, a $Ru(bipy)_3{}^{2+}$ in its triplet state was formed by the annihilation reaction (Eq. 10.27).

In principle, an excited molecule can also be produced by an electron transfer from the conduction band of a semiconductor to the oxidized species of an organic molecule (e.g. $Ru(bipy)_3{}^{3+}$). Instead of the annihilation reaction given by Eq. (10.27) we have then for the Ru complex

$$Ru^{3+} + e^-(\text{c.b.}) \rightarrow {}^{*}Ru^{2+} \qquad (10.29)$$

Such a process is possible at a semiconductor electrode where the conduction band occurs above the redox potential of the excited molecule; i.e. in terms of the usual energy scheme that $E_c > E^{*}_{F,redox}({}^{*}Ru^{2+}/Ru^{3+})$. The first experimental results were published by Gleria et al., who observed luminescence of $Ru(bipy)_3{}^{2+}$ upon cathodic

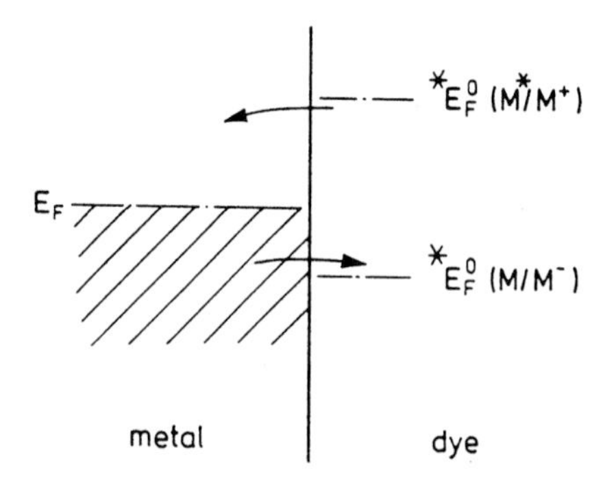

Fig. 10.30 Electron transfer between excited dye and metal electrode

polarization of an n-SiC electrode in an electrolyte containing $Ru(bipy)_3^{3+}$ [61]. This result was interpreted on the basis of the processes given in Eqs. (10.29) and (10.28).

Bard and co-workers have pointed out that it is frequently difficult to attribute the electrogenerated luminescence unambiguously to the process discussed above [62]. In several cases, instead of reaction (Eq. 10.29), the reduced species is also formed at a semiconductor electrode leading to the annihilation process (Eq. 10.27). This difficulty is caused by the fact that the reduction potential of a molecule in the dark ($E_{F,redox}(M/M^-)$) is frequently rather close to the oxidation potential of the excited molecule ($E^*_{F,redox}(M^+/M)$) (see e.g. Fig. 10.3). Luttmer and Bard found one system, rubrene, for which these two potentials are well separated. These authors observed a luminescence due to electron transfer from a ZnO electrode to the oxidized species of rubrene [62]. Another interesting example is the formation of an excited molecule by transfer of hot electrons, as already discussed in Section 7.8.

11 Applications

Some applications of semiconductor electrochemistry are already widely used, such as etching of semiconductor devices, and there are others which are promising for the future. The principles underlying these applications and some examples will be presented in this chapter.

11.1 Photoelectrochemical Solar Energy Conversion

During the 1980s and 1990s there have been efforts in many laboratories around the world to develop photoelectrochemical systems based on semiconductor materials for the utilization of solar energy. In principle, photoelectrochemical systems can be applied to the conversion of solar energy into electrical energy, as well as for the production of a chemical fuel. Some of the basic processes involved in photoelectrochemical cells are similar to those occurring in the photosynthesis of plants. The lifetime of biological systems is rather short, and the conversion efficiencies hardly exceed 1 %. In nature, this is not really a problem because plants regenerate themselves once a year. In the case of technical systems, however, an efficiency of about 10 % and stability over many years are required, which is a great challenge for scientists working in this field. Hundreds of papers on photoelectrochemical systems have been published. The basic concepts and many results have been summarized in various review articles [1–14].

In the following discussions, electrochemical photovoltaic cells and photoelectrolysis cells are treated separately.

11.1.1 Electrochemical Photovoltaic Cells

In principle, an electrochemical photovoltaic cell can be constructed very simply. It consists of a semiconductor electrode, an electrolyte containing a redox system and an inert counter electrode. The energy diagram for a cell with an n-type semiconductor and a redox system with a potential close to the valence band is shown in Fig. 11.1. In the upper part of this figure, the equilibrium condition is presented, i.e. the Fermi level is constant throughout the system and the voltage between the two electrodes is zero. Upon illumination of such a cell, holes are created and transferred to the reduced species of the redox system. The electrons reach the rear ohmic metal contact of the semiconductor electrode, traverse the external circuit, do useful work, and then reduce the oxidized component of the redox couple at the metal counter electrode (Fig. 11.1c). Since no net consumption of material in the electrolyte occurs, this cell works under regenerative conditions. The current–potential dependence of such a cell is given by Eq. (7.84); i.e. in this case

$$j = -j_0 \left[\exp\left(-\frac{eU}{nkT} \right) - 1 \right] \tag{11.1}$$

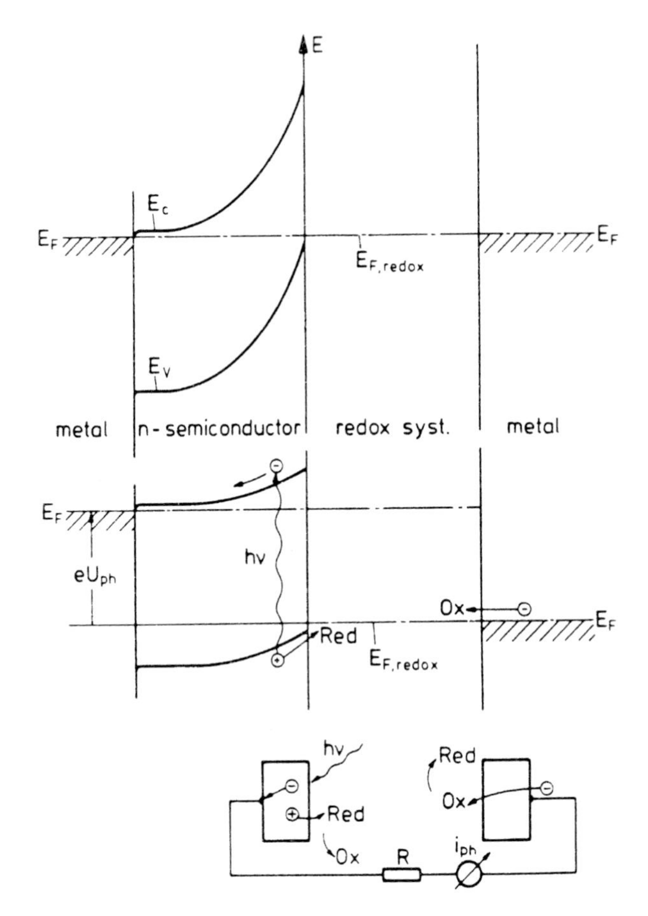

Fig. 11.1 Energy level diagram for an electrochemical photovoltaic cell using an n-type semiconductor electrode

Here the externally applied voltage between the n-type semiconductor and the metal counter electrodes is U, rather than the overvoltage η in Eq. (7.84).

Eq. (11.1) is also valid for pure solid state devices, such as semiconductor–metal contacts (Schottky junctions) and p–n junctions, as described in Chapter 2. The physics of the individual systems occurs only in j_0. The main difference appears in the cathodic forward current which is essentially determined by j_0. In this respect it must be asked whether the forward current is carried only by minority carriers (minority carrier device) or by majority carriers (majority carrier device). Using semiconductor–liquid junctions, both kinds of devices are possible. A minority carrier device is simply made by using a redox couple which has a standard potential close to the valence band of an n-type semiconductor so that holes can be transferred from the redox system into the valence band in the dark under cathodic polarization. In this case, the dark current is determined by hole injection and recombination (minority carrier device) and j_0 is given by Eq. (7.65), i.e.

$$j_0 = \frac{eDn_i^2}{N_D L_p} \tag{11.2}$$

In the case of a conduction band process, j_0 is given by the kinetics of electron transfer at the interface (see Eq. 7.54), i.e.

$$j_0 = k_c^- n_s^0 c_{ox} \tag{11.3}$$

in which n_s^0 is the carrier density at equilibrium. A current–voltage curve in the dark and under illumination is given in Fig. 11.2. The limitation of the cathodic dark current is mostly due to diffusion. The cathodic dark current is expected to increase exponentially with U. Since there are frequently deviations from an ideal slope of 60 mV/decade, a so-called quality factor n is introduced into the exponent.

Besides the photocurrent, the photovoltage of a photovoltaic cell is of importance. It is obtained for $j = 0$, i.e. $U = U_{ph}$ (see Fig. 11.2). We have then

$$U_{ph} = (nkT/e)\ \ln\left(j_{ph}/j_0 + 1\right) \tag{11.4}$$

At this voltage ($U = U_{ph}$), the anodic photocurrent and the cathodic dark current are equal. This is called the open-circuit condition as illustrated in the energy diagram (Fig. 11.1b). The conversion efficiency η of a photovoltaic cell is defined as

$$\eta = P_m/P_0 \tag{11.5}$$

in which P_0 is the power of incident sunlight whereas P_m is the maximum power output as given by (compare also with Fig. 11.2)

$$P_m = U_{ph}^m j_{ph}^m \tag{11.6}$$

The absolute values of U_{ph}^m and j_{ph}^m depend on the shape of the current–potential curve under illumination and must be selected so that the rectangle (dashed area in Fig. 11.2) is maximized. The conversion efficiency depends on the bandgap of the semiconductor as derived in Section 11.1.3. The short-circuit photocurrent amounts to about 20 mA cm^{-2} under sunlight irradiation. On the other hand, the photovoltage not only depends on j_{ph} but also on j_0. According to Eq. (11.4), j_0 should be as small as possible. Its final value depends on the combination of semiconductor and redox system and on the mechanism of the charge transfer in the dark (majority or minority carrier process).

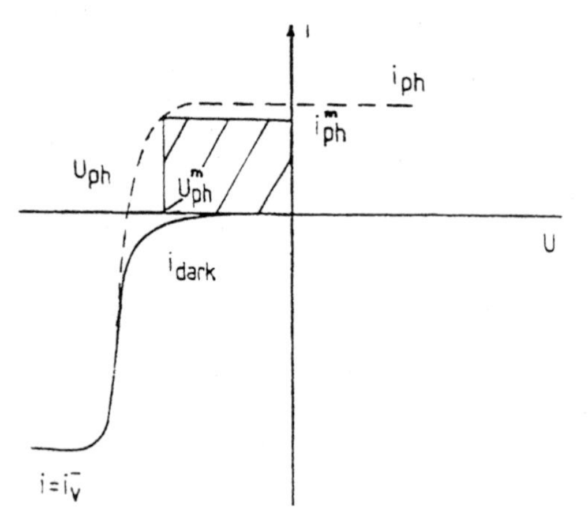

Fig. 11.2 Current–voltage curves in the dark and under illumination for an electrochemical photovoltaic cell (with n-semiconductor)

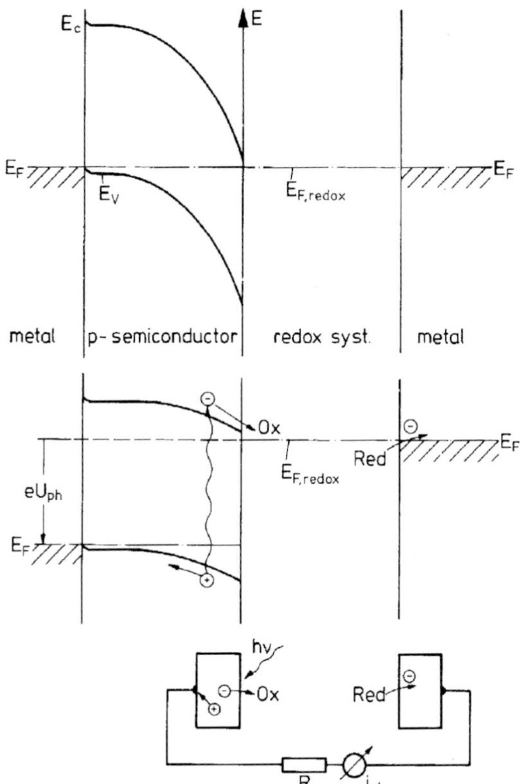

Fig. 11.3 Energy level diagram for an electrochemical photovoltaic cell using a p-type semiconductor electrode

The performance of a solar cell is further characterized by the so-called fill factor (FF) which is defined as

$$FF = \frac{j_{\mathrm{ph}}^{\mathrm{m}} U_{\mathrm{ph}}^{\mathrm{m}}}{j_{\mathrm{ph,s}} U_{\mathrm{ph}}^{0}} \tag{11.7}$$

in which $j_{\mathrm{ph,s}}$ and U_{ph}^{0} are the short-circuit current and the open-circuit voltage, respectively.

It should be mentioned that a p-type electrode can also be used in a photovoltaic cell (Fig. 11.3). In such a cell electrons are the minority carriers created by light, and a redox system with a standard potential close to the conduction band should be selected. In this case, the electrons are driven toward the semiconductor surface where they are transferred to the oxidized species of the redox system. Here, electrons are injected from the reduced species of the redox system into the counter electrode and traverse the external circuit in the opposite direction from that in the n-type electrode system (see refs. [2, 5–7, 12, 13]).

The conversion efficiencies depend on the semiconductor bandgap in semiconductor solid and solid–liquid devices, as proved by thermodynamic calculations. The maximum efficiency is about 30 % for a bandgap around 1.3 eV. Further details concerning conversion efficiencies are given in Section 11.1.1.3.

11.1.1.1 Analysis of Systems

In the 1970s and early 1980s, quite a large number of systems were proposed and investigated. Besides various semiconductors, aqueous and non-aqueous liquids have been used. Problems arise mainly in aqueous electrolytes because reactions other than charge transfer with a redox system can occur, such O_2 and H_2 evolution and, especially, anodic dissolution (see below). Although most semiconductors undergo corrosion in aqueous solutions, a number of systems have been found in which the anodic decomposition could be sufficiently suppressed in the presence of a suitable redox system. Another possibility is the use of non-aqueous electrolytes such as acetonitrile (CH_3CN) or alcohol (CH_3OH). A selection of systems is summarized in Table 11.1. Only those systems are listed here which exhibit or promise high conversion efficiency besides good stability. It is interesting to note that stable cells with n-type electrodes in aqueous solutions were mainly fabricated by using S^{2-}/S_n^{2-} (or Se^{2-}/Se_n^{2-}) and I^-/I_3^-as redox systems. The latter redox couple may be favorable with regard to stabilization because iodine adsorbs fairly strongly on surfaces. In the case of chalcogenide electrodes, one of the other redox systems is advantageous because here any elemental sulfur or selenium formed as a corrosion product is dissolved as polysulfide (S_n^{2-}) or polyselenide, leading to clean surfaces [15, 16].

Most systems, especially the redox systems, were not selected systematically. The selection was mainly based on good stability of the electrode. Quite remarkable conversion efficiencies ($\phi > 14$ %) have been obtained with several cells. Conversion efficiencies were obtained in the order of those reported for pure solid state devices. In the case of Si (numbers 10 and 11 in Table 11.1) the conversion efficiency is not limited by the surface chemistry but by the quality of the semiconductor [17–19]. The charge

Table 11.1 Electrochemical photovoltaic cell systems. Photovoltage (U_{ph}), photocurrent (j_{ph}), fill factor FF, efficiency (η)

Cell	E_g (eV)	Solvent	U_{ph} (V)	j_{ph} (mA cm^{-2})	FF	η (%)	Ref.
(1) n–CdSe/(S^{2-}/S_n^{2-})	1.7	NaOH	0.75	12	–	8	[133]
(2) n-CdSe,Te/(S^{2-}/S_n^{2-})	1.7	H_2O	0.78	22	0.65	12.5	[134]
(3) n-GaAs/(Se^{2-}/Se_n^{2-})	1.4	NaOH	0.65	20	–	12	[40, 44]
(4) n-CdS/(I^-/I_3^-)	2.5	CH_3CN	0.95	0.03	–	009.5	[135]
(5) n-$CuInSe_2$/(I^-/I_3^-)	1.01	H_2O	0.64	21	–	009.7	[136]
(6) n-$MoSe_2$/(I^-/I_3^-)	1.1	"	0.55	9	–	–	[137]
(7) n-WSe_2/(I^-/I_3^-)	1.2	"	0.63	28	–	>14	[138]
(8) n-FeS_2/(I^-/I_3^-)	0.95	"	0.25	10	–	2.8	[44]
(9) n-WSe_2/($Fephen^{2+/3+}$)	1.2	H_2SO_4	0.65	10	–	–	[27]
(10) n-Si/(Br^-/Br_2)	1.1	H_2O	0.68	22	–	>10	[24]
(11) n-Si/(Fc^{1+}/Fc)*	1.1	CH_3OH	0.67	20	>0.7	>10	[139]
(12) n-GaAs/(Fc+/Fc)*	1.4	CH_3CN	0.7	20	–	11	[140]
(13) n-GaAs/Cu^{2+}/Cu^+)	1.4	HCl	0.65	0.5	–	–	[27]
(14) p-InP(V^{3+}/V^{2+})	1.3	HCl	0.65	25	0.65	11.5	[47]

*Ferrocene

transfer processes and the limiting factors in the case of the n-Si/(Fc^0/Fc^+) cell have been studied in detail, and it was shown that this system is a minority carrier device [17]. Further details are given below. With most other systems it is not clear whether a minority or a majority carrier device is involved, because most researchers have analyzed the cells only in terms of classical parameters such as photocurrent, photovoltage, conversion efficiency and fill factor. In most cases, however, it can be assumed that the dark forward current is due to a majority carrier transfer process, i.e. electron transfer from the conduction band of an n-type semiconductor electrode to a corresponding electron acceptor in the solution. Such a majority carrier current is mostly much higher than the equivalent minority current because the recombination between electrons and holes within a semiconductor electrode occurs at a rather small rate. This has also been observed with pure solid state devices; i.e. higher forward currents and correspondingly smaller photovoltages have been found with majority carrier devices compared with minority carrier systems.

Most authors have characterized the cell performance by a so-called power plot, i.e. i_{ph} vs. U_{ph}, and not by a complete i–U characteristic in the dark and under illumination. Two examples are given in Fig. 11.4. There is only one case reported in the literature where the charge transfer processes and the limiting factors have been studied in more detail (number 11 in Table 11.1). Here n-Si electrodes were used and ferrocene derivatives (Fc^{1+}/Fc^0) as redox systems in water-free methanol [18, 20–21]. Since the current–voltage curve exhibited a slope of about 60 mV per decade in current (ideality factor of $n = 1.05$), it was concluded that any oxide formed on the Si surface must remain sufficiently thin so that electrons and holes can tunnel through the oxide layer. It has been further found that the open-circuit voltage U_{ph} increased linearly with increasing standard potential of the ferrocene derivatives as shown in Fig. 11.5. The open-circuit voltage became constant above a standard potential of $U^0_{redox} = +0.15$ V vs. saturated calomel electrode (SCE) as found, for instance, with dimethyl ferrocene (Me_2Fc). Such a dependence of the photovoltage vs. U^0_{redox} is only obtainable for an n-type electrode if

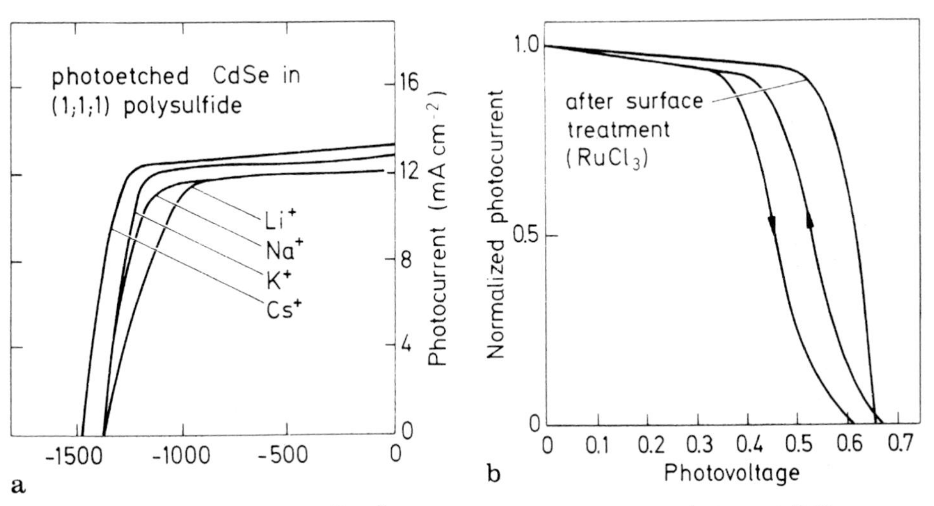

Fig. 11.4 Power plot: a) CdSe/(S^{2-}/S_n^{2-}) system in aqueous solutions (After ref. [22]); b) n-Si/($Me_2Fc^{+/0}$) system in methanol (After ref. [142])

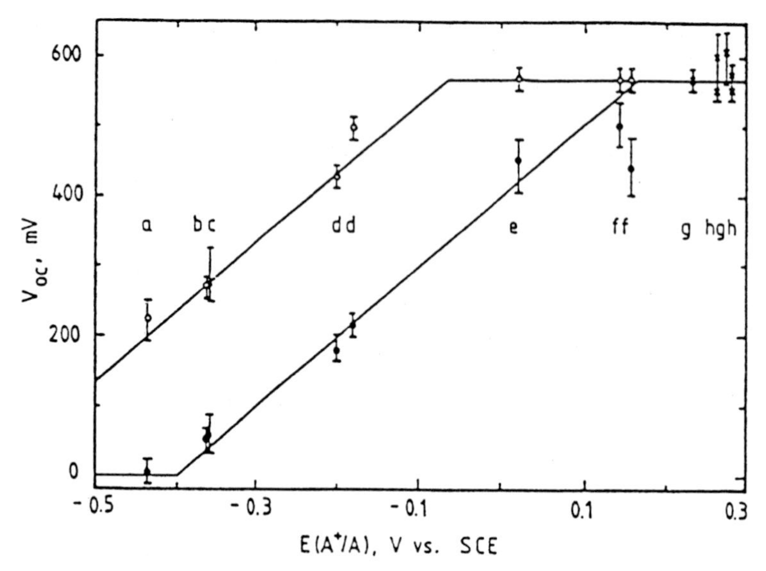

Fig. 11.5 Photovoltage vs. redox potential for an n-Si electrode in CH_3OH using various redox couples. The lower curve corresponds to the initial reading of V_{oc}, and the upper curve was obtained after the electrode was stabilized. SCE, saturated calomel electrode. (After ref. [142])

the cathodic forward current is due to an electron transfer via the conduction band. In the case of redox systems with a high standard potential, U^0_{redox} occurs close to the valence band of the n-Si electrode. Under these conditions, the cathodic forward current is due to the injection of holes into the valence band. According to Eqs. (11.1) and (11.2), the dark forward current is governed by electron–hole recombination in the bulk. Since j_0 is then constant, U_{ph} also becomes independent of U^0_{redox}. This has been verified by the response of U_{ph} to independent variation of parameters such as donor density N_D and diffusion length L_p [20, 22]. In the case of n-Si with a resistivity of 0.015 Ω cm, there is a photovoltage of 0.67 V at a photocurrent of 20 mA cm^{-2}. The same value was calculated on the basis of the Shockley model using Eqs. (11.4) and (11.2). This high photovoltage is remarkable because it is close to the thermodynamically possible value of around 0.7 V (see also Section 11.1.1.3). Such a high value has not even been achieved in simple p–n homojunctions, for which the lower photovoltage has been interpreted as recombination losses in the depletion layer [23]. None of these losses seem to occur in the n-Si/CH_3OH liquid junction. Residual losses in the short-circuit arise from optical reflectivity and absorption processes and losses in the fill factor arise from concentration overpotentials and uncompensated series resistance losses from the potentiostat [20]. Neglecting the latter losses, this cell is the first system with which the same efficiency was obtained as that found with Si homojunctions.

Another interesting investigation with Si electrodes was performed using the photoelectrochemical solar cell n-Si(Br^0/Br^-) in aqueous solutions (number 10 in Table 11.1) [24]. In this case, extremely small Pt islands were deposited on the Si electrode through which the contact to the solution was made. Here also a relatively large band bending was obtained and a high photovoltage of U_{ph} = 0.68 V, comparable to that of the other Si system. Here, however, the cell is a majority carrier device. Since the free Si surface was covered by a thicker oxide film, the corresponding electron transfer to the redox system occurred only via the Pt islands. The surface area covered by Pt islands was much smaller than the geometric surface. This caused a small j_0 value which is responsible for the high photovoltage in this majority carrier device. Details on the mechanism of the charge transfer and its catalysis by the Pt islands are given in Section 7.9.

There are two other systems in which the forward current is also due to minority carrier injection, namely n-WSe_2($Fephen_3^{2+/3+}$) [25] and n-GaAs(Cu^{1+}/Cu^{2+}) [26], both in aqueous solutions. Since here the diffusion length of hole is in the order of 1 μm, i.e. much smaller than in the case of Si ($L_p \approx 10^{-2}$ cm), the recombination of the injected holes occurs mostly within the space charge region [27]. Although this leads to rather high j_0 values (see Eq. 2.39), relatively high photovoltages were obtained (0.65 V in both cases).

In the case of layer compounds such as WSe_2, $MoSe_2$ or MoS_2, the kinetics of charge transfer have also been studied while taking into account surface recombination which plays an important role. In the presence of suitable redox systems some materials show very little corrosion [25, 28]. This is due to the morphology of the crystal surfaces, and it is generally assumed that the corrosion occurs only at steps of different crystal planes (see also Sections 8.4 and 8.5) [29]. The steps also play an important role in the fill factor as determined by surface recombination measurements [30]. It is interesting to note that Tenne and colleagues found a great improvement in the conversion efficiency of a WSe_2(I^-/I_3^-) liquid junction after photoetching the electrode [31, 32]. These authors proved that the surface recombination was reduced considerably by this treatment. In fact, they reported a high photovoltage (0.63 V) and an efficiency of >14 % (Table 11.1). This result contradicts other investigations to some extent, as this kind of etching leads to a higher density of steps and an increased corrosion [33, 24] (see also Section 8.4). Possibly the corrosion problem is less severe when the iodine couple is used as a redox system because it is strongly adsorbed on the electrode surface. According to the position of bands at the surface and of the standard redox potential of the iodine couple, it is clear that this system is a majority carrier device.

As already mentioned, many investigations have been performed with chalcogenide electrodes using S^{2-}/S_n^{2-} or Se^{2-}/Se_n^{2-} as redox couples. In particular, a group at the Weizmann Institute in Israel has studied these systems in detail (see ref. [34] and literature cited there). Since the standard potential occurs somewhere in the middle of the bandgap ($E_c - eU^0_{redox} \approx 0.8$ V, Table 11.1), the forward current is carried by electron transfer from the conduction band to the redox system. Accordingly, these systems must be majority carrier devices. It has been found furthermore that etching also plays an important role here [36]. In addition, it has been observed that various alkali cations influence the power plot of corresponding cells (Fig. 11.4a). This effect was related to ion pairing for strongly hydrogenated cations such as Li^+, which results into a decreased activity of the active (poly)sulfide at the electrode. When Cs^+ was used instead of Li^+ there was considerable improvement, not only in the electrochemical kinetics, but also in the stability [21].

In the late 1990s $CuInX_2$ electrodes (X = S or Se) have been studied extensively. These electrodes have shown much better stability in polysulfide solutions than CdX_2, but the fill factor was rather poor in corresponding solar cells [34]. On the other hand, in iodide solutions a high fill factor and good conversion efficiency were determined, but $CuInSe_2$ was found to be unstable in I_3^- solutions [37, 38]. Surprisingly, the stability was greatly enhanced by adding Cu^{1+} ions to the electrolyte [37, 39]. It has been shown that the stabilization is caused by the formation of a passivating layer of p-type $CuInSe_3$-Se^0 [39]. Accordingly, a p–n heterojunction is formed which is actually the active part of the solar cell: that is, we have a solid state solar cell in contact with an electrolyte.

Another cell which has been studied in some detail is the n-GaAs(Se^{2-}/Se_n^{2-}) liquid junction [40, 41]. In this system, the oxidation of Se^{2-} can compete sufficiently with the anodic decomposition of GaAs in alkaline solutions. Since the standard potential of the (Se^{2-}/Se_n^{2-}) couple occurred here near the middle of the bandgap, the forward dark current must be due to an electron transfer from the conduction band of the n-GaAs electrode to the Se^{2-} ions and it is controlled by the surface kinetics (majority carrier device). Considerable overvoltages for the onset of a photocurrent with respect to the flatband potential have frequently been observed (see e.g. Section 7.3). This effect leads to rather poor power plots and a very low fill factor. Several authors have shown that the fill factor improves considerably after some metal atoms, such as Ru on GaAs [41] and Cu on CdSe, have been deposited on the electrode [42]. This effect has been interpreted by Nelson et al. by assuming a decrease of surface recombination due to strong interaction between Ru and the GaAs surface atoms, resulting in a splitting of surface states into new states which are not active any more [43]. Abrahams et al. made the same observation with GaAs using a metal ion treatment. They interpreted the increase of the slope of the photocurrent potential dependence and of the fill factor by assuming that the metal deposited on the surface catalyses the redox reaction [20]. Catalytic effects which play an important role in the photoelectrochemical reactions at semiconductor electrodes and particles have already been treated in Sections 7.9 and 9.2.

In the search for new semiconducting material FeS_2 (pyrite) has attracted considerable attention. FeS_2 has a direct bandgap of 0.95 eV. Therefore, the corresponding absorption coefficient is very high ($\alpha = 5 \cdot 10^5\ cm^{-1}$ for $h\nu = 1.3$ eV). These properties makes FeS_2 interesting as an absorber material for thin film solar cells [44]. Despite these favorable physical properties, the conversion efficiencies of pyrite never exceeded 2.7 % [44]. The photovoltages of all FeS_2 junctions were considerably smaller than the theoretically expected value of about 0.5 V. Surface states and bulk defects were attributed to the significant sulfur deficiency found in many natural and synthetic crystals [45].

The most severe problem is the stability of the semiconductor electrode. Most semiconductors, having a bandgap lower than about 2 eV, undergo anodic decomposition in aqueous solutions. In order to get more insight into this problem, the mechanism of the anodic dissolution and the competition between anodic decomposition and redox reactions have been studied in greater detail, as already discussed in Section 8.5. These investigations have shown that these processes are controlled by thermodynamic as well as by kinetic parameters. As already mentioned, only in the case of a few selected redox systems could the decomposition be suppressed to a very low value [27]. However, it is difficult to obtain long-term stability with semiconductor materials which undergo anodic corrosion in aqueous solutions. The transition metal chalcogenides such as WSe_2 and others, which form layer crystals, comprise a distinct class. As already discussed in Section 8.1.4., a high stability was expected because the electronic states of the highest valence band are formed by non-bonding d-electron states of the metal [9]. Certainly, a higher stability against anodic decomposition was found when the basal planar surfaces were exposed to the aqueous solution because the metal was shielded from the surface. However, according to calculations by Gerischer [3], MoS_2 and $MoSe_2$ should not be thermodynamically stable. In this case, excellent stability can only be achieved in the presence of a suitable redox system for kinetic reasons [25].

Most investigations have been performed with n-type semiconductor electrodes and only a few with p-type materials. In principle, the application of p-type electrodes should be more favorable because electrons created by light excitation are transferred from the conduction band to the redox system. Accordingly, we have here a cathodic photocurrent. This would be advantageous with respect to the stability of the electrode because most semiconductors do not show cathodic decomposition [46]. Usually H_2 is formed during cathodic polarization and a redox reaction has to compete with this process. Unfortunately, large overvoltages for the onset of photocurrent with respect to the flatband potential have been found with most p-electrodes, which has been interpreted as having been caused by strong surface recombination and trapping. In one case the overvoltage was reduced to a very low level by deposition of a very thin rhodium film on a p-InP electrode [47]. Using such an p-InP electrode and the redox couple (V^{3+}/V^{2+}) a reasonably high efficiency was obtained (see number 11 in Table 11.1) [47].

Several other attempts have been made by various authors to avoid anodic corrosion at n-type electrodes and surface recombination at p-type electrodes, by modifying the surface or by depositing a metal film on the electrode in order to catalyse a reaction. It has been frequently overlooked that the latter procedure leads to a semiconductor–metal junction (Schottky junction) which by itself is a photovoltaic cell (see Section 2.2) [14, 27]. In the extreme case, then only the metal is contacting the redox solution. We have then a pure solid state photovoltaic system which is contacting the solution via a metal. Accordingly, catalysis at the semiconductor electrode plays a minor role under these circumstances.

11.1.1.2 Dye-Sensitized Solar Cells

In the late 1990s sensitizers have been used in regenerative solar cells. The sensitization effect is based on the excitation of a dye molecule adsorbed onto the surface of a semiconductor electrode, followed by an electron injection into the conduction band of an n-type electrode, as already described in detail in Chapter 10. The energy diagram of a complete cell is illustrated in Fig. 11.6. In the presence of a suitable electron donor, such as I^-/I_3^-, the oxidized dye molecule is reduced back to its original state (I_3^-). The current in the solution is carried by the redox system which is then reduced at the metal or carbon counter electrode (Fig. 11.6). In principle, this is a majority carrier device. Here the role of the semiconductor is simply to transport the electrons from the semiconductor–electrolyte interface toward the rear ohmic contact; that is, electrons injected from the dye into an n-type semiconductor electrode after light excitation (Fig. 11.6). The advantage of this method is that a large bandgap semiconductor, such as n–type TiO_2, can be used, which is not involved in the photoelectrochemical reaction; in other words, the semiconductor electrode remains stable.

Several research groups have studied the sensitization processes, and the mechanism is fairly well understood (see Chapter 10). Applications to solar cells have been suggested at quite an early stage [48]. Although the quantum yield of this process can

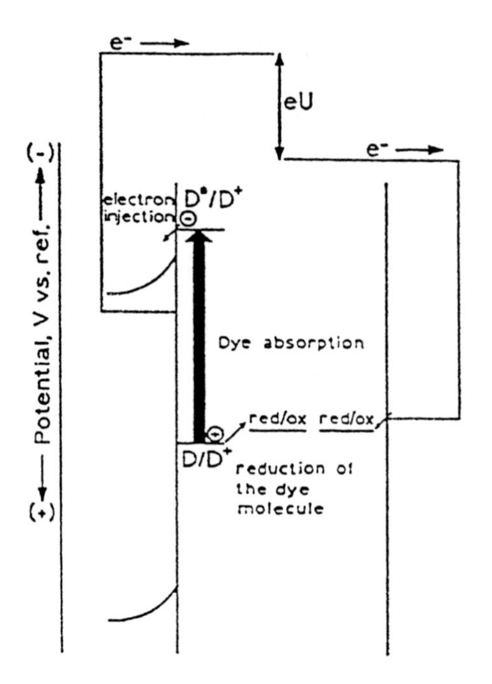

Fig. 11.6 Electrochemical photovoltaic cell based on dye sensitization of a wide bandgap semiconductor

reach more than 90 %, the photocurrent efficiency with respect to the incident light was less than 1 % of the light. A few attempts have been made to increase the surface area by using sintered electrodes, without much success, however [49, 50]. A breakthrough became possible after Stalder et al. succeeded in preparing highly porous TiO_2 electrodes [51]. Using a ruthenium complex as a dye, high current efficiencies were obtained with these electrodes [52]. During the late 1990s considerable improvements were achieved by utilizing very porous nanocrystalline TiO_2 [53, 54] (see also Section 10.2.7). Conversion efficiencies of about 7 % have been routinely obtained, with some values higher than 10 % being reported. This system has generated great interest, and various research groups have started to work on it. One of the problems to be solved for dye-sensitized systems is the long-standing issues of the stability of the dye itself because the oxidized dye formed in the first reaction step may also react with the solvent.

11.1.1.3 Conversion Efficiencies

As shown in the previous sections and in Chapter 7, a photocurrent can flow through a solar cell if a certain band bending below the semiconductor surface exists. In order to obtain a high power output, a high photovoltage is also required. The photovoltage, however, is mainly determined by the exchange current, j_0. Since this current can be relatively small if it is controlled by the surface kinetics, high photovoltages can be expected for electrochemical cells (see Eqs. 11.3 and 11.4). The question arises, however, concerning the highest photovoltages and, consequently, conversion efficiencies which are theoretically obtainable. A thermodynamic limit has been derived by Ross and co-workers [55, 56]. Their theory, applicable for all kinds of photochemical solar energy conversion systems, yields a lower limit of a recombination rate which cannot be passed. The basic concept of the theory is as follows.

At equilibrium in the dark, the recombination rate for radiative transitions is equal to the radiation emitted by a blackbody, i.e.

$$j_{\mathrm{r,dark}} = j_{\mathrm{BB}} = \int 8\pi n^2 \lambda^{-4} \left[\exp\left(\frac{hc}{\lambda kT}\right) - 1\right] \sigma(\lambda) \mathrm{d}\lambda \tag{11.8}$$

in which n is the refractive index of the absorber, c the light velocity, λ the wavelength and $\sigma(\lambda)$ the absorption probability. It is assumed that all photons of an energy above the bandgap energy are completely absorbed whereas those of lower energy are not absorbed. Assuming further that the radiative recombination rate is proportional to $(np - n_i^2)$ during illumination (non equilibrium), then the recombination rate is given by

$$j_{\mathrm{r}} = \exp\left(\frac{\Delta E_{\mathrm{F}}}{kT}\right) j_{\mathrm{BB}} \tag{11.9}$$

in which ΔE_F is the difference between the quasi-Fermi levels of electrons and holes. Eq. (11.9) is always valid because it can be derived from the first and second laws of thermodynamics [57, 58]. The excitation rate of the illuminated absorber (semiconductor) is given by

$$j_{\mathrm{e}} = \int I(\lambda) E_{\mathrm{ph}} \sigma(\lambda) \, \mathrm{d}\lambda \tag{11.10}$$

in which $I(\lambda)$ is the differential light intensity and E_{ph} the photon energy. The maximum difference between the two quasi-Fermi levels $\Delta E_{F,max}$ is obtained when $j_e = j_r$. Then we have with Eq. (11.9)

$$\Delta E_{\mathrm{F,max}} = kT \, \ln\left(\frac{j_{\mathrm{e}}}{j_{\mathrm{r}}}\right) \tag{11.11}$$

This difference of quasi-Fermi levels is identical to the maximum photovoltage which can be obtained with a solar cell. This is illustrated in Fig. 11.7 for a photoelectrochemical cell. Evaluating Eq. (11.11) one can prove that the maximum photovoltage is always considerably smaller than the bandgap; for instance, for GaAs (E_g = 1.4 eV) $\Delta E_{F,max}$ = 0.9 eV. Using in this case a redox couple with a standard potential which is located near the valence band, then under illumination a band bending of about 0.5 eV remains. The maximum photovoltages obtainable for semiconductors of different bandgaps are given in Fig. 11.8. These considerations lead to a very important conclusion: although it was shown in the previous section that the photovoltage can be relatively large (see Eq. 11.4), there is a thermodynamic limit.

Further, it should be emphasized that the output power of a solar cell and not the photovoltage should be maximized. In this case, the photovoltage will be lower, i.e. $\Delta E_{F,max}$ must be reduced to $\Delta E_{F,s}$. According to Ross et al., one obtains for the maximum power [55]

$$P = j_{\mathrm{e}} \left(1 - \frac{kT}{\Delta E_{\mathrm{F,max}}}\right) \Delta E_{\mathrm{F,s}} \tag{11.12}$$

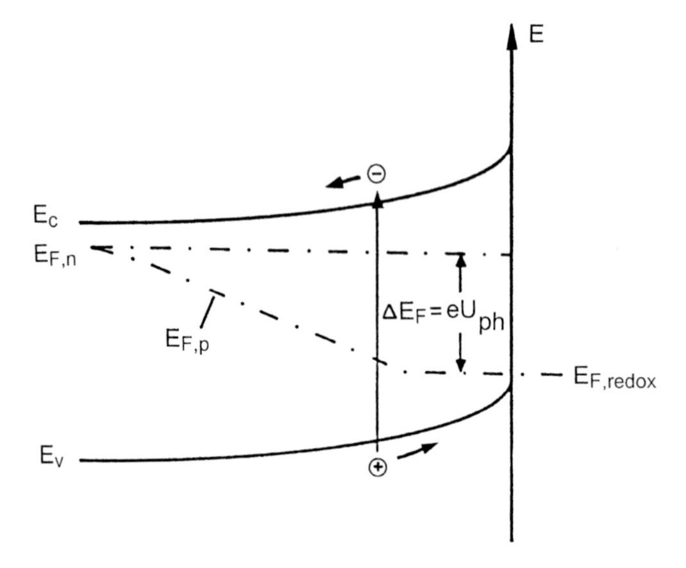

Fig. 11.7 Energy band model and photovoltage in terms of differences in the quasi-Fermi levels

with

$$\Delta E_{F,s} = \Delta E_{F,max} - kT \ln\left(\frac{\Delta E_{F,max}}{kT}\right) \tag{11.13}$$

Assuming a quantum efficiency of unity, then j_e is the photocurrent of the cell. The maximum conversion efficiency is defined by Eq. (11.5). It can be calculated for semiconductors of different bandgaps from Eqs. (11.5) and (11.12). The results are presented in Fig. 11.9. The highest efficiency is 28 % at $E_g = 1.2$ eV. This calculation is valid for solid state photovoltaic devices (p–n junction, Schottky junction) as well as for a photoelectrochemical photovoltaic cell.

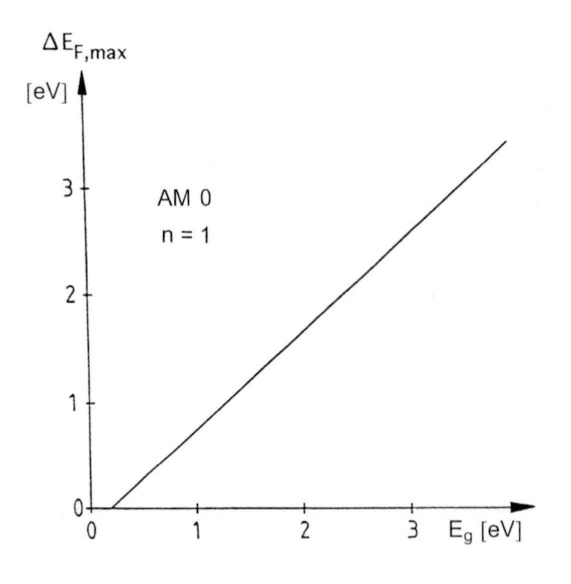

Fig. 11.8 Maximum difference between quasi-Fermi levels (photovoltage) vs. bandgap. (After ref. [27])

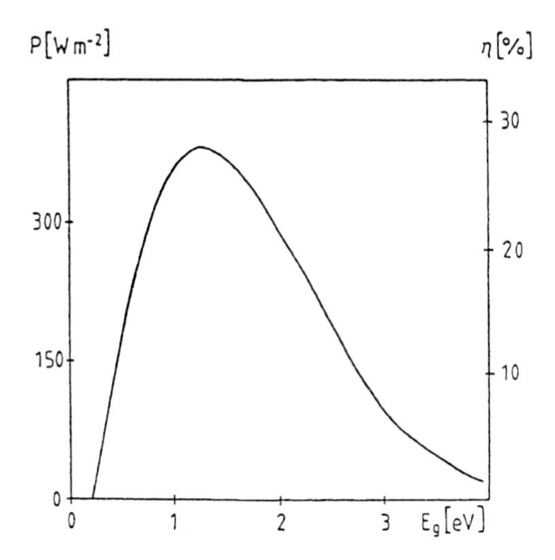

Fig. 11.9 Theoretical conversion efficiency vs. bandgap for a photovoltaic cell. (After ref. [27])

The theory of Ross et al. derived for radiative transitions can be extended to radiationless recombinations. It is assumed here that at equilibrium electron–hole pairs are also created by electron–phonon interactions. The same type of interaction is considered for the recombination process. On the basis that its rate increases exponentially with the difference between the quasi-Fermi levels – an assumption which does not follow from the first principles in thermodynamics – Ross and Collins derived a recombination current as given by [56]

$$j_r = \zeta \exp\left(\frac{\Delta E_F}{kT}\right) j_{BB} \tag{11.14}$$

in which ζ is the ratio of radiationless to radiative transitions. This effect can lead to a considerable decrease of photovoltage and of efficiency.

The application of this theory to regenerative photoelectrochemical cells has consequences. In principle, a photocurrent can flow across a semiconductor–electrolyte interface when the energy bands of an n-type semiconductor are bent upward, i.e. its onset is expected at the flatband potential as indicated by curve a in Fig. 11.10. In order to obtain a high photovoltage and, consequently, a large conversion efficiency, one would select a redox couple of a rather positive standard potential which is located near the valence band of an n-type semiconductor electrode. According to Eq. (11.4), a photovoltage in the order of the bandgap $U_{ph} \approx E_g/e$ could be expected if the cathodic partial current, determined essentially by Eq. (11.3), is extremely low as illustrated by curve b in Fig. 11.10. However, this result is in conflict with the thermodynamic derivation, according to which the photovoltage or the maximum difference of the quasi-Fermi levels $\Delta E_{F,max}$ is always considerably smaller than E_g/e (Fig. 11.8). Since the thermodynamic conditions must be fulfilled, excited electron–hole pairs have to recombine at lower band bending [27], i.e. the photocurrent onset occurs at more anodic potentials as shown by c in Fig. 11.10. Such an additional recombination has been observed in various experiments. It has been interpreted as corresponding transitions in the space charge region or via surface states [59, 60]. In some cases a shift of the flatband potential has also been observed during illumination [27, 141].

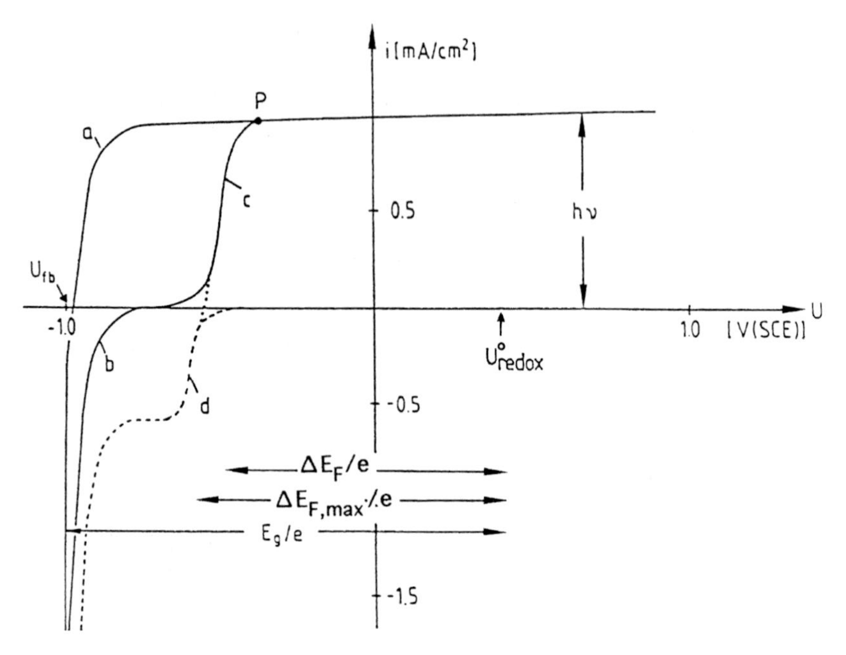

Fig. 11.10 Curve a, photocurrent for a redox couple with a relatively negative standard potential; curve b, dark current (without redox system; curve c, photocurrent in the presence of the reduced species of a redox system with a rather positive standard potential (see U^0_{redox}); curve d, dark current in the presence of the oxidized species of the redox couple. (After ref. [27])

Concerning recombination via energy states in the space charge region or at the surface, it is important to note that it also affects the cathodic dark current at an n-type electrode. For instance, using a redox couple of a standard potential which is very close to the valence band, holes are injected. Since these holes recombine via the same surface states, the cathodic dark current rises at the same potential at which the photocurrent occurs [62], as shown by curve d in Fig. 11.10. Accordingly, a current–potential dependence as given by curve b is very unlikely.

The optimal conditions are presented by the potential dependence of the total current (the dotted curve in Fig. 11.10, sum of curves c and d). The highest conversion efficiency is obtained at point P with a photovoltage of $U_{\mathrm{ph}} = \Delta E_{\mathrm{F,s}}/e$ [27]. According to Fig. 11.10, it is not necessary to select a redox couple of a standard potential near the valence band. Theoretically it is only necessary to have $U^0_{\mathrm{redox}} - U_{\mathrm{fb}} \geq \Delta E_{\mathrm{F,max}}/e$. The application of this theory to the production of a chemical fuel will be discussed in Section 11.1.2.

11.1.2 Photoelectrolysis

As already mentioned, the photoelectrolysis of water into hydrogen and oxygen has been an objective of many researchers. The work done since the early 1970s has been reviewed by many authors [4, 12, 63–67]. Several approaches to photoelectrolysis are possible. One approach is to use semiconductor–liquid junctions to produce the internal electric fields required to efficiently separate the electron–hole pairs created by the absorption of light in the semiconductor, with the holes subsequently oxidizing water at an anode region and the electrons reducing water at a cathode region. The anode and cathode may be separate electrodes [4, 12, 63–65], or combined into monolithic structures called photochemical diodes [68]. Simple dispersions or semiconductor particles as single-phase or multiphase materials may also achieve photoelectrolysis if the bandgap and flatband potential are appropriate (see below).

A second possible approach, based on semiconductor–liquid junctions, is to adsorb dye molecules onto the semiconductor surface which, upon light absorption, will inject electrons (into n-type semiconductors) or holes (into p-type semiconductors) from the excited state of the dye molecule into the semiconductor. In principle, the photo-oxidized (or photoreduced) dye can then oxidize (reduce) water, and the complementary redox process can occur at the counter electrode in the cell. However, this approach has never been demonstrated experimentally.

A third approach is to use solid state p–n or Schottky junctions to produce the required internal fields for efficient charge separation and the production of a sufficient photovoltage to decompose water. The solid state photovoltaic structure could be external to the electrolysis device, or it could be configured as a monolithic structure and simply immersed into aqueous solution.

11.1.2.1 Two-Electrode Configurations

For photoelectrolysis using semiconductor–liquid junctions in which light is absorbed in the semiconductor, various cells are possible. The first type consists of an n-type semiconductor electrode and a metal electrode and is configured similarly to a photoelectrochemical photovoltaic cell. The appropriate energy diagram is illustrated in Fig. 11.11a. The two electrodes, n-semiconductor and metal counter electrode, are short-circuited by an external wire. In the case of an n-type electrode, the holes created by light excitation must react with H_2O resulting in O_2 formation, whereas at the counter electrode H_2 is produced (see e.g. [27, 69]. The electrolyte can be described by two redox potentials, $E^0(H_2O/H_2)$ and by $E^0(H_2O/O_2)$, which differ by 1.23 eV. At equilibrium (left side of Fig. 11.11), i.e. in the dark, the electrochemical potential (Fermi level) is constant in the whole system and occurs somewhere between the two standard energies $E^0(H_2O/H_2)$ and $E^0(H_2O/O_2)$. Its position depends very sensitively on the relative concentrations of H_2 and O_2. The two reactions, O_2 formation at the n-type semiconductor and H_2 formation at the counter electrode, can obviously only occur if the bandgap is >1.23 eV, the conduction band being above (negative) of $E^0(H_2O/H_2)$ and the valence band below (positive) $E^0(H_2O/O_2)$. Since multi-electronic steps are involved in the reduction and oxidation of H_2O, certain overvoltages occur for the individual processes which lead to losses. Accordingly, the bandgap of the semiconductor must sufficiently exceed the minimum energy of 1.23 eV. The same conditions hold for a p-type electrode. The corresponding energetics are presented in Fig. 11.11b. In this case, H_2 is formed at the semiconductor electrode and O_2 at the counter electrode [27].

It is clear from the cell design that only semiconductor electrodes which are initially stable can be used. There are several oxide semiconductors available which show suffi-

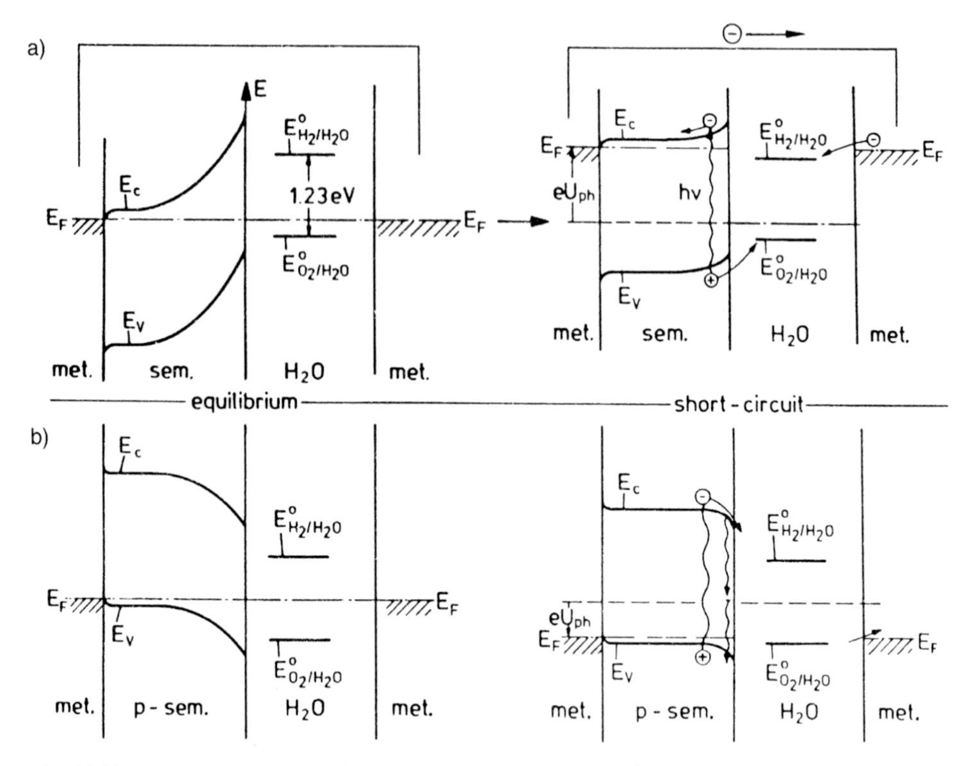

Fig. 11.11 Photocleavage of H_2O at n- and p-type electrodes (energy diagram)

cient stability. However, only a few oxides, such as $SrTiO_3$, $KTaO_3$ and ZrO_2, meet the energetic conditions discussed above. Water photocleavage at $SrTiO_3$ has been studied in detail. Unfortunately, this compound has, like most other stable oxides a relatively large bandgap (3.4–3.5 eV), which results in a very low solar absorptivity; hence they are inefficient (<1 %) in solar energy conversion systems [70]. Actually this is the only photoelectrochemical system, so far, with which photocleavage of H_2O has been realized without any additional voltage. Other oxides such as WO_3 and Fe_2O_3 which would be of interest because of a lower bandgap (around 2 eV), do not fulfil the energetic requirements because the conduction band occurs below $E^0(H_2O/H_2)$. In general, it can be stated that it is usually easy to produce hydrogen with n-type semiconductor electrodes; the real problem is the oxidation of H_2O.

Many investigations have shown that it is fairly easy to produce hydrogen at an n-type semiconductor electrode. The main problem is the oxidation of H_2O in which four elementary steps are involved. The question remains, however, whether other semiconducting materials besides oxides can be found which absorb light in the visible range and which are also stable in aqueous solutions in the absence of a redox couple; in other words, materials with which O_2 can be produced from water. Tributsch followed the strategy of searching for semiconducting transition metal compounds which possess valence bands derived from transition metal d-states. Besides the layer compounds discussed above, compounds of pyrite structure, such as RuS_2 or FeS_2, and cluster compounds are of special interest. The electrochemical and stability behavior

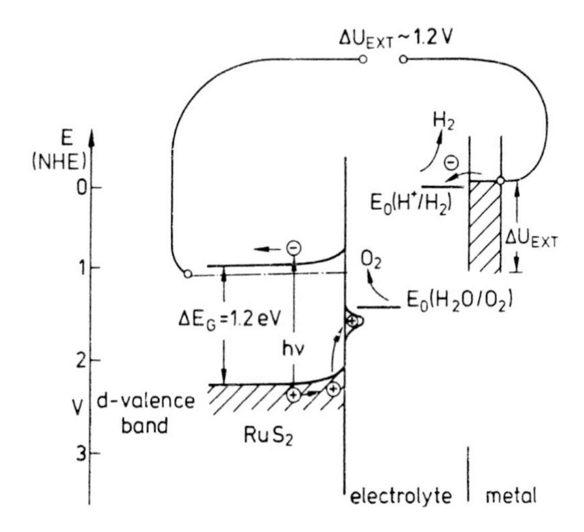

Fig. 11.12 Photocleavage of H_2O at n-RuS_2 under external bias. (After ref. [72])

of these two materials are very different. For instance, n-RuS_2 (E_g = 1.25 eV) showed a very good stability and the formation of O_2 has been observed. As discussed in Chapter 8, one or two layers of RuO_2 were formed by the interaction of RuS_2 with water [71, 72]. This led to additional energy states at the electrode surface. According to investigations using impedance measurements, the Fermi level remains pinned at the level of these RuO_2 states at the RuS_2 surface upon variation of the electrode potential [73] (for details see Section 5.3.5). Accordingly, the position of the conduction and valence band becomes unpinned. During light excitation even oxygen evolution was found, which is not surprising because RuO_2 is a good catalyst for oxygen formation. Although FeS_2 also has a pyrite structure, its stability against anodic decomposition is rather poor because its electronic structure as well as its response to crystal structure distortion is almost entirely determined by the sulfur ligand 3p states [79].

Although a RuS_2 electrode is stable due to the formation of an oxide layer on the surface, this material does not fulfil the energetic requirements as in the dark the valence band occurs above $E^0(H_2O/O_2)$. However a large downward shift of the energy bands, by 1.8 eV, was found upon illumination [73], similar to that reported for WSe_2 (see Section 5.3.5). The final position of the valence band at the surface of RuS_2 then occurred below $E^0(O_2/H_2O)$ as shown in Fig. 11.12. Oxygen evolution was found under illumination during anodic polarization. On the other hand, photocleavage of H_2O was not possible under short-circuit conditions. It can only be achieved in a cell under an appropriate external bias as illustrated in Fig. 11.12.

In principle, it would be more interesting to use p-type semiconductors as photocathodes because the stability problem is less severe. An appropriate two-electrode configuration is given in Fig. 11.11b. The same energetic conditions must be fulfilled as for n-type electrodes. Here, H_2 is formed at the p-electrode under illumination and O_2 at the metal counter electrode. Since at the p-electrode only a cathodic reaction occurs (H_2 formation), the anodic decomposition reaction can be avoided [14]. Unfortunately, however, only a few p-type materials are available which fulfil the energetic conditions. In addition, all p-type photocathodes show a large overvoltage for the onset of the cathodic photocurrent. This is a general problem which has not been solved satisfactorily. The highest efficiency reported for photoelectrolysis is based on a system containing a p-InP photocathode ($E_g = 1.3$ eV). The high overvoltage of the photocurrent with respect to the flatband potential and the resulting recombination losses were considerably reduced by depositing islands of noble metals such as rhodium, on the InP surface [74, 75]. Since the bandgap of InP is not sufficiently large, water cleavage was only found under external bias (solar-assisted electrolysis). A relatively high efficiency of 12 % was reported for this system [74].

In another type of cell, both electrodes consist of semiconducting materials, i.e. one is n-type and the other is p-type (Fig. 11.13) [69]. This configuration is of special interest because the available electron–hole potential for driving chemical reactions in the electrolyte is enhanced when both electrodes are illuminated. In the cell, two photons must be absorbed (one in each electrode) to produce one net electron–hole pair for the cell reaction. This electron–hole pair consists of the minority hole and minority elec-

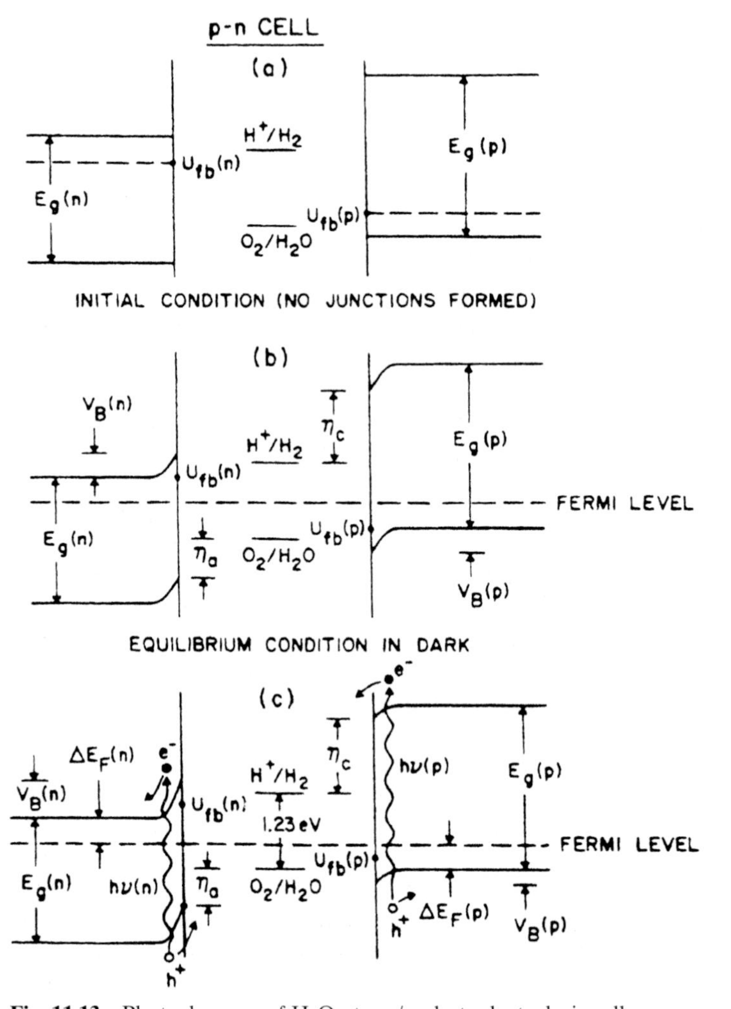

Fig. 11.13 Photocleavage of H_2O at a p/n photoelectrolysis cell

tron from the n-type and p-type electrodes, respectively, and it has a potential energy greater than that available from the absorption of one photon. An important advantage of the double-electrode cell is that, for a given cell reaction, it may allow the use of smaller bandgap semiconductors [69]. Since the maximum photocurrent available from sunlight increases rapidly with decreasing bandgap, higher conversion efficiencies can be produced. Various attempts have been made to produce suitable cells. One example is a configuration of an n-TiO_2 ($E_g = 3.1$ eV) and a p-GaP ($E_g = 2.25$ eV) electrode. Photoelectrolysis of water into H_2 and O_2 was achieved with simulated sunlight [76, 77]. This is also an interesting example insofar as photocleavage of water was obtained at zero external bias. This would not be possible in a simple cell configuration with an n-TiO_2 electrode and a metal counter electrode because the conduction band of TiO_2 is just at the standard hydrogen potential $E^0(H_2O/H_2)$. Severe stabilty problems did not occur here because the O_2 formation occurred at a stable oxide electrode. Various attempts have also been made with other combinations of semiconductors. In all these cases the stabilty problem was avoided by using a large bandgap oxide semiconductor (for further information see [69]).

11.1.2.2 Photochemical Diodes

The elimination of bias requirements for photoelectrolysis by the use of double electrode systems (n- and p-electrode, Fig. 11.13) leads to a very interesting configurational variation. This configuration called the "photochemical diode" [78], comprises photoelectrolysis cells that are collapsed into monolithic particles (compare also with Chapter 9) containing no external wires. In one simple form, a photochemical diode consists of a sandwich of either a semiconductor and a metal or an n-type and a p-type semiconductor, connected through ohmic contacts as illustrated by the energy diagrams in Fig. 11.14 [69]. To generate water cleavage, photochemical diodes are simply immersed into the aqueous solution, and the semiconductor faces are illuminated. A comparison of the energy level diagrams in Fig. 11.14 with that for biological photosynthesis reveals very interesting analogies: both systems require the absorption of two photons to produce one useful electron–hole pair. The potential energy of this electron–hole pair is enhanced so that chemical reactions requiring energies greater than that available from one photon can be driven. The n-type semiconductor is analogous to photosystem II, the p-type semiconductor is analogous to photosystem I, and the recombination of majority carriers at the ohmic contacts is analogous to the recombination of the excited electron from excited pigment II with the hole in photosystem I. The size of a photochemical diode is arbitrary; when semiconductor particles are used, their size may approach colloidal or perhaps macromolecular dimensions.

The platination of semiconductor powders is a method for producing semiconductor–metal type photochemical diodes with an energy level scheme as shown in Fig. 11.14a. This was demonstrated for the first time with platinized TiO_2 powders which showed excellent photocatalytic activity for the photodecarboxylation of acetate (the photo-Kolbe reaction), a process which has already been discussed in detail in Section 9.2.3 [80]. Various attempts have also been made to photocleave water by using semiconductor powders on which a catalyst such as Pt or RuO_2 has been deposited. The relevant experiments usually failed, either because the semiconductor was

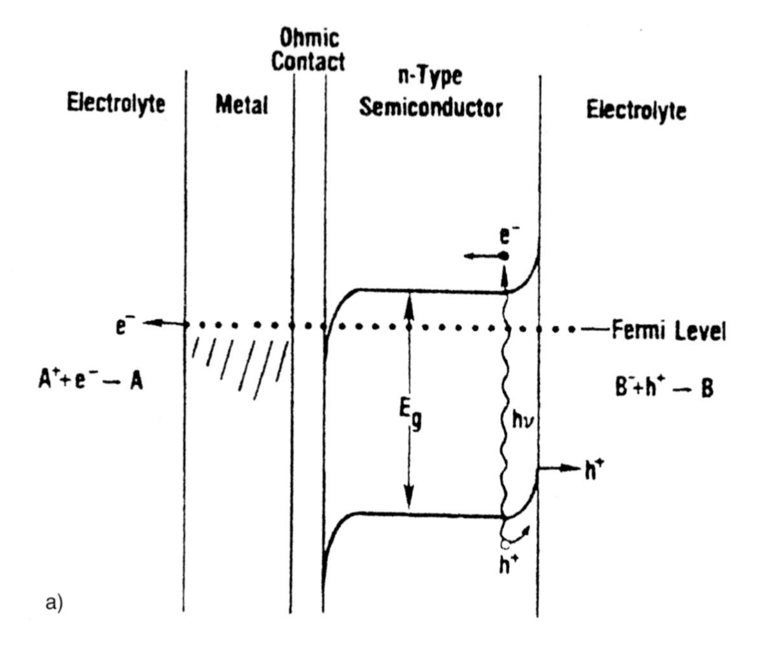

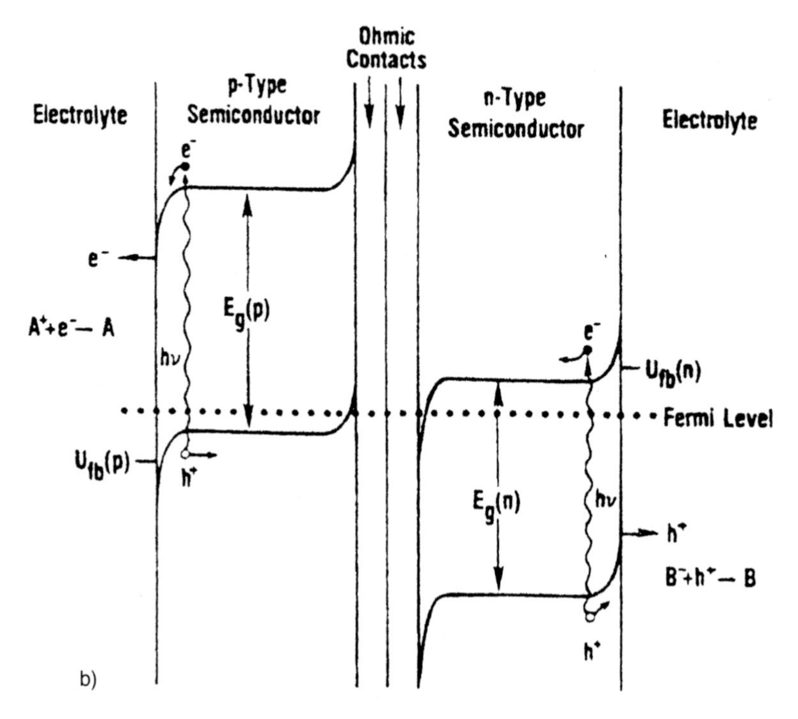

Fig. 11.14 Energy level diagrams for: a) semiconductor–metal type photochemical diode, b) p/n type photochemical diode. (After ref. [69])

not photostable or the energetic requirements were not fulfilled (for details see e.g. refs. [14, 27]).

There are some further aspects which must be considered when photochemical diodes are used. When a metal (catalyst) is deposited on a semiconductor, then frequently a Schottky barrier instead of an ohmic contact is formed at the semiconductor–metal interface. In this case, the latter junction behaves as a photovoltaic system by itself, which may determine or essentially change the properties of the photochemical diode. The consequences have been discussed in detail in refs. [14, 27]. Frequently, colloidal semiconducting particles have been used, their size being much smaller than the thickness of the space charge region expected . Since then no space charge exists, also no electric field is available for separating electron–hole pairs at the semiconductor–liquid interface or at a Schottky junction. In this case, the reaction rates are entirely determined by the kinetics of the charge transfer, which may reduce the efficiency. On the other hand, the use of powders loaded with a catalyst is a very convenient method for the photocleavage of water. A disadvantage is, however, that H_2 and O_2 are produced in the same vessel. This can be avoided by using so-called "monograin membranes", a technique where semiconductor particles are fixed in a thin polymer membrane and where each particle is in contact with different electrolytes on both sides of the membrane, as illustrated in Fig. 9.14. In this case each product is formed on the corresponding side of the membrane as quantitatively illustrated for the photocleavage of H_2S at a CdS monograin membrane. Details of this technique and of the reactions were discussed in Section 9.2.1.

11.1.2.3 Photoelectrolysis Driven by Photovoltaics

As considerable problems have occurred with direct photoelectroysis the question arises concerning whether it would be more feasible to convert solar energy into electrical energy using a separate solid state photovoltaic system, the latter being connected to a standard electrolysis cell [69]. This question is of interest because cascade-type or tandem multijunction semiconductor systems are actively pursued in photovoltaic research to produce high-efficiency solar-to-electrical power conversion [81]. Recently, a 30 % efficient device was reported, based on a two-junction system comprising GaAs and $GaInP_2$ [82]. An important practical question is how photoelectrolysis compares with such high-efficiency photovoltaic cells which are coupled to dark electrolysis. Since the efficiency of dark electrolysis can easily be in the range of 80 %, a coupled photovoltaic electrolysis system could show efficiencies as high as 24 %. Such a coupled system could either be two separate devices electrically connected or an integrated monolithic device. The band diagram for such a monolithic photovoltaic electrolysis cell is shown in Fig. 11.15. This system should be compared with the photochemical diode for which the energy diagram is shown in Fig. 11.14b.
There are several significant differences between these two systems, as follows:

(1) The photovoltaic electrolysis cell has twice as many semiconductor layers as the photochemical diode.
(2) In the photovoltaic electrolysis cell the photoactive junctions are p–n junctions between two semiconductors, while in the photochemical diode the photoactive junctions are between semiconductors and aqueous solutions.

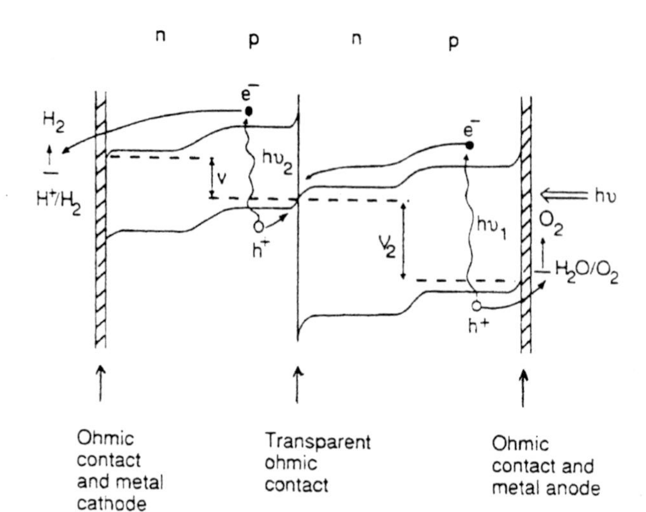

Fig. 11.15 Energy band diagram for two-junction photovoltaic device. (After ref. [69])

(3) In the photovoltaic electrolysis cell, the n-type region of the device, covered with a metal layer, becomes a cathode while the p-type region covered with a metal layer becomes an anode (i.e. it behaves like a majority carrier device); in the photochemical diode, the opposite is true, i.e. it is a minority carrier device with the n-type region acting as anode and the p-type region acting as a cathode.

The photovoltaic electrolysis cell must be covered on the illuminated side with a transparent conductor which forms an ohmic contact and is catalytic for the relevant gas evolution reaction. Metallic coatings consisting of small metal islands may also serve to stabilize the photoelectrodes against corrosion, catalyse the H_2 evolution, and produce efficient photoelectrolysis.

An integrated monolithic $GaInP_2$/GaAs pn tandem cell device, illustrated in Fig. 11.16, has recently been investigated [83]. The solid state tandem cell consists of a pn GaAs bottom cell connected to a p-$GaInP_2$ top layer through a tunnel diode interconnection. The p-$GaInP_2$ forms a semiconductor–liquid junction with the aqueous solution. Accordingly, the complete cell is a two-photon system. The $GaInP_2$ top layer (E_g = 1.83 eV) absorbs the visible portion of the solar spectrum, and the bottom pn-GaAs junction (E_g = 1.4 eV) absorbs the near-infrared portion of the spectrum transmitted through the top junction. The GaAs is connected to a Pt electrode via an external wire (Fig. 11.16). Concerning the production of hydrogen, a conversion efficiency of 12.4 % is obtained [83], which is remarkable compared with the maximum realizable efficiency of 16 % for this type of cell (see next section). The key to making this system work so well appears to be the requirement that the bottom cell should be the limiting electron provider. Since these tandem systems operate by requiring two photons (one per junction) to produce one electron in the external circuit, great care was taken in the solid state system to match the photon absorption characteristics so that equal numbers of photocarriers are generated in the top and in the bottom cells.

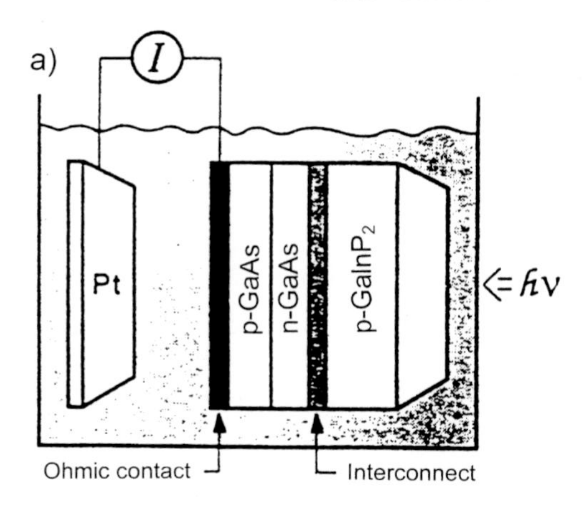

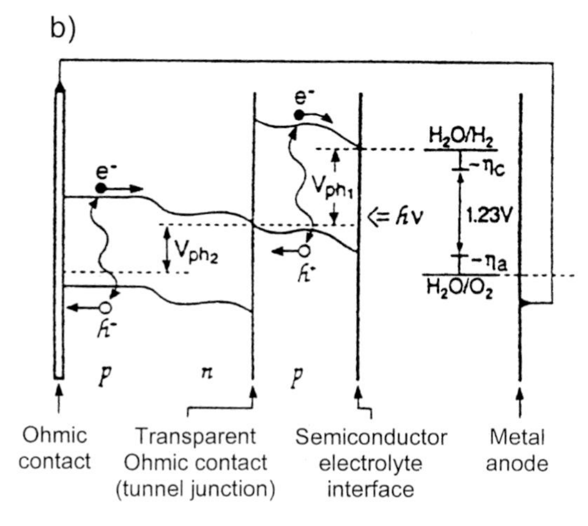

Fig. 11.16 a) Schematic of the monolithic combination of a photoelectrochemical/photovoltaic (PEC/PV) device. b) Idealized energy level diagram for the monolithic PEC/PV photoelectrolysis device. (After ref. [83])

11.1.2.4 Efficiency

Conversion efficiencies for electrochemical photovoltaic cells have already been derived in Section 11.1.1.3, on the basis of the theory developed by Ross et al. [58]. The maximum efficiencies were calculated on the basis of Eq. (11.12). If energy should be stored in a chemical fuel, further losses must be encountered, i.e. overpotentials, η_{ox} for the oxidation and η_{red} for the reduction, have to be considered. The latter are defined as the difference between the quasi-Fermi level and the redox potentials. Using Eq. (11.12), the conversion efficiency for the production of a chemical fuel is then given by

$$\eta = \frac{P_{stor}}{P_e} = \frac{j_{ph}}{P_e}\left(1 - \frac{kT}{\Delta E_{F,max}}\right)\Delta E_{stor} \tag{11.15}$$

where ΔE_{stor} is the storable energy. Here it is assumed that

$$\Delta E_{F,s} \geq \Delta E_{stor} + e\eta_{ox} + e\eta_{red} \tag{11.16}$$

On other hand, we have $P_{stor} = 0$ for

$$\Delta E_{stor} + e\eta_{ox} + e\eta_{red} \geq \Delta E_{F,max} \tag{11.17}$$

In the small range between $\Delta E_{F,s}$ and $\Delta E_{F,max}$ we have [27]

$$P_{stor} = j_{ph}\left[1 - \exp\left(\frac{\Delta E_{stor} + e\eta_{ox} + e\eta_{red} - \Delta E_{F,max}}{kT}\right)\right]\Delta E_{stor} \tag{11.18}$$

if the condition

$$\Delta E_{F,s} \leq \Delta E_{stor} + e\eta_{ox} + e\eta_{red} \leq \Delta E_{F,max} \tag{11.19}$$

is fulfilled. Eq. (11.18) was derived [14] on the basis of the thermodynamic model published by Bolton et al. [84]. A detailed description of the thermodynamic model as derived by Bolton et al. would be beyond the scope of this chapter. In the range where Eq. (11.18) is valid, the difference of quasi-Fermi levels ΔE_F should be as small as possible in order to avoid recombination losses, i.e. $\Delta E_F = \Delta E_{stor} + e\eta_{ox} + e\eta_{red}$. Fig. 11.17 shows the stored power P and the corresponding efficiency η for the photocleavage of water as a function of bandgap for different overpotentials, assuming $\Delta E_{stor} = 1.23$ eV. The curves in Fig. 11.17a were calculated for a cell consisting of a semiconductor and a metal counter electrode (energy diagram in Fig. 11.11), both being short-circuited. The data given in Fig. 11.17b were obtained for a system in which H_2O is cleaved by reduction at a p-electrode and by oxidation at an n-electrode (two-photon process; see energy diagram in Fig. 11.13). Both sets of curves exhibit a distinct maximum. The efficiencies decrease with increasing overpotentials. It is interesting to note that in both cases a maximum theoretical efficiency of about 27 % is obtained for $\eta_{ox} = \eta_{red} = 0$. This is a surprisingly large value compared with the maximum theoretical efficiency of 31 % for the conversion of sunlight into electrical energy (see dotted curve in Fig. 11.17). A higher conversion efficiency can only be expected for case b in Fig. 11.17 if the bandgaps of two semiconductors were to be different. Concerning the conversion of solar radiation into electrical energy, the maximum efficiency can be increased up to

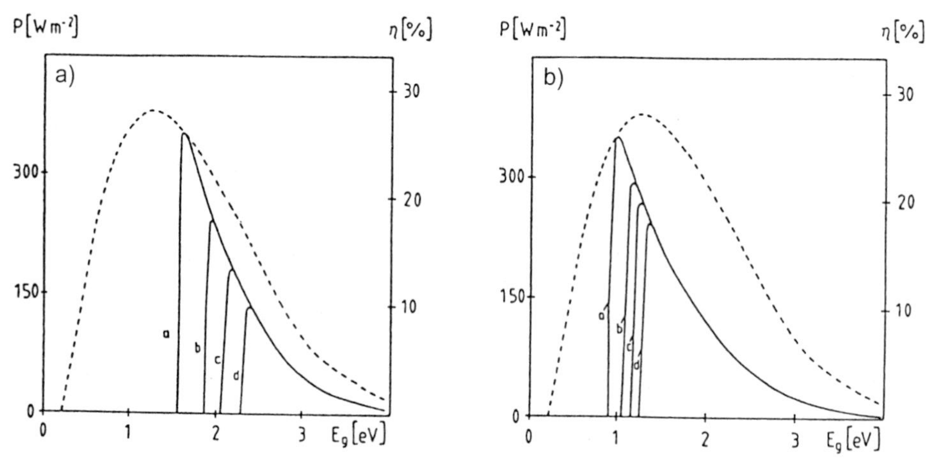

Fig. 11.17 Theoretical conversion efficiencies for photoelectrochemical H_2O-splitting: a) n-type electrode combined with a metal counter electrode; b) n-type electrode short-circuited with a p-electrode of equal bandgap; calculated for different overvoltages of: a = 0 V; b = 0.3 V; c = 0.5 V; d = 0.7 V. (After ref. [27])

42 % for a two-bandgap system. As the number of bandgaps in a tandem multiphoton device increases, so also does the efficiency; the ultimate limit is 67 % with an infinite number of tandem layers. The limiting efficiency is quickly approached after four or five layers, so that for practical reasons only systems with two or perhaps three were investigated; the three-layer system would have a theoretical efficiency of 52 %. For the photoelectrolysis of H_2O with sunlight, the maximum realizable efficiencies, taking into consideration all possible losses, have been estimated to be about 10 % for a single-bandgap and 16 % for a two-bandgap system [85].

11.1.3 Production of Other Fuels

11.1.3.1 Photoelectrolysis of H_2S

As discussed in the previous section, great difficulties occur in the photocleavage of H_2O. In the case of H_2S photocleavage, the situation is much simpler because the oxidation of S^{2-} ions can compete sufficiently fast with anodic decomposition. This process was first verified by Nozik [78] using CdS/Pt photodiodes. Further investigations were performed with CdS suspensions, the particles being loaded with RuO_2 [86] or Pt [87] as catalysts. The effect of RuO_2 was originally attributed to the catalysis of hole transfer from the valence band of CdS to the S^{2-} ions in the solution [86]. Later, with the use of monograin membranes it was found that the electron rather than the hole transfer is catalysed by RuO_2 because H_2 evolution was found at this catalyst [88], as already discussed in detail in Section 9.2.1 (Fig. 9.15). Quantum efficiencies of up to 30 % have been obtained. The result was interpreted as the formation of a Schottky barrier at the CdS/RuO_2 suspension which could be increased considerably by addition of sulfite because the latter served as a sink for sulfur produced during the photoreaction [89].

11.1.3.2 Photoelectrolysis of Halides

The photoelectrolysis of halides was studied using a cell containing a p-type InP and a Pt counter electrode [90]. A thin layer of Rh was deposited on the semiconductor electrode in order to avoid any oxide formation and a high overvoltage with regard to the onset potential of the cathodic photocurrent (see also Section 11.1.2.1 and Section 7.9). Under short-circuit conditions and illumination, hydrogen is produced at the p-InP photocathode and iodine at the Pt counter electrode as shown in Fig. 11.18. This figure is in principle identical to Fig. 11.11b (Section 11.1.2.1). According to Fig. 11.18a, considerable currents are obtained. Under short-circuit conditions, the cathodic photocurrent at the p-InP electrode (H_2 formation) and the anodic current at the Pt electrode (formation of I_3^-) are equal. One can easily recognize that the partial currents are rather high under these conditions.

This is an interesting system insofar as the products can be stored. When this photocell is combined with a fuel cell system then the fuels can be used again for producing electricity. Several years ago, researchers at Texas Instruments developed a similar kind of device. This system consisted of a panel of small p–n spheres of silicon circuited

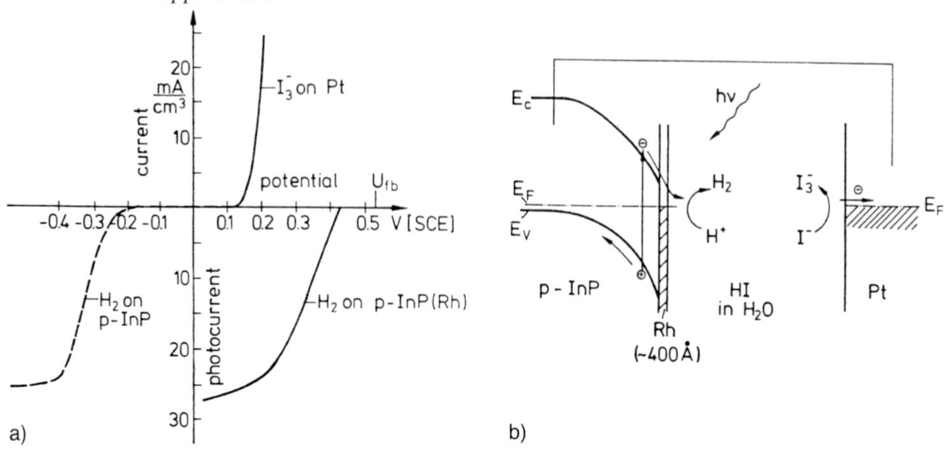

Fig. 11.18 Photoelectrolysis of HI at p-InP loaded with Rh: a) partial currents; b) energy diagram. (After ref. [90])

in series [91]. This system of photocells was connected with an electrolysis cell and the photovoltage was sufficient to drive the decomposition of HBr. Such a device is actually a solid state device integrated in a electrolysis cell.

11.1.4 Photoreduction of CO_2

It is a very difficult task to reduce CO_2 to a useful fuel such as methanol or methane by electrochemical methods because six electrons per molecule are required for the production of methanol and as many as eight electrons for methane. Another difficulty is that high energy intermediates are involved in most steps, imposing high kinetic barriers. Hence, most electrochemical reduction experiments have yielded only formic acid as a product, according to the reaction

$$CO_2 + 2H^+ + 2e^- \rightarrow HCOOH \qquad E^0_{redox} = 0.2V \text{ vs. SCE} \tag{11.20}$$

It may be surprising that this reaction works rather well at metal electrodes, although the first step in this reaction ($CO_2 + e^- \rightarrow COO^{\bullet -}$) already requires a potential of $E = -1.6$ V (SCE) [92]. In principle, there are better chances with semiconductor electrodes because the conduction band can be sufficiently negative [93]. However, formic acid is also the only product with semiconductor electrodes generally. One example is a colloidal solution of ZnS with sulfite as a hole scavenger [94]. Interestingly, the formation of various hydrocarbons has been found, especially at Cu electrodes, as mainly studied by Hori and his group [95]. The relative concentrations of the products depend on the composition of the solution and on the potential. Considerable current yields of the order of some 10 % were obtained as shown in Fig. 11.19. Hydrocarbons were also obtained at some noble metal electrodes such as rhodium and ruthenium, but at much lower concentrations.

Frese and Canfield have carried out some very interesting investigations with GaAs electrodes [92]. They have observed that CO_2 is selectively reduced to CH_3OH at the

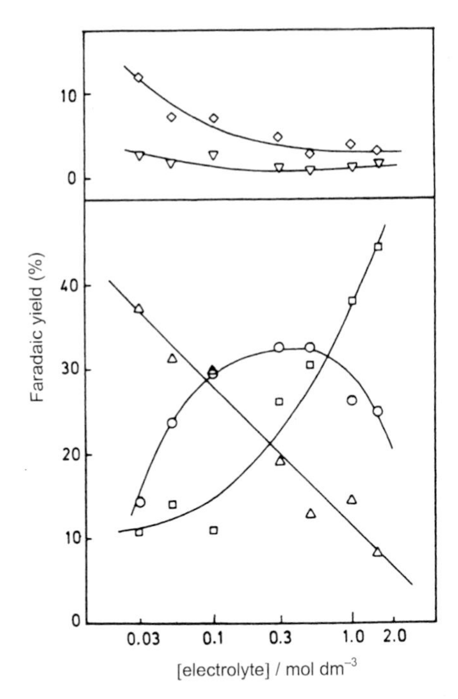

Fig. 11.19 Faradaic yields of the products in the electrochemical reduction of CO_2 at Cu-electrodes in $KHCO_3$ aqueous solutions of various concentrations: △ C_2H_4; O, CH_4, □, H_2; ◇ EtOH; ▽ PrOH. (After ref. [95])

(111) As face during cathodic polarization. Current efficiencies of up to 100 % were obtained. It is important to note that this result was only found by using reagent grade salts and distilled H_2O. In the case of much purer chemicals (99.999 % $1.7 \times 10^7\ \Omega$ cm H_2O) practically no CH_3OH was obtained. This surprising result may be due to the deposition of traces of impurities (Zn, Cu, As, and Ru) which may catalyse the CH_3OH formation. The authors proposed a reaction mechanism as follows:

$$nH^+ + ne^- \rightarrow nH_{ad} \tag{11.21}$$

$$2H_{ad} \rightarrow H_2 \tag{11.22}$$

$$CO_2 + e^- + H^+ \rightarrow COOH_{ad} \tag{11.23}$$

$$H_{ad} + COOH_{ad} \rightarrow I + CH_3OH \tag{11.24}$$

in which I represents some unknown intermediate. Other authors [96] have mentioned that CH_3OH is also formed by illuminating p-GaAs electrodes under open-circuit conditions [96]. The "impurities" mentioned above, have probably acted as surface catalysts. There are various other examples. The charge transfer excited state of a titanium oxide species, highly dispersed within zeolites, plays a significant role in the reduction of CO_2 with H_2O with a high selectivity for the formation of CH_3OH, while the catalyst involving the aggregated octahedrally coordinated titanium oxide species shows high selectivity for producing CH_4 [97]. Another example is the selective formation of CH_3OH from CO_2 by photoexcitation of TiO_2 in propylene solution [98].

Plants reduce CO_2 by a photochemical process on a very large scale, although the efficiency of this process is less than 1 %. It is a general question, however, whether the photochemical or photoelectrochemical reduction of CO_2 to alcohol or any other hydrocarbon on a technical scale makes any sense, because of the large entropy factor for collecting CO_2 from air. It is certainly more advantageous to avoid the production of CO_2 and to produce electrical energy directly instead of burning coal or oil.

11.2 Photocatalytic Reactions

In the first part of Chapter 11 systems have been described in which solar energy is used for producing electrical energy or a storable fuel. Both processes lead to an increase of free energy, i.e. $\Delta G > 0$ (uphill reaction). On the other hand, a photocatalytic reaction is a downhill reaction ($\Delta G < 0$), where light excitation is only used to speed up a reaction which is thermodynamically possible in the dark but is kinetically inhibited. Examples are reactions involved in pollution control or in the synthesis of some organic compounds; these are discussed below.

11.2.1 Photodegradation of Pollutants

During the 1990s, the photocatalytic oxidation of organic compounds using semiconductor particles has been of considerable interest for environmental applications, particularly the degradation of hazardous waste. Classes of compounds which have been degraded, include alkanes, haloalkanes, aliphatic alcohols, carboxylic acids, alkenes, aromatics, polymers, surfactants, herbicides, pesticides and dyes. Many of the results are summarized in various review articles [99–105]. As with fuel production, certain energetic requirements must be met; i.e. the valence band must be positive with respect to the oxidation potential of the pollutant and the conduction band negative relative to the reduction potential of the electron acceptor. In order to conserve electroneutrality, both conditions must be fulfilled simultaneously. TiO_2 particles in aqueous solutions have mostly been used because it is a stable material.

The degradation of organic pollutants, especially chlorinated hydrocarbons, was the subject of many investigations. Only two examples will be given here, which are investigated in more detail. One is the degradation of chloroform. The overall reaction is given by

$$2CHCl_3 + O_2 + H_2O + h\nu(TiO_2) \rightarrow 2CO_2 + 6H^+ + 6Cl^- \tag{11.25}$$

in which $h\nu(TiO_2)$ symbolizes the photon absorption in TiO_2. This overall reaction was quantitatively confirmed by measuring the chloride production rate [106], the formation of carbonates [99, 106], the rate of hydroxide consumption at constant pH [99] and the depletion of O_2 [99]. In addition, the photodegradation of various chlorophenols has been investigated extensively [107]. The second example is the complete mineralization of pentachlorophenols during illumination. The overall reaction is given by

$$HO–C_6Cl_5 + 4\tfrac{1}{2}O_2 + H_2O + h\nu(TiO_2) \rightarrow 6CO_2 + 5HCl \tag{11.26}$$

In the latter case, the photodegradation was measured at various semiconductor particles as shown in Fig. 11.20 [108]. According to these results, TiO_2 is the best catalyst

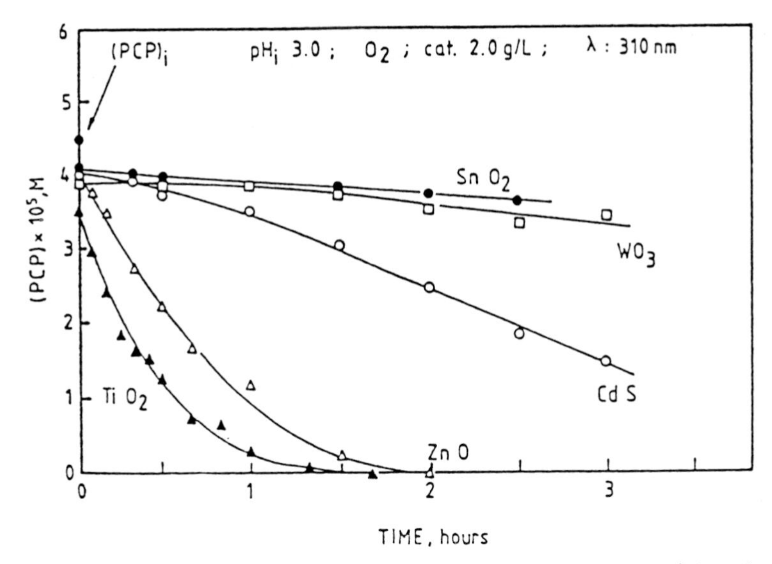

Fig. 11.20 Concentration of parachlorophenol vs. irradiation time (photodegradation). (After ref. [108])

alongside ZnO. This has been found by many other researchers. In order to compare different catalysts the activity was defined as the ratio of the photocatalytic degradation rate (in units mol l^{-1} s^{-1}) and the incident light intensity [110]. The photocatalytic activity was found to depend on the conditions under which TiO_2 was prepared. There is no general rule; in one case better activity was found for one type of organic contamination, and for another there was higher activity with a differently prepared TiO_2 catalyst [109]. Most researchers have applied P25, a TiO_2 catalyst prepared by Degussa. It consists of a 70:30 mixture of non-porous anatase and rutile, with a BET surface area of about 55 m^2 g^{-1} and crystallite sizes of 30 nm in 0.1-μm diameter aggregates. The reason for its outstanding catalytic properties is not known.

In Eqs. (11.25) and (11.26) the overall reactions are presented which do not give any information on single steps. The complete oxidation of an organic molecule generally involves many reaction steps. Rather little is known, concerning the mechanism of the photodegradation. The organic compound is primarily oxidized by hole transfer via the valence band. Since the valence band of TiO_2 occurs at a very positive energy (E_v = +2.7 eV vs. normal hydrogen electrode (NHE) at pH 7), the holes in the valence band have a very high oxidation power (see also Section 9.2). It is generally assumed, and there is some vague experimental evidence, that an $OH^{\bullet}$ surface radical is formed in the first step which in turn oxidizes the organic compound. The role of oxygen is twofold. On the one hand, O_2 is used as an electron acceptor. On the other hand, O_2 itself or some intermediates formed by its reduction, are also involved in the oxidation of the organic molecules. This has recently been illustrated for reactions of aliphatic compounds at TiO_2 particles [111]. Here, the photogenerated electrons react with O_2 to form a superoxide radical. The latter combines with the organoperoxy radical, formed by the organic radical which was generated by the hole reaction then reacting with molecular oxygen. The product of the combination of the organoperoxy radical and

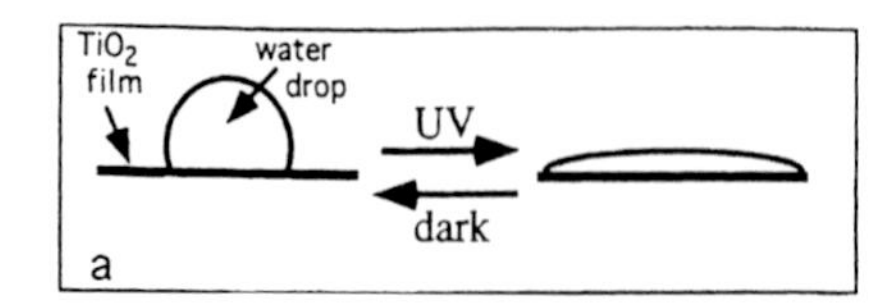

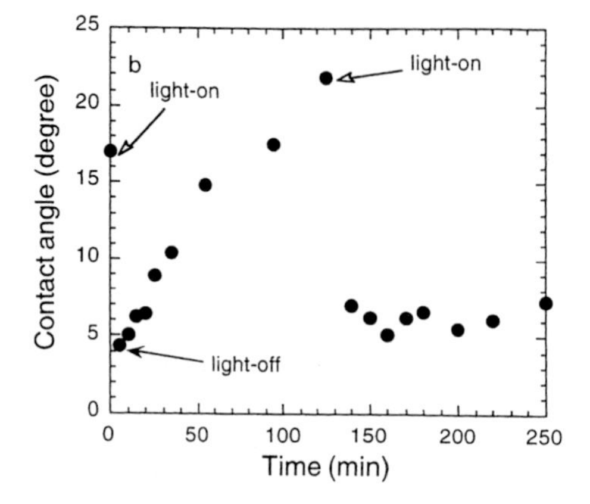

Fig. 11.21 a) Schematic representation of the superhydrophilicity phenomenon. b) Water contact angle on TiO_2-coated glass as dependent on time. (After ref. [112])

the superoxide radical is an unstable organotetroxide which decomposes to further products. In most other cases the reactions seem to be even more complicated. Nevertheless, several pathways for the various reactions have been suggested [109]. This method is also applicable for gas phase treatments [109] as shown, for example, in the oxidation of aldehyde [105, 112].

The technique of photocatalytic purification of water is especially suitable for small concentrations of pollutants. Compared with other techniques, obviously here no other dangerous compounds like dioxine are formed. Another interesting application is the photocatalytic treatment of oil slicks [113]. In order to keep the photocatalytic material on the surface, TiO_2 was deposited on hollow microbeads of aluminosilicate.

In all cases where sunlight is used for detoxification processes the efficiency is rather small because TiO_2 absorbs light only in the UV ($\lambda \leq 400$ nm). Accordingly, various attempts have been made to increase the activity. One way of doing this is the use of TiO_2 particles loaded with Pt [114].

As a further application, the deposition of TiO_2 on tiles for light-induced sterilization purposes has recently been suggested [115]. In addition it has been shown that illumination of TiO_2 powder can lead to the killing of T-24 human bladder cancer cells [105, 116]. Interestingly, as a kind of side-effect it was found that tiles or glass windows covered by a thin layer of TiO_2, stay much cleaner than uncovered glasses [105, 112]. A detailed analysis of this effect has shown that an illumination of these TiO_2 films leads to a decrease of the contact angle between a water droplet and the substrate (Fig. 11.21) [112]. This was also found with oily liquids, i.e. the surface becomes hydrophilic and oleophilic [117]. The oleophilic areas are oxidized by the photocatalytic reaction so that dust does not stick to the surface and contaminants are easily washed away by rain. In addition, besides this self-cleaning effect, windows do not appear to fog because no water droplets are formed [117].

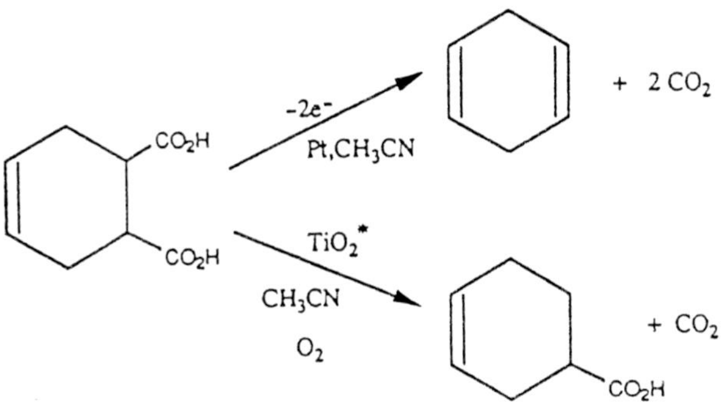

Fig. 11.22 Reaction scheme for the oxidation of cyclohexene dicarboxylate. (After ref. [119])

11.2.2 Light-Induced Chemical Reactions

Suspensions of semiconductor particles (mainly TiO_2) have been used for the selective photo-oxidation of organic molecules or functional groups. To provide electroneutrality within the particle, oxygen served as an electron acceptor in organic solvents. The advantages of this method compared with pure chemical reactions and conventional electrochemical routes are threefold [108]: (1) restriction to single-electron transfer routes; (2) control of the environment in which the radical ion is generated, and (3) preferential adsorption effects.

Concerning the first point, it should be mentioned that two-electron oxidation occurs with organic substrates in conventional electrochemical routes when an intermediate generated in the primary electrochemical step is itself oxidized more easily than its parent substrate. At a given electrode potential the intermediate is thus further oxidized as fast as it is formed. In the case of excited particles, however, only one hole is transferred to the organic substrate because it may take a rather long time before the same particle absorbs another photon (see Section 9.2.6). Thus oxidizable intermediates can accumulate or react by alternative routes. One example investigated by Fox and her group, is the oxidation of cyclohexene dicarboxylate [118, 119]. The conventional electrochemical oxidation leads to 1.4-cyclohexadiene as shown by the upper path of the reaction scheme in Fig. 11.22.

In the case of irradiated semiconductor suspensions, the monoacid is formed in a one-hole transfer reaction as given by the lower path in the reaction scheme. This reaction is of great interest because the monoacid described here is difficult to synthesize by ordinary chemical methods. Many other reactions have been investigated [118].

11.3 Etching of Semiconductors

Etching of semiconductors, in the dark and under illumination, plays an important role in device fabrication. Since Si and GaAs are mostly used in devices, research about etching processes concentrates on these materials. One must distinguish between purely chemical and electrochemical etching. In the case of chemical etching, the semiconductor material is usually oxidized as, for example, Si by HNO_3 and the resulting

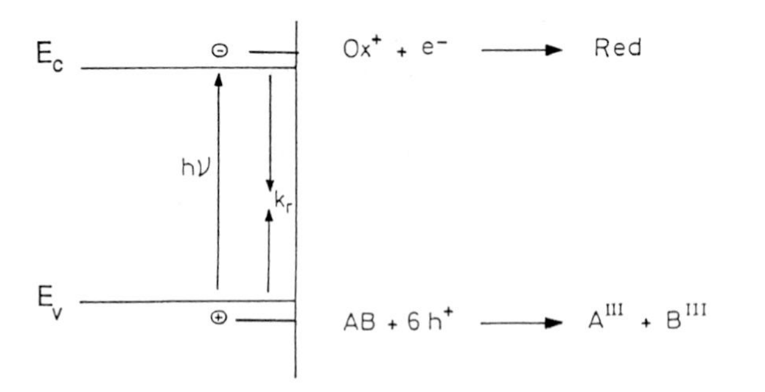

Fig. 11.23 Model for photoetching at open-circuit potential. (After ref. [122])

SiO_2 is dissolved by HF. The electrochemical etching is characterized by hole transfer. Taking GaAs as an example, the dissolution at low pH can be described by [120]:

$$GaAs + 3H_2O + 6h^+ \rightarrow Ga^{3+} + H_3AsO_3 + 3H^+ \tag{11.27}$$

When etching is carried out at open-circuit potentials, these holes must be supplied by an oxidizing agent of a fairly positive standard potential such as Ce^{4+} [121]. We have then

$$6Ce^{4+} \rightarrow 6Ce^{3+} + 6h^+ \tag{11.28}$$

Since this process occurs via the valence band it works for a p-type electrode as well as for an n-type material. This is an etching process which occurs at metal electrodes in the same way.

Photoetching is of special interest, because the semiconducting material can be locally etched away by focussing a light beam on a certain spot (or by using a laser beam). To achieve enhanced etching under illumination at open-circuit potentials, it is necessary to involve both majority and minority carriers, in the dissolution mechanism. The principle of photoetching is illustrated for an n-type semiconductor in Fig. 11.23. During light excitation, the anodic decomposition occurs via hole consumption (valence band process), whereas electrons are transferred to the Ox form of a corresponding redox system (conduction band process). This principle can be applied to n- as well as to p-type semiconductors. In the case of a p-type electrode, the anodic dissolution can occur in the dark, but the electron transfer reaction requires light excitation. With n-type, the reduction of the redox system occurs in the dark, but holes have to be excited for the dissolution process. Accordingly, corresponding anodic and cathodic partial currents are required under open-circuit conditions. Suitable redox couples ought to be those where the standard potential occurs somewhat below the conduction band. Investigations with redox couples such as $Eu^{2+/3+}$, $V^{2+/3+}$ or $Cr^{2+/3+}$ have shown, however, that no etching takes place at GaAs [121]. This result is due to extremely small exchange currents at the open-circuit potential. In particular, the small rate of electron transfer from the conduction band to the redox system is responsible for this effect.

Interestingly, some rather complex redox systems are much more effective with respect to photoetching. A typical example for GaAs is H_2O_2. Here, the reduction current in the presence of H_2O_2 occurs at a very high rate at n-type as well at p-type GaAs electrodes; the overvoltages for this reduction process are surprisingly low. Gerischer has suggested that these rates can be possible for reactions via surface states, if the charge transfer from one of the bands into the surface state is rate-limiting [123]. However, this would require an even faster rate for the transfer from the surface state to the redox system for which the same maximum rate should be valid. Accordingly, this cannot really explain the unusually high rates. Interestingly, a high rate has also been found for the reduction of H_2O_2 at n-GaAs [124]. In this case, surface radicals are formed by pure chemical etching according to the reaction

$$\begin{array}{lll} & & \quad HO^- \;\bullet OH \\ -Ga\bullet\,\bullet As- & + \; H_2O_2 \;\rightarrow & -Ga^+\bullet As- \\ \;\;| \quad\;\; | & & \quad\;\; | \quad\;\; | \end{array} \tag{11.29}$$

Accordingly, a surface state (surface radical) is formed which acts as an effective electron trap, i.e.

$$\begin{array}{lll} HO^- \;\bullet OH & & \qquad\quad \bullet OH \\ -Ga^+\,\bullet As- & + \; e^- \;\rightarrow & -Ga\bullet\;\bullet As{-}OH^- \\ \;\;| \qquad\; | & & \quad\;\; | \qquad\; | \end{array} \tag{11.30}$$

The $^\bullet OH$ radical is further reduced by hole injection into the valence band:

$$\bullet OH \rightarrow OH^- + p^+ \tag{11.31}$$

This is actually a detailed description of the current-doubling process found for the reduction of H_2O_2 as already discussed in Section 7.6. Since the electron transfer from the conduction band into the surface state (Eq. 11.30) can be rather fast and the corresponding rate may be determined by the thermal velocity of electrons toward the surface, it has to be assumed that the initial chemical etching reaction (Eq. 11.29) is even faster.

According to this reaction mechanism, the surface radical obviously plays a key role in the etching process [122]. Other redox systems, such as Br_2 and BrO_3^{2-}, behave similarly. Kelly and co-workers have studied the corresponding reaction mechanisms in detail. They proposed a unified model which they applied for all three redox systems [124, 125].

Finally it should be mentioned that this technique can also be applied for etching small and deep grooves into n-type semiconductors as illustrated in Fig. 11.24. The

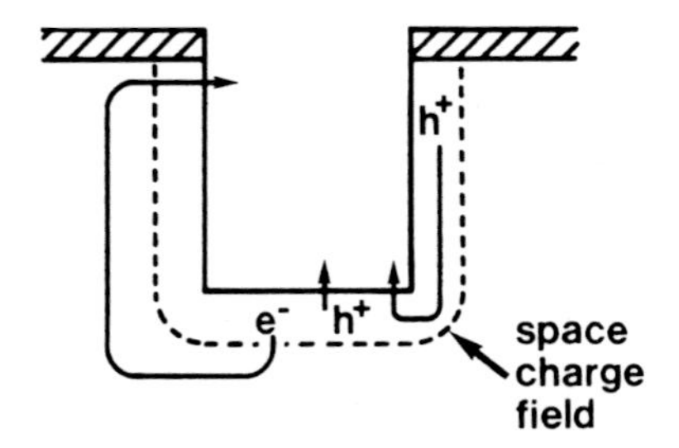

Fig. 11.24 Mechanism of charge migration in a photoetching structure. On the left, migration of majority carriers (electrons) occurs from the bottom to the less intensely illuminated side walls. There, they are used in reduction of an oxidizing agent in solution or recombination with an injected hole. On the right, minority holes generated or injected at the walls migrate towards the base through the two-dimensional majority carrier depleted space charge field. (After ref. [127])

focussed light beam produces electron–hole pairs at the bottom of the growing groove where the holes are consumed for the anodic decomposition of the semiconductor. The electrons, available everywhere in the n-type material, can be transferred anywhere, i.e. also at the top of the groove. Accordingly, there is no diffusion limitation for the redox ions within the groove [126, 127].

11.4 Light-Induced Metal Deposition

Selective metal deposition is of interest in several applications such as the formation of conduction patterns for integrated circuits and semiconductor devices. Instead of depositing a complete metal film and producing the pattern by selective etching, there is the interesting goal of forming the pattern directly by photodeposition. The basic concept of the procedure was already developed 25 years ago [113, 128, 129], but the results were not sufficiently reproducible. The situation has recently improved because of a better understanding of the primary reaction steps. The principles of the photodeposition are as follows:

When an n-type semiconductor which is in contact with a metal ion-containing electrolyte is illuminated, then two equal partial currents occur under open-circuit conditions (Fig. 11.25a). The anodic photocurrent is due to O_2 formation in H_2O, whereas the cathodic partial current corresponds to the reduction of the metallic ions. Since the holes cannot diffuse very far, most of them collect at the illuminated interface. In the case of an n-type semiconductor, sufficient electrons are availabe everywhere, so that metal deposition should occur at illuminated as well as at dark surface sites (Fig. 11.25b), according to which conclusion, selective deposition would be impossible. Experimentally, however, selective metal deposition has been observed, e.g. at CdS at illuminated surfaces [130] and at TiO_2 at the dark sites [131]. In the case of CdS, this phenomenon was interpreted as a downward shift of the energy bands at the illuminated surface which is more favorable to an electron transfer there [130]. The result obtained with TiO_2 has been explained by strong internal and external recombination

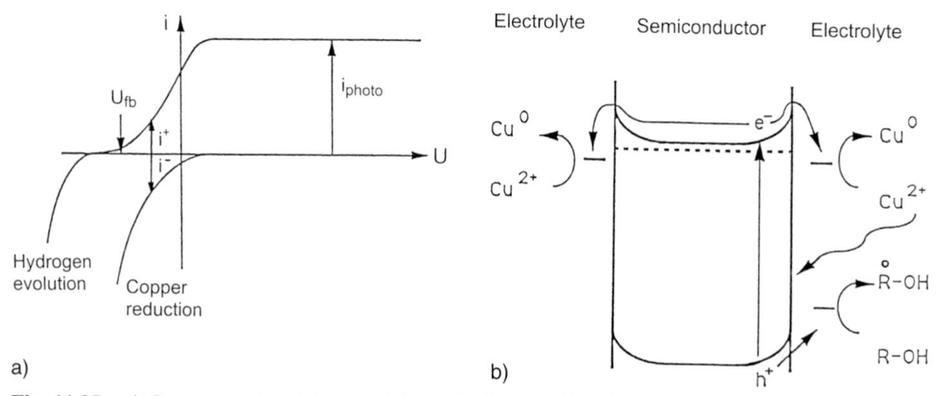

Fig. 11.25 a) Current–potential curve (theoretical) of an illuminated TiO_2 electrode in the presence of metal ions in the electrolyte. b) Energy model for metal deposition. (After ref. [132])

[131, 132]. Since TiO_2 is more suitable for application, much effort has been devoted to influencing the surface sites where deposition occurs. Many attempts have been made in applying different surface treatments or using other hole acceptors in the solution. For instance, the result described above, was only obtained with very well polished and etched TiO_2 surfaces, whereas with rough surfaces Cu deposition was found everywhere on the crystal. A real improvement was finally obtained by using another hole acceptor, such as methanol or formic acid. In this case, the metal deposition was found almost entirely on the illuminated side, and the deposition rate was considerably enhanced. This result was explained as follows.

In solutions without Cu^{2+}, the anodic photocurrent is increased upon addition of CH_3OH due to the current-doubling effect (see Section 7.6), i.e.

$$CH_3OH + h^+ \rightarrow \bullet CH_2OH + H^+ \tag{11.32}$$

$$\bullet CH_2OH \rightarrow H_2CO + H^+ + e^- \tag{11.33}$$

According to this reaction scheme, only one hole created by light excitation is required for the first reaction step. In the second step, an electron is injected from the radical into the conduction band, which proceeds without any light excitation. Upon addition of Cu^{2+} ions, the current-doubling effect disappears. Accordingly, the second reaction (Eq. 11.33) does not take place anymore. It has been concluded from this result that the radical must be capable of reducing Cu^{2+}, i.e.

$$2\bullet CH_2OH + Cu^{2+} \rightarrow H_2CO + 2H^+ + Cu^0 \tag{11.34}$$

Since the radicals are formed by hole transfer only at the illuminated sites, Cu^0 must also be deposited there [132].

Another way of getting metal deposition only at the illuminated sites is the use of very highly ohmic TiO_2 material where the initial electron density is extremely low. The only reasonable electron density is then produced by light excitation. In this case both electrons and holes exist only at the illuminated areas so that the reduction of Cu^{2+} can also only occur at these sites. This requires, however, very careful production of the TiO_2 layer.

Appendices

Appendix A
List of Major Symbols

Symbol	Meaning	Usual dimension
A	Electrode area	cm^2
A_R	Richardson constant	$A\ cm^{-2}$
a	Activity coefficient	
C_H	Differential Helmholtz capacity	$F\ cm^{-2}$
C_{sc}	Differential space charge capacity	$F\ cm^{-2}$
c_j	Concentration of species *j* in solution	$M\ cm^{-3}$
D_j	Diffusion constant of species *j* in the solid or in the electrolyte	$cm^2\ s^{-1}$
$\mathscr{D}$	Electric displacement	$C\ cm^{-2}$
D_{redox}	Density of states of a redox system	cm^{-3}
d_{sc}	Thickness of space charge region	cm^{-1}
E	Electron energy	eV
E^o	Standard electron energy of a redox system	eV
E_g	Bandgap energy	eV
E_A	Energy of electron acceptors in semiconductor	eV
E_D	Energy of electron donors in semiconductor	eV
E_c	Energy of the lower edge of the conduction band	eV
E_v	Energy of the upper edge of the valence band	eV
E_F	Fermi level energy	eV
E_t	Electron energy in surface states	eV
E_{fb}	Flatband energy	eV
$E^*(M/M^*)$	Electron energy of a redox system in its excited state	eV
$\mathscr{E}$	Electric field strength	$V\ cm^{-1}$
e	Elementary charge	A s
F	Faraday constant	C
f	Fermi distribution	
G	Gibbs free energy	eV
ΔG	Gibbs free energy change in a chemical process	kJ, eV
ΔG^0	Standard Gibbs free energy change in a chemical process	kJ, eV
$\Delta G^{\#}$	Standard Gibbs free energy of activation	kJ/mol, eV
ΔH	Enthalpy change in a chemical process	kJ, eV
h	Planck constant	J s
j	Current density	$A\ cm^{-2}$
j_c^+	Anodic current density via the conduction band of a semiconductor	$A\ cm^{-2}$
j_c^-	Cathodic current density via the conduction band	$A\ cm^{-2}$
j_v^+	Anodic current density via the valence band	$A\ cm^{-2}$
j_v^-	Cathodic current density via the valence band	$A\ cm^{-2}$
j_{rec}	Recombination current within the semiconductor	$A\ cm^{-2}$

Symbol	Meaning	Usual dimension
j_{lim}	Diffusion limiting current	A cm^{-2}
j_F	Faraday current density	A cm^{-2}
j_{ph}	Photocurrent density	A cm^{-2}
j_0	Exchange current density	A cm^{-2}
K	Equilibrium constant	depends on order
kT	Thermal energy	eV
k_j^i	Rate constant	depends on order
k	Boltzmann constant	J/K
L_D	Thickness of Debye layer	cm
L_n	Diffusion length of electrons in a semiconductor	cm
L_p	Diffusion length of holes in a semiconductor	cm
m_e	Free electron mass	Kg
m_e^*	Effective electron mass	Kg
m_h^*	Effective hole mass	Kg
N_c	Density of states at the lower edge of the conduction band	cm^{-3}
N_v	Density of states at the upper edge of the valence band	cm^{-3}
N_D	Density of donor states in the semiconductor	cm^{-3}
N_A	Density of acceptor states	cm^{-3}
N_t	Density of surface states	cm^{-2}
n_0; n_s	Electron density in the bulk and at the surface of a semiconductor	cm^{-3}
n_i	Intrinsic electron density	cm^{-3}
n	Ideality factor	
p_0; p_s	Hole density in the bulk and at the surface of a semiconductor	cm^{-3}
Q_{sc}	Space charge below the semiconductor surface	A s cm^{-2}
R	(a) Gas constant	J mol^{-1} K^{-1}
	(b) Resistance	Ω
R_{ct}	Charge transfer resistance	Ω
R_s	Series resistance	Ω
R_{th}	Thermionic resistance	Ω
ΔS	Entrpy change in a chemical process	kJ K^{-1}
ΔS^0	Standard entropy change in a chemical reaction	kJ K^{-1}
s	Surface recombination velocity	cm s^{-1}
T	Absolute temperature	
t	Time	
U	Applied voltage	V
U_E	Electrode potential	V
U_{fb}	Flatband potential	V
U^0_{redox}	Standard redox potential	V
$U_{1/2}$	Halfwave potential	V
Z	Impedance	Ω
Z'	Real part of impedance	Ω
Z''	Imaginary part of impedance	Ω
Z_F	Impedance of Faradaic process	Ω
Z_w	Warburg impedance	Ω

Symbol	Meaning	Usual dimension
Greek Symbols		
α	(1) Absorption coefficient	cm^{-1}
	(2) Transfer coefficient	
δ_N	Nernst diffusion layer thickness	cm
ε	Dielectric constant	
ε_o	Permitivity of free space	$F\ cm^{-1}$
η	Overpotential	V
λ	(1) Wavelength	nm
	(2) Reorganization energy	eV
Λ_f	Equivalent ionic conductivity	$cm^2\ \Omega^{-1} eq^{-1}$
μ	Mobility of ions	$cm^2 V^{-1} s^{-1}$
μ_n	Mobility of electrons in a semiconductor	$cm^2 V^{-1} s^{-1}$
μ_p	Mobility of holes in a semiconductor	$cm^2 V^{-1} s^{-1}$
$\bar{\mu}_{e,\ redox}$	Electrochemical potential of electrons in a redox system	eV
ϱ	Resistivity	Ωcm
σ	Conductivity	$(\Omega cm)^{-1}$
ϕ	Electrostatic potential	V
ϕ_b	Barrier height	V
$\Delta\phi_{sc}$	Potential across the space charge layer	V
$\Delta\phi_H$	Potential across the Helmholtz layer	V
χ	Electron affinity	eV

Appendix B
Physical Constants

Quantity	Symbol	Value
Elementary charge	e	$1{,}6 \times 10^{-9}$ A s
Electron volt	eV	$1\ eV = 1.6 \times 10^{-19}$ VA s
Faraday constant	F	$1\ F = 9.65 \times 10^4$ As /equiv
Planck constant	h	6.63×19^{-34} J s
Boltzmann constant	k	$1.38 \times 10^{-23}\ J\ K^{-1}$
Avogadro constant	N_{avo}	$6.02 \times 10^{23}\ mol^{-1}$
Gas constant	R	$8.31\ J\ mol^{-1} K^{-1}$
Permitivity in vacuum	ε_o	$8.85 \times 10^{-14}\ F\ cm^{-1}$
Wavelength of 1 eV quantum	λ	1.24 μm

Appendix C
Lattice Parameters of Semiconductors

Semiconductor material	Crystal structure
C	diamond
Si	diamond
Ge	diamond
SiC	wurtzite
	zincblende
GaP	zincblende
GaAs	zincblende
InP	zincblende
ZnO	rocksalt
ZnS	zincblende
	wurtzite
CdS	zincblende
	wurtzite
CdSe	wurtzite
TiO_2	rutile
	anatase
SnS_2	layered dichalcogenide
MoS_2	layered dichalcogenide
WSe_2	layered dichalcogenide
FeS_2	pyrite

Appendix D
Properties of Important Semiconductors[a)]

Bandgap (eV)			Mobility ($cm^2V^{-1}s^{-1}$)		Effective mass (m^*/m_o)	
			electrons	holes	electrons	holes
C	5.47	(i)	1800	1200	0.2	0.25
Si	1.12	(i)	1500	450	0.98	0.16
Ge	0.66	(i)	3900	1900	1.64	0.04
SiC	3.0–3.2	(i)	400	50	0.6	1.0
GaP	2.26	(i)	110	75	0.82	0.62
GaAs	1.42	(d)	8500	400	0.067	0.082
InP	1.35	(d)	4600	150	0.077	0.64
ZnS	3.68	(d)	165	5	0.4	
ZnO	3,35	(d)	200	180	0.27	
CdS	2.42	(d)	340	50	0.21	0.8
CdSe	1.7	(d)	800		0.13	0.45
TiO_2	3.1	(i)	100			
SnS_2	2.2	(d)				
MoS_2	1.23[b)]	(i)	≈ 200 (parallel to			
WSe_2	1.2[b)]	(i)	≈ 100 basal plane)			
FeS_2	0.95[b)]	(d)	≈ 100			

[a)] Most data are taken from Sze, S.M., Physics of Semiconductor Devices, 2nd edition, John Wiley & Sons, New York 1981

[b)] Taken from W. Jägermann and H. Tributsch, Progr. Surf. Sci. 29, 1 (1988)

Appendix E
Effective Density of States and Intrinsic Carrier Densities

Material	E_g (eV)	Density of states in conduction band (cm^{-3})	Density of states in valence band (cm^{-3})	Carrier density intrinsic (cm^{-3})
Ge	0.66	1.04×10^{19}	6×10^{18}	1.5×10^{13}
Si	1.12	2.8×10^{19}	1.04×10^{19}	3.2×10^{10}
InP	1.35	5.8×10^{17}	1.4×10^{19}	5.4×10^{6}
GaAs	1.42	4.7×10^{17}	7.0×10^{18}	8.4×10^{5}
GaP	2.26	2.0×10^{19}	1.25×10^{19}	3.7×10^{-1}
CdS	2.42	2.6×10^{18}	1.9×10^{19}	6.7×10^{-3}
SiC	3.1	1.25×10^{19}	2.7×10^{19}	2.2×10^{-8}
ZnO	3.35	2.6×10^{18}		$\approx 2 \times 10^{-11}$
ZnS	3.68	6.8×10^{18}		$\approx 8 \times 10^{-14}$
C	5.47	2.4×10^{18}	3.3×10^{18}	2.4×10^{-29}

Appendix F
Major Redox Systems and Corresponding Standard Potentials

a) Aqueous Solutions

Reaction	Potential, V
$Cr_2O_7^{2-} + 14\ H^+ + 6e^- \Leftrightarrow 2\ Cr^{3+} + 7\ H_2O$	–1.33
$S + 2e^- \Leftrightarrow S^{2-}$	–0.51
$Eu^{3+} + e^- \Leftrightarrow Eu^{2+}$	–0.44
$Cr^{3+} + e^- \Leftrightarrow Cr^{2+}$	–0.43
$V^{3+} + e^- \Leftrightarrow V^{2+}$	–0.27
$Cu^{2+} + e^- \Leftrightarrow Cu^{1+}$ (in 6M HCl)	0.16
$[Fe(CN)_6]^{3-} + e^- \Leftrightarrow [Fe(CN)_6]^{4-}$	0.36
$I_3^- + 2e^- \Leftrightarrow 3\ I^-$	0.54
p–benzoquinone + $2H^+ + 2e^- \Leftrightarrow$ hydroquinone	0.70
$Fe^{3+} + e^- \Leftrightarrow Fe^{2+}$	0.77
$[Fe(phen)_3]^{3+} + e^- \Leftrightarrow [Fe(phen)_3]^{2+}$	1.06
$Br_2 + 2e^- \Leftrightarrow 2\ Br^-$	1.07
$Ru(bipy)_3^{3+} + e^- \Leftrightarrow Ru(bipy)_3^{2+}$	1.24
$Cl_2 + 2e^- \Leftrightarrow 2\ Cl^-$	1.36
$Ce^{4+} + e^- \Leftrightarrow Ce^{3+}$	1.44
$H_2O_2 + 2e^- \Leftrightarrow 2\ OH^-$	1.78
$S_2O_8^{2-} + 2e^- \Leftrightarrow 2\ SO_4^{2-}$	2.01

b) In Acetonitrile (vs. Ag/AgCl)

Reaction	Potential, V
Cobaltocene: $Co(Cp)_2^{1+} + e^- \Leftrightarrow Co(Cp)_2$	–1.3
Methylviologene: $MV^{2+} + e^- \Leftrightarrow MV^{1+}$	–1.21
Dimethylferrocene: $DMFe(Cp)_2^{1+} + e^- \Leftrightarrow DMFe(Cp)_2$	–0.1
Ferrocene: $Fe(Cp)_2^{1+} + e^- \Leftrightarrow Fe(Cp)_2$	0.03

Appendix G
Potentials of Reference Electrodes

Reference electrode	Potential vs. NHE (Volts)
Hg/Hg_2SO_4, K_2SO_4 (saturated)	+0.64
Hg/Hg_2Cl_2, KCl (0.1 M)	+0.33
Hg/Hg_2Cl_2, KCl (1 M), NCE	+0.28
Hg/Hg_2Cl_2, KCl (saturated), SCE	+0.24
Ag/AgCl, KCl (saturated)	+0.20

References

References for Chapter 1

[1] Kittel, C., Introduction to Solid State Physics, 5th ed. John Wiley & Sons, Inc. New York 1976
[2] Smith, R.A., Semiconductors, 2nd ed. Cambridge University Press, London 1979
[3] Moss, T.S., Handbook on Semiconductors, Vol.1–4, North Holland, Amsterdam 1986
[4] Pankove, J.I., Optical Processes in Semiconductors, Dover Publications Inc., New York 1971
[5] Madelung, O., Physics of III-V Compounds, John Wiley, New York 1964
[6] Sutton, A.P., Electronic Structure of Materials, Oxford Science Publications, Clarendon Press, Oxford 1993
[7] Sze, S.M., Physics of Semiconductor Devices, John Wiley & Sons, New York 1981
[8] Dexter, K.L. and Knox, R.S., Excitons, John Wiley, New York 1965
[9] Gutman, F. and Lyons, L.E., Organic Semiconductors, John Wiley, New York 1967
[10] van der Pauw, L.J., Philips Res. Rep. 13, 1 (1958)
[11] Chelikowsky, J.R. and Cohen, M.L., Phys. Rev. B14, 556 (1976)

References for Chapter 2

[1] Jägermann, W., in: Modern Aspects of Electrochemistry, Vol. 30 (eds. E. White et al.) Plenum Press, New York 1996, p.1
[2] Hering, C. and Nichols, M.H., Rev. Mod. Phys. 21, 185 (1949)
[3] Duke, C.B., Adv. Sol. State Phys. 33, 1 (1994)
[4] Kalm, A., Surf. Sci. Rep. 3 193 (1983)
[5] Many, A., Goldstein, Y. and Grover, N.B., Semiconductor Surfaces, North Holland Publishing Company, Amsterdam 1965
[6] Morrison, S.R., The Chemical Physics of Surfaces, Plenum Press, New York 1977
[7] Lüth, H., Surfaces and Interfaces of Solids, Springer Verlag, Berlin 1993
[8] Forstmann, F., in: Photoemission and the Electronic Properties of Surfaces (eds.: B. Feuerbacher et al.), John Wiley, Chichester 1978
[9] Jones, R.O., in Surface Physics of Semiconductors and Phosphors, Academic Press, New York 1975
[10] Garcia-Moliner, F. and Flores, F., Introduction to the Theory of Solid Surfaces, Cambridge University Press, Cambridge, UK 1979
[11] Desjonqueres, M.C. and Spanjaard, D., Concepts in Surface Physics, Springer Verlag, Berlin 1993
[12] Brillson, L.J., Surf. Sci. Reports 2, 123 (1982)
[13] Archer, R.J. and Atalla, G., Am. Acad. Sci. New York 101, 697 (1963)
[14] Traum, M.M., Rowe, J.E., and Smith, N.V., J. Vac. Sci. Technol.12, 298 (1975)
[15] Himpsel, F.J. and Eastman, D.E., J. Vac. Sci. Technol. 16, 1297 (1979)
[16] Sze, S.M., Physics of Semiconductor Devices, 2nd edition, John Wiley and Sons, New York 1981
[17] Kurtin, S., McGill, T.C. and Mead, C.A., Phys. Rev. Lett. 22, 1433 (1969)
[18] Mead, C.A., Appl. Phys. Lett. 6, 103 (1965)

[19] Mönch, W., Semiconductor Surfaces, Springer Verlag, Berlin 1993
[20] Rhoderick, E.H. and Williams, R.H., Metal-Semiconductor Contacts, 2nd ed., Oxford Science Publication, Oxford 1988
[21] Spicer, W.E., Lindau, I., Skeath, P. and Su, C.Y., J. Vac. Sci. Technol. 17, 1019 (1980); Phys. Rev. Lett. 44, 420 (1980)
[22] Lindau, I. and Kendelwicz, T., Crit. Rev. Sol. State Mater. Sci. 13, 27 (1986)
[23] Flores, F. and Tejedor, C., J. Phys. C (Sol. State Phys.) 20, 145 (1985)
[24] Aspnes, D.E. and Heller, A., J. Vac. Sci. Technol. B1, 602 (1983)
[25] Aspnes, D.E. and Heller, A., J. Phys. Chem. 87, 919 (1983)
[26] Reineke, R. and Memming, R. Surf. Sci. 192, 66 (1987)
[27] Bethe, H.A., MIT Radiat. Lab. Rep. 42–12 (1942)
[28] Chang, C.Y. and Sze, S.M., Solid State Electron. 13, 727 (1970)
[29] Shockley, W., Electrons and Holes in Semiconductors, van Norstrand Comp., New York 1950
[30] Lewerenz, H.J. and Jungblut, H., Photovoltaic, Springer Verlag, Berlin 1995
[31] Gärtner, W.W., Phys. Rev. 116, 84 (1959)
[32] Shockley, W. and Read, W.T., Phys. Rev. 87, 835 (1952)
[33] Stevenson, D.T. and Keyes, R.J., Physica 20, 1041 (1954)
[34] Morrison, R.S., Electrochemistry at Semiconductor and Oxidized Metal Electrodes, Plenum Press 1980
[35] Uhlendorf, I., Reineke-Koch, R. and Memming, R. J. Phys. Chem. 100, 4930 (1996)

References for Chapter 3

[1] Lyons, E.H., Introduction in Electrochemistry, D.C. Heath, Boston 1967
[2] Bockris, J. O'M. and Reddy, A., Modern Electrochemistry, Plenum Press, New York 1970
[3] Bard, A.J. and Faulkner, L.R. Electrochemical Methods, John Wiley, New York 1980
[4] Bockris, J. O'M. and Khan, S.U.M., Surface Electrochemistry, Plenum Press, New York 1993
[5] Hamnett, A., Hamann, C.H. and Vielstich, W., Electrochemistry, John Wiley and Sons, New York 1997
[6] Atkins, P.W., Physical Chemistry, Oxford University Press, Oxford, UK.1978
[7] Bard, A.J., Memming, R. and Miller, B., Pure & Appl. Chem. 63, 589 (1991)
[8] Gerischer, H., in: Physicall Chemistry (eds.: M. Eyring et al.) Academic Press, New York 1970
[9] Lohmann, F., Z. Naturforsch. 22a, 843 (1967)
[10] Advances in Electrochemistry and Electrochemical Engineering, Vol. 10 (eds.: Gerischer, H. and Tobias, C.W.) John Wiley, New York 1977, p. 213
[11] Cash, B., in: Reactions of Molecules at Electrodes (ed. N.S. Hush) Wiley Interscience, New York 1971, p. 46
[12] Randles, J.B.E., Discuss. Faraday Soc. 24, 194 (1957)
[13] Gomer, R., J. Chem. Phys. 66, 4413 (1977)
[14] Kötz, R., Neff, H. and Müller, K., J. Electroanal. Chem. 215, 331 (1986).

References for Chapter 4

[1] Delahay, P., New Instrumental Methods in Electrochemistry, Interscience Publishers, Inc., New York 1954
[2] Vetter, K., Electrochemical Kinetics, Academic Press, New York 1967

[3] Bockris, J.O'M. and Reddy, A., Modern Electrochemistry, Plenum Press, New York 1970
[4] Sawyer, D.T. and Roberts, J.L., Experimental Electrochemistry for Chemists, John Wiley, New York 1974
[5] Greef, R., Peat,R., Peter, L.M., Pletcher,D. and Robinson, J., Instrumental Methods in Electrochemistry, Ellis Horwaood, Chichester, 1985.
[6] Bruckenstein, S. and Miller, B. Acc.Chem. Res. 10, 54 (1977)
[7] Pleskov, Yu.,V. and Filinovskii, The Rotating Disc Electrode, Consultants Bureau, New York 1976
[8] Memming, R., Ber. Bunsenges. Phys. Chem. 81, 732 (1977)
[9] Wipf, D.O. and Bard, A.J., J. Electrochem. Soc. 138, 469 (1991)
[10] Harten, H.U. and Schultz, W., Ztschr. f. Phys. 141, 319 (1955)
[11] Brattain, W.H. and Garrett, C.G.B., Bell Syst. Tech. J. 34, 129 (1955)
[12] Pleskov, Yu.V., Dokl. Akad. Nauk SSSR 129,111 (1959)
[13] Gabrielli, C., Identification of Electrochemical Processes by Frequency Response Analysis, Schlumberger Technologies (004183) 1983
[14] McDonald, J.R., Impedance Spectroscopy, John Wiley & Sons, New York 1987
[15] Vanmaeckelbergh, D., Electrochim. Acta 42, 1135 (1987)
[16] Li, J. and Peter, L.M., J. Electroanal. Chem. 199, 1 (1986)
[17] Harrick, N.J., Internal Reflection Spectroscopy, Interscience Publ. Comp., New York (1967)
[18] Beckmann, K.H., Surf. Sci. 3, 14 (1965)
[19] Harrick, N.J. and Beckmann, K.J., in: Characterization of Solid Surfaces (eds. P.F. Kane and A.B. Larrabee) Chapter 10, Plenum Press, New York 1974, p. 243
[20] Rao, A.V., Chazalviel, J.N. and Ozanam, F., J. Appl. Phys. 60, 696 (1986)
[21] Rao, A.V. and Chazalviel, J.N., J. Electrochem. Soc. 134, 2777 (1987)
[22] Sherwood, P.A.M., Chem. Soc. Reviews 14, 1 (1985)
[23] Jägermann, W., in Modern Aspects of Electrochemistry, Vol. 30 (eds. E. White et al.) Plenum Press, New York 1996, p.1
[24] Jägermann, W., in: Photoelectrochemistry and Photovoltaics of Layered Semiconductors (ed. R. Arnchamy), Kluwer, Dordrecht, NL 1992
[25] Uosaki, K. and Koinuma, M., in: Electrochemical Nanotechnology (eds. W.L. Lorenz and W. Plieth), Wiley-VCH, Weinheim, Germany 1998
[26] Dakkouri, A.S., Dietterle, M. and Kolb, D.M., Festkoerperprobleme/ Adv. Sol. State Phys. Vol. 36 (ed. R. Helbig), Vieweg, Braunschweig/Wiesbaden, 1997, p.1
[27] Scheeweiß, M.A. and Kolb, D.M., Das Rastertunnelmikroskop in der Elektrochemie, Chemie in unserer Zeit, in press
[28] Itaya, K. and Tosnita, E., Surf. Sci. 219, 2515 (1989)
[29] Allongue, P. Brune,H. and Gerischer, H., Surf. Sci. 275, 414 (1982)
[30] Allongue, P., Costa-Kieling, V. and Gerischer, H., J. Electrochem. Soc. 140, 1009 (1993)
[31] Itaya, K., Prog. Surf. Sci. 58, 121 (1998)

References for Chapter 5

[1] Hoffmann, M.R., Martin, S.T., Choi, W. and Bahnemann, D.W., Chem. Rev. 95, 69 (1995)
[2] Pettinger, B., Schöppel, H.R., Yokoyama, T. and Gerischer, H., Ber. Bunsenges. Phys. Chem. 78, 1024 (1974)
[3] Memming, R. and Schwandt, G., Surf. Sci. 4, 109 (1966)
[4] Bard, A.J. and Faulkner, L.R. Electrochemical Methods, John Wiley, New York 1980
[5] Memming, R., in Electroanalytical Chemistry (ed. A.J. Bard.) Vol.11, Marcel Dekker, New York 1979, p.1

[6] Gomes, W.P. and Cardon, F., Progr. Surf. Sci. 12, 155 (1979)
[7] Pleskov, Yu.V., in: Progress in Surface and Membrane Science, Vol.7, Academic Press, New York 1973, p. 57.
[8] Pleskov, Yu.V., in Comprehensive Treatise of Electrochemistry, Vol.1 (eds. J.O'M. Bockris et al.) Plenum Press, New York 1980, p. 291
[9] Memming, R., in Comprehensive Treatise of Electrochemistry, Vol.7 (eds. B.E. Conway et al.) Plenum Press, New York 1983, p. 529
[10] Myamlin, V.A. and Pleskov, Yu.V., Electrochemistry of Semiconductors, Plenum Press, New York 1967.
[11] Morrison, S.R., Electrochemistry at Semiconductor and Oxidized Metal Electrodes, Plenum Press, New York 1980
[12] Delahay, P., Double Layer and Electrode Kinetics, John Wiley, New York 1965
[13] Memming, R. and Schwandt, G., Ang. Chem. Int. Ed. 79, 833 (1967)
[14] Kingston, R.H. and Neustadter, S.F., J.Appl. Phys. 26, 718 (1955)
[15] Bohnenkamp, K. and Engell, H.J., Ztschr. Elektrochem. Ber. Bunsenges. Phys. Chem. 61, 1184 (1957)
[16] Brattain, W.H. and Boddy, P.J., J. Electrochem. Soc. 109, 574 (1962)
[17] Harten, H.U. and Memming, R., Phys. Lett. 3, 95 (1962)
[18] Hoffmann-Perez, M. and Gerischer, H., Ztschr. Elektrochem., Ber. Bunsenges. Phys. Chem. 65, 77 (1961)
[19] Boddy, P.J. and Brattain, W.H., J. Electrochem. Soc. 110, 570 (1963)
[20] Gerischer, H., Maurer, A. and Mindt, W., Surf. Sci. 4, 431 (1966)
[21] Gerischer, H. and Mindt, W., Surf. Sci. 4, 440 (1966)
[22] Memming, R. and Neumann, G., Surf. Sci. 10, 1 (1968)
[23] Memming, R. and Neumann, G., Electroanal. Chem. 21, 295 (1969)
[24] Memming, R., Surf. Sci. 2, 436 (1964)
[25] Harten, H.U. Z. f. Naturforsch. 16a, 459 (1961)
[26] Memming, R. and Schwandt, G., Surf. Sci. 5, 97 (1966)
[27] Pomykal, K.E., Fajardo, A.M. and Lewis, N.S., J. Phys. Chem. 100, 3652 (1996)
[28] Meissner, D., Memming, R. and Kastening, B., J. Phys. Chem. 92, 3476 (1988)
[29] Lewerenz, H.J., Gerischer, H. and Lübke, M., J. Electrochem. Soc. 131, 100 (1984)
[30] McEvoy, A.J., Etman, M. and Memming, R., Electroanal. Chem. 190, 225 (1985)
[31] Mayer, T., Klein, A., Lang, O., Pettenkofer, C. and Jägermann, W., Surf. Sci. 269/270, 909 (1992)
[32] Memming, R., J. Electrochem. Soc. 116, 785 (1969)
[33] Bard, A.J., Bocarsly, A.B., Fan, F., Walton, E.W. and Wrighton, M.S., J. Am. Chem. Soc. 102, 3671 (1980)
[34] Ennaoui, A. and Tributsch, H., J. Electroanal. Chem. 204, 185 (1986)
[35] Mishra, K.K. and Osseo-Asare, K., J. Electrochem. Soc. 139, 749 (1992)
[36] Fantini, M.C.A., Shen, W.M., Tomkiewicz, M. and Gambino, J.P., J. Appl. Phys. 65, 4884 (1989)
[37] Ba, B., Fotouhi, B.B., Gabouze, N., Gorochov, O. and Cachet,H., J. Electroanal. Chem. 334, 263 (1992)
[38] Meier, A., Selmarten, D.C., Siemoneit, K., Smith, B.B. and Nozik, A.J., J. Phys. Chem. B 103, 2122 (1999)
[39] Siemoneit, K., Meier, A., Reineke-Koch, R. and Memming, R., Electrochim. Acta, (2000) in press
[40] Schröder, K. and Memming, R., Ber. Bunsenges. Phys. Chem. 89, 385 (1985)
[41] Rosenwaks, Y., Thacker, B.R., Nozik, A.J., Shapira, Y. and Huppert, D., J. Phys. Chem. 97, 1042 (1993)

[42] Vanmaeckelbergh, D., Gomes, W.P. and Cardona, F., Ber. Bunsenges. Phys. Chem. 89, 994 (1985)
[43] Memming, R. and Kelly, J.J., in: Photochemical Conversion and Storage of Solar Energy (ed. J.S. Connally), Academic Press, New York, 1980, p. 243
[44] Kelly, J.J. and Memming, R., J. Electrochem. Soc. 129, 730 (1982)
[45] Memming, R., in Topics in Current Chemistry, Springer Verlag, Vol. 169, 1994, p. 105
[46] Sinn, Chr., Meissner, D. and Memming, R., J. Electrochem. Soc. 137, 168 (1990)
[47] Kühne, H.M. and Tributsch, H., J. Electroanal. Chem. 201, 263 (1986)
[48] Jägermann, W. and Tributsch, H., Prog. Surf. Sci. 29, 1 (1988)
[49] Schefold, J. and Kühne, H.M., J. Electroanal. Chem. 300, 211 (1991)
[50] Meissner, D. and Memming, R., in: Photocatalytic Production of Energy-rich Compounds (eds. G. Grassi and D.O. Hall), Elsevier, London 1988, p. 138

References for Chapter 6

[1] Marcus, R.A., J. Chem. Phys. 24, 966 (1956)
[2] Marcus, R.A., Ann. Rev. Phys. Chem.15, 155 (1964)
[3] Hush, N.S., Trans. Faraday Soc. 57, 557 (1961)
[4] Dogonadze, R.R., Kuznetsov, A.M. and Chizmadzhev, S.C., Russ. J. Phys. Chem. 38, 652 (1964)
[5] Levich, V.G., in: Advances in Electrochemistry and Electrochemical Engineering (eds. P. Delahay and C.W. Tobias), Vol. 4, Wiley Interscience, New York 1966, p. 249
[6] Gerischer, H., Ztschr. Phys. Chem. N.F. 26, 223 (1960); 26, 325 (1960); 27, 48 (1961)
[7] Gerischer, H., in: Physical Chemistry (eds. M. Eyring et al.), Vol. 4, Academic Press, New York 1970, p. 463
[8] Newton, M.D. and Sutin, N., Ann. Rev. Phys. Chem. 35, 437 (1984)
[9] Marcus, R.A., J. Phys. Chem. 94, 4152 (1990); 95, 2010 (1991)
[10] Marcus, R.A., J.Phys. Chem. 94, 1050 (1990)
[11] Smith, B.B. and Koval, C.A., J. Electroanal. Chem. 277, 43 (1990)
[12] Morrison, S.R., Electrochemistry at Semiconductor and Oxidized Metal Electrodes, Plenum Press, New York 1980
[13] Memming, R., in: Comprehensive Treatise of Electrochemistry, Vol.7 (eds. B.E. Conway et al.) Plenum Press, New York 1983, p. 529
[14] Cohen-Tannoudji, C., Din, B. and Laloe, F., Quantum Mechanics, Vol.1, John Wiley & Sons, New York, 1977
[15] Bockris, J.O'M. and Khan, S.U.M., Quantum Electrochemistry, Plenum Press, New York 1979
[16] Smith, B.B. and Hynes, J., J. Chem. Phys. 99, 6517 (1993)
[17] Landau, L., Phys. Z. Sowjet. 1, 89 (1932); 2, 46 (1932)
[18] Zener, C., Proc. Roy. Soc. A 137, 696 (1932)
[19] Schmickler, W., Interfacial Electrochemistry, Oxford University Press, London 1996; J. Electroanal. Chem. 204, 31 (1986)
[20] Anderson, P.W., Phys. Rev. Lett. 18, 1049 (1967)
[21] Muscat, J.P. and Newns, D.M., Prog. Surf. Sci. 9, 1 (1978)
[22] Kittel, C., Introduction to Solid State Physics, 5th ed. John Wiley & Sons, Inc. New York 1976
[23] Smith, B.B. and Nozik, A.J., Chem. Phys. 205, 47 (1996)
[24] Smith, B.B. and Nozik, A.J., J. Phys. Chem. B 101, 2459 (1997)
[25] Smith, B.B., private communication

References for Chapter 7

[1] Vetter, K.J., Electrochemical Kinetics, Academic Press 1967
[2] Rieger, P.H., Electrochemistry, Prentice-Hall International Editions, Englewood Cliffs, N.J. 1992
[3] Bockris, J.O'M. and Khan, S.U.M., Quantum Electrochemistry, Plenum Press, New York 1979
[4] Miller, C. and Grätzel, M., J. Phys. Chem. 95, 5225 (1991)
[5] Miller, C., Cuendet, P. and Grätzel, M., J. Phys. Chem. 95, 877 (1991)
[6] Bard, A.J. and Faulkner, L.R. Electrochemical Methods, John Wiley, New York 1980
[7] Albery, W.J., Electrode Kinetics, Oxford University Press, London 1975
[8] Levich, V.G., Physicochemical Hydrodynamics, Prentice Hall, Englewood, NJ 1962
[9] Pleskov, Yu.V. and Filinovskii, G., The Rotating Disc Electrode, Consultants Bureau, New York 1976
[10] Heyrovsky, J. and Ilcovic, J., Coll. Czech. Chem. Comm. 7, 198 (1935)
[11] Nicholson, R.S. and Shain, J., Anal. Chem. 36, 706 (1964)
[12] Myamlin, V.A. and Pleskov, Yu.V., Electrochemistry of Semiconductors, Plenum Press, New York 1967.
[13] Jägermann, W. and Tributsch, H., Prog. Surf. Sci. 29, 1 (1988)
[14] Memming, R., in Electroanalytical Chemistry (ed. A.J. Bard) Vol.11, Marcel Dekker, New York 1979, p.1
[15] Morrison, S.R., Electrochemistry at Semiconductor and Oxidized Metal Electrodes, Plenum Press, New York 1980
[16] Gomes, W.P. and Cardon, F., Prog. Surf. Sci. 12, 155 (1982)
[17] Pleskov, Yu.V., in: Comprehensive Treatise of Electrochemistry, Vol.1 (eds. J.O'M. Bockris et al.) Plenum Press, New York 1980, p. 291
[18] Memming, R., in: Comprehensive Treatise of Electrochemistry, Vol.7 (eds. B.E. Conway et al.) Plenum Press, New York 1983, p. 529
[19] Memming, R., in: Topics of Current Chemistry, Vol.169, Springer Verlag, Berlin 1994, p.106
[20] Miller, R.J.D., McLendon, G.L., Nozik, A.J., Schmickler, W. and Willig, F. Surface Electron Transfer Processes, VCH New York 1995
[21] Ba, B., Fotouhi, B.B., Gabouze, N., Gorochov, O. and Cachet,H., J. Electroanal. Chem. 334, 263 (1992)
[22] Gerischer, H., in: Physical Chemistry (eds. M. Eyring et al.), Vol. 4, Academic Press, New York 1970, p. 463
[23] Reichman, J., Appl. Phys. Lett. 36, 574 (1980)
[24] Gärtner, W.W., Phys. Rev. 116, 84 (1959)
[25] Butler, M.A., J. Appl. Phys. 48, 1914 (1977)
[26] Wilson, R.H., J. Appl. Phys. 48, 4292 (1977)
[27] Morrison, S.R., Surf. Sci. 15, 363 (1969)
[28] Memming, R. and Möllers, F., Ber. Bunsenges. Phys. Chem. 76, 475 (1972)
[29] Rimmasch, J. and Meissner, D., unpublished results
[30] Pomykal, K.E. and Lewis, N.S., J. Phys. Chem. B 101, 2476 (1997)
[31] Fajardo, A.M. and Lewis, N.S., J. Phys. Chem. B 101, 11136 (1997); Science 274, 969 (1996)
[32] Meier, A., Kocha, S.S., Hanna, M.C., Nozik, A.J., Siemoneit, K., Reineke-Koch, R. and Memming, R., J. Phys. Chem. B 101, 7038 (1997)
[33] Meier, A., Selmarten, D.C., Siemoneit, K., Smith, B.B. and Nozik, A.J., J. Phys. Chem. B 103, 2122 (1999)
[34] Diol, S.J., Poles, E., Rosenwaks, Y. and Miller, R.J.D., J. Phys. Chem. B 102, 6193 (1998)
[35] Reineke, R. and Memming, R., J. Phys. Chem. 96, 1317 (1992)

[36] Uhlendorf, I., Reineke-Koch, R. and Memming, R., Ber. Bunsenges. Phys. Chem. 99, 1082 (1995)
[37] Schröder, K. and Memming, R., Ber. Bunsenges. Phys. Chem. 89, 385 (1985)
[38] Howard, J.N. and Koval, C.A., Anal. Chem. 63, 2777 (1991)
[39] Wipf, D.O. and Bard, A.J., J. Electrochem. Soc. 138, 469 (1991)
[40] Lewis, N.S., Ann. Rev. Phys. Chem. 42, 543 (1991)
[41] Smith, B.B., Halley, J.W. and Nozik, A.J., Chem. Phys. 205, 245 (1996)
[42] Smith, B.B. and Nozik, A.J., J. Phys. Chem. B 101, 2459 (1997)
[43] Bard, A.J., Bocarsly, A.B., Fan, F., Walton, E.W. and Wrighton, M.S., J. Am. Chem. Soc. 102, 3671 (1980)
[44] Nagasubramanian, G., Wheeler, B.L. and Bard, A.J., J. Electrochem. Soc. 130, 1680 (1983)
[45] Van den Berghe, R.A.L., Cardon, F. and Gomes, W.P., Surf. Sci. 39, 368 (1973)
[46] Gerischer, H., J. Electroanal. Chem. 150, 553 (1983)
[47] Kautek, W. and Willig, W., Electrochim. Acta 26, 1709 (1981)
[48] Meissner, D. and Memming, R., Electrochim. Acta 37, 799 (1992)
[49] McEvoy, A.I., Etman, M. and Memming, R., J. Electroanal. Chem. 190, 225 (1985)
[50] Kelly, J.J. and Memming, R., J. Electrochem. Soc. 129, 730 (1982)
[51] Rosenwaks, Y., Thacker, B.R., Bertness, K. and Nozik, A.J., J. Phys. Chem. 99, 7871 (1995)
[52] Sze, S., Physics of Semiconductor Devices, 2nd edition, John Wiley, New York (1981)
[53] Becka, A.M. and Miller, C.J., J. Phys. Chem. 96, 2657 (1992)
[54] Reineke, R. and Memming, R., J. Phys. Chem. 96, 1310 (1992)
[55] Vanmaekelbergh, D., Gomes, W.P., Cardon, F., J. Electrochem. Soc. 129, 564 (1982)
[56] Lu Shou Yun, Vanmaekelbergh, D. and Gomes, W.P., Ber. Bunsenges. Phys. Chem. 91, 390 (1987)
[57] Vanmaekelbergh, D., Lu Shou Yun and Gomes, W.P., J. Electroanal. Chem. 221, 187 (1987)
[58] Notten, P.H.L., Electrochim. Acta 32, 575 (1987)
[59] Notten, P.H.L. and Kelly, J.J., J. Electrochem. Soc. 134, 444 (1987)
[60] McCann, J.F. and Haneman, D., J. Electrochem. Soc. 129, 1134 (1982)
[61] Haneman, D. and McCann, J.F., Phys. Rev. B 25, 1241 (1982)
[62] Gregg, B.A. and Nozik, A.J., J. Phys. Chem. 97, 13441 (1993)
[63] Shreve, G.A. and Lewis, N.S., J. Electrochem. Soc. 142, 112 (1995)
[64] Ming X. Tan, Kenyon, C.N., Krüger, O. and Lewis, N.S., J. Phys. Chem. B 101, 2830 (1997)
[65] Smith, B.B. and Koval, C.A., J. Electroanal. Chem. 277, 43 (1990)
[66] Sutin, N., Ann. Rev. Nucl. Sci. 12, 285 (1962)
[67] Nielson, R.M., McManis, G.E.,Safford, L.K. and Weaver, M.J., J. Phys. Chem. 93, 2152 (1989)
[68] Memming, R., J. Electrochem. Soc. 116, 785 (1969)
[69] Memming, R. and Möllers, F., Ber. Bunsenges. Phys. Chem. 76, 609 (1972)
[70] Li, J. and Peter, L.M., J.Electroanal. Chem. 182, 399 (1985)
[71] Vetter, K.J., Z. Elektrochemie 56, 797 (1952)
[72] Bridge, N.V. and Porter, G., Proc. Roy. Soc. A224, 259 (1958)
[73] Minks, B.P. PhD thesis, University of Utrecht (Holland) 1991
[74] Notten, P.H.L., J. Electroanal. Chem. 119, 41 (1987)
[75] Morrison, S.R. and Freund, T., J. Chem. Phys. 47, 1543 (1967); Electrochim. Acta 13, 1343 (1968)
[76] Micka, K. and Gerischer, H., J. Electroanal. Chem. 38, 87 (1972)
[77] Lee, J., Kato, T., Fujishima, A. and Honda, K., Bull. Chem. Soc. Jap. 57, 1179 (1984)
[78] Yamagata, S., Nakabayashi, T., Sancier, K. M. and Fujishima, A., Bull. Chem. Soc. Jap. 61, 3429 (1990)
[79] Chmiel, G. and Gerischer, H., J. Phys. Chem. 94, 1612 (1990)
[80] Meissner, D., Memming, R. and Kastening, B., J. Phys. Chem. 92, 3476 (1988)

[81] Smandek, B., Chmiel, G. and Gerischer, H., Ber. Bunsenges. Phys. Chem. 93, 1094 (1989)
[82] Rosenwaks, Y., Thacker, B.R., Ahrenkiel, R.K. and Nozik, A.J., J. Phys. Chem. 96, 10096 (1992)
[83] Rosenwaks, Y., Thacker, B.R., Nozik, A.J., Shapira, Y. and Huppert,D. J. Phys. Chem. 97, 10421 (1993)
[84] Rosenwaks, Y., Shapira, V. and Huppert, D., Phys. Rev. B 45, 9108 (1992)
[85] Decker, F., Pettinger, B. and Gerischer, H., J. Electrochem. Soc. 130, 1335 (1983)
[86] Meulenkamp, E.A., PhD thesis, University of Utrecht (Holland) 1993
[87] Schoenmakers, R., Waagenaar, R. and Kelly, J.J., J. Electrochem. Soc. 142, L60 (1995)
[88] Edelstein, D.C., Tang,C.L. and Nozik, A.J., Appl. Phys. Lett. 51, 48 (1987)
[89] Boudreaux, D.S., Williams, F. and Nozik, A.J., J. Appl. Phys. 51, 2158 (1980)
[90] Lugli, P. and Goodnick, S.M., Phys. Rev.Lett. 59, 716 (1987)
[91] Williams, F. and Nozik, A.J., Nature 271, 137 (1978)
[92] Ross, R.T. and Nozik, A.J., J. Appl. Phys. 53, 3813 (1982)
[93] Cooper, G., Turner, J.A., Parkinson, B.A. and Nozik, A.J., J. Appl. Phys. 54, 6463 (1983)
[94] Min, L. and Miller, R.J.D., Appl.Phys. Lett. 56, 524 (1990)
[95] Turner, A.J.,and Nozik, A.J., Appl.Phys. Lett. 41, 101 (1982)
[96] Koval, C.A. and Segar, P.R., J. Am. Chem. Soc. 111, 2004 (1989)
[97] Sung, Y.E., Galliard, F. and Bard, A.J., J. Phys. Chem. B 102, 9797 (1998)
[98] Sung, Y.E. and Bard, A.J., J. Phys. Chem. B 102, 9806 (1998)
[99] Smith, B.B., private communication
[100] Nozik, A.J., private communication
[101] Memming, R., in Topics of Current Chemistry, Vol. 143, Springer Verlag, Berlin 1988, p.79
[102] Heller, A., Acc. Chem. Res. 14, 5 (1981)
[103] Tan, M.X., Newcomb, C. Kunnar,A., Lunt, S.R., Sailor, M.J., Tufts, B.J. and Lewis, N.S., J. Phys. Chem. 95, 10133 (1991)
[104] Lewerenz, H.J., Aspnes, D.E., Miller, B., Malm, D.L. and Heller,A., J. Am. Chem. Soc. 104, 38 (1982)
[105] Tsubomura, H. and Nakato, Y., Nouv. J. Chim. 11, 167 (1987)
[106] Nakato, Y., Ueda, K., Yano, H. and Tsubomura, H., J. Phys.Chem. 92, 2316 (1988)
[107] Meier, A., Uhlendorf, I. and Meissner, D., Electrochim. Acta 40, 1523 (1995)
[108] Meier, A., Uhlendorf, I., Meissner, D. and Memming, R., J. Phys. Chem., in press
[109] Hope, G.A. and Bard, A.J., J. Phys. Chem. 87, 1979 (1983)
[110] Spitz, Ch., Reineke-Koch,R. and Memming, R., unpublished data
[111] Hale, J.M., in: Reactions of Molecules at Electrodes (ed. N.S. Hush), Wiley Interscience, New York 1971, p. 229
[112] Beckmann, K.H. and Memming, R., J. Electrochem. Soc. 116, 368 (1969)

References for Chapter 8

[1] Dewald, J.F., in: Semiconductors (ed. N.B. Hannay),Reinhold Publishing Corporation, New York 1959
[2] Garrett, C.G.B. and Brattain, W.H., Phys. Rev. 99, 376 (1955)
[3] Brattain, W.H. and Garrett, C.G.B., Bell Syst. Techn. J. 34, 129 (1955)
[4] Pleskov, Yu.V. Dokl. Akad. Nauk SSSR, 132,1360 (1969)
[5] Beck, F. and Gerischer, H., Z. Elektrochem. 63, 500 (1959)
[6] Gerischer, H., in: Advance in Electrochemistry and Electrochemical Engineering (ed. P. Delahay) Vol.1, Interscience Publisher, New York 1961, p. 139
[7] Beck, F. and Gerischer, H., Z. Elektrochem. 63, 943 (1959)
[8] Memming, R. and Schwandt, G., Surf. Sci. 4, 109 (1966)

[9] Turner, D.R., J. Electrochem. Soc. 105, 402 (1958)
[10] Uhlir, A., Bell Syst. Tech. J. 35, 333 (1956)
[11] Levich, V.G., Physiochemical Hydrodynamics, Prentice Hall, Englewood Cliffs, NJ 1962
[12] Eddowes, J.A., J. Electroanal. Chem. 280, 297 (1990)
[13] Beckmann, K.H., Surf. Sci. 3, 314 (1965)
[14] Ubara, H., Imura, T. and Hitaki, A., Sol. State Comm. 59, 673 (1984)
[15] Yablonovich, E., Allara, D.L., Chang, C.C., Gmitter, T. and Bright, T.B., Phys. Rev. Lett. 57, 249 (1986)
[16] Chabal, Y.J., Higashi, G.S., Raghavachari, K. and Burrows, V.A., J. Vac. Sci. Technol. A7, 2104 (1989)
[17] Jacobs, P. and Chabal, Y.J. J. Chem. Phys. 95, 2897 (1991)
[18] Blackwood, D. J., Borazio, A., Greef, R., Peter, L.M. and Stumper, J., Electrochim Acta 37, 889 (1992)
[19] Gerischer, H., Allongue, P. and Kieling, V.C., Ber. Bunsenges. Phys. Chem. 97, 753 (1993)
[20] Cattarin, S., Peter, L.M. and Riley, D.J., J. Phys. Chem. B 101, 4071 (1997)
[21] Matsumura, M. and Morrison, S.R., J. Electroanal. Chem. 144, 113 (1983); 147, 157 (1983)
[22] Lewerenz, H.J., Stumper, J. and Peter, L.M., Phys. Rev. Lett. 61, 1989 (1988)
[23] Kooj, E.S., PhD-thesis, University of Utrecht (Holland) 1997
[24] Brus, L., J. Phys. Chem. 98, 3575 (1994)
[25] Lehmann, V. and Gösele, U., Appl. Phys. Lett. 58, 856 (1991)
[26] Canham, L.T., Appl. Phys. Lett. 57, 1046 (1990)
[27] Bressers,, M.M.C., Knapen, J.W.J., Meulenkamp, E.A. and Kelly, J.J., Appl. Phys. Lett. 61, 108 (1992)
[28] Brandt, M.S., Fuchs, H.D., Stutzmann, M., Weber, J. and Cardona, M., Sol. State. Commun. 81, 307 (1992)
[29] Petrova-Koch, V., Muschile, T., Kux, A., Meyer, B.K. and Koch, F., Appl. Phys. Lett. 61, 943 (1992)
[30] Friedman, S.L., Marcus, M.A., Adler, D.L., Xie, Y.H., Harris, T.D. and Citrin, P.H., Appl. Phys. Lett. 62, 1934 (1993)
[31] Valence, A. Phys. Rev. B 52, 8323 (1995)
[32] Lauermann, I., Memming, R. and Meissner, D., J. Electrochem. Soc. 144, 73 (1997)
[33] Hirayama, H., Kawabuko, T., Goto, A. and Kaneko, T., J. Am. Ceram. Soc. 72, 2049 (1989)
[34] Shor, J.S., Grimberg, I., Weiss, B.Z. and Kurtz, A.D., Appl. Phys. Lett. 62, 2836 (1993)
[35] Gerischer, H., Ber. Bunsenges. Phys. Chem. 69, 578 (1965)
[36] Minks, B.P. and Kelly, J.J., J. Electroanal. Chem. 273, 119 (1989)
[37] Reineke, R. and Memming, R., J. Phys. Chem. 96, 1310 (1992)
[38] Morrison, S.R., Electrochemistry at Semiconductor and Oxidized Metal Electrodes, Plenum Press, New York 1980
[39] Notten, P.H.L. and Kelly, J.J., J. Electrochem. Soc. 134, 444 (1987)
[40] Memming, R. and Schwandt, G., Electrochim. Acta 13, 1299 (1968)
[41] Madou, M.J., Cardon, F. and Gomes, W.P., Ber. Bunsenges. Phys. Chem. 81, 1186 (1977)
[42] Pourbaix, M., Atlas d'Equilibres Electrochimique, Ganthier-Villars, Paris 1963
[43] Meissner, D., Sinn, Ch., Memming, R., Notten, P.H.L. and Kelly, J.J., in: Homogeneous and Heterogeneous Photocatalysis (eds.: E. Pelizzetti and N. Serpone), D. Reidel, Dordrecht (Holland) 1986, p. 343
[44] Erné, B.H., Vanmaeckelbergh, D. and Vermeir, I.E., Electrochim. Acta 38, 2559 (1993)
[45] Meissner, D., Memming, R. and Kastening, B., J. Phys. Chem. 92, 3476 (1988)
[46] Meissner, D., Benndorf, C. and Memming, R., Appl. Surf. Sci. 27, 423 (1987)
[47] de Wit, A.R.,, Vanmaeckelbergh, D. and Kelly, J.J., J. Electrochem. Soc. 139, 2508 (1992)
[48] Tributsch, H. and Bennett, J.C., J. Electroanal. Chem. 81, 97 (1977)
[49] Tributsch, H., in: Structure and Bonding, Vol. 49, Springer Verlag, Berlin, 1982, p.129

[50] Jägermann, W. and Tributsch, H., Prog. Surf. Sci. 29, 1 (1988)
[51] Gerischer, H., in: Topics of Applied Physics, Vol. 31, Springer Verlag, Berlin, 1979, p.115
[52] McEvoy, A.J., Etman, M. and Memming, R., J. Electroanal. Chem. 190, 225 (1985)
[53] Sinn, Ch., Meissner, D. and Memming, R., J. Electrochem. Soc. 137, 168 (1990)
[54] Lewerenz, H.J., Gerischer, H. and Lübke, M., J. Electrochem. Soc. 131, 100 (1984)
[55] Gerischer, H. and Wallem-Mattes, J., Z. Phys. Chem. 64, 187 (1969)
[56] Kelly, J.J. and Reynders, A.C., Appl. Surf. Sci. 29, 149 (1987)
[57] Minks, B.P., Oskam, G., Vanmaeckelbergh, D. and Kelly, J.J., J. Electroanal. Chem. 273, 119 (1989); 273, 133 (1989)
[58] Gerischer, H., J. Vac. Sci. Technol. 15, 1422 (1978)
[59] Bard, A.J. and Wrighton, M.S., J. Electrochem. Soc. 124, 1706 (1977)
[60] Memming, R., J. Electrochem. Soc. 125, 117 (1978)
[61] Bard, A.J., Parson, R. and Jordan, J., Standard Potentials in Aqueous Solutions, Marcel Dekker Inc., New York 1985
[62] Inoue, T., Watanabe, T., Fujishima, A., Honda, K. and Kohayakawa, P., J. Electrochem. Soc. 124, 719 (1977)
[63] Memming, R., Ber. Bunsenges. Phys. Chem. 81,732 (1977)
[64] Memming, R., Topics of Current Chemistry, Vol.169, Springer Verlag, Berlin (1994), p. 106
[65] Frese, K., Madou, M.J. and Morrison, S.R., J. Phys. Chem. 84, 3172 (1980); J. Electrochem. Soc. 128, 1528 (1981)
[66] Vanmaekelbergh, D.and Gomes, W.P., Ber. Bunsenges. Phys.Chem. 89, 987 (1985); J. Phys. Chem. 94, 1571 (1990)
[67] Allongue, P., Blonkowsky, S. and Lincot, D., J. Electroanal. Chem. 300, 261 (1991)
[68] Cardon, F., Gomes, W.P., Kerchove, F., Vanmaekelbergh, D. and Overmeire, F., Faraday Discussion 70, 153 (1980)
[69] Gerischer, H. and Lübke, M., Ber. Bunsenges. Phys. Chem. 87, 123 (1983)
[70] Memming, R., Topics of Current Chemistry, Vol. 143, Springer Verlag, Berlin 1988, p. 79
[71] Memming, R., in: Photoelectrochemistry, Photocatalysis and Photoreactors (ed. M. Schiavello), D. Reidel, Dordrecht (Holland) 1980, p. 107
[72] Reineke, R. and Memming, R., J. Phys. Chem. 96, 1317 (1992)

References for Chapter 9

[1] Henglein, A. in Topics of Current Chemistry 143 (ed. E. Steckhan), Springer Verlag Berlin (1988) p. 113
[2] Miller, R.J.D., McLendon, G.L., Nozik, A.J., Schmickler, W. and Willig, F. in: Surface Electron Transfer Processes, VCH New York, Chapter 6 (1995), p. 311
[3] Henglein, A., Chem. Rev. 89, 1861 (1989)
[4] Bawendi, M.G., Steigerwald, M.L. and Brus, L.E., Ann. Rev. Phys. Chem. 41, 477 (1990)
[5] Memming, R. in: Topics of Current Chemistry 169 (ed. J. Mattay), Springer Verlag Berlin (1994) p. 105
[6] Nozik A.J. and Memming R., J. Phys. Chem. 100, 13061 (1996)
[7] Weller, H., Ang. Chem. Int. Ed. 32, 41 (1993)
[8] Alivisatos, A.P., J. Phys. Chem. 100, 13226 (1996)
[9] Henglein, A. Ber. Bunsenges. Phys. Chem. 99, 903 (1995)
[10] Fox, M.A. in: Topics of Current Chemistry 142, Springer Verlag Berlin (1987), p. 42
[11] Ekimov, A.I. and Onushchenko, A.A., Sov. Phys. Semicond. 16, 775 (1982)
[12] Efros, A.I.L. and Efros, A.L., Sov. Phys. Semicond. 16, 772 (1982)
[13] Brus, L.E., J. Chem. Phys. 79, 5566 (1983); 80, 4403 (1984);
[14] Rossetti, R., Hull, R., Gibson, J.M. and Brus, L.E., J. Chem. Phys. 82, 552 (1985)

[15] Weller, H.,Schmidt, H.M., Koch, U., Fojtik, A., Baral, S., Henglein, A., Kunath, W., Weiss, K. and Dieman, E., Chem. Phys. Lett. 124, 557 (1986)
[16] Zunger, A., MRS Bulletin, February 1998, p. 35
[17] Krishna, M.V.R. and Friesner, R., J. Chem. Phys. 95, 8309 (1991)
[18] Lippens, P.E. and Lannoo, M., Phys. Rev. B 39, 10935 (1989); B41, 6079 (1990)
[19] Fu, H. and Zunger, A., Phys. Rev. B 56, 1496 (1997)
[20] Wang, L.W. and Zunger, Phys. Rev. B 53, 9579 (1996)
[21] Vossmeyer, T., Katsikas, L., Giersig, M., Popovic, I.G., Diesner, K., Chemseddine, A. Eychmüller, A. and Weller, H., J. Phys. Chem. 98, 7665 (1994)
[22] Mičič, O.I., Curtis, C.J., Jones, K.M., Sprague, J.R. and Nozik, A.J. J. Phys. Chem. 98, 4966 (1994)
[23] Mičič, O.I., CHeong, H.M., Fu, H., Zunger, A., Sprague, J.R., Mascarenhas, A. and Nozik, A.J., J. Phys. Chem. B 101, 4904 (1997)
[24] Mičič, O.I., Sprague, J.R., Lu, Z. and Nozik, Appl. Phys. Lett. 68, 3150 (1996)
[25] Kumar, A., Janata, E. and Henglein, A., J. Phys. Chem. 92, 2587 (1988)
[26] Brus, L.E., Szajowski, P.F., Wilsom, W.L., Harris, T.D., Schuppler, S. and Citrin, P.H., J. Am. Chem. Soc. 117, 2915 (1995)
[27] Calcott, P.D.J., Nash, K.J., Canham, L.T., Kane, M.J. and Brumhead, D.J., J. Lumin. 57, 257 (1993)
[28] Suemoto, T., Tanaka, K., Nakajima, A. and Itakura, T., Phys. Rev. Lett. 70, 3659 (1993)
[29] Bastard, G., Phys. Rev. B 24, 5693 (1981); B 25, 7594 (1982)
[30] Altarelli, M.J., J. Lumin. 30, 472 (1985)
[31] Bastard, G. and Brum, J.A., Quantum Electron. 22, 1625 (1986)
[32] Dingle, R., Gossard, A.C. and Wiegmann, W., Phys. Rev. Lett. 34, 1327 (1975)
[33] Dingle, R., Wiegmann, W. and Henry, C.H., Phys. Rev. Lett. 33, 827 (1974)
[34] Dingle, R., Festkörperprobleme 15, 21 (1975)
[35] Hodes, G., Israel. J. Chem. 33, 95 (1993)
[36] Hodes, G., Albu-Yaron, A., Decker, F. and Motisuke, P., Phys. Rev. B 36, 4215 (1987)
[37] Grätzel, M. and Frank, A.J., J. Phys. Chem. 86, 2964 (1982)
[38] Henglein, A., Ber. Bunsenges. Phys. Chem. 86, 301 (1982)
[39] Spanhel, L., Haase, M., Weller, H. and Henglein, A., J. Am. Chem. Soc. 109, 56 (1987)
[40] Henglein, A., J. Phys. Chem. 86, 2291 (1982)
[41] Fojtik, A., Weller, H., Koch, U. and Henglein, A., Ber.Bunsenges. Phys. Chem. 88, 969 (1984)
[42] Dimitrijevič, N.M., Li, S. and Grätzel, M., J. Am. Chem. Soc. 106, 6565 (1984)
[43] Meissner, D., Memming, R., Kastening, B. and Bahnemann, D., Chem. Phys. Lett. 127, 419 (1996)
[44] Bahnemann, D., Henglein, A., Lilie, J. and Spanhel, L., J. Phys. Chem. 88, 709 (1984)
[45] Bahnemann, D., Hilgendorff, M. and Memming, R., J. Phys. Chem. B 101, 4265 (1997)
[46] King, B. and Freund, F., Phys. Ref. B 29, 5814 (1984)
[47] Serpone, N., Lawless, D., Khairutdinov, R. and Pelizzetti, E., J. Phys. Chem. 99, 16655 (1995)
[48] Colombo, D.P. and Bowman, R.M., J. Phys. Chem. 99, 11752 (1995)
[49] Chen, G., Zen, J.M., Fan, F.R.F. and Bard, A.J., J. Phys. Chem. 95, 3682 (1991)
[50] Memming, R., in: Topics of Current Chemistry 143 (ed. E. Steckhan), Springer Verlag Berlin, 1988, p. 79
[51] Meissner, D., Memming, R. and Kastening, B., Chem. Phys. Lett. 96, 34 (1983)
[52] Alfassi, Z., Bahnemann, D. and Henglein, A., J. Phys. Chem. 86, 4656 (1982)
[53] Gurevich, Yu.Ya., Pleskov, Yu.V. and Rotenberg, Z.A., Photoelectrochemistry, Consultants Bureau New York (1988)
[54] Haase, M., Weller, H. and Henglein, A., J. Phys. Chem., 92, 4706 (1988)

[55] Briggs , D. (Ed.): Practical surface analysis. 1. Auger and X-ray Photoelectron Spectroscopy, 2. ed., Wiley, Chichester, 1990; 2 . Ion and Neutral Spectroscopy, 2. ed., Wiley, Chichester, 1992
[56] Kraeutler, B. and Bard, A.J., J. Am. Chem. Soc. 99, 7729 (1977)
[57] Kraeutler, B. and Bard, A.J., J. Am. Chem. Soc. 100, 5985 (1978)
[58] Hoffmann, M.R., Martin, S.T., Choi, W. and Bahnemann, D., Chem. Rev. 95, 69 (1995)
[59] Bahnenmann, D. in Photochemical Conversion and Storage of Solar Energy (eds. E. Pelizzetti, M. Schiavello), Kluver The Netherlands (1991) p. 251
[60] Kormann, C., Bahnemann, D. and Hoffmann, M.R., Langmuir 6, 555 (1990)
[61] Henglein, A. and Gutierrez, M., Ber. Bunsenges. Phys. Chem. 87, 852 (1983)
[62] Müller, B.B., Majoni, S., Memming, R. and Meissner, D., J. Phys. Chem. B 101, 2501 (1997)
[63] Henglein, A., Kumar, A., Janata, E. and Weller, H., Chem. Phys. Lett. 132, 133 (1986)
[64] Hoyer, P. and Weller, H., Chem. Phys. Lett. 221, 379 (1994)
[65] Burstein, E., Phys. Rev. 93, 632 (1954)
[66] Nedeljkovic, J.M., Nenadovic, M.T., Mičič, O.I. and Nozik, A.J., J. Phys. Chem. 90, 12 (1986)
[67] Dimitrijevic, D., Savic, O.I., Mičič, O.I. and Nozik, J. Phys. Chem. 88, 5827 (1984)
[68] Spanhel, L., Weller, H. and Henglein, A., J. Am. Chem. Soc. 109, 6632 (1987)
[69] Spanhel, L., Henglein, A. and Weller, H., Ber. Bunsenges. Phys. Chem. 91, 1359 (1987)
[70] Kietzmann, R., Willig, F., Weller, H., Vogel, R., Nath, D.N., Eichberger, R., Liska, P. and Lehnert, J., Mol. Cryst. 194, 169 (1991)
[71] Nozik, A.J., Appl. Phys. Lett. 30, 567 (1977)
[72] Vogel, R., Hoyer, P. and Weller, H., J. Phys. Chem. 98, 3183 (1994)
[73] Boudreaux, D.S., Williams, F. and Nozik, A.J., J. Appl. Phys. 51, 2158 (1980)
[74] Lugli, P. and Goodnick, S.M., Phys. Rev. Lett. 59, 716 (1987)
[75] Lyon, S.A., J. Lumin. 51, 35 (1987)
[76] Rosker, M.J., Wise, F.W. and Tang, C.L., Appl. Phys. Lett. 49, 1726 (1986)
[77] Edelstein, D.C., Tang, C.L. and Nozik, A.J., Appl. Phys. Lett. 51, 48 (1987)
[78] Rosenwaks, Y., Hanna, M.C., Levi, D.H., Szmyd, D.M., Ahrenkiel, R.K. and Nozik, A.J., Phys. Rev. B 48, 14675 (1993)
[79] Parsons, C.A., Peterson, M.W., Thacker, B.R., Turner, J.A. and Nozik, A.J., J. Phys. Chem. 94, 3381 (1990)
[80] Parsons, C.A., Thacker, B.R., Szmyd, D.A., Peterson, M.W., McMahon, W.E. and Nozik, A.J., J. Chem. Phys. 93, 7706 (1990)
[81] Diol, S.J., Poles, E., Rosenwaks, Y. and Miller, R.J.D., J. Phys.Chem. B. 102, 6193 (1998)
[82] Meier, A., Selmarten, D.C., Siemoneit, K., Smith, B.B. and Nozik, A.J., J. Phys. Chem. B. 103, 2122 (1999)
[83] O'Reagen, B. and Grätzel, M., Nature 353, 737 (1991)
[84] Bahnemann, D., Kormann, C. and Hoffmann, M.R., J. Phys. Chem. 91, 3789 (1987)

References for Chapter 10

[1] Vogel, H.W., Ber. Deut. Chem. Ges. 6,1302 (1873)
[2] West,W. and Carrol, B.H., in: The Theory of Photographic Processes (eds. C.E.K. Mees and T.H. James), McMillan, New York 1966, p. 233
[3] Berriman, R.W. and Gilman, P.B., Photogr. Sci. Eng. 17, 235 (1973)
[4] Weller, A., in: Fast Reactions and Primary Processes in Chemical Kinetics (ed Claesson, S.), Almquist and Wiksell 1967, p. 413
[5] Stevens, R., Sharpe, R.R. and Bingham, W.S.W., Photochem. Photobiol. 6, 83 (1967)

[6] Turro, N.J., Modern Molecular Photochemistry, Benjamin Cummings Publ. Comp., Menlo Park, Cal. 1978
[7] Loutfy, R.O. and Sharp, J.H., Photogr. Sci. Eng. 20, 165 (1976)
[8] Spitler, M.T., J. Electroanal. Chem. 228, 69 (1987)
[9] Memming, R., Progr. Surf. Sci. 17, 7 (1984)
[10] Tokel-Takooryan, N.F., Hemmingway, R.E. and Bard, A.J., J. Am. Chem. Soc. 95, 6582 (1973)
[11] Balzani, V., Bolletta, F., Gandolfi, M.T. and Maestri, M., in: Topics of Current Chemistry 75, Springer Verlag Berlin 1978, p. 1
[12] Sutin, N. and Creutz, C., Adv. Chem. Ser. 168, Inorg. Organomet. Photochemistry 1978, p.1
[13] Bock, C.R., Meyer, T.J. and Whitten, D.G., J. Am. Chem. Soc. 97, 2911 (1975)
[14] Gleria, M. and Memming, R., Z. Phys. Chem. N.F. 98, 303 (1975)
[15] Clark, W.D.K. and Sutin, N., J. Am. Chem. Soc. 99, 4676 (1977)
[16] Memming, R. Surf. Sci. 101, 551 (1980)
[17] Gerischer, H. and Tributsch, H., Ber. Bunsenges. Phys. Chem. 72, 437 (1968)
[18] Gerischer, H., Michel-Beyerle, M.E., Rebentrost, F. and Tributsch, H., Electrochim. Acta 13, 1509 (1968)
[19] Fujishima, A., Watanabe, T., Tatsuoki, O. and Honda, K., Chem. Lett. 13 (1975)
[20] Memming, H. and Tributsch, H., J. Phys. Chem.75, 562 (1971)
[21] Parkinson, B.A., Langmuir 4, 967 (1988)
[22] Parkinson, B.A. and Spitler, M.T., Electrochim. Acta 37, 943 (1992)
[23] Yamase, T., Gerischer, H., Lübke, M. and Pettinger, B., Ber. Bunsenges. Phys. Chem. 82,1041 (1978)
[24] Tribursch, H., Ber. Bunsenges. Phys. Chem. 73, 582 (1969)
[25] Scheibe, G., Angew. Chemie 52, 633 (1933)
[26] Memming, R. Photochem. Photobiol. 16, 325 (1972)
[27] Spitler, M.T., Lübke, M. and Gerischer, H., Chem. Phys. Lett. 56, 577 (1983)
[28] Gerischer, H. and Willig, F., Topics of Current Chemistry, Vol. 61, Springer Verlag, Berlin 1976, p. 33
[29] Kuhn, H., Möbius, D. and Bücher, H., in: Physical Methods of Chemistry, Vol.1 (ed. A. Weissberger and B.W. Rossiter), John Wiley, New York 1972, p. 577
[30] Anderson, S., Constable, E.C., Dare-Edwards M.P., Goodenough, J.B., Hamnett, A., Seddon, K.R. and Wright, R.D., Nature 280, 571 (1979)
[31] Hamnett, A., Dare-Edwards, M.P., Wright, R.D., Seddon, K.R. and Goodenough, J.B., J. Phys. Chem. 83, 3280 (1979)
[32] Kudo, A., Steinberg, M., Bard, A.J., Campion, A., Fox, M.A., Mallouk, T.E., Webber, S.E. and White, J.M., J. Electrochem. Soc. 137, 3846 (1990)
[33] Memming, R., Far. Discussions Chem. Soc. 58, 261 (1974)
[34] Arden, W. and Frommherz, P., Ber. Bunsenges. Phys. Chem. 82, 868 (1978)
[35] Miyasaka, T., Watanabe, T., Fujishima, A. and Honda, K., J. Am. Chem. Soc. 100, 6657, (1978)
[36] Miyasaka, T. and Honda, K., Surf. Sci. 101, 541 (1980)
[37] Memming, R. and Schröppel, R., Chem. Phys. Lett. 62, 207 (1979)
[38] Miyasaka, T., Watanabe, T., Fujishima, A. and Honda, K., Photochem. Photobiol. 32, 217 (1980)
[39] Arbour, C., Nebesney, K.W., Lee, P.A., Lai-Kwan Chan, Armstrong, N.R. and Parkinson, B.A., J. Am. Chem. Soc., in press
[40] Desilvestro, J., Grätzel, M., Kavan, L., Moser, J. and Augustynski, J., J. Am. Chem. Soc. 107, 2988 (1985)
[41] Fessenden, R.W. and Kamat, P.V., J. Phys. Chem. 99, 12902 (1995)

[42] Deppe, J., Meissner, D. and Memming, R., unpublished results
[43] Tributsch, H. and Gerischer, H., Ber. Bunsenges. Phys. Chem. 73, 251, 850 (1969)
[44] Adelmann, A.H. and Oster, G., J. Am. Chem. Soc. 78, 3977 (1956)
[45] Watanabe, T., Fujishima, A. and Honda, K., Ber. Bunsenges. Phys. Chem. 79, 1213 (1975)
[46] Kim, H. and Laitinen, H.A., J. Electrochem. Soc. 122, 53 (1975)
[47] Spitler, M.T. and Parkinson, B.A., Langmuir 2, 549 (1986)
[48] Lanzafame, J.M., Miller, R.J.D. Miller, Muenter, A.A. and Parkinson, B.A., J. Phys. Chem. 96, 2820 (1991)
[49] Pettinger, B., Schöppel, H.R. and Gerischer, H., Ber. Bunsenges. Phys. Chem. 77, 960 (1973)
[50] O'Reagan, B. and Grätzel, M., Nature 353, 737 (1991)
[51] Hannappel, T., Burfeindt, B., Storck, W. and Willig, F., J. Phys. Chem. B 101, 6799 (1997)
[52] Tachibana, Y., Moser, J.E., Grätzel, M., Klug, D.R. and Durrant, J.R., J. Phys. Chem. 100, 20056 (1996)
[53] Ellingson, R.J., Asbury, J.B., Ferrere, S., Ghosh, H.N., Sprague, J.R., Lian, T. and Nozik, A.J., J. Phys. Chem. B 102, 6455 (1998)
[54] Argazzi, R., Bignozzi, C.A., Heimer, T.A., Castellano, F.N. and Meyer, G., J. Inorg. Chem. 33, 5741 (1994)
[55] Kamat, P.V., Bedja, I., Hotchandani, S. and Patterson, L.K., J. Phys. Chem. 100, 4900 (1996)
[56] Kuhn, H,. Naturwissenschaften 54, 429 (1967)
[57] Memming, R. and Kürsten, G., Ber. Bunsenges. Phys. Chem. 76, 4 (1972)
[58] Eisenberg, M. and Silvermann, H.P., Electrochim. Acta 5, 1 (1961)
[59] Gleria, M. and Memming, R., Z. Phys. Chem. N.F. 101, 171 (1976)
[60] Luttmer, J.D. and Bard, A.J., J. Electrochem. Soc. 125, 1423 (1978)

References for Chapter 11

[1] Harris, L.A. and Wilson, R.H., Annu. Rev. Mater. Sci. 1978, p. 99
[2] Gerischer, H., Topics in: Applied Physics, Vol. 31, Springer Verlag, Berlin (1979), p. 115
[3] Memming, R., Phil. Tech. Rev. 38, 160 (1978/79)
[4] Nozik, A.J., Annu. Rev. Phys. Chem. 29, 189 (1978)
[5] Rajeshwar, K., Singh, P. and Dubow, J., Electrochim. Acta 23, 1117 (1978)
[6] Butler, M.A. and Ginley, D.S., J. Mater. Sci. 15, 1 (1980)
[7] Memming, R., Electrochim. Acta 25, 77 (1980)
[8] Heller, A. Accts. Chem. Res. 14, 154 (1981)
[9] Tributsch, H., in: Structure and Bonding, Vol. 49 (eds. M.J. Clarke et al.), Springer Verlag 1982, p.127
[10] Bard, A.J., Science 207, 139 (1980)
[11] Grätzel, M., in Structure and Bonding, Vol.49 (eds. Clarke, M.J. et al.), Springer Verlag 1982, p.37
[12] Lewis, N.S., Annu. Rev. Mater. Sci., 14, 95 (1984)
[13] Pleskov, Yu.V., Prog. Surf. Sci. 15, 401 (1984)
[14] Memming, R., Topics in Current Chemistry 143, 81 (1987)
[15] Meissner, D., Memming, R. and Kastening, B. J. Phys. Chem. 92, 3476 (1988)
[16] Meissner, D., Benndorf, K. and Memming, R., Appl. Surf. Sci. 27, 423 (1987)
[17] Nozik, A.J. and Memming, R., J. Phys. Chem. 100, 13068 (1996)
[18] Rosenbluth, M.L. and Lewis, N.S., J. Am. Chem. Soc. 108, 4689 (1986)
[19] Nakato, Y., Ueda, K. Yano, H. and Tsubomura, H., J. Phys. Chem. 92, 2316 (1988)
[20] Abrahams, J.L., Casagrande, L.G., Rosenblum, M.D., Rosenbluth, M.L., Santangelo, P.G., Tufts, B.J. and Lewis, N.S., Nouv. J. Chim. 11, 157 (1987)

[21] Fantini, M.C.A., Shen, W.M., Tomkiewicz, M. and Gambino, J.P., J. Appl.Phys. 65, 4884 (1989)
[22] Licht, S., Tenne, R., Flaisher, H. and Manassen, J., J. Electrochem. Soc. 133, 52 (1986)
[23] Fahrenbruch, A.L. and Bube, R.H., Fundamentals of Solar Cells, Academic Press, New York, 1983
[24] Nakato, Y., Yano, H. and Tsubomura, Chem. Lett. 987 (1986)
[25] Sinn, Ch., Meissner, D. and Memming, R., J. Electrochem. Soc. 137, 168 (1990)
[26] Reineke, R. and Memming, R., J. Phys. Chem. 96, 1310 (1992)
[27] Memming, R., in: Photochemistry and Photophysics, Vol. 2 (ed. J.F. Rabek), CRC Press, Boca Raton, Fl. 1990, p.143
[28] Kautek, W. and Gerischer, H., Electrochim. Acta 27, 355 (1982)
[29] Lewerenz, H.J., Gerischer, H. and Lübke, M., J. Electrochem. Soc. 131, 100 (1984)
[30] Lewerenz, H. J., Heller, A. and Disalvo, F.J., J. Am. Chem. Soc. 102,1877 (1980)
[31] Tenne, R., Spahni, W. Calzaferri, G. and Wold, A., J. Electroanal. Chem. 189, 247 (1985)
[32] Tenne, R. and Wold, A., Ber. Bunsenges. Phys. Chem. 90, 545 (1986)
[33] McEvoy, A.J., Etman, M. and Memming, R., J. Electroanal. Chem. 190, 225 (1985)
[34] Hodes, G., J. Photochem. 29, 243 (1985)
[35] Hodes, G., in: Energy Resources through Photochemistry and Catalysis (ed. Grätzel, M.), Academic Press, New York 1983, p. 521
[36] Licht, S. and Manassen, J., J. Electrochem. Soc. 132, 1076 (1985)
[37] Cahen, D. and Chen, Y.W., Appl. Phys. Lett. 45, 746 (1984)
[38] Menezes, S., Lewerenz, H.J. and Bachmann, K.J., Nature, 305, 615 (1983)
[39] Lewerenz, H.J. and Kötz, E.R., J. Appl. Phys. 60, 1430 (1986)
[40] Chang, K.C., Heller, A., Schwartz, B., Menezes, S. and Miller, B., Science, 196, 1097 (1977)
[41] Parkinson, B.A., Heller, A. and Miller, B., J. Electrochem. Soc. 126, 954 (1979)
[42] Hodes, G, Manassen, J. and Cahen, D., J. Electrochem. Soc. 127, 544 (1980)
[43] Nelson, R.J., Williams, J.S., Leamy, H.J., Miller, B., Casey, H.C., Parkinson, B.A. and Heller, A., Appl. Phys. Lett. 36, 76 (1980)
[44] Ennaoui, A., Fiechter, S., Pettenkofer, Alonse-Vante, N., Büker, K., Bronold, M., Höpfner, C. and Tributsch, H., Solar Energy Mater. Sol. Cells 29, 289 (1993)
[45] Fiechter, S., Birkholz, M., Hartmann, A., Dulski, P., Giersig, M., Tributsch, H. and Tilley, R.J.D., J. Mater. Res. Bull. 21, 1481 (1996)
[46] Memming, R., in: Comprehensive Treatise of Electrochemistry, Vol.7 (eds. B.E. Conway et al.), Plenum Press, New York 1983, p. 529
[47] Heller, A., Miller, B. and Thiel, F., Appl. Phys. Lett. 38, 282 (1981)
[48] Tributsch, H. and Calvin, M., Photochem. Photobiol. 16, 261 (1972)
[49] Matsumura, N., Nomura, Y. and Tsubomura, H, Bull. Chem. Soc. Jap. 59, 2533 (1977)
[50] Alonso, N., Beley, V.M., Chartier, P. and Erns, N., Rev. Phys. Appl. 16, 5 (1981)
[51] Stalder, C. and Augustynski, J., Electrochem. Soc. 126, 2007 (1979)
[52] Desilvestro, L., Grätzel, M., Kavau, L., Moser, L. and Augustynski, J., J. Am. Chem. Soc. 107, 2988 (1985)
[53] Nazeedruddin, L. and Grätzel, M., J. Am. Chem. Soc. 115, 6382 (1993)
[54] O'Reagan, B. and Grätzel, M., Nature 353, 737 (1991)
[55] Ross, R.T. and Hsiao, T.L., J. Appl. Phys. 48, 4783 (1977)
[56] Ross, R.T. and Collins, J.M., J. Appl. Phys. 51, 4504 (1980)
[57] Bolton, J.R., Haught, A.F. and Ross, R.T., in: Photochemical Conversion and Storage of Solar Energy (ed. J.S. Connolly) Academic Press, New York 1980, p. 297
[58] Ross, R.T., J. Chem. Phys. 45,1 (1966); 46, 4590 (1967)
[59] Handschuh, M. and Lorenz, W., Z. Phys. Chem. (Leipzig) 264, 15 (1983)

[60] Vanmaeckelbergh, D., Gomes, W.P. and Cardon, F., Ber. Bunsenges. Phys. Chem. 89, 987 (1985)
[61] Allongue, P., Cachet, H. and Horowitz, G., J. Electrochem. Soc. 130, 2352 (1981)
[62] Memming, R., in: Photoelectrochemistry, Photocatalysis and photoreactors (ed. M. Schiavello), D.Reidel, Dordrecht (Holland) 1980, chapter 9
[63] Bard, A.J. and Fox, M.A., Acc. Chem. Res. 28, 141 (1995)
[64] Tributsch, H., Modern Aspects of Electrochemistry, Plenum Press, New York, (1986)
[65] Parkinson, B.A., J. Chem. Educ. 60, 338 (1993)
[66] Koval, C.A. and Howard, J.N., Chem. Rev. 92, 411 (1992)
[67] Memming, R., Topics in Current Chemistry,Vol. 169, Springer Verlag, Berlin, 1994, p. 105
[68] Nozik, A.J., Appl. Phys. Lett. 30, 567 (1977)
[69] Nozik, A.J. and Memming, R., J. Phys. Chem. 100, 13061 (1996)
[70] Mavroides, J.G., Kafalos, J.A. and Kolesar, D.F., Appl. Phys. Lett. 28, 241 (1976)
[71] Jägermann, W. and Tributsch, H., Prog. Surf. Sci. 29, 1 (1988)
[72] Tributsch, H. J. Photochem. 29, 89 (1985)
[73] Kühne, H.M. and Tributsch, H., Electroanal. Chem. 201, 263 (1986)
[74] Heller, A. and Vadimsky, R.G., Phys. Rev. Lett. 46, 1153 (1981)
[75] Aharon-Shalom, E. and Heller, A., J. Electrochem. 129, 2865 (1982)
[76] Nozik, A.J., Appl. Phys. Lett. 29, 150 (1976)
[77] Yoneyama, H., Sakamoto, H. and Tamura, H., Electrochim. Acta 20, 341 (1975)
[78] Nozik, A.J., Appl. Phys. Lett. 30, 567 (1977)
[79] Eyert, V., Höck, K.H., Fiechter, S. and Tributsch, H., Phys. Rev. B 57, 6350 (1998)
[80] Kraeutler, B. and Bard, A.J., J. Am. Chem. Soc. 100, 5895; 100, 2239 (1978)
[81] Olson, J.M., Kurtz, S.R., Kibbler, A.E. and Faine, P., Appl. Phys. Lett. 56, 623 (1990)
[82] Bertness, K.A., Kurtz, S.R., Frieman, D.J., Kibbler, A.E., Kramer, C. and Olson, J.M., Appl. Phys. Lett. 65, 989 (1994)
[83] Khaselev, O. and Turner, J.A., Science 280, 425 (1998)
[84] Archer, M.D. and Bolton, J.R., J. Phys. Chem. 94, 8028 (1990)
[85] Bolton, J.R., Strickler, S.J. and Connolly, J.S., Nature 316, 495 (1985)
[86] Borgarello, E., Kalyanasundaram, K. and Grätzel, M., Helv. Chim. Acta 65, 243 (1982)
[87] Matsumura, M., Saho, Y. and Tsubomura, H., J. Phys. Chem. 87, 3807 (1983)
[88] Meissner, D., Memming, R. and Kastening, B., Chem. Phys. Lett. 96, 34 (1983)
[89] Thewissen, D.H.M.W., Tinnemans, A:H.A., Eenwhorst Reinten, M., Timmer, K. and Mackor, A., Nouv. J. Chim. 7, 191 (1983)
[90] Levy-Clement, C., Heller, A., Bonner, W.A. and Parkinson, B.A., J. Electrochem. Soc. 129, 1701 (1982)
[91] Johnson, E.L., in: Electrochemistry in Industry (eds. V. Landan et al.), Plenum Press, New York, 1982, p. 299
[92] Frese, K.W. and Canfield, D., J. Electrochem. Soc. 130, 1772 (1983); 131, 2518 (1984)
[93] Honda, K., Fujishima, A., Watanabe, T., in: Solar Hydrogen Energy Systems (ed. T. Ohta), Pergamon Press, Oxford (England) 1981, p.137
[94] Henglein, A., Pure Appl. Chem. 56, 1215 (1984)
[95] Hori, Y., Murata, A. and Takahashi, A., J. Chem. Soc. Faraday Trans. 85, 2309 (1989)
[96] Sears, W.M. and Morrison, S.R., J. Phys. Chem. 89, 3295 (1985)
[97] Yamashita, H., Fujii, Y., Ichihashi, Y., Zhang, S.G., Ikene, K., Park, R.P., Koyano, K., Tatsumi, T. and Ampo, M., Catalysis Today 45, 221 (1998)
[98] Kuwabata, S., Uchida, H., Ogawa, A., Hirao, S. and Yoneyama, H., J.Chem. Soc. Chem. Commun. 1995, p. 829
[99] Hoffmann, M.R., Martin, S.T., Choi, W. and Bahnemann, D.W., Chem. Rev. 95, 69 (1995)
[100] Venkatardi, R. and Peters, R.W., Hazard Waste & Hazard Mat. 10, 107 (1993)
[101] Kamat, P.V., Chem. Rev. 93, 341 (1993)

[102] Fox, M.A. and Dulay, M.T., Chem. Rev. 93, 341 (1993)
[103] Legrini, O., Oliveros, E. and Braun, A.M., Chem. Rev. 93, 671 (1993)
[104] Pelizzetti, E., and Minero, C., Electrochim. Acta 38, 47 (1993)
[105] Fujishima, A., Hashimoto, K. and Watanabe, T., TiO_2 Photocatalysis, Fundamentals and Applications, BKS, Inc. Tokyo, Japan 1990
[106] Pruden, A.L. and Ollis, D.F., Environ. Sci. Technol. 17, 628 (1983)
[107] Borgarello, E., Serpone, N., Barbeni, N., Minero, C., Pelizzetti, E. and Pramauro, E., Chim. Ind. Milan. 68, 52 (1986)
[108] Barbeni, M., Pramauro, E., Pelizzetti, E., Borgarello, E., Grätzel, M. and Serpone, N., Nouv. J. Chim. 8, 547 (1984)
[109] Bahnemann, D., in Handbook of Environmental Chemistry, Part L. (ed.: Boule, P.), Springer Verlag, Berlin (1999), p. 285
[110] Mills, A., Davies, R.H. and Worsley, D., Chem. Soc. Rev. 22, 417 (1993)
[111] Schwitzgebel, J., Ekerdt, J.G., Gerischer, H. and Heller, A., J. Phys. Chem. 99, 5633 (1995)
[112] Fujishima, A. and Rao, T.N., Proc. In. Acad. Sci. 109, 471 (1997)
[113] Jackson, N.B., Wang, C.M., Luo, Z., Schwitzgebel, J., Ekerdt, J.G., Brock, J.R. and Heller, A., J. Electrochem. Soc. 138, 3660 (1991)
[114] Bahnemann, D.W., Hilgendorff, M. and Memming, R., J. Phys. Chem. B 101, 4265 (1997)
[115] Watanabe, T., Kitamura, A., Kojima, E., Nakayama, C., Hashimoto, K. and Fujishima, A., Photocatalytic Purification and Treatment of Water and Air, Elsevier Science Publisher B.V., Amsterdam (Holland) 1993
[116] Kubota, Y., Shuin, T., Kawasaki, C., Hosaka, M., Kitamura, H., Cai, R.C., Sakai, H., Hashimoto, K. and Fujishima, A., J. Cancer 70, 1107 (1994)
[117] Wang, R., Hashimoto, K., Fujishima, A., Chikusie, M., Kojima, E., Kitamura, A., Shimohigoshi, M. and Watanabe, T., Nature 388, 431 (1997)
[118] Fox, M.A., in Topics in: Organic Chemistry (eds. A.J. Fry and W.E. Brittan), Plenum press, New York 1986, p.177
[119] Fox, M.A., Topics in Current Chemistry 143, 72 (1987)
[120] Ostermayer, Kohl, P.A. and Burton, R.H., Appl. Phys. Lett. 43, 642 (1983)
[121] Kelly, J.J. and Notton, P.H.L., Electrochim. Acta 29, 589 (1984)
[122] Kelly, J.J., van den Meeraker, J.E.A.M. and Notten, P.H.L., in: Grundlagen von Elektrodenreaktionen (ed. H. Behrens), Dechema Monographien Vol. 102, Verlag Chemie, Weinheim 1986, p. 453
[123] Gerischer, H., J. Phys. Chem. 95, 1356 (1991)
[124] Minks, B.P., Oskam, G., Vanmaeckelbergh, D. and Kelly, J.J., J. Electroanal. Chem. 273, 119 (1989)
[125] Minks, B.P., Vanmaeckelbergh, D. and Kelly, J.J., J. Electroanal. Chem. 273, 133 (1989)
[126] v. d. Ven, J. and Nabben, H.J.P., J. Appl. Phys. 67, 7572 (1990)
[127] v. d. Ven, J. and Nabben, H.J.P., J. Electrochem. Soc. 138, 144 (1991)
[128] Möllers, F., Tolle, H.J. and Memming, R., J. Electrochem. Sic. 121, 1160 (1974)
[129] Jacobs, J.W.M., J. Phys. Chem. 90, 6507 (1986)
[130] Lauermann, I., Meissner, D. and Memming, R., in Proc. Vol. 88–14 of the Electrochemical Society, Pennington, N.J. 1988, p.190
[131] Kobayashi, T., Taniguchi, Y., Yoneyama, H. and Tamura, H., J. Phys. Chem. 87, 768 (1983)
[132] Richter, W., Rimmasch, J., Kastening, B., Memming, R. and Meissner, D., in: Proc. Vol. 91–3 of the Electrochem. Soc., Pennington, N.J. 1991, p. 149
[133] Wang, Y. and Herron, N., J. Phys. Chem. 91, 257 (1987)
[134] Dingle, R., in: Advances in Solid State Physics, Vol. 53 (ed. H.I. Queisser), Pergamon Vieweg, Braunschweig 1975, p. 136
[135] Möllers, F. and Memming, Ber. Bunsenges. Phys. Chem. 76, 469 (1976)

[136] Memming, R. and Möllers, F., Ber. Bunsenges. Phys. Chem. 76, 475 (1976)
[137] Lohmann, F., Ber. Bunsenges. Phys. Chem. 70, 428 (1966)
[138] Bolts, I.M. and Wrighton, M.S., J. Phys. Chem. 80, 2641 (1976)
[139] Gerischer, H. and Mindt, W., Surf. Sci. 4, 440 (1966)
[140] Gerischer, H., Hoffmann-Perez and Mindt, W., Ber. Bunsenges. Phys. Chem. 69, 130 (1965)
[141] Van den Meerakker, J.E.A.M., Kelly, J.J. and Notten, P.H.L., J. Electrochem. Soc. 132, 638 (1985)
[142] Rosenbluth, M. and Lewis, N.S., J. Am. Chem. Soc. 108, 4689 (1986)

Subject Index